SECOND EDITION

VISUALIZING
PHYSICAL
GEOGRAPHY

VISUALIZING
PHYSICAL
GEOGRAPHY

SECOND EDITION

TIM FORESMAN, PhD

INTERNATIONAL CENTER FOR
REMOTE SENSING EDUCATION

UNIVERSITY OF MARYLAND

ALAN STRAHLER, PhD

DEPARTMENT OF GEOGRAPHY
AND ENVIRONMENT

BOSTON UNIVERSITY

WILEY
VISUALIZING™

WILEY

VICE PRESIDENT AND EXECUTIVE PUBLISHER   Jay O'Callaghan
EXECUTIVE EDITOR   Ryan Flahive
DIRECTOR OF DEVELOPMENT   Barbara Heaney
MANAGER, PRODUCT DEVELOPMENT   Nancy Perry
PRODUCT DESIGNER   Beth Tripmacher
EDITORIAL OPERATIONS MANAGER   Lynn Cohen
EDITORIAL ASSISTANTS   Darnell Sessoms, Brittany Cheetham
DIRECTOR, MARKETING COMMUNICATIONS   Jeffrey Rucker
SENIOR MARKETING MANAGER   Margaret Barrett
SENIOR CONTENT MANAGER   Micheline Frederick
SENIOR PRODUCTION EDITOR   William Murray
CREATIVE DIRECTOR   Harry Nolan
COVER DESIGN   Harry Nolan
INTERIOR DESIGN   Jim O'Shea
SENIOR PHOTO EDITOR   Mary Ann Price
PHOTO RESEARCHER   Ellinor Wagner
SENIOR ILLUSTRATION EDITOR   Sandra Rigby
PRODUCTION SERVICES   Furino Production

This book was set in New Baskerville by Precision Graphics, printed and bound
by Quad/Graphics. The cover was printed by Quad/Graphics.

ISBN 13: 978-0-470-62615-3
BRV ISBN: 978-1-118-12658-5

Printed in the United States of America
10 9 8 7 6 5 4 3 2 1

Preface

How Is Wiley Visualizing Different?

Wiley Visualizing is based on decades of research on the use of visuals in learning (Mayer, 2005).[1] The visuals teach key concepts and are pedagogically designed to **present, explain,** and **organize** new information. The figures are tightly integrated with accompanying text; the visuals are conceived with the text in ways that clarify and reinforce major concepts, while allowing students to understand the details. This commitment to distinctive and consistent visual pedagogy sets Wiley Visualizing apart from other textbooks.

The texts offer an array of remarkable photographs, maps, media, and film from photo collections around the world, including those of National Geographic. Wiley Visualizing's images are not decorative; such images can be distracting to students. Instead, they are purposeful and the primary driver of the content. These authentic materials immerse the student in real-life issues and experiences and support thinking, comprehension, and application.

Together these elements deliver a level of rigor in ways that maximize student learning and involvement. Wiley Visualizing has proven to increase student learning through its unique combination of text, photographs, and illustrations, with online video, animations, simulations and assessments.

(1) Visual Pedagogy. Using the Cognitive Theory of Multimedia Learning, which is backed up by hundreds of empirical research studies, Wiley's authors create visualizations for their texts that specifically support students' thinking and learning—for example, the selection of relevant materials, the organization of the new information, or the integration of the new knowledge with prior knowledge.

(2) Authentic Situations and Problems. *Visualizing Physical Geography 2e* benefits from National Geographic's more than century-long recording of the world and offers an array of remarkable photographs, maps, media, and film. These authentic materials immerse the student in real-life issues in environmental science, thereby enhancing motivation, learning, and retention (Donovan & Bransford, 2005).[2]

(3) Designed with Interactive Multimedia. *Visualizing Physical Geography 2e* is tightly integrated with *WileyPLUS,* our online learning environment that provides interactive multimedia activities in which learners can actively engage with the materials. The combination of textbook and *WileyPLUS* provides learners with multiple entry points to the content, giving them greater opportunity to explore concepts and assess their understanding as they progress through the course. *WileyPLUS* is a key component of the Wiley Visualizing learning and problem-solving experience, setting it apart from other textbooks whose online component is mere drill-and-practice.

Wiley Visualizing and the *WileyPLUS* Learning Environment are designed as natural extensions of how we learn

To understand why the Visualizing approach is effective, it is first helpful to understand how we learn.

1. Our brain processes information using two main channels: visual and verbal. Our *working memory* holds information that our minds process as we learn. This "mental workbench" helps us with decisions, problem-solving, and making sense of words and pictures by building verbal and visual models of the information.

2. When the verbal and visual models of corresponding information are integrated in working memory, we form more comprehensive, lasting, mental models.

3. When we link these integrated mental models to our prior knowledge, stored in our *long-term memory*, we build even stronger mental models. When an integrated (visual plus verbal) mental model is formed and stored in long-term memory, real learning begins.

The effort our brains put forth to make sense of instructional information is called *cognitive load*. There are two kinds of cognitive load: productive cognitive load, such as when we're engaged in learning or exert positive effort to create mental models; and unproductive cognitive load, which occurs when the brain is trying to make sense of needlessly complex content or when information is not presented well. The learning process can be impaired when the information to be processed exceeds the capacity of working memory. Well-designed visuals and text with effective pedagogical guidance can reduce the unproductive cognitive load in our working memory.

---

[1]Mayer, R.E. (Ed) (2005). The Cambridge Handbook of Multimedia Learning. Cambridge University Press.

[2]Donovan, M.S., & Bransford, J. (Eds.) (2005). How Students Learn: Science in the Classroom. The National Academy Press. Available at http://www.nap.edu/openbook.php?record_id=11102&page=1

Wiley Visualizing is designed for engaging and effective learning

The visuals and text in *Visualizing Physical Geography 2e* are specially integrated to present complex processes in clear steps and with clear representations, organize related pieces of information, and integrate related information with one another. This approach, along with the use of interactive multimedia, minimizes unproductive cognitive load and helps students engage with the content. When students are engaged, they are reading and learning, which can lead to greater knowledge and academic success.

Research shows that well-designed visuals, integrated with comprehensive text, can improve the efficiency with which a learner processes information. In this regard, SEG Research, an independent research firm, conducted a national, multisite study evaluating the effectiveness of Wiley Visualizing. Its findings indicate that students using Wiley Visualizing products (both print and multimedia) were more engaged in the course, exhibited greater retention throughout the course, and made significantly greater gains in content area knowledge and skills, as compared to students in similar classes that did not use Wiley Visualizing.[3]

The use of *WileyPLUS* can also increase learning. According to a white paper titled "Leveraging Blended Learning for More Effective Course Management and Enhanced Student Outcomes" by Peggy Wyllie of Evince Market Research & Communications, studies show that effective use of online resources can increase learning outcomes. Pairing supportive online resources with face-to-face instruction can help students to learn and reflect on material, and deploying multimodal learning methods can help students to engage with the material and retain their acquired knowledge.

[3]SEG Research (2009). Improving Student-Learning with Graphically-Enhanced Textbooks: A Study of the Effectiveness of the Wiley Visualizing Series.

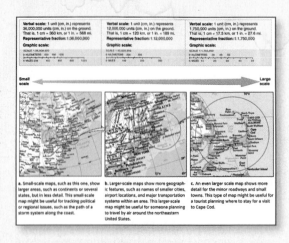

Figure 1: Map scale (Fig. 1.18) Through a logical progression of visuals and graphic features such as the arrow and circles, this illustration directs learners' attention to the underlying concept. Textual and visual elements are physically integrated. This eliminates split attention (when we must divide our attention between several sources of different information).

Figure 2: Monitoring change through satellite scanning (Fig. 1.23) Images are paired so that students can compare and contrast them, thereby grasping the underlying concept. Adjacent captions eliminate split attention.

Figure 3: Environmental temperature lapse rate (Fig. 3.10) From abstraction to reality: Linking points on the graph to a photo of landscapes illustrates how the data on the graph translates to environmental differences in ecosystems.

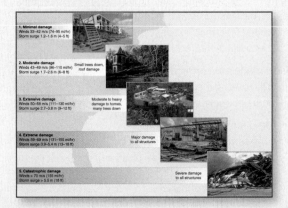

Figure 4: Saffir-Simpson scale of tropical cyclone intensity (Fig. 6.14) This matrix visually organizes abstract information to reduce cognitive load.

How Are the Wiley Visualizing Chapters Organized?

Student engagement is more than just exciting videos or interesting animations—engagement means keeping students motivated to keep going. It is easy to get bored or lose focus when presented with large amounts of information, and it is easy to lose motivation when the relevance of the information is unclear. The design of *WileyPLUS* is based on cognitive science, instructional design, and extensive research into user experience. It transforms learning into an interactive, engaging, and outcomes-oriented experience for students.

Each Wiley Visualizing chapter engages students from the start

Chapter opening text and visuals introduce the subject and connect the student with the material that follows.

Chapter Introductions illustrate key concepts in the chapter with intriguing stories and striking photographs.

Chapter Outlines anticipate the content.

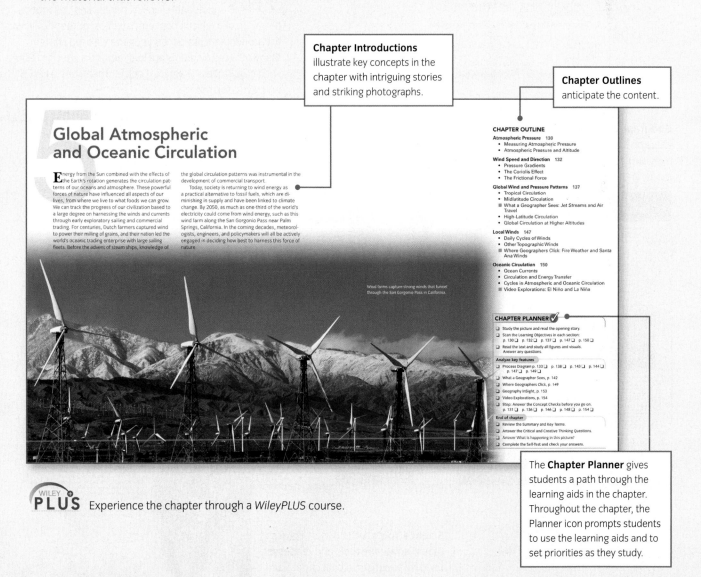

Global Atmospheric and Oceanic Circulation

Energy from the Sun combined with the effects of the Earth's rotation generates the circulation patterns of our oceans and atmosphere. These powerful forces of nature have influenced all aspects of our lives, from where we live to what foods we can grow. We can track the progress of our civilization based to a large degree on harnessing the winds and currents through early exploratory sailing and commercial trading. For centuries, Dutch farmers captured wind to power their milling of grains, and their nation led the world's oceanic trading enterprise with large sailing fleets. Before the advent of steam ships, knowledge of

the global circulation patterns was instrumental in the development of commercial transport.

Today, society is returning to wind energy as a practical alternative to fossil fuels, which are diminishing in supply and have been linked to climate change. By 2050, as much as one-third of the world's electricity could come from wind energy, such as this wind farm along the San Gorgonio Pass near Palm Springs, California. In the coming decades, meteorologists, engineers, and policymakers will all be actively engaged in deciding how best to harness this force of nature.

Wind farms capture strong winds that funnel through the San Gorgonio Pass in California.

CHAPTER OUTLINE

Atmospheric Pressure 130
- Measuring Atmospheric Pressure
- Atmospheric Pressure and Altitude

Wind Speed and Direction 132
- Pressure Gradients
- The Coriolis Effect
- The Frictional Force

Global Wind and Pressure Patterns 137
- Tropical Circulation
- Midlatitude Circulation
- ■ What a Geographer Sees: Jet Streams and Air Travel
- High-Latitude Circulation
- Global Circulation at Higher Altitudes

Local Winds 147
- Daily Cycles of Winds
- Other Topographic Winds
- ■ Where Geographers Click: Fire Weather and Santa Ana Winds

Oceanic Circulation 150
- Ocean Currents
- Circulation and Energy Transfer
- Cycles in Atmospheric and Oceanic Circulation
- ■ Video Explorations: El Niño and La Niña

CHAPTER PLANNER ✓
- ☐ Study the picture and read the opening story.
- ☐ Scan the Learning Objectives in each section:
  p. 130 ☐   p. 132 ☐   p. 137 ☐   p. 147 ☐   p. 150 ☐
- ☐ Read the text and study all figures and visuals. Answer any questions.

Analyze key features
- ☐ Process Diagram p. 133 ☐   p. 138 ☐   p. 143 ☐   p. 144 ☐ p. 147 ☐   p. 149 ☐
- ☐ What a Geographer Sees, p. 142
- ☐ Where Geographers Click, p. 149
- ☐ Geography InSight, p. 153
- ☐ Video Explorations, p. 154
- ☐ Stop: Answer the Concept Checks before you go on. p. 131 ☐   p. 136 ☐   p. 146 ☐   p. 148 ☐   p. 154 ☐

End of chapter
- ☐ Review the Summary and Key Terms.
- ☐ Answer the Critical and Creative Thinking Questions.
- ☐ Answer What is happening in this picture?
- ☐ Complete the Self-Test and check your answers.

WILEY **PLUS** Experience the chapter through a *WileyPLUS* course.

The **Chapter Planner** gives students a path through the learning aids in the chapter. Throughout the chapter, the Planner icon prompts students to use the learning aids and to set priorities as they study.

Wiley Visualizing guides students through the chapter

The content of Wiley Visualizing gives students a variety of approaches—visuals, words, interactions, video, and assessments—that work together to provide a guided path through the content.

Learning Objectives at the start of each section indicate in behavioral terms the concepts that students are expected to master while reading the section.

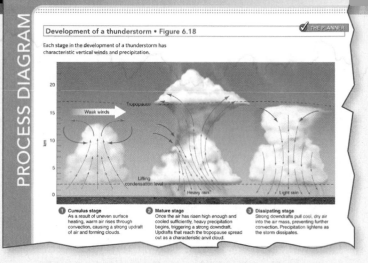

PROCESS DIAGRAM

Development of a thunderstorm • Figure 6.18

Each stage in the development of a thunderstorm has characteristic vertical winds and precipitation.

1 Cumulus stage As a result of uneven surface heating, warm air rises through convection, causing a strong updraft of air and forming clouds.

2 Mature stage Once the air has risen high enough and cooled sufficiently, heavy precipitation begins, triggering a strong downdraft. Updrafts that reach the tropopause spread out as a characteristic anvil cloud.

3 Dissipating stage Strong downdrafts pull cool, dry air into the air mass, preventing further convection. Precipitation lightens as the storm dissipates.

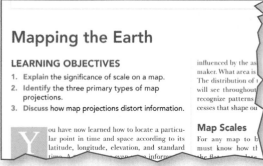

Mapping the Earth

LEARNING OBJECTIVES

1. Explain the significance of scale on a map.
2. Identify the three primary types of map projections.
3. Discuss how map projections distort information.

Y ou have now learned how to locate a particular point in time and space according to its latitude, longitude, elevation, and standard time.

influenced by the as maker. What area is The distribution of will see throughout recognize patterns cesses that shape ou

Map Scales

For any map to b must know how t

Process Diagrams provide in-depth coverage of processes correlated with clear, step-by-step narrative, enabling students to grasp important topics with less effort.

Interactive Process Diagrams provide additional visual examples and descriptive narration of a difficult concept, process, or theory, allowing the students to interact and engage with the content. Many of these diagrams are built around a specific feature such as a Process Diagram. Look for them in *WileyPLUS* when you see this icon.

WILEY **PLUS** Animation

Geography InSight features are multipart visual sections that focus on a key concept or topic in the chapter, exploring it in detail or in broader context using a combination of photos, diagrams, maps, and data.

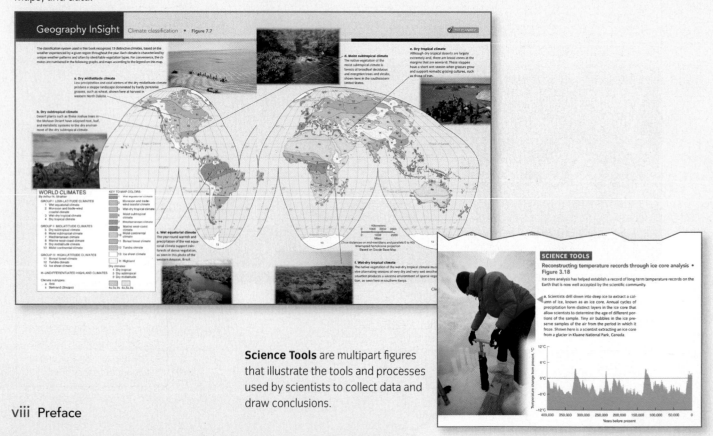

Geography InSight   Climate classification • Figure 7.7

Science Tools are multipart figures that illustrate the tools and processes used by scientists to collect data and draw conclusions.

WHAT A GEOGRAPHER SEES

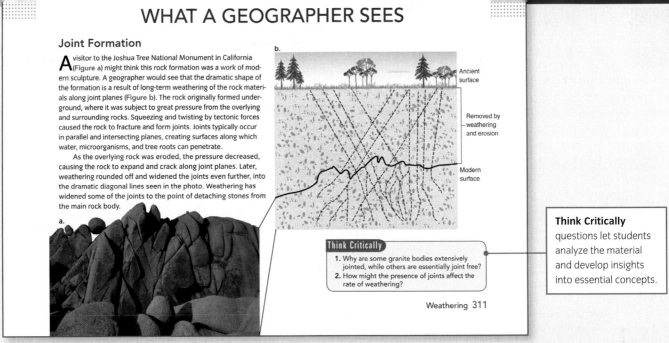

Joint Formation

A visitor to the Joshua Tree National Monument in California (Figure a) might think this rock formation was a work of modern sculpture. A geographer would see that the dramatic shape of the formation is a result of long-term weathering of the rock materials along joint planes (Figure b). The rock originally formed underground, where it was subject to great pressure from the overlying and surrounding rocks. Squeezing and twisting by tectonic forces caused the rock to fracture and form joints. Joints typically occur in parallel and intersecting planes, creating surfaces along which water, microorganisms, and tree roots can penetrate.

As the overlying rock was eroded, the pressure decreased, causing the rock to expand and crack along joint planes. Later, weathering rounded off and widened the joints even further, into the dramatic diagonal lines seen in the photo. Weathering has widened some of the joints to the point of detaching stones from the main rock body.

Think Critically

1. Why are some granite bodies extensively jointed, while others are essentially joint free?
2. How might the presence of joints affect the rate of weathering?

Weathering 311

Think Critically questions let students analyze the material and develop insights into essential concepts.

What a Geographer Sees highlights a concept or phenomenon that would stand out to geographers. Photos and figures are used to improve students' understanding of the usefulness of a geographic perspective and to develop their observational skills.

Video Explorations
The Power of Water

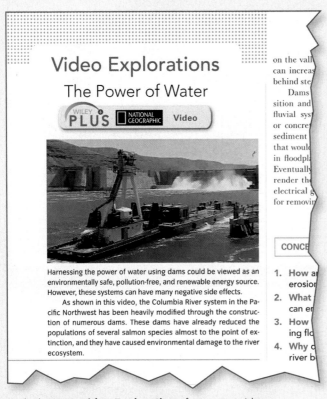

Harnessing the power of water using dams could be viewed as an environmentally safe, pollution-free, and renewable energy source. However, these systems can have many negative side effects.

As shown in this video, the Columbia River system in the Pacific Northwest has been heavily modified through the construction of numerous dams. These dams have already reduced the populations of several salmon species almost to the point of extinction, and they have caused environmental damage to the river ecosystem.

Each chapter's **Video Explorations** feature provides visual context for key concepts, ideas, and terms through a rich collection of videos, including more than 30 National Geographic videos from their award-winning collection.

Where Geographers CLICK
Global Monitoring of Climate Change

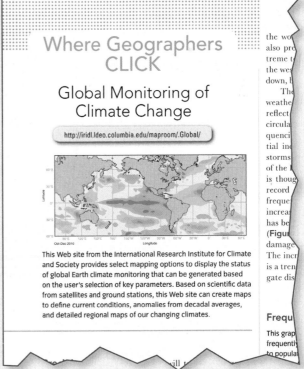

http://iridl.ldeo.columbia.edu/maproom/.Global/

This Web site from the International Research Institute for Climate and Society provides select mapping options to display the status of global Earth climate monitoring that can be generated based on the user's selection of key parameters. Based on scientific data from satellites and ground stations, this Web site can create maps to define current conditions, anomalies from decadal averages, and detailed regional maps of our changing climates.

Where Geographers Click showcases a Web site that professionals use and encourages students to try out its tools.

 Streaming videos are available to students in *WileyPLUS*.

In concert with the visual approach of the book, **www.ConceptCaching.com** is an online collection of photographs that explores places, regions, people, and their activities. Photographs, GPS coordinates, and explanations of core geographic concepts are "cached" for viewing by professors and students alike. Professors can access the images or submit their own by visiting the Web site.

GeoDiscoveries Media Library is an interactive media source of animations, simulations, and interactivities allowing instructors to visually demonstrate key concepts in greater depth. Assets from the library are included in the *WileyPLUS* course.

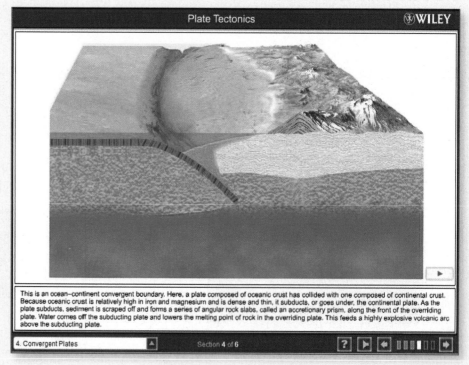

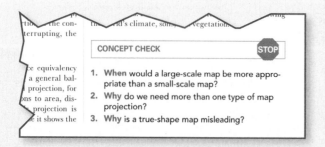

CONCEPT CHECK — STOP

1. When would a large-scale map be more appropriate than a small-scale map?
2. Why do we need more than one type of map projection?
3. Why is a true-shape map misleading?

Coordinated with the section-opening **Learning Objectives**, at the end of each section **Concept Check** questions allow students to test their comprehension of the learning objectives.

Student understanding is assessed at different levels

Wiley Visualizing with *WileyPLUS* offers students lots of practice material for assessing their understanding of each study objective. Students know exactly what they are getting out of each study session through immediate feedback and coaching.

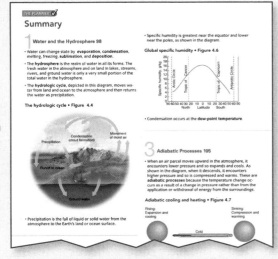

The **Summary** revisits each major section, with informative images taken from the chapter. These visuals reinforce important concepts.

Critical and Creative Thinking Questions

1. How does air pressure change as you walk up a mountain, and what is the implication for the amount of oxygen you absorb with each breath?

2. What type of weather systems dominate the area where you live? Do you notice a consistent pattern of cyclones and anticyclones? Is your weather influenced by prevailing westerly winds? Are you subject to any local winds (such as mountain breezes or Santa Ana winds)?

3. An island directly on the equator might experience a particular annual pattern of wet and dry weather. There would be two rainy seasons (spring and fall) and two dry seasons (summer and winter). How could the migration of the ITCZ explain this pattern?

4. An airline pilot is planning a nonstop flight from Los Angeles to Sydney, Australia. What general wind conditions can the pilot expect to find in the upper atmosphere as the airplane travels? What jet streams will be encountered? Will they slow or speed the aircraft on its way?

5. Predict some changes in weather that might occur on land in South America caused by changes in ocean temperatures associated with El Niño conditions.

6. On the map, notice the latitude of Europe compared to that of North America. Why does England have a relatively warm climate compared to cities at similar latitudes in Canada (for example, Winnipeg)?

Critical and Creative Thinking Questions challenge students to think more broadly about chapter concepts. The level of these questions ranges from simple to advanced; they encourage students to think critically and develop an analytical understanding of the ideas discussed in the chapter.

What is happening in this picture?

This spectacular photo, taken by astronauts on the Gemini XI mission, shows the southern tip of the Red Sea and the southern tip of the Arabian Peninsula.

Think Critically

1. How do you explain the fact that these two land masses almost seem like jigsaw puzzle pieces that could fit back together?
2. What plate boundary would you expect to find here?
3. What would you expect to happen in this region in the future?

What is happening in this picture? presents a photograph that is relevant to a chapter topic and illustrates a situation students are not likely to have encountered previously.

Think Critically questions ask students to apply what they have learned in order to interpret and explain what they observe in the image.

Self-Test

(Check your answers in the Appendix.)

1. Label the following regions on this diagram showing the zones of subsurface water: (a) the saturated zone, (b) the soil-water belt, (c) the unsaturated zone, and (d) the water table.

2. The water table is at its highest _____.
   a. under the lowest areas of land surface
   b. under the highest areas of land surface
   c. adjacent to perennial streams
   d. in lakes

3. A layer of rock or sediment that contains abundant, freely flowing ground water is known as a(n) _____.
   a. aquiclude
   b. infiltration
   c. aquifer
   d. water table

4. Geographers apply the term _____ to the topography of any limestone area where sinkholes, as shown in this photograph, are numerous and small surface streams are nonexistent.
   a. travertine
   b. karst
   c. carbonate
   d. dolomite

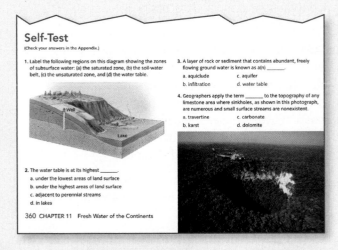

360 CHAPTER 11 Fresh Water of the Continents

Visual end-of-chapter **Self-Tests** pose review questions that ask students to demonstrate their understanding of key concepts.

Why *Visualizing Physical Geography 2e?*

With the help of modern technologies, such as remote sensing and GIS, geographers are able to "see" our world in new ways, from the top layers of the Earth's atmosphere to the ice caps of the poles and the forests of the farthest continents. In this Second Edition of *Visualizing Physical Geography*, we take students to these new frontiers, highlighting the tools and methods scientists use to collect data, test hypotheses, and build theories.

One of these scientific frontiers is the ongoing effort to understand how and why our climate is changing. Recognizing the challenges posed by ongoing changes to our climate, we have expanded our coverage of climate change throughout. In particular, we focus on *how* scientists have established evidence for climate change and how that evidence can be used to model climate outcomes.

As part of this enhanced focus on the process of science, we also want students to learn to think more like scientists themselves. Throughout every chapter, new questions prompt students to think critically about the topics at hand and extend their knowledge to the world around them.

While we hope that students become excited by the cutting-edge science of geography, our primary goal remains to help students understand the many common natural phenomena that they experience every day, from the formation of a thunderstorm to the building of a sand dune. We also aim to excite them about the diversity of the Earth's environments, with the hope that they will be motivated to experience at first hand as much of the Earth's physical geography as they can.

Our treatment builds on basic science principles from meteorology, climatology, geology, geomorphology, soil science, ecology, and biogeography, weaving them together to describe the physical processes that shape the varied landscapes of our planet. Each chapter begins by establishing fundamental principles and processes and then guides students through more specific phenomena with the aid of numerous visual tools.

This book is an introductory text aimed primarily at undergraduate nonscience majors, and we assume little prior knowledge of physical geography. The accessible format of *Visualizing Physical Geography 2e* allows students to move easily from jumping-off points in the early chapters to more complex concepts covered later. Chapters are designed to be as self-contained as possible, so that instructors can choose chapter orderings and emphases that suit their needs. Our book is appropriate for use in one-semester physical geography courses offered by a variety of departments, including geography, earth sciences, environmental studies, biology, and agriculture.

Organization

Physical geography is concerned with processes resulting from two vast energy flows—a stream of solar electromagnetic radiation that drives surface temperatures and flows of surface and atmospheric gases and fluids, and a stream of heat from the Earth's interior that manifests itself in the slow motion of earth materials in the upper layers of the Earth's crust. These flows interact at the Earth's surface, which is the province of the physical geographer. At this interface, the biological realm exploits these flows, adding a third element to the landscape. Following this schema, we present physical geography beginning with weather and climate, then turn to earth materials, geomorphology, and finally ecology and biogeography.

The book's opening three chapters cover core physical principles and lay the groundwork for understanding climate, surface features, and global biodiversity discussed in later chapters.

- **CHAPTER 1** introduces the science and tools of geography, including the scientific method, maps, and remote sensing technologies.

- **CHAPTER 2** examines the effects of the Earth's rotation and revolution about the Sun on changing patterns of insolation around the globe. This chapter also introduces the greenhouse effect.

- **CHAPTER 3** builds on this foundation to explain daily and annual air temperature patterns at different locations. This chapter also introduces global warming and climate change — both describing the most recent evidence for this trend and discussing its possible impact in the future.

- **CHAPTERS 4, 5, and 6** investigate weather systems, including the development of precipitation and global wind patterns. The concepts are presented with references to examples covering the globe, including a discussion of the causes and consequences of some of the most dramatic recent weather events seen within the United States and around the world.

- **CHAPTER 7** combines the foundational concepts from the core chapters on insolation and air temperature with the information on weather patterns from Chapters 4, 5, and 6 in a detailed discussion of climate regions. Climate change is revisited with this new understanding.

- **CHAPTERS 8 to 15** deal with the formation of landforms and features on the Earth's surface. We begin with a simple yet thorough introduction to rock types, plate tectonics, and weathering processes. We then discuss volcanic and tectonic landforms, as well as features created by mass wasting, running water, waves and wind, and moving glaciers. Also provided is a detailed discussion of soil formation and soil types around the globe.

- **CHAPTERS 16 and 17** turn to ecological, historical, and global biogeography. Here we show how the climates, surface features, and environments discussed in earlier chapters can affect living organisms now and in the future.

New to this edition

This Second Edition of *Visualizing Physical Geography* is dedicated to further enhancing the student learning experience through several new and unique features, including

- **New emphasis on climate change.** Discussion of the impacts of climate change is woven throughout the book, including the most current research and predictions.

- **Expanded coverage of important topics.** Throughout, this edition contains new or expanded discussions of important and relevant topics.

   Chapter 1 includes new coverage of geography as a science, the scientific method, remote sensing technologies, and GIS.

   Chapter 6 includes new coverage of the Saffir-Simpson scale and how and why lightning occurs. It also features new illustrations of the formation of thunderstorms and tornadoes. Coverage of thunderstorms has been moved here from Chapter 4.

   Chapter 7 features updated climate change coverage, expanded to investigate the methods of climate modeling.

   Chapter 9 features new figures on earthquake focus and epicenter, and new coverage of the 2011 earthquake and tsunami in Japan. Coverage of plate tectonics has been moved to this chapter from Chapter 8.

   Chapter 10 now includes expanded discussion of forces that act on a slope.

Chapter 11 features expanded coverage of subsidence and groundwater pollution.

Chapter 12 has been expanded with new coverage of downcutting and rejuvenation, as well as nickpoint migration, and expanded coverage of migration of meanders and landforms associated with meandering streams.

Chapter 13 features expanded coverage and new illustrations of dune formation, wave formation, wave refraction, and barrier coast landforms. Tsunami coverage has been moved here from Chapter 9.

Chapter 14 includes expanded coverage and new illustrations for types of alpine glaciers, processes of glacier formation and movement, and glacial landforms. Permafrost coverage has been moved to this chapter from Chapter 10.

Chapter 15 features expanded coverage and new or revised figures for gravitational, capillary and hygroscopic water, and soil pH.

- **Enhanced visuals.** Throughout the text, photos, figures, and diagrams have been carefully examined and revised to increase their diversity and overall effectiveness as aids to learning. In particular, the size of world maps has been increased for improved legibility.

- **New *Science Tools* feature.** In each chapter, a new *Science Tools* figure demonstrates the tools and methods scientists use to learn about the Earth.

- **New *Where Geographers Click* feature.** Every chapter includes a new *Where Geographers Click* feature, which showcases a Web site that professionals use and encourages students to try out its tools.

- **New *Video Explorations* feature.** In each chapter, a *Video Explorations* feature showcases a relevant video, frequently from the award-winning National Geographic Society collection. Linked to the text, the videos provide visual context for key concepts.

- **More opportunities for critical thinking.** In every chapter, select figures are paired with critical thinking questions. *Ask Yourself* questions test student comprehension. *Put It Together* questions ask students to combine information from the figure with previously learned material. *Think Critically* questions require students to apply what they've learned to new situations.

Also available

Earth Pulse 2e. Utilizing full-color imagery and National Geographic photographs, *EarthPulse* takes you on a journey of discovery covering topics such as *The Human Condition, Our Relationship with Nature, and Our Connected World*. Illustrated by specific examples, each section focuses on trends affecting our world today. Included are extensive full-color world and regional maps for reference. *EarthPulse* is available only in a package with *Visualizing Physical Geography 2e*. Contact your Wiley representative for more information or visit www.wiley.com/college/earthpulse.

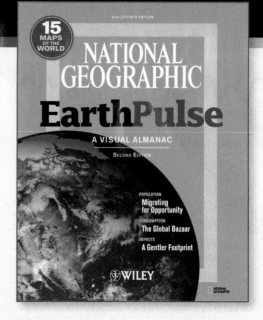

How Does Wiley Visualizing Support Instructors?

Wiley Visualizing Site

WILEY VISUALIZING™

The Wiley Visualizing site hosts a wealth of information for instructors using Wiley Visualizing, including ways to maximize the visual approach in the classroom and a white paper titled "How Visuals Can Help Students Learn," by Matt Leavitt, instructional design consultant. Visit Wiley Visualizing at www.wiley.com/college/visualizing.

Wiley Custom Select

WILEY *Custom*
LEARNING SOLUTIONS

Wiley Custom Select gives you the freedom to build your course materials exactly the way you want them. Offer your students a cost-efficient alternative to traditional texts. In a simple three-step process create a solution containing the content you want, in the sequence you want, delivered how you want. Visit Wiley Custom Select at http://customselect.wiley.com.

Videos

Each chapter's **Video Explorations** provide visual context for key concepts, ideas, and terms through a rich collection of videos, including more than 30 National Geographic videos from their award-winning collection.

Streaming videos are available to students in the context of *WileyPLUS* and accompanying assignments that can be graded online and added to the instructor gradebook.

Book Companion Site

www.wiley.com/college/foresman

All instructor resources (the Test Bank, Instructor's Manual, PowerPoint presentations, and all textbook illustrations and photos in jpeg format are housed on the book companion site (www.wiley.com/college/foresman). Student resources include self quizzes and flashcards.

PowerPoint Presentations

(available in *WileyPLUS* and on the book companion site)

A complete set of highly visual PowerPoint presentations—one per chapter—by Jonathan Little, Monroe Community College, is available online and in *WileyPLUS* to enhance classroom presentations. Tailored to the text's topical coverage and learning objectives, these presentations are designed to convey key text concepts, illustrated by embedded text art. We offer three different types of PowerPoint presentations for each chapter: PowerPoint's with just the text art, PowerPoint's with text art and presentation notes and Media-Integrated PowerPoint's with links to videos and animations.

Test Bank

(available in *WileyPLUS* and on the book companion site)

The visuals from the textbook are also included in the Test Bank by Thomas Orf, Las Positas College. The Test Bank has a diverse selection of test items including multiple-choice and essay question, with at least 20 percent of them incorporating visuals from the book. The test bank is available online in MS Word files as a Computerized Test Bank, and within *WileyPLUS*. The easy-to-use test-generation program fully supports graphics, print tests, student answer sheets, and answer keys. The software's advanced features allow you to produce an exam to your exact specifications. We also offer Pre-Lecture Clicker questions, by Lee Stocks, University of North Carolina- Pembroke, which can be used in any clicker question software.

Instructor's Manual

(available in *WileyPLUS* and on the book companion site)

The Instructor's Manual includes creative ideas for in-class activities, discussion questions, and lecture transitions by David Sallee, Tarrant County College. It also includes answers to Critical and Creative Thinking questions and Concept Check questions.

Guidance is also provided on how to maximize the effectiveness of visuals in the classroom.

1. **Use visuals during class discussions or presentations.** Point out important information as the students look at the visuals, to help them integrate separate visual and verbal mental models.

2. **Use visuals for assignments and to assess learning.** For example, learners could be asked to identify samples of concepts portrayed in visuals.

3. **Use visuals to encourage group activities.** Students can study together, make sense of, discuss, hypothesize, or make decisions about the content. Students can work together to interpret and describe the diagram, or use the diagram to solve problems, conduct related research, or work through a case study activity.

4. **Use visuals during reviews.** Students can review key vocabulary, concepts, principles, processes, and relationships displayed visually. This recall helps link prior knowledge to new information in working memory, building integrated mental models.

5. **Use visuals for assignments and to assess learning.** For example, learners could be asked to identify samples of concepts portrayed in visuals.

6. **Use visuals to apply facts or concepts to realistic situations or examples.** For example, a familiar photograph, such as the Grand Canyon, can illustrate key information about the stratification of rock, linking this new concept to prior knowledge.

Image Gallery

(available in *WileyPLUS* and on the book companion site)

All photographs, figures, maps, and other visuals from the text are online and in *WileyPLUS* and can be used as you wish in the classroom. These online electronic files allow you to easily incorporate images into your PowerPoint presentations as you choose, or to create your own handouts.

In addition to the text images, you also have access to **ConceptCaching**, an online database of photographs that explores what a region looks like and what it feels like to live in that region. Photographs and GPS coordinates are "cached" for viewing by core geographical concept and by region. Professors can access the images or submit their own by visiting www.conceptcaching.com.

Wiley Faculty Network

The Wiley Faculty Network (WFN) is a global community of faculty, connected by a passion for teaching and a drive to learn, share, and collaborate. Their mission is to promote the effective use of technology and enrich the teaching experience. Connect with the Wiley Faculty Network to collaborate with your colleagues, find a mentor, attend virtual and live events, and view a wealth of resources all designed to help you grow as an educator. Visit the Wiley Faculty Network at www.wherefacultyconnect.com.

How Has Wiley Visualizing Been Shaped by Contributors?

Wiley Visualizing and the *WileyPLUS* learning environment would not have come about without lots of people, each of whom played a part in sharing their research and contributing to this new approach.

Academic Research Consultants

Richard Mayer, Professor of Psychology, UC Santa Barbara. His Cognitive Theory of Multimedia Learning provided the basis on which we designed our program. He continues to provide guidance to our author and editorial teams on how to develop and implement strong, pedagogically effective visuals and use them in the classroom.

Jan L. Plass, Professor of Educational Communication and Technology in the Steinhardt School of Culture, Education, and Human Development at New York University. He co-directs the NYU Games for Learning Institute and is the founding director of the CREATE Consortium for Research and Evaluation of Advanced Technology in Education.

Matthew Leavitt, Instructional Design Consultant. He advises the Visualizing team on the effective design and use of visuals in instruction and has made virtual and live presentations to university faculty around the country regarding effective design and use of instructional visuals.

Independent Research Studies

SEG Research, an independent research and assessment firm, conducted a national, multisite effectiveness study of students enrolled in entry-level college Psychology and Geology courses. The study was designed to evaluate the effectiveness of Wiley Visualizing. You can view the full research paper at www.wiley.com/college/visualizing/huffman/efficacy.html.

Instructor and Student Contributions

Throughout the process of developing the concept of guided visual pedagogy for Wiley Visualizing, we benefited from the comments and constructive criticism provided by the instructors and colleagues listed below. We offer our sincere appreciation to these individuals for their helpful reviews and general feedback:

Visualizing Reviewers, Focus Group Participants, and Survey Respondents

James Abbott, Temple University
Melissa Acevedo, Westchester Community College
Shiva Achet, Roosevelt University
Denise Addorisio, Westchester Community College
Dave Alan, University of Phoenix
Sue Allen-Long, Indiana University Purdue
Robert Amey, Bridgewater State College
Nancy Bain, Ohio University
Corinne Balducci, Westchester Community College
Steve Barnhart, Middlesex County Community College
Stefan Becker, University of Washington – Oshkosh
Callan Bentley, NVCC Annandale
Valerie Bergeron, Delaware Technical & Community College
Andrew Berns, Milwaukee Area Technical College
Gregory Bishop, Orange Coast College
Rebecca Boger, Brooklyn College
Scott Brame, Clemson University
Joan Brandt, Central Piedmont Community College
Richard Brinn, Florida International University
Jim Bruno, University of Phoenix
William Chamberlin, Fullerton College

Oiyin Pauline Chow, Harrisburg Area Community College
Laurie Corey, Westchester Community College
Ozeas Costas, Ohio State University at Mansfield
Christopher Di Leonardo, Foothill College
Dani Ducharme, Waubonsee Community College
Mark Eastman, Diablo Valley College
Ben Elman, Baruch College
Staussa Ervin, Tarrant County College
Michael Farabee, Estrella Mountain Community College
Laurie Flaherty, Eastern Washington University
Susan Fuhr, Maryville College
Peter Galvin, Indiana University at Southeast
Andrew Getzfeld, New Jersey City University
Janet Gingold, Prince George's Community College
Donald Glassman, Des Moines Area Community College
Richard Goode, Porterville College
Peggy Green, Broward Community College
Stelian Grigoras, Northwood University
Paul Grogger, University of Colorado
Michael Hackett, Westchester Community College
Duane Hampton, Western Michigan University

Thomas Hancock, Eastern Washington University
Gregory Harris, Polk State College
John Haworth, Chattanooga State Technical Community College
James Hayes-Bohanan, Bridgewater State College
Peter Ingmire, San Francisco State University
Mark Jackson, Central Connecticut State University
Heather Jennings, Mercer County Community College
Eric Jerde, Morehead State University
Jennifer Johnson, Ferris State University
Richard Kandus, Mt. San Jacinto College District
Christopher Kent, Spokane Community College
Gerald Ketterling, North Dakota State University
Lynnel Kiely, Harold Washington College
Eryn Klosko, Westchester Community College
Cary T. Komoto, University of Wisconsin – Barron County
John Kupfer, University of South Carolina
Nicole Lafleur, University of Phoenix
Arthur Lee, Roane State Community College
Mary Lynam, Margrove College
Heidi Marcum, Baylor University
Beth Marshall, Washington State University
Dr. Theresa Martin, Eastern Washington University
Charles Mason, Morehead State University
Susan Massey, Art Institute of Philadelphia
Linda McCollum, Eastern Washington University
Mary L. Meiners, San Diego Miramar College
Shawn Mikulay, Elgin Community College
Cassandra Moe, Century Community College
Lynn Hanson Mooney, Art Institute of Charlotte
Kristy Moreno, University of Phoenix
Jacob Napieralski, University of Michigan - Dearborn
Gisele Nasar, Brevard Community College, Cocoa Campus
Daria Nikitina, West Chester University
Robin O'Quinn, Eastern Washington University
Richard Orndorff, Eastern Washington University
Sharen Orndorff, Eastern Washington University
Clair Ossian, Tarrant County College
Debra Parish, North Harris Montgomery Community College District
Linda Peters, Holyoke Community College

Robin Popp, Chattanooga State Technical Community College
Michael Priano, Westchester Community College
Alan "Paul" Price, University of Wisconsin – Washington County
Max Reams, Olivet Nazarene University
Mary Celeste Reese, Mississippi State University
Bruce Rengers, Metropolitan State College of Denver
Guillermo Rocha, Brooklyn College
Penny Sadler, College of William and Mary
Shamili Sandiford, College of DuPage
Thomas Sasek, University of Louisiana at Monroe
Donna Seagle, Chattanooga State Technical Community College
Diane Shakes, College of William and Mary
Jennie Silva, Louisiana State University
Michael Siola, Chicago State University
Morgan Slusher, Community College of Baltimore County
Julia Smith, Eastern Washington University
Darlene Smucny, University of Maryland University College
Jeff Snyder, Bowling Green State University
Alice Stefaniak, St. Xavier University
Alicia Steinhardt, Hartnell Community College
Kurt Stellwagen, Eastern Washington University
Charlotte Stromfors, University of Phoenix
Shane Strup, University of Phoenix
Donald Thieme, Georgia Perimeter College
Pamela Thinesen, Century Community College
Chad Thompson, SUNY Westchester Community College
Lensyl Urbano, University of Memphis
Gopal Venugopal, Roosevelt University
Daniel Vogt, University of Washington – College of Forest Resources
Dr. Laura J. Vosejpka, Northwood University
Brenda L. Walker, Kirkwood Community College
Stephen Wareham, Cal State Fullerton
Fred William Whitford, Montana State University
Katie Wiedman, University of St. Francis
Harry Williams, University of North Texas
Emily Williamson, Mississippi State University
Bridget Wyatt, San Francisco State University
Van Youngman, Art Institute of Philadelphia
Alexander Zemcov, Westchester Community College

Student Participants

Karl Beall, Eastern Washington University
Jessica Bryant, Eastern Washington University
Pia Chawla, Westchester Community College
Channel DeWitt, Eastern Washington University
Lucy DiAroscia, Westchester Community College
Heather Gregg, Eastern Washington University
Lindsey Harris, Eastern Washington University
Brenden Hayden, Eastern Washington University
Patty Hosner, Eastern Washington University

Tonya Karunartue, Eastern Washington University
Sydney Lindgren, Eastern Washington University
Michael Maczuga, Westchester Community College
Melissa Michael, Eastern Washington University
Estelle Rizzin, Westchester Community College
Andrew Rowley, Eastern Washington University
Eric Torres, Westchester Community College
Joshua Watson, Eastern Washington University

Reviewers of *Visualizing Physical Geography 2e*

John All, Western Kentucky University
Thomas Davinroy, Metropolitan State College of Denver
Bryant Evans, Houston Community College
John Frye, Kutztown University
Sarah Goggin, Cypress College
Paul Hanson, University of Nebraska
Susan Hartley, Lake Superior College
Gary Haynes, Ohio University
Tyler Huffman, Eastern Kentucky University
Ryan Kelly, Kentucky Bluegrass Community and Technical College
Laura Lewis, University of Maryland
Lee Lines, Rollins College

Jonathan Little, Monroe Community College
Ingrid Luffman, Eastern Tennessee State University
Armando Mendoza, Cypress College
Heidi Natel, Monroe Community College
Shouraseni Sen Roy, University of Miami
Sherman Silverman, Prince Georges Community College
Erica Smithwick, The Pennsylvania State University
Lee Stocks, University of North Carolina
John Van de Grift, Metropolitan State College of Denver
James Vaughan, University of Texas
Mark Welford, Georgia Southern University
Kenneth Yanow, Southwestern College

Student Feedback/Class Testing

To make certain that *Visualizing Human Geography* met the needs of current students, we asked several instructors to class-test a chapter. The feedback that we received from students and instructors confirmed our belief that the visualizing approach taken in this book is highly effective in helping students to learn. We wish to thank the following instructors and their students who provided us with helpful feedback and suggestions:

Michael Binford, University of Florida
Zachary Bortolot, James Madison University
Ralph Dubayah, University of Maryland
Scott Greene, University of Oklahoma
Christine Hallman, Northeastern State University
Susan Hartley, Lake Superior College
Ryan Kelly, Bluegrass Community College

Mohsen Khani, Sinclair Community College
Kevin Law, Marshall Univeristy
Ingri Luffman, East Tennessee State University
Richard Miller, Florida State University
Tom Orf, Las Positas College
Steve Stadler, Oklahoma State
Forrest Wilkerson, Minnesota State University. Mankato

Reviewers of *Visualizing Physical Geography 1e*

Christiana Asante, Grambling State University
Philip Chaney, Auburn University
James Duvall, Contra Costa College
Kenneth Engelbrecht, Metropolitan State College of Denver
Curtis Holder, University of Colorado-Colorado Springs
Steve LaDochy, California State University-Los Angeles
Michael Madsen, Brigham Young University-Idaho

Herschel Stern, Mira Costa College
John Teeple, Cañada College
Alice Turkington, University of Kentucky-Lexington

Student Feedback/Class Testing

Dorothy Sack, The Ohio State University

Focus Group and Telesession Participants

A number of professors and students participated in focus groups and telesessions, providing feedback on the text, visuals, and pedagogy. Our thanks to the following participants for their helpful comments and suggestions.

Sylvester Allred, Northern Arizona University
David Bastedo, San Bernardino Valley College
Ann Brandt-Williams, Glendale Community College
Natalie Bursztyn, Bakersfield College
Stan Celestian, Glendale Community College
O. Pauline Chow, Harrisburg Area Community College
Diane Clemens-Knott, California State University, Fullerton
Mitchell Colgan, College of Charleston
Linda Crow, Montgomery College
Smruti Desai, Cy-Fair College
Charles Dick, Pasco-Hernando Community College
Donald Glassman, Des Moines Area Community College
Mark Grobner, California State University, Stanislaus
Michael Hackett, Westchester Community College
Gale Haigh, McNeese State University
Roger Hangarter, Indiana University
Michael Harman, North Harris College
Terry Harrison, Arapahoe Community College
Javier Hasbun, University of West Georgia
Stephen Hasiotis, University of Kansas
Adam Hayashi, Central Florida Community College
Laura Hubbard, University of California, Berkeley

James Hutcheon, Georgia Southern University
Scott Jeffrey, Community College of Baltimore County, Catonsville Campus
Matther Kapell, Wayne State University
Arnold Karpoff, University of Louisville
Dale Lambert, Tarrant County College NE
Arthur Lee, Roane State Community College
Harvey Liftin, Broward Community College
Walter Little, University at Albany, SUNY
Mary Meiners, San Diego Miramar College
Scott Miller, Penn State University
Jane Murphy, Virginia College Online
Bethany Myers, Wichita State University
Terri Oltman, Westwood College
Keith Prufer, Wichita State University
Ann Somers, University of North Carolina, Greensboro
Donald Thieme, Georgia Perimeter College
Kip Thompson, Ozarks Technical Community College
Judy Voelker, Northern Kentucky University
Arthur Washington, Florida A&M University
Stephen Williams, Glendale Community College
Feranda Williamson, Capella University.

Dedication

We dedicate the Second Edition of *Visualizing Physical Geography* to Dr. John E. Estes, one of the founding geographers of the University of California at Santa Barbara's Department of Geography. Jack, as he was known to one and all, introduced modern remote sensing to the world of geography. Through Jack's pioneering work, remote sensing rapidly advanced as a major tool for geographers to map, measure, and monitor the dynamics of our planet. Jack was a close friend to both of us, mentoring Tim as a graduate student and researcher and Alan as a young faculty member as we learned the trade of remote sensing and its application to a wide range of areas within physical geography. He managed to hook us both on a field that has proven to be exciting, challenging, and satisfying beyond our wildest expectations. Jack passed away in 2001 and life for us has never been quite the same. But as we thumb the pages of our new edition, we are constantly reminded of all we learned from Jack and of how much of that knowledge we have been able to pass along to you.

TF
AS

About the Authors

Tim Foresman earned his Ph.D. degree in Geography from the University of California at Santa Barbara. He is adjunct Professor of Geography at University of Maryland and president of the International Center for Remote Sensing Education. He has over 100 professional articles and books on environmental protection and sustainable development using the applied tools of geography, including remote sensing and geographic information systems (GIS). His pioneering work in spatial information systems and geographic information systems includes a seminal publication on the history of GIS and a Lifetime Achievement Award from ESRI. His active research and field studies have included tenure with the US Department of Defense, US Environmental Protection Agency, NASA, and the United Nations Environment Programme. His international recognition includes service as adjunct professor at Coventry University in England and Qinghai University in China, as well as visiting professor at Keio University in Japan. He was elected as Fellow of Sigma Xi, the scientific research society.

Alan Strahler earned his Ph.D. degree in Geography from Johns Hopkins in 1969, and is presently Professor of Geography at Boston University. He has published over 250 articles in the refereed scientific literature, largely on the theory of remote sensing of vegetation, and has also contributed to the fields of plant geography, forest ecology, and quantitative methods. In 2011, he received the William T. Pecora Award for outstanding contributions to understanding the Earth by remote sensing, sponsored by the Department of the Interior and National Aeronautics and Space Administration. He has also been awarded the Association of American Geographers/Remote Sensing Specialty Group Medal for Outstanding Contributions to Remote Sensing. With Arthur Strahler, he is a coauthor of seven textbook titles with twelve revised editions on physical geography and environmental science. He holds the honorary degree D.S.H.C. from the Université Catholique de Louvain, Belgium, and is a Fellow of the American Association for the Advancement of Science.

Special Thanks

Visualizing Physical Geography, Second Edition, has been a great experience for the authors. With the guidance of our skilled and experienced Developmental Editor, Rebecca Heider, we achieved not only a remarkable visual presentation of the concepts and processes of physical geography, but also a better internal structure and organization to enhance the learning process. We couldn't have done it without her. We also owe Nancy Perry, development manager for the Wiley Visualizing Imprint, a great debt for putting our ideas on the page in a way that adds visual excitement and makes our text easy to read and learn from. We greatly appreciate the help that editorial assistants Brittany Cheetham and Darnell Sessoms provided to this project.

Our new edition is also greatly enhanced by new photos found by Senior Photo Editor Mary Ann Price and Photo Researcher Elle Wagner, in addition to many new selections from the National Geography library. Sandra Rigby, Senior Illustration Editor, helped us execute the many new and revised art pieces that are a part of our Second Edition. Jeanine Furino of Furino Production led our production team with great skill on the straight and narrow path from final chapter manuscripts to bound books, with support from Micheline Frederick and Bill Murray. Our compositor and illustration studio, Precision Graphics, laid out every page to our team's most exacting specifications while producing amazing art pieces and striking page spreads. Our Executive Editor Ryan Flahive, along with Vice-President and Publisher Jay O'Callaghan and Senior Marketing Manager Margaret Barrett, provided the strategy for bringing the book to our users. Beth Tripmacher and Darnell Sessoms brought together the extensive Web support package that students and instructors rely on for adding to the learning experience.

Special thanks to all the members of our great team!

Special thanks from Tim to his wife, Joyce, who was most supportive over the many months of this production process.

Tim Foresman
Alan Strahler

1 Discovering the Earth's Dimensions

2 The Earth's Global Energy Balance

3 Air Temperature

4 Atmospheric Moisture and Precipitation

5 Global Atmospheric and Oceanic Circulation

6 Weather Systems

9 Plate Tectonics, Earthquakes, and Volcanoes

10 Weathering and Mass Wasting

11 Fresh Water of the Continents

12 Landforms Made by Running Water

13 Landforms Made by Wind and Waves

14 Glacial and Periglacial Landforms

15 Global Soils

16 Biogeographic Processes

17 Global Biogeography

A series or combination of figures and photos that describe and depict a complex process

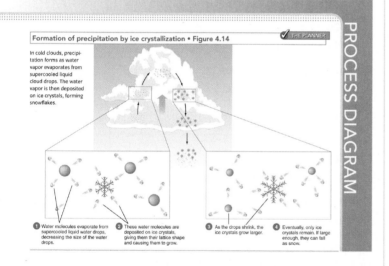

Formation of precipitation by ice crystallization • Figure 4.14

Multipart visual presentations that focus on a key concept or topic in the chapter

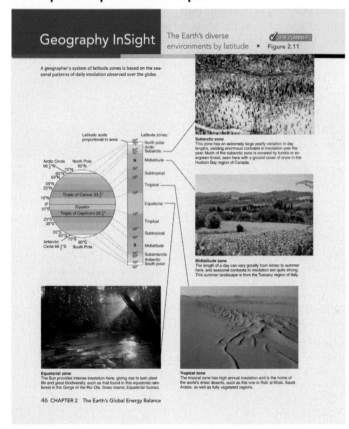

Chapter 1
Geometry of map projections

Chapter 2
The Earth's diverse environments by latitude

Chapter 3
Net radiation and air temperature by latitude
Consequences of global warming

Chapter 4
Global precipitation

Chapter 5
Cyclic patterns of oceanic circulation

Chapter 6
Source regions of air masses

Chapter 7
Factors controlling global precipitation
Climate classification

Chapter 8
Geologic time
Global landforms

Chapter 9
Lithospheric plates and their motions
Types of volcanoes

Chapter 10
Types of mass wasting

Chapter 11
Water access and use

Chapter 12
Erosional and depositional landforms

Chapter 13
Types of sand dunes

Chapter 14
Types of alpine glaciers

Chapter 15
Soil orders and global agriculture

Chapter 16
Threatened biodiversity

Chapter 17
How climate influences terrestrial biomes

Multipart figures that illustrate the tools and processes used by scientists to collect data and draw conclusions

Chapter 1
Monitoring change through satellite scanning
Constellation of Earth-monitoring satellites
GIS basics

Chapter 2
Calibrating sensors for atmospheric ozone monitoring

Chapter 3
Measuring the Earth's temperature
Reconstructing temperature records through ice core
    analysis

Chapter 4
Measuring precipitation

Chapter 5
Radiosonde

Chapter 6
Conditions inside a hurricane

Chapter 7
Collecting climate data

Chapter 8
Chalk cliffs provide a record of climate change
Undersea topography

Chapter 9
Space-based instruments imaging Earth movements

Chapter 10
Mapping slope stability

Chapter 11
Land subsidence mapping

SCIENCE TOOLS

GIS basics • Figure 1.25

GIS allows for the integration of a variety of different spatial data sources and formats into a common digital geocoded database that maintains the spatial locations of all database elements. GIS data can be viewed, using this shared geography, as a series of data layers, covering the same geographic area, which can then be used for modeling, analysis, and display.

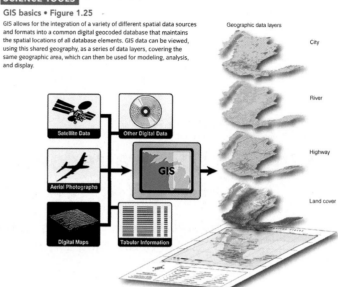

Chapter 12
Delta dynamics

Chapter 13
Transformation of the Sahara

Chapter 14
Monitoring the effect of climate change on glaciers

Chapter 15
Paving paradise

Chapter 16
Carbon budget in boreal forest ecosystem

Chapter 17
Deforestation in Borneo

WILEY PLUS

www.wileyplus.com

WileyPLUS is a research-based online environment for effective teaching and learning.

WileyPLUS builds students' confidence because it takes the guesswork out of studying by providing students with a clear roadmap:

- what to do
- how to do it
- if they did it right

It offers interactive resources along with a complete digital textbook that help students learn more. With *WileyPLUS*, students take more initiative so you'll have greater impact on their achievement in the classroom and beyond.

1 Discovering the Earth's Dimensions

In ancient Egypt during the 3rd century BCE, a traveling scholar in the southern town of Syene (modern Aswan) observed on the summer solstice, the longest day of the year, that vertical objects—towers or poles—cast no shadows at noon, because the Sun was directly overhead. This man, a Greek named Eratosthenes, was the librarian for Alexandria, a northern city, where he had previously noted that shadows were cast during the summer solstice.

As the world's first geographer, Eratosthenes used his observations of differences in the Sun's noon shadows for his classic scientific investigation of the Earth's size. He combined his knowledge of geometry, astronomy, and geography to calculate the Earth's circumference in present units as 39,690 kilometers (24,662 miles), with an error of less than 2%. Although his assumptions contained numerous errors, his sound approach to the problem helped establish the scientific principles in use in modern geography.

What we know today about our planet and its interrelated systems is predicated on the same scientific method that Eratosthenes used: establishing questions, gathering data, and sharing the results of our discoveries. After 2300 years, geographers continue to explore and map the world.

To reach Syene, now Aswan, from Alexandria, Eratosthenes might have sailed up the Nile River on a felucca, an ancient type of Egyptian sailboat, shown here in a present-day photograph.

CHAPTER OUTLINE

CHAPTER PLANNER ✓

❏ Study the picture and read the opening story.

❏ Scan the Learning Objectives in each section:
   p. 4 ❏   p. 10 ❏   p. 11 ❏   p. 14 ❏   p. 20 ❏   p. 25 ❏

❏ Read the text and study all figures and visuals. Answer any questions.

Analyze key features

❏ Process Diagrams p. 6 ❏   p. 8 ❏

❏ Geography InSight, p. 22

❏ Where Geographers Click p. 25 ❏   p. 30 ❏

❏ Video Explorations, p. 27

❏ What a Geographer Sees, p. 29

❏ Stop: Answer the Concept Checks before you go on.
   p. 9 ❏   p. 11 ❏   p. 14 ❏   p. 19 ❏   p. 25 ❏   p. 30 ❏

End of chapter

❏ Review the Summary and Key Terms.

❏ Answer the Critical and Creative Thinking Questions.

❏ Answer What is happening in this picture?

❏ Complete the Self-Test and check your answers.

The World of Physical Geography

LEARNING OBJECTIVES

1. **Distinguish** between physical and human geography.
2. **Describe** the Earth's four systems.
3. **Outline** the steps in the scientific method.

Over the next 40 years, the Earth's population is projected to rise from about 7 billion to 9 billion people. The entirety of our combined scientific knowledge will be required to address the challenge of feeding and sheltering this growing population without depleting or irreparably damaging the Earth's life-sustaining systems. The science of physical geography plays a valuable role in helping us better understand the planet and address issues of sustainability for all life forms in the twenty-first century.

The Science of Geography

Geography is an ancient discipline devoted to the study of the Earth. Geographers are concerned with the interesting features and places on the Earth's surface, the distribution of phenomena on the Earth, and how all these things change over time. They take into account both spatial considerations (related to physical space) and temporal considerations (related to changes with the passage of time). Geographers sometimes refer to these five essential themes of geography:

1. *Location*—Geography helps identify a specific location on a map, such as your home address or your position as located by a Global Positioning System (GPS).
2. *Place*—Geography takes into consideration not just map coordinates but a combination of characteristics

Geography's major fields and subfields • Figure 1.1

Geographers engage across a variety of fields, using spatial technologies and methods, to examine, map, and better understand the physical world and its spatial interactions over time. Geography's two major domains, physical and human geography, cover a range of subfields, only a portion of which are shown here.

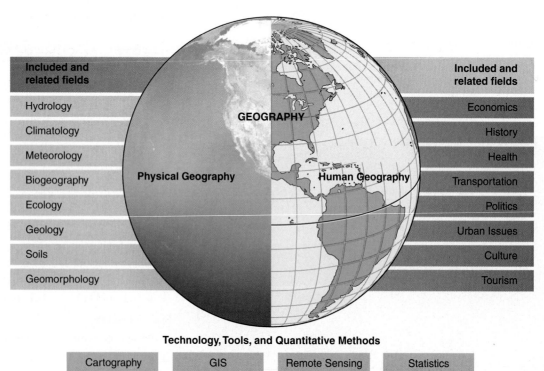

that makes a place unique, including landscape, climate, language, and culture.

3. *Region*—Geographers group areas with similar characteristics in common, such as climate or industry, into regions to provide context for comparing similarities and differences as well as changes over time.

4. *Human-Earth relationships*—The study of human-Earth relationships helps geographers understand how humans shape our Earth and exploit its resources and how the environment influences humans.

5. *Movement*—Geographers study movement to understand how people, goods, ideas, and energy travel across the globe.

Geography was once the domain of explorers and philosophers, whose privileged position within a ruler's court allowed them to document with maps their discoveries of the known world. As geography evolved from its foundations in astronomy and mathematics, multiple offshoots of the discipline have grown into separate disciplines, ranging from geology to ecology, political science to medical science, transportation to migration.

Today, geographers divide their discipline into two main subfields. **Physical geography** addresses the study of the features of the Earth's living and nonliving systems, such as its landscapes and the natural processes that formed them, as well as its planetary origins. **Human geography** focuses on spatial interactions and patterns related to human activity, including social, cultural, and economic topics (**Figure 1.1**).

Physical geographers help refine and quantify our knowledge about the Earth, especially its surface and systems. Such diverse topics as climate change, earthquakes, resource extraction, and the dynamic Earth systems that support the human species are all components of modern physical geography. By understanding and applying modern tools of physical geography, students will discover the wonders of our natural world and the processes that continue to reshape them.

A World of Systems

Physical geography examines the Earth in terms of four major systems: the atmosphere, hydrosphere, lithosphere, and biosphere (**Figure 1.2**). The **atmosphere** is composed of gases that envelop and protect the planet from harmful solar radiation. The **hydrosphere** includes all the water, in all its forms, including the oceans, surface waters of the lands, ground water, water held in the atmosphere, and precipitation. The **lithosphere** is the

The Earth's four major systems • Figure 1.2

Constant interactions of energy and material between the Earth's four major systems create the life-supporting conditions on our unique planet.

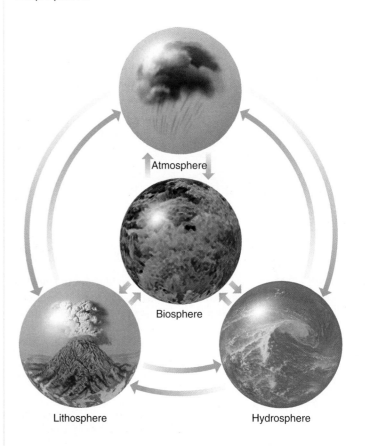

Atmosphere

Biosphere

Lithosphere

Hydrosphere

structural rock foundation of the planet's surface. The **biosphere** represents the Earth's living plants and organisms, which use the hydrosphere for moisture, the atmosphere for life-supporting gases, and the lithosphere for nutrients.

The natural processes that continue to shape our planet involve dynamic interactions between these four spheres. For example, water molecules in the atmosphere play a central role in the hydrosphere, affecting the weather and providing critical moisture—from rain, streams, lakes, and the ocean—for the biosphere. Rain and water flowing over land continuously reshape the lithosphere, eroding mountains and plains. The constant flow of energy and matter among these four principal systems defines the main interrelationships within the Earth's dynamic systems. How these systems operate and the influence each system has on the others will be a recurring focus of the coming chapters.

Methods and Tools for Geography

The tools of geography have evolved over the many centuries since the time of Eratosthenes's first global measurements, mentioned in the chapter opener. Space-age technologies, including remote sensing, geographic information systems (GIS), GPS, and the Internet, have added many quantitative spatial techniques to the geographer's toolbox. Advances in software mapping programs and Web-based, three-dimensional visualization tools, such as Google Earth, have made the mapping tools of professional geographers accessible to billions of citizens on the planet. Such Web tools provide a means for uniting our global and local perspectives for a better-informed society. With unprecedented accuracy and timeliness, anyone who has access to the Internet can ask and answer questions about the Earth's geography.

Computer technologies alone, however, do not foster good science. Like all other scientists, regardless of their tools, geographers follow the conventional steps of the **scientific method** to ensure that their theories and hypotheses are confirmed by reproducible

> **scientific method**
> The formal process that a scientist uses to solve a problem, which involves first observing and formulating a hypothesis and then testing and evaluating results.

PROCESS DIAGRAM

Steps of the scientific method • Figure 1.3

Each step of the scientific method must be carefully documented to enable other scientists to replicate the process and independently confirm the findings. A scientist may test and revise numerous hypotheses before making a scientific conclusion.

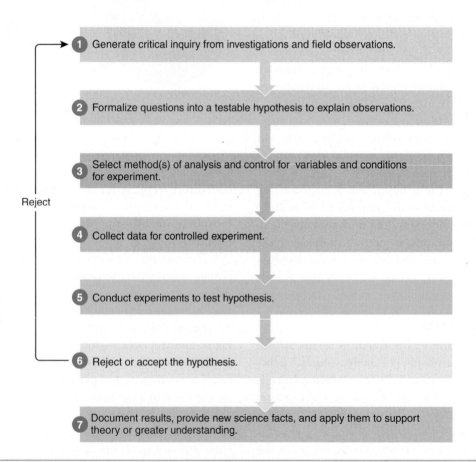

1. Generate critical inquiry from investigations and field observations.
2. Formalize questions into a testable hypothesis to explain observations.
3. Select method(s) of analysis and control for variables and conditions for experiment.
4. Collect data for controlled experiment.
5. Conduct experiments to test hypothesis.
6. Reject or accept the hypothesis.
7. Document results, provide new science facts, and apply them to support theory or greater understanding.

Reject

experimental and observational results. This ensures that each scientific advance builds on the solid foundation of previous scientific work.

A geographer's best tool is an inquisitive mind. The first step in the scientific method is to use field observations or literature reviews to stimulate questions and ideas. If a process or spatial pattern presents itself, the geographer then frames a key question or set of questions into a testable **hypothesis**. The geographer then designs an approach to evaluate the hypothesis, tests the hypothesis, and, based on the results, either accepts the hypothesis as valid or returns to the formulation step to create alternatives for further testing (**Figure 1.3**). By documenting and reporting the results, geographers are able to share their discoveries to advance our common understanding of how the world works.

> **hypothesis** A logical explanation for a process or phenomenon that allows prediction and testing by experiment.

Now let's take a closer look at how a group of geographers might apply the scientific method to their research (**Figure 1.4** on the next page).

Step 1. *Generate critical inquiry from investigations and field observations.* In this example, geographers who study volcanoes along the Hawaiian archipelago observe a pattern of vegetative growth that is different on the west side of each of the Hawaiian Islands than on the east side.

Step 2. *Formalize questions into a testable hypothesis to explain observations.* The geographers have three reasonable guesses about what factor is causing the differences in vegetation they witnessed in the field: rock type, temperature, or rainfall. These guesses form the basis for three hypotheses, which they will test one at a time.

Step 3. *Select method(s) of analysis and control for variables and conditions for experiment.* The geographers use the big island of Hawaii to map the specific 16 × 16 km (10 × 10 mi) square test areas on the west and east sides of the island.

Step 4. *Collect data for controlled experiment.* The geographers collect existing data about the four factors they are investigating. They obtain vegetation maps from NASA satellites. They also obtain geologic maps, which are readily available, both in paper and digital formats, from the U.S. Geological Survey (USGS) and state agencies. They get temperature data from maps published by the National Oceanic and Atmospheric Agency (NOAA) and the USGS. They collect rainfall data from NOAA's weather service.

Step 5. *Conduct experiments to test hypothesis.* The geographers begin by overlaying the rock type map on the vegetation map to look for similar patterns, or correlation, between the rock types and the vegetation within the square grids. If they reject this hypothesis, they will return to this step to test the next variable.

Step 6. *Reject or accept the hypothesis.* The comparison of the rock type map and the temperature map reveals no correlation with vegetation patterns, so the geographers reject these hypotheses. In the test of rainfall patterns, however, the geographers find a positive correlation with the vegetation pattern. They therefore accept this hypothesis and conclude that differences in vegetation patterns can be explained by differences in the rainfall patterns on the west and east sides of the island.

Step 7. *Document results, provide new science facts, and apply them to support theory or greater understanding.* Documentation of the geographers' experiments enables others to review the experiments for verification. When sufficient research has been conducted, for example, on each of the Hawaiian Islands, the hypothesis may be elevated to a theory. A **theory** represents a hypothesis that through rigorous validation obtains the status of a recognized scientific principle or conclusion. The word

> **theory** A hypothesis that has been tested and is strongly supported by experimentation, observation, and scientific evidence.

theory is often used in casual speech when we mean hypothesis. But in the scientific community, theories receive a much higher level of respect and are rare, whereas hypotheses are considered the basic tools for scientific exploration and advancement of knowledge.

Geographic use of the scientific method • Figure 1.4

 THE PLANNER

The scientific method is used to investigate spatial vegetation patterns and discover which alternative hypothesis best explains the observed variations from the west coasts and east coasts of the Hawaiian Islands.

1 Generate critical inquiry. Scientists observe different vegetation patterns on the east and west sides of the Hawaiian Islands and speculate about what causes those differences.

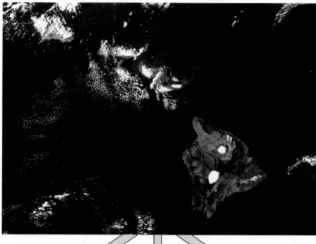

2 Formalize questions into a testable hypothesis. They formulate three hypotheses to explain the vegetative growth patterns observed in the field.

Hypothesis 1
Vegetation patterns are explained by rock type.

Hypothesis 2
Vegetation patterns are explained by temperatures.

Hypothesis 3
Vegetation patterns are explained by rainfall.

3 Select method(s) of analysis. They define boundaries for the specific area that will be sampled for each variable.

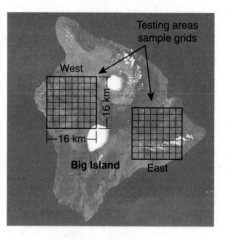

4 Collect data. They collect data on each variable.

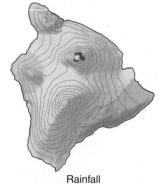

Rock type

Temperature

Rainfall

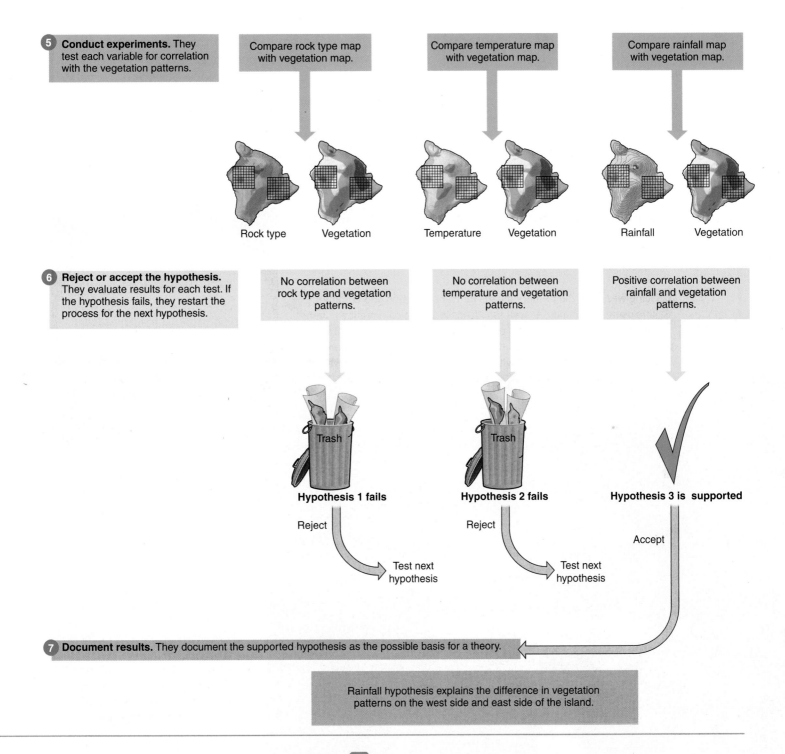

5 **Conduct experiments.** They test each variable for correlation with the vegetation patterns.

Compare rock type map with vegetation map.

Compare temperature map with vegetation map.

Compare rainfall map with vegetation map.

Rock type   Vegetation   Temperature   Vegetation   Rainfall   Vegetation

6 **Reject or accept the hypothesis.** They evaluate results for each test. If the hypothesis fails, they restart the process for the next hypothesis.

No correlation between rock type and vegetation patterns.

No correlation between temperature and vegetation patterns.

Positive correlation between rainfall and vegetation patterns.

Trash

Trash

Hypothesis 1 fails

Hypothesis 2 fails

Hypothesis 3 is supported

Reject

Reject

Accept

Test next hypothesis

Test next hypothesis

7 **Document results.** They document the supported hypothesis as the possible basis for a theory.

Rainfall hypothesis explains the difference in vegetation patterns on the west side and east side of the island.

CONCEPT CHECK   STOP

1. **Which** subfield of geography is interested in population distribution around the globe?

2. **What** role does water play in each of the Earth's four spheres?

3. **What** is the difference between a hypothesis and a theory?

The Shape of the Earth

LEARNING OBJECTIVES

1. **Explain** how we know the Earth is approximately spherical.

2. **Describe** how and why the Earth's actual shape deviates from a perfect sphere.

Geography, as we have just discussed, is concerned with the study of places on the Earth, so the most fundamental place to begin our study of geography is with the entire Earth itself. What do we actually know about the Earth's shape?

People have known for thousands of years that the Earth is spherical, based on mathematical calculations along with observations of the Sun and sky (**Figure 1.5**).

Curvature of the Earth • Figure 1.5

The Earth's curvature was evident to ancient scholars and seamen from their keen observations of the shapes of the Moon and Sun, ships sailing over the horizon, and sunsets drifting over the horizon. Today, space photographs provide ample evidence that the Earth is indeed round.

a. Sailing was once the primary means for transportation and commerce. Sailors using telescopes, as well as citizens in sailing communities, were accustomed to seeing the mast of ships first on the horizon as ships approached ports of call. As ships left port and sailed over the horizon, witnesses were soon acquainted with the concept of the Earth's curvature.

b. As you watch the last minutes of the sunset from the ground, the Sun lies below the horizon, no longer illuminating the land around you. But at the height of the clouds, the Sun has not yet dipped below the horizon, so it still bathes the clouds in red and pink rays. As the Sun washes over the clouds, the red band of light slowly moves farther toward the horizon. In this dramatic sunset photo, the far distant clouds are still directly illuminated by the Sun's last rays. For the clouds directly overhead, however, the Sun has left the sky.

c. Today we have the advantage of observing the Earth from the vantage point of space. This photo of Skylab, taken by astronauts leaving the ship, shows the Earth's curved horizon.

axis An imaginary straight line through the center of the Earth around which the Earth rotates.

pole One of the two points on the Earth's surface where the axis emerges.

The Earth is not, however, as perfectly round as is suggested by most world globes, due to the centrifugal forces acting on the Earth as it spins constantly around its **axis**. As the Earth spins, the outward force of rotation causes it to bulge slightly around the middle, or equator, and flatten at the **poles (Figure 1.6)**. The difference is very small—about 0.3%—but strictly speaking, the Earth's squashed shape is closer to what is technically known as an oblate spheroid, not a perfect sphere.

CONCEPT CHECK **STOP**

1. **How** can observations of the sky lead to the conclusion that the Earth is round?
2. **What** is the effect of the Earth's rotation on its shape?

Shape of the Earth • Figure 1.6

The diameter of the Earth is slightly larger at the equator than it is between the poles.

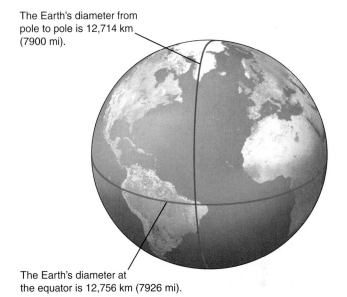

The Earth's diameter from pole to pole is 12,714 km (7900 mi).

The Earth's diameter at the equator is 12,756 km (7926 mi).

Global Location

LEARNING OBJECTIVES

1. **Describe** the geographic grid system.
2. **Explain** how we determine position on the globe.

Locations are essential to the study of geography, and to discuss locations, we need a universal system to identify a position on the Earth. Cartography, navigation on sea and land, and geographic studies are all based on the application of this universal system.

The Geographic Grid

It is impossible to lay a flat sheet of paper over a sphere without creasing, folding, or cutting it—as you know if you have tried to gift wrap a ball. This simple fact has challenged mapmakers for centuries, as they have grappled with expressing the three-dimensional world on two-dimensional parchment. Because the Earth's surface is curved, we cannot divide it into a rectangular grid any more than we could smoothly wrap a globe in a sheet of graph paper. Instead, we divide the Earth into a geographic grid, which is made up of a system of imaginary circles.

Meridians Imagine slicing the Earth in half through its axis of rotation, just as you might slice an apple along its core. The cut edge of the Earth forms a circle touching the poles that has the Earth's center as the circle's center. This circle belongs to the family of **great circles**, which are the largest circles that can be traced on the globe. All great circles bisect the Earth and have the center of the Earth as their center. Although great circles can be drawn along any plane that splits the globe evenly, the north–south lines that make up the geographic grid, called **meridians**, involve only those great circles that pass through both poles. Each meridian is actually only one-half of a great circle, connecting one pole to the other.

meridian A north–south line on the Earth's surface that connects the poles.

The paths of the great circles also represent the shortest distance between two points on the globe. For example, airplanes on long international flights often follow routes that arc toward the poles, along the paths of the great circles, because those are the shortest routes between two places.

The geographic grid • Figure 1.7

The geographic grid is a network of imaginary circles that segments the surface of the Earth.

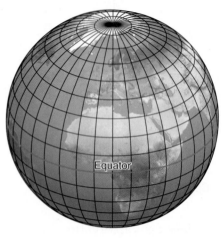

a. Meridians are halves of great circles that divide the globe vertically from pole to pole.

b. Parallels are small circles that lie parallel to the equator and divide the globe crosswise into rings. (The equator is actually a great circle, since the equatorial plane passes through the Earth's center.)

c. The intersection of meridians and parallels makes up the geographic grid.

Parallels Imagine cutting the globe just as you might slice an onion to make onion rings. Lay the globe on its side, so that the axis runs parallel to your imaginary chopping board, and begin to slice. Each cut creates a circular outline that passes right around the surface of the globe. These **small circles** are smaller at the poles and grow to their largest point at the **equator**. Like great circles, small circles can be traced on any plane, but the only small circles that make up the geographic grid are those that run east and west, parallel to the equator, called **parallels**. The equator is the only parallel that is also a great circle, because its center is also the Earth's center. When meridians and parallels are traced on the globe, they create a grid of intersecting lines (**Figure 1.7**).

> **equator** The parallel that lies midway between the Earth's poles.
>
> **parallel** An east–west circle on the Earth's surface, lying in a plane parallel to the equator.

Meridians and parallels define geographic directions. When you walk directly north or south, you follow a meridian; when you walk east or west, you follow a parallel. An infinite number of parallels and meridians can be drawn on the Earth's surface, just as an infinite number of positions exist on the globe. Every point on the Earth is associated with a unique combination of one parallel and one meridian. The position of a point is defined by such an intersection.

Latitude and Longitude

We label parallels by describing their distance north or south of the equator, or their **latitude** (**Figure 1.8a**). "Latitudes are like ladders," is a memory device that describes the manner in which parallels climb up the globe from south to north. The equator divides the globe into two equal portions, or hemispheres. All parallels in the northern hemisphere are given as north latitude, and all points south of the equator are given as south latitude (N or S). So the equator is at 0°, while the North Pole is at 90° N and the South Pole is at 90° S.

> **latitude** An angular distance for a point north or south of the equator, as measured from the Earth's center.

Meridians are identified by **longitude**, which is an angular measure of how far eastward or westward the meridian is from an arbitrary reference meridian (**Figure 1.8b**). International agreement, in 1884, established the reference meridian, known as

> **longitude** The angular distance for a point east or west of the prime meridian (Greenwich), as measured from the Earth's center.

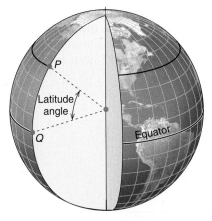

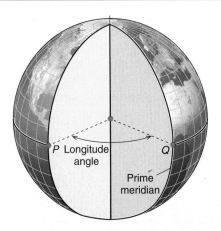

a. The latitude of a parallel is the angular distance between a point on the parallel (*P*) and a point on the equator along the same meridian (*Q*) as measured from the Earth's center.

b. The longitude of a meridian is the angular distance between a point on that meridian at the equator (*P*) and a point on the prime meridian at the equator (*Q*) as measured from the Earth's center.

the **prime meridian**, as the one that passes through the old Royal Observatory at Greenwich, near London, England (**Figure 1.9**). Until the early twentieth century, other reference meridians, such as the Paris Meridian of French cartography, were in simultaneous usage, especially for maps of West Africa. Today, Greenwich's prime meridian is firmly established as the international standard. It has a value of 0° longitude.

The longitude of a meridian on the globe is measured eastward or westward from the prime meridian, depending on which direction gives the smaller angle. So longitude ranges from 0° to 180° east or west (E or W).

Prime meridian • Figure 1.9

This stripe in the forecourt of the old Royal Observatory at Greenwich, England, marks the prime meridian. Many international conventions were held from the late 1800s to mid-1900s to formulate treaties confirming this global standard. Today, all navigational charts and electronic systems adhere to this line among the cobblestones of Greenwich as marking the universal longitude reference meridian.

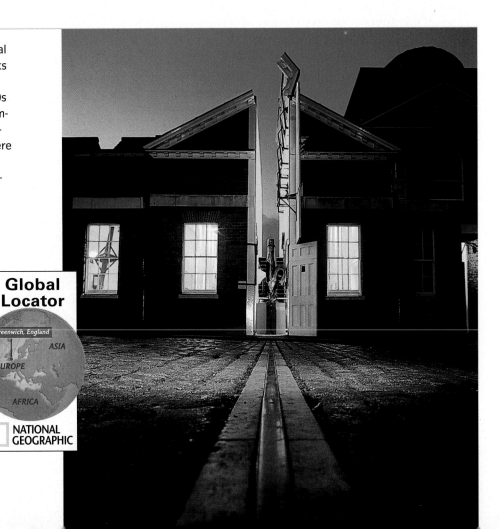

Latitude and longitude of a point • Figure 1.10

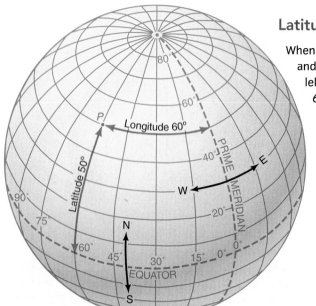

When both the latitude and longitude of a point are known, it can be accurately and precisely located on the geographic grid. The point *P* lies on the parallel of latitude at 50° north (50° from the equator) and on the meridian at 60° west (60° from the prime meridian). Its location is therefore lat. 50° N, long. 60° W.

Together, the intersection of latitude and longitude pinpoint locations (in theory to within a few meters) on the geographic grid (**Figure 1.10**). Fractions of latitude or longitude angles are described using minutes and seconds. A minute is 1/60 of a degree, and a second is 1/60 of a minute, or 1/3600 of a degree. So, the latitude 41°, 27 minutes (′), and 41 seconds (″) north (lat. 41°27′41″ N) means 41° north plus 27/60 of a degree plus 41/3600 of a degree. This cumbersome system has now been replaced almost entirely by decimal notation. In this example, the latitude 41°27′41″ N translates to 41.4614° N.

A third axis exists for defining locations. Altitude—that is, height or elevation—is denoted by *z* to complement

the *x* and *y* calculated from latitude and longitude. Altitude is usually measured from either sea level or from the center of the Earth. GPS devices allow the user to define *x*, *y*, and *z* positions to within less than 1 meter, which is considerably better than could be achieved using latitudes and longitudes from a map. Three-dimensional mapping and analysis rely on the use of altitude to create terrain maps and building profiles and to represent the Earth more realistically, as discussed in greater detail later in this chapter.

CONCEPT CHECK

1. **How** are meridians and parallels related to the geographic grid?
2. **How** are latitude and longitude related to meridians and parallels?

Global Time

LEARNING OBJECTIVES

1. **Describe** how the Sun's position regulates global time.
2. **Discuss** the need for world time zones.
3. **Explain** why we need the international date line and daylight saving time.

Now we've seen the universal means of identifying a location on the globe. But we also want to be able to pinpoint a location at a specific time, so we need a universal measure of time. The Earth's geographic grid and the rotation of the Earth help define global time.

Solar Time

Our most basic measure of time is based on the Earth's rotation. The Earth spins slowly through the day and night, making one full turn with respect to the Sun every day. We define a solar day as one rotation, and for centuries we

Solar timekeeping • Figure 1.11

This version of a sundial at the University of Arizona works because a shadow cast on the ground changes its angle as the position of the Sun in the sky changes through the day. In the morning, the user's shadow falls to the user's left. At solar noon, it points straight ahead, to the north. In the afternoon, the shadow falls to the user's right. This sundial also factors in how the noon elevation of the Sun changes according to time of year. The user selects a standing position based on the day of the year and then reads the time from the position of the shadow.

have chosen to divide the solar day into exactly 24 units, called hours, and devised clocks to keep track of hours in groups of 12.

As the Earth rotates, the Sun appears to move across the sky from east to west. In the morning, the Sun is low on the eastern horizon, and as the day progresses, it rises higher, until at solar noon it reaches its zenith, the highest point in the sky. Solar noon is distinct from 12:00 on timepieces, as will be discussed next. After solar noon, the Sun's elevation in the sky decreases. By late afternoon, the Sun hangs low in the sky, and at sunset it rests on the western horizon. Noting the position of the Sun in the sky is the most basic way to determine time (**Figure 1.11**).

Because time is referenced to the Sun's position in the sky, while we are waking up, people on the other side of the planet are going to sleep. This is because the Sun only shines on half of the Earth at one time. As the Earth rotates, the Sun shines on eastern regions first and gradually illuminates westward regions as the day progresses. Relative local time is determined primarily by longitude because only one meridian can be directly under the Sun and experience solar noon at a given moment (**Figure 1.12**).

Time is determined by longitude • Figure 1.12

When it is noon in Chicago, locations on meridians to the east of Chicago, such as New York, have passed solar noon, and locations to the west of Chicago, such as Portland, have not yet reached solar noon. Even though Mobile is about 1600 km (1000 mi) away from Chicago, because these two cities have nearly the same longitude, they experience solar noon at approximately the same time.

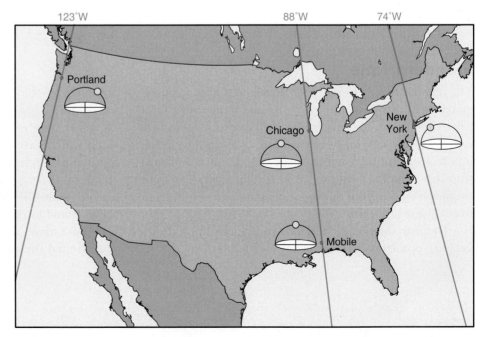

This map shows standard meridians radiating out from the North Pole. This scheme provides the framework for world time zones, although commerce and political consideration have generated variations to this scheme.

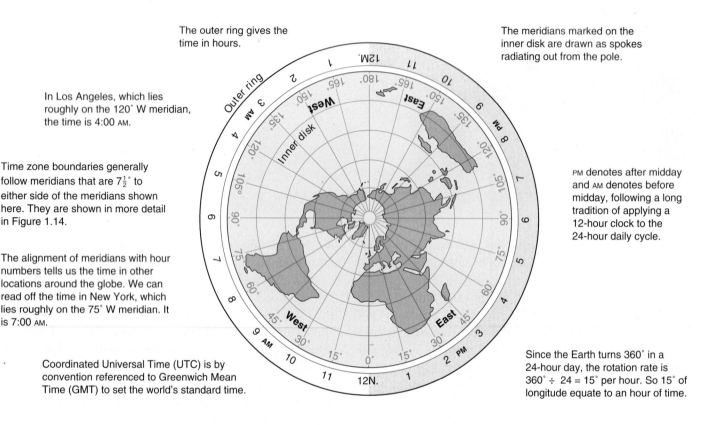

The outer ring gives the time in hours.

The meridians marked on the inner disk are drawn as spokes radiating out from the pole.

In Los Angeles, which lies roughly on the 120° W meridian, the time is 4:00 AM.

Time zone boundaries generally follow meridians that are $7\frac{1}{2}°$ to either side of the meridians shown here. They are shown in more detail in Figure 1.14.

The alignment of meridians with hour numbers tells us the time in other locations around the globe. We can read off the time in New York, which lies roughly on the 75° W meridian. It is 7:00 AM.

Coordinated Universal Time (UTC) is by convention referenced to Greenwich Mean Time (GMT) to set the world's standard time.

PM denotes after midday and AM denotes before midday, following a long tradition of applying a 12-hour clock to the 24-hour daily cycle.

Since the Earth turns 360° in a 24-hour day, the rotation rate is 360° ÷ 24 = 15° per hour. So 15° of longitude equate to an hour of time.

The prime meridian is directly under the Sun—that is, at this moment, the 0° meridian is directly on the 12:00 noon mark—the highest point of its path in the sky in Greenwich, England.

Standard Time

The Sun has been a timekeeper throughout history, but as global commerce became more prevalent, the use of local solar time began to cause problems. For example, as new railroads were developed in the United States during the nineteenth century, each railroad used its own standard time, based on the local time at its headquarters—so train timetables were often incompatible and, as a result, train wrecks were a hazard.

The introduction of standard time solved these time-keeping problems by designating 24 standard meridians around the globe, at equal intervals from the prime meridian (**Figure 1.13**).

Each standard meridian forms the basis for a time zone, which extends roughly 7 1/2° to each side of the meridian. People within a time zone all keep the same standard time, even though solar noon is experienced at somewhat different times within the zone—earlier to the east and later to the west. If we know the time at the prime meridian, we can determine the time in every time zone around the globe (**Figure 1.14**).

Time zones of the world • Figure 1.14

World time zones are labeled at the bottom of the figure by the number of hours of difference between that zone and Greenwich mean time. So the number –7 tells us that local time is seven hours behind Greenwich time, while a +3 indicates that local time is three hours ahead of Greenwich time. 15° meridians are dashed lines, while the 7 1/2° meridians, which form many of the boundaries between zones, are bold lines. Adjacent zones appear in different colors. The top line shows the time of day in each zone when it is noon at the Greenwich meridian.

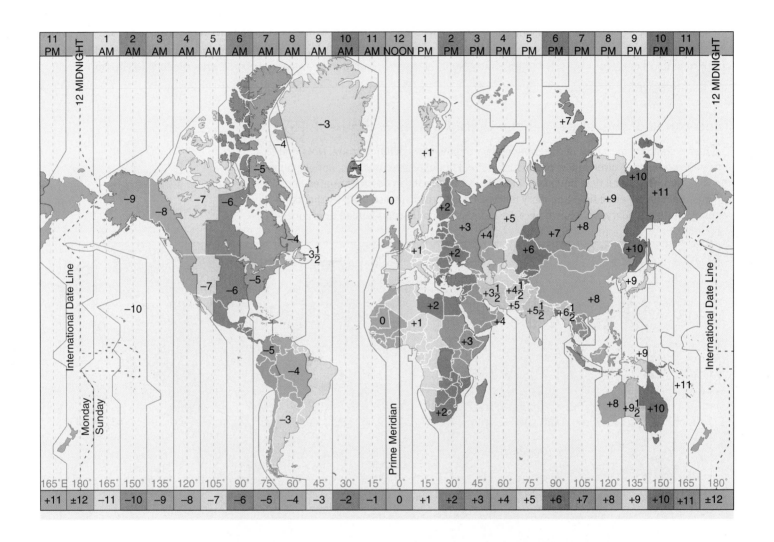

Put It Together

Review Figure 1.13 and answer this question:
As you go farther toward the poles, the east–west distance between time zones _____.

a. gets bigger
b. gets smaller
c. stays the same
d. does not change in a systematic way

<hr>

Some large countries span numerous time zones, and they typically modify the boundaries of the zones to accommodate political and economic considerations. From east to west, Russia spans 11 zones, but they are grouped into 8 standard time zones. China covers 5 time zones but runs on a single national time, using the standard meridian of Beijing. The United States and its Caribbean possessions are covered by 7 time zones (**Figure 1.15**).

A few countries, such as India and Iran, keep time using a meridian that is positioned midway between standard meridians, so that their clocks depart from those of their neighbors by 30 or 90 minutes. Some regions within countries also keep time by 7 1/2° meridians, such as the Canadian province of Newfoundland and the interior Australian states of South Australia and Northern Territory.

Coordinated universal time Since the 1950s, the most accurate time has been kept using atomic clocks, which are based on the frequency of microwave energy emission from atoms of the element cesium cooled to near absolute zero. These very accurate clocks keep atomic time, which is a universal standard that is not related to the Earth's rotation. Civil time sources use **coordinated universal time (UCT)**, which is derived from atomic time and provides a day of 86,400 seconds (24 hours) in length to match the Earth's mean rotation rate with respect to the Sun.

Time zones of the conterminous United States • Figure 1.15

The name, standard meridian, and number code are shown for each time zone. Note that time zone boundaries often follow preexisting natural or political boundaries. For example, the Eastern time–Central time boundary line follows Lake Michigan down its center, and the Mountain time–Pacific time boundary follows a ridge-crest line also used by the Idaho–Montana state boundary. Not shown here are the Alaska and Hawaii-Aleutian time zones.

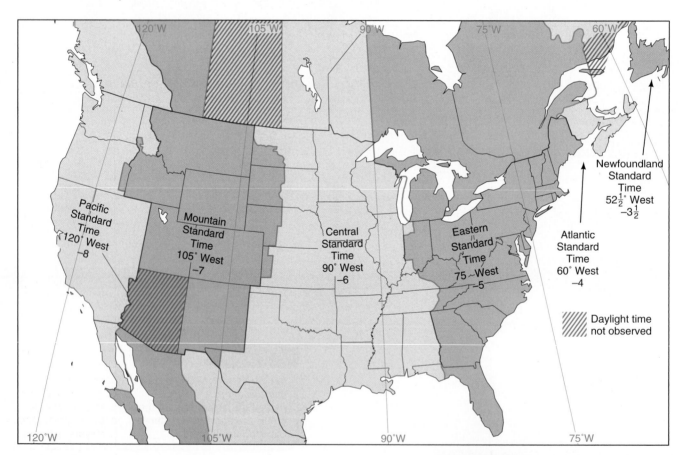

Ask Yourself

If it is 3:00 PM in New York City, what time is it in:

a. Atlanta?
b. Dallas?
c. Los Angeles?

The international date line follows the 180th meridian, with notable exceptions where it passes through the Aleutian Islands to the north and the island nation of Kiribati to the south. The Aleutians all keep the date to the east of the 180th meridian, while all the island groups within Kiribati keep the date to the west.

Ask Yourself

If you wanted to be among the first people in the world to celebrate New Year's Eve, would you travel to the Aleutian Islands or to Kiribati?

The International Date Line

Take a world map or globe with 15° meridians. Start at the Greenwich 0° meridian and count along the 15° meridians in an eastward direction. You will find that the 180° meridian is number 12 and that the time at this meridian is therefore 12 hours later than Greenwich standard time. Counting in a similar manner westward from the Greenwich meridian, we find that the 180° meridian is again number 12 but that the time is 12 hours earlier than Greenwich time. We seem to have a paradox: How can the same meridian be both 12 hours ahead of Greenwich time and 12 hours behind it? The answer is that each side of this meridian is experiencing a different day.

Imagine that you are on the 180° meridian on June 26. At the exact instant of midnight, the same 24-hour calendar day covers the entire globe. Stepping east will place you in the very early morning of June 26, while stepping west will place you very late in the evening of June 26. You are in the same calendar day on both sides of the meridian but 24 hours apart in time.

Doing the same experiment an hour later, at 1:00 AM, stepping east you will find that you are in the early morning of June 26. But if you step west, you will find that midnight of June 26 has passed, and it is now the early morning of June 27. So on the west side of the 180th meridian, it is also 1:00 AM, but it is one day later than on the east side. For this reason, the 180th meridian serves as the **international date line** (**Figure 1.16**). This means that if you travel westward across the date line, you must advance your calendar by one day. If you are traveling eastward, you set your calendar back by a day.

Air travelers on Pacific routes between North America and Asia cross the date line. On an eastward flight from Tokyo to San Francisco, you may actually arrive the day before you take off, due to the date change!

Daylight Saving Time

The United States and many other countries observe some form of **daylight saving time**, in which clocks are set ahead by an hour (sometimes two) for part of the year. Although it was once thought that adding daylight hours to the end of the workday would save electricity and reduce traffic accidents and crime, the evidence now shows that the primary effects are economic—allowing more retail shopping and recreation, for example, or reducing television watching. Although something of a mixed blessing, daylight saving time is now a part of normal life in most places.

CONCEPT CHECK	

1. **How** might a clock help you determine your longitude?

2. **Why** are standard meridians 15 degrees apart?

3. **Why** and in what circumstances do we routinely reset dates and times?

Mapping the Earth

LEARNING OBJECTIVES

1. **Explain** the significance of scale on a map.
2. **Identify** the three primary types of map projections.
3. **Discuss** how map projections distort information.

You have now learned how to locate a particular point in time and space according to its latitude, longitude, elevation, and standard time. A **map** gives us even more information about a location, by representing its spatial relationship with a surrounding area (**Figure 1.17**). Cartography is the subfield of geography that focuses on representing the Earth through maps.

Representing our expansive three-dimensional planet on a small, two-dimensional scale is a process that is heavily

map A graphic, scaled representative view of the Earth, or any portion of the Earth, as viewed from above, depicting various features of interest.

influenced by the assumptions and intentions of the mapmaker. What area is shown? What information is included? The distribution of the Earth's features on a map, as you will see throughout this book, can allow geographers to recognize patterns and draw conclusions about the processes that shape our planet.

Map Scales

For any map to be useful, we must know how the shapes on the flat page relate to the three-dimensional world around us. **Map scale** tells us how to convert between the distances drawn on our map and true Earth distances.

map scale The relationship between distance on a map and distance on the ground, given as a fraction or a ratio.

A verbal scale explains this through words, such as 1 cm = 1 km, which means that every centimeter of distance on the map equals 1 km of distance on the Earth. A representative fraction expresses the map scale in the form of a ratio. The ratio 1:50,000, for example, means that 1 unit

Basic features of a map • Figure 1.17

Basic features for good maps have stood the test of time. The standard features include title, scale and projection, date, location, author, and legend, although not all features are included on every map.

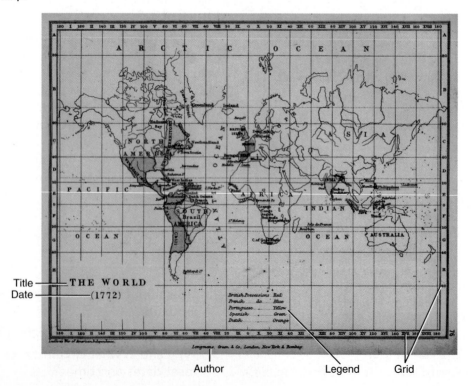

Title — THE WORLD
Date — (1772)

Author   Legend   Grid

Map scale • Figure 1.18

The usefulness of a certain map scale depends on the type of information you are searching for on the map. In each of these maps, the red circle shows the same area of Cape Cod, in eastern Massachusetts, displayed at a different scale.

Verbal scale: 1 unit (cm, in.) represents 36,000,000 units (cm, in.) on the ground. That is, 1 cm = 360 km, or 1 in. = 568 mi.
Representative fraction: 1:36,000,000
Graphic scale:

SCALE 1:36,000,000
0 KILOMETERS 600 800 1000
0 MILES 200 400 600 800 1000

Verbal scale: 1 unit (cm, in.) represents 12,000,000 units (cm, in.) on the ground. That is, 1 cm = 120 km, or 1 in. = 189 mi.
Representative fraction: 1:12,000,000
Graphic scale:

SCALE 1:12,000,000
0 KILOMETERS 200 300
0 MILES 100 200 300

Verbal scale: 1 unit (cm, in.) represents 1,750,000 units (cm, in.) on the ground. That is, 1 cm = 17.5 km, or 1 in. = 27.6 mi.
Representative fraction: 1:1,750,000
Graphic scale:

SCALE 1:1,750,000
0 KILOMETERS 30 40 50
0 MILES 10 20 30 40 50

Small scale ←—————————————————————————→ Large scale

a. Small-scale maps, such as this one, show larger areas, such as continents or several states, but in less detail. This small-scale map might be useful for tracking political or regional issues, such as the path of a storm system along the coast.

b. Larger-scale maps show more geographic features, such as names of smaller cities, airport locations, and major transportation systems within an area. This larger-scale map might be useful for someone planning to travel by air around the northeastern United States.

c. An even larger scale map shows more detail for the minor roadways and small towns. This type of map might be useful for a tourist planning where to stay for a visit to Cape Cod.

of map distance equals 50,000 units of the same distance units on the Earth. That is, 1 cm on the map equals 50,000 cm or 500 m or 0.5 km on the ground. A graphic scale shows the scale on a bar, with the relationship between distances marked. A graphic scale is useful because it remains accurate even if the map is reproduced smaller or larger than the original.

Whether a small scale or a large scale is chosen for a map is determined by the map's purpose (**Figure 1.18**). We use small-scale maps for large geographic areas and large-scale maps for small areas.

map projection
A system of parallels and meridians representing the Earth's curved surface drawn on a flat surface.

Map Projections

A globe shows a fairly accurate representation of distance relationships across the Earth's curved surface. But a globe is limited by its scale and its cumbersome shape. If we want to examine portions of the Earth at a larger scale, or if we want to have a map that can be reproduced in a book or on a computer screen, the globe is no longer practical, so we translate the geographic grid onto a flat surface, in the form of a **map projection**. **Figure 1.19** (on the next page) explores the three main ways that map projections are made.

Cartographers can project the Earth's curved surface onto flat paper by using the geometry of planes, cones, or cylinders. However, only at the point where these surfaces touch the globe is the flat map completely accurate. The farther from these points of contact, the more distorted (stretched or squeezed) features become.

a. Cylindrical projections

A cylindrical projection wraps a cylinder around the globe so that the paper touches the globe at the equator. The equator is the standard line for a cylindrical map.

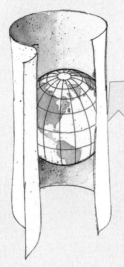

In this cylindrical projection (a Mercator projection) the meridians are evenly spaced, but the spacing between parallels increases at higher latitude so that the spacing at 60° is double that at the equator, which distorts landmasses, as seen with Greenland.

b. Plane projections

A plane projection is produced by projecting a map from a single point, usually at the Earth's center or on the opposite pole, onto a piece of paper touching the globe at any point.

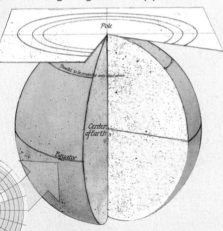

Mapmakers can choose any center point, from which directions and distances are true, but in outer areas, shapes and sizes are distorted. On this projection, Antarctica, the Arctic Ocean, and several continents appear.

c. Conic projections

In a conic projection, a cone sits atop the globe like a dunce cap, with the point of the cone typically situated over one of the poles. Accuracy is greatest along the circle it touches—the standard parallel.

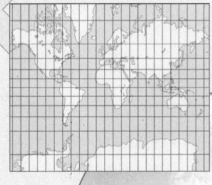

This conic equal-area map is a good format for mapping mid-latitude regions that are larger east to west than north to south.

d. Elliptical projections

In most elliptical projections (also called oval projections), the central meridian and all parallels are straight lines, with relative sizes represented accurately, but shapes are distorted at the edges. Elliptical projections are often used for thematic or political maps. This elliptical equal-area projection was devised by the German mathematician Carl B. Mollweide.

Put It Together

Review Figure 1.18 and answer this question:
As the scale of a map gets larger, the distortions due to the projection _____.

a. increase
b. decrease
c. stay the same
d. do not change in a systematic way

Distortions in Map Projections

Map projections are more convenient and versatile than a globe, but they are also less accurate. Because we cannot project a curved surface onto a flat sheet without some distortion, the scale of a map will be true only for one point or for one or two single lines. Away from that point or line, the scale will be distorted. This distortion is greatest for maps that show large regions, such as continents or hemispheres. A mapmaker must decide what features are most important to display accurately and choose a type of projection accordingly.

Conformity Map projections have two primary areas of distortion: shape and size. **True-shape projections** prioritize conformity, which means they show the true shape of areas but distort their size. The Mercator projection, originated in 1569, is one of the most commonly used true-shape projections. The Mercator projection is useful for navigation because a straight line on a Mercator map is a line of constant compass bearing, called a rhumb line, that can be used to plot a course between two points.

Because the Mercator projection shows the true compass direction of any straight line on the map, it is used to show many types of straight-line features. These include wind and ocean current flow lines, directions of crustal features (such as chains of volcanoes), and lines of equal values, such as lines of equal air temperature or equal air pressure.

The drawback of Mercator's map is that it distorts the relative size of areas, with areas near the pole appearing larger than they really are. For example, on the Mercator projection, Greenland appears 17 times as large as it really is. The Mercator projection can also falsely make the shortest distance between two points seem much longer than the compass line joining them (**Figure 1.20**).

Distortion on the Mercator projection • Figure 1.20

The Mercator projection shows true shape but distorts size and can give a false impression of the shortest distance between two points.

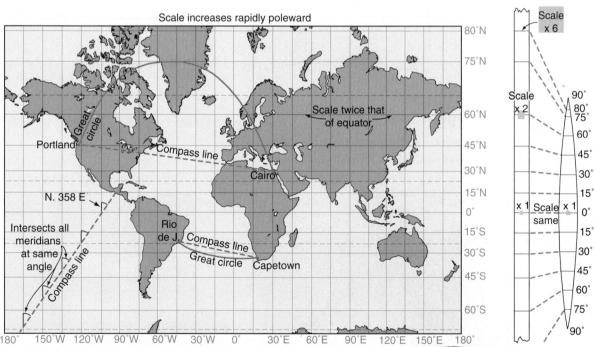
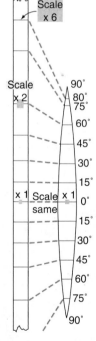

a. The compass line connecting two locations, such as Portland and Cairo, shows the compass bearing of a course directly connecting them. However, the shortest distance between them lies on a great circle, which is a longer, curving line on this map projection.

b. On a globe, it is easy to see that the great circle route is shorter.

Think Critically

The Mercator projection has also been criticized because its distortions reinforce the cultural biases of the European explorers. Describe one example that supports this point.

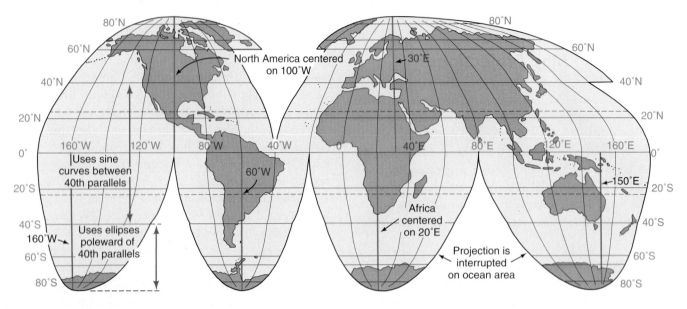

The Goode projection • Figure 1.21

This **interrupted projection** minimizes distortion of scale and shape while maintaining equal area. The straight, horizontal parallels make it easy to scan across the map at any given level to compare regions most likely to be similar in climate.

Equivalence In contrast to true-shape projections, **equal-area projections** prioritize equivalence, which means they show the size of different areas on the map accurately but distort the shapes of those areas. For example, the Goode projection indicates the true size of regions on the Earth's surface, so we can easily scan over regions to get a feel for the relative size of places affected by different features (**Figure 1.21**). This makes the Goode map ideal for showing the world's climates, soils, and vegetation.

The Goode map preserves the relative surface area of different regions, but it distorts the shapes of places,

The Winkel Tripel projection • Figure 1.22

This projection has parallels that are nearly straight, curving slightly toward the edges of the map. The meridians are increasingly curved farther from the central meridian.

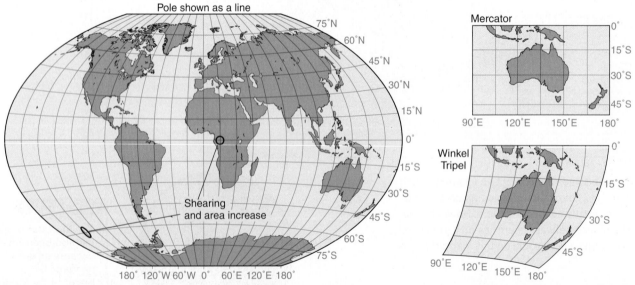

a. The circle in the lower left is distorted by the curving meridians compared to a circle around the same area at the equator.

b. The shape of Australia is distorted on the Winkel Tripel map compared to the Mercator map, but the Winkel Tripel map shows the area of Australia more accurately.

particularly in high latitudes and at the far right and left edges. For example, Greenland, which is located along the edge of the North American part of the map, is strongly sheared to the left. The distortion of the continents is minimized by breaking, or interrupting, the map in oceanic areas.

Compromises Some projections sacrifice equivalency and conformity for the sake of portraying a general balance between the two. The Winkel Tripel projection, for example, minimizes the sum of distortions to area, distance, and direction (**Figure 1.22**). This projection is very useful for displaying world maps because it shows the shapes and areas of countries and continents with only small distortions compared to the Mercator or other global projections. It is an ideal choice for world maps showing the world's climate, soils, and vegetation.

CONCEPT CHECK STOP

1. **When** would a large-scale map be more appropriate than a small-scale map?
2. **Why** do we need more than one type of map projection?
3. **Why** is a true-shape map misleading?

Frontiers in Mapping Technologies

LEARNING OBJECTIVES

1. **Explain** how remote sensing has changed perspectives for geographers.
2. **Describe** the basic features of GIS.
3. **Identify** the components and applications for GPS.
4. **Explain** how Web-based geobrowsers are expanding geographers' tools.

The collection and management of geographic data has changed radically since the time of the earliest mapmakers. The modern geographer's toolkit now includes powerful digital technologies, such as remote sensing, geographic information systems, and 3-D geobrowsers.

Remote-Sensing Tools

Geographers have been mapping the surface of the Earth for thousands of years, but until recently, their view of the world was limited to places where humans could travel and observe the environment around them. Only in the past 100 years or so have geographers been able to view vast areas of the Earth at great distances, first from aerial photographs taken from airplanes and later from space. The use of technology to record observations from a distance is called **remote sensing** (see *Where Geographers Click*).

Where Geographers CLICK

Remote Sensing

http://rst.gsfc.nasa.gov

This Website provides an overview of the wide range of remote-sensing technologies that have been developed over the past four decades. This site is an excellent resource for novices and experts interested in types of sensors and fields of study that apply remotely sensed data.

Monitoring change through satellite scanning • Figure 1.23

Scientists use comparisons of satellite data over time to show incremental and drastic changes to the Earth's surface.

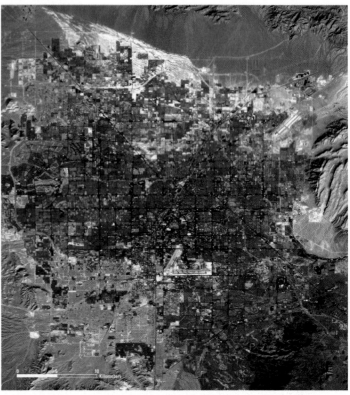

a. In 1973, approximately 500,000 people lived in the desert city of Las Vegas.

b. By 2006, the population had grown to 1.5 million. Las Vegas sprawled in every direction, especially to the northwest and south. This unprecedented rate of growth has put a strain on water and energy supplies.

Scientists are using remote-sensing technologies to investigate many of the environmental and land use issues facing our world and affecting our future. In fact, remote-sensing technologies have become so commonplace that many people are now accustomed to viewing satellite images when planning vacations and routing trips on their home computers.

The science of remote sensing relies on the physics of electromagnetic energy and the fact that the Earth reflects, absorbs, and emits energy waves, which we will discuss in greater detail in Chapter 2. Handheld, aircraft, and satellite-mounted sensors capture radio waves (including radar in many variations), microwaves, infrared energy (including thermal or heat energy), and visible and near-infrared wavelengths.

The resolution of satellite images determines the amount of detail that is shown. *Spatial resolution* refers to the size of the smallest area, or pixel, imaged. Modern satellite imagers can have a spatial resolution of less than 1 square meter. *Spectral resolution* refers to the range of wavelengths captured by the detectors in the sensor systems. Today's sensor technologies can image data across the spectrum from ultraviolet to thermal infrared wavelengths.

The launch of the first Landsat satellite by NASA in 1972 provided nearly complete coverage of the Earth's surface for the first time in history, acquiring simultaneous images drawn from four wavelength bands in the visible and near-infrared parts of the spectrum. Since that time, six additional Landsat imagers have contributed to a record of four decades of images of our planet. Remotely sensed data from Landsat and other satellites are used across a range of geographic studies, from crop analysis to transportation corridor planning to sea ice measurements to mapping of the expansion of deserts. With long records of satellite data, scientists and policymakers can study changes on the Earth's surface to evaluate how humans are changing our planet and what our environmental response to those changes should be (**Figure 1.23**).

SCIENCE TOOLS

Constellation of Earth-monitoring satellites • Figure 1.24

The Earth Observing System (EOS) satellites depicted in this image highlight a portion of NASA's many remote-sensing research programs that target daily events and long-term changes for our planet, including biodiversity protection, natural resources depletion, human effects on the planet, and climate changes. Together these satellites provide continuous data about all of the Earth's systems.

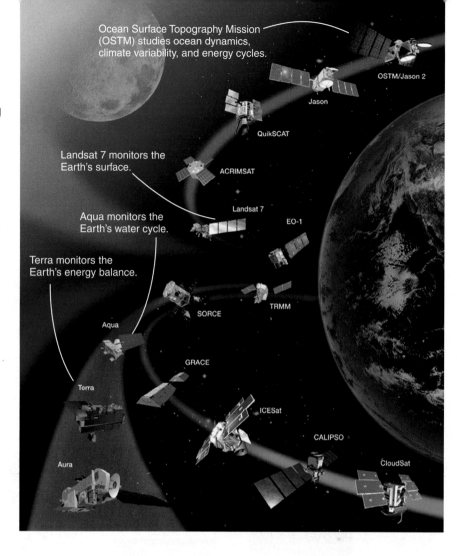

Ocean Surface Topography Mission (OSTM) studies ocean dynamics, climate variability, and energy cycles.

Landsat 7 monitors the Earth's surface.

Aqua monitors the Earth's water cycle.

Terra monitors the Earth's energy balance.

Satellites in geosynchronous orbit remain over the same spot on the Earth's surface at all times. Weather satellites in high-altitude geosynchronous orbits (at altitudes of 35,800 km [22,300 mi]) provide constant monitoring of meteorological systems for full disc views of the planet from a fixed location by imaging thermal and visible wavelengths through the atmosphere. Weather satellites record ozone levels and sea-surface temperatures, as well as keep track of storms. NASA, NOAA, and a growing number of nations are actively operating constellations of satellites for monitoring the Earth's systems (**Figure 1.24**). Much as x-ray and scanning technologies have advanced health services, remote sensing has revolutionized the investigations into the health of the planet. Scientists and those who make three-dimensional maps continue to create innovative uses for remotely sensed data (see *Video Explorations*).

Geographic Information Systems (GIS)

As sensing technology advances, we have access to more and more computer-compatible geographic data about our planet. However, the complexity of integrating large amounts of data could easily overwhelm our ability to understand and

Video Explorations
Solid Terrain Modeling

WILEY PLUS | NATIONAL GEOGRAPHIC | Video

Maps that render landforms in three dimensions provide geographers with information about the terrain that can be lost in a flat map. This video explores a new form of mapping called solid terrain modeling, which combines satellite imagery and aerial photography with computer modeling to cut precise models of the Earth in dense polyurethane foam. These models give the user the perspective of someone looking down from a plane 1524 meters (5000 feet) in the air—but they also allow users to zoom in and touch the land with their hands.

GIS basics • Figure 1.25

GIS allows for the integration of a variety of different spatial data sources and formats into a common digital geocoded database that maintains the spatial locations of all database elements. GIS data can be viewed, using this shared geography, as a series of data layers, covering the same geographic area, which can then be used for modeling, analysis, and display.

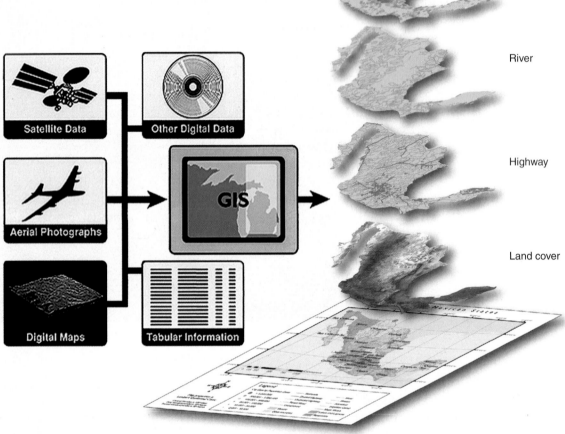

Geographic data layers

City

River

Highway

Land cover

geographic information system (GIS) A combination of software, data, and operational organization that provides the capacity to capture and communicate spatial relationships among geographic features, values, and objects in digital databases.

manage the Earth's resources. Using **geographic information systems (GIS)** is an important method of organizing and analyzing the spatial data we have available from a variety of sources (**Figure 1.25**).

GIS are designed to answer questions involving spatial information and connections. For example, what is the best route to deliver packages? Which wells will be polluted by underground aquifer contamination? What property values will suffer from loan defaults in one neighborhood as compared to another? These questions are answered by merging conventional data with their geographic locations in a process called *geocoding*. GIS provide powerful tools to display,

analyze, and manipulate geocoded data to answer questions about what is surrounding an object, or what is near an object, or what is contained within an object, and where other objects that share interesting characteristics are located.

Visualization is one of the powerful features of GIS. Once spatial information is collected in the database, layers of data collected from different sources can be examined and overlaid to create a comprehensive map showing all data relevant to the topic under investigation. Figure 1.25 shows different data layers that might be used to guide planning for regional development in Mexico. Visualization of complex environmental systems using GIS technology has become a standard method for scientists and government agencies to work together and share a common understanding of complex environmental issues. GIS maps have become a common feature for public hearings due to their authoritative data and facility for visual communications.

Global Positioning System (GPS)

The **Global Positioning System (GPS)** now provides everyone with the ability to determine geographic location rapidly and accurately. The GPS consists of a constellation of 24 or more satellites, plus 21 or more spares, that orbit the Earth at an altitude of 17,550 km (10,900 mi). The system was originally created by the U.S. Department of Defense to triangulate positions and check precision timing for military purposes. Now extended to public use, GPS technology is widely available in handheld and car-mounted receivers and is even routinely integrated into mobile phones.

GPS receivers work by measuring and triangulating the relative time delay of signals from a minimum of three, but usually four or more, GPS satellites, each of which contains onboard atomic clocks. While accuracy can be in the centimeter range, inexpensive GPS receivers can be accurate to about a meter under the best conditions.

Handheld devices and phones automatically enable GPS data to be recorded, along with photographs and notes. This has led to a new cartographic phenomenon in which roads and other map features are being digitally created and supported over the Internet. The Open Street Map project (www.openstreetmap.org) represents a creative approach to citizen-led data collection for mapping the world's streets and making this information freely available over the Internet. This new facility for citizen-contributed mapping has raised concerns for government agencies and private businesses that previously maintained tight control on publishing such data. Regardless of nations' mapping policies, these new citizen-mapping trends are changing the field of cartography.

Geobrowsers and 3-D Mapping

In the past few years, a tremendous growth in Internet mapping has been augmented with the advent of three-dimensional (3-D) mapping. The realistic perspective of a 3-D globe has demonstrated new value to the mapping community. Millions of average citizens also enjoy its ease of access and use. Creative applications of Web-based mapping technology have yielded a number of valuable uses, ranging from reporting on humanitarian conditions to tracking eco-disasters across the planet (see *What a Geographer Sees*).

WHAT A GEOGRAPHER SEES

Visualizing the Earth

This view of the Earth, without country outlines or lines of latitude and longitude, looks like a photograph of a desktop globe. But a geographer will recognize this as a synthetic Earth view from NASA's Earth visualization model, known as the Blue Marble. This model is actually a collection of satellite-based observations, recorded over many months, that have been stitched together to provide a versatile 3-D visualization tool. This view of the Blue Marble shows details of the land surface collected by the MODIS (Moderate Resolution Imaging Spectroradiometer) instrument located on the Terra and Aqua satellites. A seamless mosaic of every square kilometer of our planet, collected by MODIS multiple times each year, enables Earth scientists to study changes in the planet's forests, glaciers, farms, and cities. Visualization facilitates our capacity to view and understand complex data. The Blue Marble includes other satellite data (not shown in this view) to visualize the major patterns of oceans, sea ice, and clouds.

Such visualization tools are helping geographers measure, map, and model our planet's systems to better understand the dynamic processes of natural and human-induced changes and to ask and answer key questions. For example, geographers can analyze satellite data from South America, shown here, to determine the rate and effects of deforestation. They can determine whether deforested areas are expanding, carbon emissions from fires are increasing, desertification of arid land is expanding, the area of irrigated agriculture is declining, and growing seasons are getting drier.

Think Critically

Brazil has an area of over 8.5 million km² (3.28 million mi²). If the resolution of Blue Marble is 1 square kilometer, how can geographers keep track of 8.5 million points to discern changes in land use?

Where Geographers CLICK

Geobrowsers in Action

http://worldwind.arc.nasa.gov/java/

A variety of popular geobrowsers are freely available on the Internet, such as NASA's World Wind software system. These virtual Earth software programs allow a user to place data linked by geo-coordinates on the 3-D globe for analysis and display purposes. A collection of field data or photographs can be readily linked to locations on the virtual globe and made accessible over the Internet to any selected audience. This capacity enhances the ability of geographers and non-geographers alike to create, publish, and share spatial data visualizations for any topic and generate Web maps. Geobrowsers facilitate the public's ability to combine remote-sensing, GIS, and GPS information into an integrated system that is proving to be both easy to use and attractive to a growing user community.

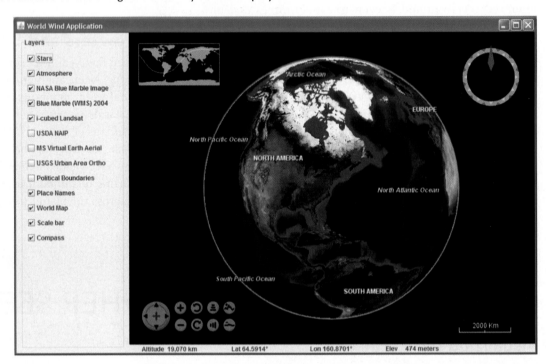

Geobrowsers are computer programs that access and query geographic data draped over a computer-generated globe. These browsers are offered by a host of commercial and government virtual globe developers and offer functions that capture and present data as three-dimensional images. Google Earth is the most widely recognized geobrowser. What distinguishes geobrowsers from traditional GIS is their simplified Web-based visualization tools (see *Where Geographers Click*). These popular 3-D mapping tools use an attractive virtual spinning Earth interface, and their use requires little or no technical training. Individuals and organizations use Web-based geobrowsers to rapidly build and share hundreds of thousands of applications.

Physical geographers and other scientists are increasingly reporting the results of their research and creating engaging visual presentations by linking their data to geobrowsers. The enhanced understanding gained from viewing maps and images in three dimensions is proving invaluable as scientists work to communicate with nontechnical audiences. In addition, these mapping systems enable us all to visualize what is underground, underwater, and inside buildings. As the old adage says, a picture is worth a thousand words. Three-dimensional mapping and widespread use around the world is adding a thousand-fold increase to our mapping and appreciation of the Earth's systems.

CONCEPT CHECK

1. **How** does remote sensing use electromagnetic energy?

2. **How** does geocoding contribute to GIS?

3. **How** do GPS satellites pinpoint locations on the ground?

4. **What** is the key feature of geobrowsers that promotes their use?

Summary

1 The World of Physical Geography 4

- **Geography** is the study of the Earth, focusing on spatial issues and spatial relationships.

- Four major systems are used to describe the Earth—**atmosphere**, **hydrosphere**, **lithosphere**, and **biosphere**—as shown in the diagram.

The Earth's four major systems • Figure 1.2

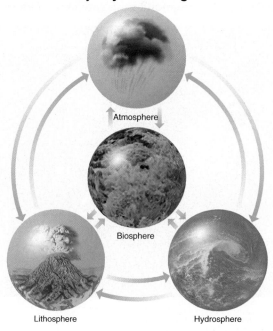

Atmosphere

Biosphere

Lithosphere

Hydrosphere

2 The Shape of the Earth 10

- From ancient measurements as well as precision space images, such as this one, we know the Earth's shape approximates a sphere.

Curvature of the Earth • Figure 1.5

- The Earth's shape is not perfectly round. Rather, it is wider around the middle because of the Earth's rotation.

3 Global Location 11

- The intersection of the **axis** with the Earth's surface marks the Earth's **poles**.

- The **geographic grid**, which consists of **meridians** and **parallels**, helps us mark location on the globe.

- Geographic location is labeled using **latitude** and **longitude**, as shown in the diagram. Altitude provides for three-dimensional locations. The **equator** and the **prime meridian** act as reference lines.

Latitude and longitude of a point • Figure 1.10

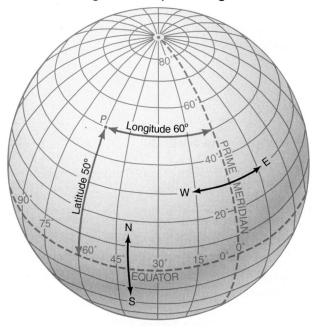

4 Global Time 14

- Our most basic form of timekeeping is the changing position of the Sun in the sky as the Earth rotates over the course of a day.

(Continued on next page.)

- The system of standard time divides the globe into 24 time zones, as shown in the diagram. We keep time according to standard meridians that are normally 15° apart, so clocks around the globe usually differ by whole hours.

Standard time • Figure 1.13

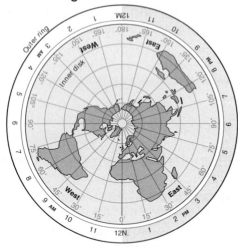

- International timekeeping is based on synchronized atomic clocks.
- As you cross the **international date line**, the calendar day changes.
- **Daylight saving time** advances the clock by one hour from spring until fall.

5 Mapping the Earth 20

- A **map** is a representation of spatial relationships between features of the Earth.
- **Map scale** indicates the relationship between distances on a map and actual distances on the Earth's surface.
- **Map projections** display the Earth's curved surface on a flat page, such as the Goode projection, shown here. The main types of projections are conical, cylindrical, plane, and elliptical projections.

The Goode projection • Figure 1.21

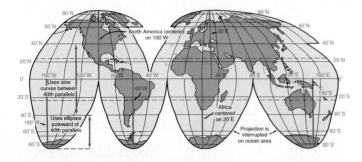

- Map projections can accurately represent the true shape of features or the true size of features, but not both.

6 Frontiers in Mapping Technologies 25

- **Remote sensing** offers global coverage of the Earth through satellites and sensors, such as the NASA satellites depicted here.

Constellation of Earth-monitoring satellites • Figure 1.24

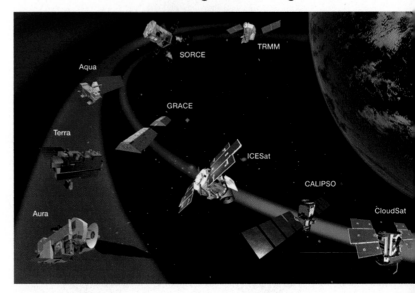

- **GIS** connects spatial attributes of mapping features for monitoring, modeling, and visualizing mapping data.
- **GPS** has unleashed the collection of geolocation information by citizens using popular handheld and car navigation systems.
- Web access to geobrowsers has launched a new era in citizen-accessible and citizen-created maps, including 3-D maps on virtual Earth globes.

Key Terms

- atmosphere 5
- axis 11
- biosphere 5
- coordinated universal time (UTC) 18
- daylight saving time 19
- equal-area projection 24
- equator 12
- geographic grid 11
- geographic information system (GIS) 28
- geography 4
- Global Positioning System (GPS) 29
- great circles 11
- human geography 5
- hydrosphere 5
- hypothesis 7
- international date line 19
- interrupted projection 24
- latitude 12
- lithosphere 5
- longitude 12
- map 20
- map projection 21
- map scale 20
- meridian 11
- parallel 12
- physical geography 5
- pole 11
- prime meridian 13
- remote sensing 25
- scientific method 6
- small circles 12
- theory 7
- true-shape projection 23

Critical and Creative Thinking Questions

1. How can mapmakers influence the way citizens view the planet?

2. A geographer is tasked with providing maps for regional planners so they can formulate policies with the aim of reducing urban sprawl. Name the disciplines of three scientists with whom collaboration would greatly assist the geographer's work. Give the reasons these disciplines would be so helpful.

3. What map projection would you select to map and measure the rate of global deforestation? Why?

4. Explain how space-age technologies have affected the field of geography. What do you think we can expect from these technologies in the future?

5. How has the idea of map scale changed in the digital age?

What is happening in this picture?

The Earth's city lights were captured by a satellite that images at night.

Think Critically

1. What information about towns and cities does this image tell us?
2. How do you know that all of the satellite images used for this picture were taken at different times?

Self-Test

(Check your answers in the Appendix.)

1. Eratosthenes used _____ to calculate the Earth's circumference.

   a. the first telescope

   b. ancient maps

   c. the scientific method

   d. circumnavigation of the globe

2. Which of the following is not an essential theme of geography?

   a. location

   b. measurement

   c. region

   d. place

3. Physical geography is distinguished from human geography based on its _____.

   a. focus on time

   b. use of technology

   c. attention to social issues

   d. study of the Earth's components and processes

4. The shape of the Earth is best described as a(n) _____.

   a. perfect sphere

   b. ellipsoid

   c. oblate spheroid

   d. spherical ellipsoid

5. Describe the position of point *P*, shown here on the geographic grid, in terms of its latitude and longitude.

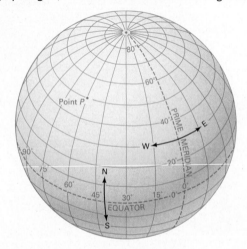

6. Which travel route between Cairo and Portland represents the quickest route and why?

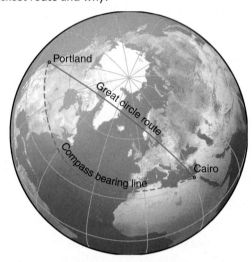

7. Greenwich, England, is important to geography because _____.

   a. it is the only major city that falls naturally on the prime meridian

   b. it is the birthplace of the telescope

   c. sunrise begins each day directly over the city

   d. international treaties are based on its location

8. _____ time is based on 24 _____-wide time zones.

   a. Polar, 7 1/2°

   b. World, 15°

   c. Standard, 7 1/2°

   d. Standard, 15°

9. Chicago's longitude is approximately 87° W, and the meridian associated with its time zone is 90° W. If daylight saving time is in effect in Chicago and in Greenwich, England, what time is it in Chicago if it is 3:00 PM on July 4 in Greenwich?

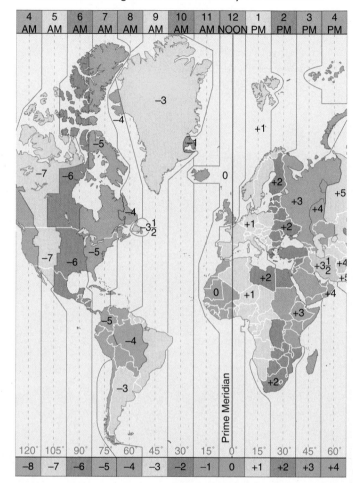

10. Currently, the most accurate time is determined by _____.

a. astronomical observations of the Sun

b. atomic clocks based on cesium emissions

c. lasers that track the surface of the Moon

d. the microwave vibrations of water

11. Distortions associated with maps are caused by _____.

a. representing Earth features digitally

b. using different scales on the same map

c. projecting a curved Earth onto a flat map

d. imperfect understanding of geographers

12. The _____ projection is a rectangular grid of meridians and parallels depicted as straight vertical and horizontal lines, respectively.

a. polar

b. Winkel Tripel

c. Mercator

d. conic

13. The _____ projection portrays the shapes and sizes of countries and continents with low distortion.

a. Mercator

b. Winkel Tripel

c. conic

d. polar

14. Which of the following technologies is not always based on data from satellites?

a. Geographical Information Systems (GIS)

b. Earth Observing System (EOS)

c. Global Positioning System (GPS)

d. Blue Marble

15. Geobrowsers display 3-D virtual globes, creating _____.

a. a universal system for all nations to adopt

b. a platform for enhanced visualization

c. a citizen-accessible system or mapping

d. both b and c

THE PLANNER ✓

Review your Chapter Planner on the chapter opener and check off your completed work.

2

The Earth's Global Energy Balance

Every second, the Sun showers our planet with enough electromagnetic energy to meet the world's energy requirements for 10 days. Theoretically, harnessing 1 minute of the Sun's incoming energy, which takes about 8.3 minutes to reach us from the Sun's surface, would provide an equivalent of just under 2 years' worth of the energy consumed by all the world's citizens.

Currently, our world economy is driven by burning fossil fuels that have been sequestered underground for millions of years. Considerations for sustainable development and climate change require that we begin to balance our global energy use with cleaner energy sources.

Today, solar energy collection is a fast-growing industry, with some people installing solar thermal collectors on their roofs to drive heating systems. Photovoltaic panels are used to power a variety of devices, from calculators to outdoor lights. Solar power plants focus sunlight on fluid-filled tubes or reservoirs to produce steam that runs turbines to generate electricity.

As society becomes increasingly concerned about our dependence on nonrenewable fossil fuels, it seems likely that the Sun will become an even more important energy provider. The Nevada Solar One power plant, shown here, uses trough-like parabolic mirrors to focus solar radiation onto pipes containing synthetic oil that carries heat into a central power plant, where it generates electric power. It is an example of one of many new solar technologies that are under development.

Generating 64 megawatts of solar thermal energy is the Nevada Solar One operation, about 48 kilometers (30 miles) north of Las Vegas, Nevada.

CHAPTER OUTLINE

CHAPTER PLANNER ✓

❏ Study the picture and read the opening story.

❏ Scan the Learning Objectives in each section:
p. 38 ❏   p. 41 ❏   p. 47 ❏   p. 52 ❏

❏ Read the text and study all figures and visuals. Answer any questions.

Analyze key features

❏ Geography InSight, p. 46

❏ Video Explorations, p. 49

❏ Process Diagram p. 50 ❏   p. 53 ❏

❏ What a Geographer Sees, p. 56

❏ Where Geographers Click, p. 59

❏ Stop: Answer the Concept Checks before you go on.
p. 40 ❏   p. 47 ❏   p. 52 ❏   p. 59 ❏

End of chapter

❏ Review the Summary and Key Terms.

❏ Answer the Critical and Creative Thinking Questions.

❏ Answer What is happening in this picture?

❏ Complete the Self-Test and check your answers.

Electromagnetic Radiation

LEARNING OBJECTIVES

1. **Describe** different types of electromagnetic radiation.
2. **Relate** wavelength to temperature.
3. **Define** the solar constant.

Most natural phenomena on the Earth's surface—from the accumulation of rain clouds to the movement of a sand dune to the growth of a forest—are driven by the Sun, either directly or indirectly. The Sun powers the hydrosphere, the atmosphere, and the biosphere. Only the lithosphere is powered without the Sun's energy. The Sun is the power source for wind, waves, weather, rivers, ocean currents, and essentially all life forms, as we will see in later chapters.

The Earth receives energy from the Sun in the form of **electromagnetic radiation**. All surfaces—from the fiery Sun in the sky to the skin covering our bodies—constantly emit radiation.

> **electromagnetic radiation** A wave form of energy radiated by any substance possessing internal energy; it travels through space at the speed of light.

You can think of electromagnetic radiation as a collection of energy waves that travel away from the surface of an object. Electromagnetic energy occurs across a wide range of wavelengths. Light, radio waves, and infrared heat waves are all forms of electromagnetic radiation of different wavelengths.

Wavelength is the measured distance separating one wave crest from the next wave crest. For wavelengths not too far from those of visible light, we measure wavelength in units of micrometers. A micrometer is one-millionth of a meter (10^{-6} m). The tip of your little finger is about 15,000 micrometers wide. We abbreviate micrometer using μm; the first letter of the abbreviation is the Greek letter μ, or mu. For very short wavelengths, we use the unit of the nanometer (nm), which is one-billionth of a meter (10^{-9} nm). For longer wavelengths, we use the familiar units of centimeters and meters.

The Electromagnetic Spectrum

The **electromagnetic spectrum** (**Figure 2.1**) defines the entire range of wavelengths for all energy. Radiation at the end of the spectrum with the shortest wavelengths includes gamma rays and X-rays, which exhibit high-energy states

The electromagnetic spectrum • Figure 2.1

Electromagnetic radiation can exist at any wavelength. Visible light begins at about 0.4 μm with the color violet. Colors then gradually change through blue, green, yellow, orange, and red until the end of the visible spectrum, at about 0.7 μm.

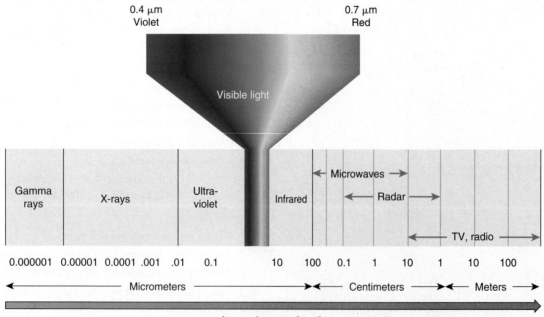

Spectra of solar and Earth radiation • Figure 2.2

The Earth radiates less energy than the Sun. This figure plots both the shortwave radiation from the Sun (left side) and the longwave radiation emitted by the Earth's surface and atmosphere (right side). We have not taken the scattering of radiation by the atmosphere into account in this illustration.

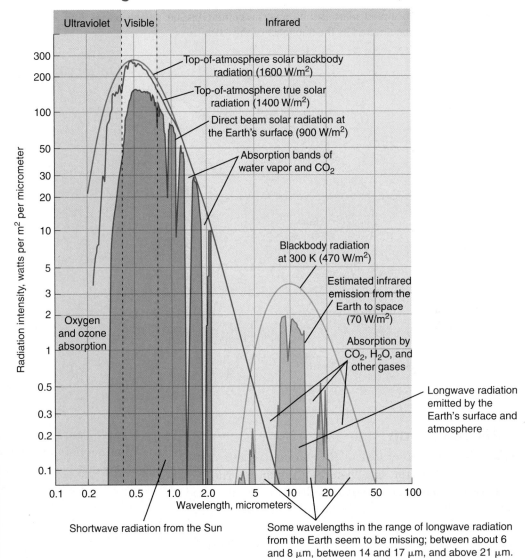

Top-of-atmosphere solar blackbody radiation (1600 W/m^2)

Top-of-atmosphere true solar radiation (1400 W/m^2)

Direct beam solar radiation at the Earth's surface (900 W/m^2)

Absorption bands of water vapor and CO_2

Oxygen and ozone absorption

Blackbody radiation at 300 K (470 W/m^2)

Estimated infrared emission from the Earth to space (70 W/m^2)

Absorption by CO_2, H_2O, and other gases

Longwave radiation emitted by the Earth's surface and atmosphere

Shortwave radiation from the Sun

Some wavelengths in the range of longwave radiation from the Earth seem to be missing; between about 6 and 8 μm, between 14 and 17 μm, and above 21 μm. These wavelengths are absorbed by the atmosphere.

shortwave radiation Electromagnetic energy in the range of 0.2 to 3 μm.

longwave radiation Electromagnetic energy in the range of 3 to 5 μm.

and can be hazardous to health. The middle of the spectrum, referred to as **shortwave radiation,** includes ultraviolet light, the visible light spectrum, and near- and shortwave-infrared radiation. Very hot objects, such as the Sun or a light bulb filament, give off radiation that is nearly all in the form of shortwave radiation generating light. **Longwave radiation** includes a portion of infrared radiation, as well as microwaves, radar, and wavelengths conventionally associated with communications transmissions, such as radio and television. Longwave infrared radiation emitted by the Earth is important to climate change science, as we will discuss later in the chapter.

Radiation and Temperature

There are two important physical principles related to the emission of electromagnetic radiation. The first is that hot objects radiate much more energy than cooler objects. The flow of radiant energy from the surface of an object, or its outgoing energy flux, is directly related to the absolute temperature of the surface. Even a small increase in temperature can mean a large increase in the rate at which radiation is given off by an object or a surface. For example, water at room temperature emits about one-third more energy than when it is at the freezing point.

The second principle is that the hotter the object, the shorter the wavelengths of energy it radiates. This inverse relationship between wavelength and temperature means that very hot objects such as the Sun emit radiation at short wavelengths. Because the Earth is a much cooler object, it radiates energy at longer wavelengths (**Figure 2.2**).

Rays of solar radiation spread apart as they move away from the Sun, so only a small fraction of the radiation emitted by the Sun reaches the Earth.

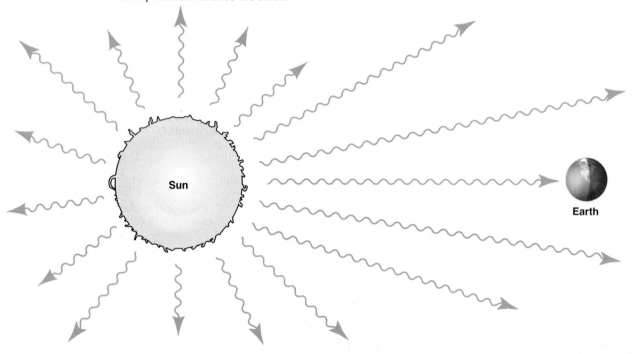

Solar Radiation

Now let's take a closer look at the electromagnetic radiation produced by the Sun. The Sun produces primarily shortwave radiation, consisting of ultraviolet, visible, and infrared radiation. Figure 2.2 shows that the Sun's output peaks in the visible part of the spectrum. Human vision is adjusted to the wavelengths where solar light energy is highest. Butterflies and some other animal species have adapted to view the world in the ultraviolet and infrared in addition to the visible portions of the spectrum.

Our Sun is a fiery ball of constantly churning gas. This average-sized star is the largest object in our solar system, containing more than 98% of the solar system's mass. At its core, the temperature reaches 15 million °C (27 million °F), and the pressure is 340 billion times the Earth's air pressure at sea level—conditions intense enough to cause the nuclear reactions that fuel the Sun's emission of electromagnetic radiation.

Because of our average distance of 150 million km (93 million mi) from the Sun, only a small fraction of the Sun's energy reaches the Earth (**Figure 2.3**). Energy that is twice as far from the Sun will be one-quarter as intense. The distance of the Earth's orbit from the Sun is seen as optimal for life due to its acceptable levels of solar radiation.

The flow rate of solar energy received by Earth is nearly constant. Its average value is known as the **solar constant**. Scientists measure the solar constant just beyond the outer limits of the Earth's atmosphere, before any energy has been lost in the atmosphere. Variations in the solar constant include sunspots, which are caused by intense magnetic activity on the Sun's surface. These dark spots, when present, can reduce the Sun's radiant output by about 0.1%.

We are familiar with the use of the watt (W) to describe the power, or rate of energy flow, of a lightbulb or other home appliance. We use these same units when we talk about the intensity of received (or emitted) radiation. For the Sun's radiative energy, we must take into account both the power of the radiation and the surface area of the planet being hit by (or giving off) energy. So we use units of watts per square meter (W/m^2). The solar constant has a value of about 1367 W/m^2.

CONCEPT CHECK

1. **What** are the similarities and differences between gamma rays, X-rays, and visible light?

2. **Why** do the Sun and the Earth emit radiation at different wavelengths?

3. **How** is the solar constant related to the Earth's distance from the Sun?

Insolation over the Globe

LEARNING OBJECTIVES

1. **Identify** the factors that affect daily insolation.

2. **Explain** why the seasons occur.
3. **Describe** how insolation varies by latitude.

The amount of solar radiation that reaches specific locations on the Earth changes over the course of a day, as well as over the course of a year, and it varies by latitude. These differences have a profound impact on regional climates around the globe.

Daily Insolation

insolation The flow rate of incoming solar energy, as measured at the top of the atmosphere.

Incoming solar radiation is known as **insolation**. It is measured in units of watts per square meter (W/m^2). At any given location, the average insolation over a 24-hour day depends on two factors: (1) the angle of sunlight and (2) day length. Insolation is measured at the top of the atmosphere, so it is unaffected by clouds and weather.

Angle of sunlight You saw in Chapter 1 that the angle of sunlight changes over the course of a day as the Earth rotates on its axis. Near noon, the Sun at its zenith is high above the horizon, whereas early in the morning and late in the evening, the Sun is low in the sky, just above the horizon.

Insolation is greatest when the Sun is directly overhead, and it decreases when the Sun is low in the sky, because the same amount of solar energy is spread out over a greater area of ground surface (**Figure 2.4**).

Because the surface of the Earth is curved, however, the noon Sun is directly overhead at only one point on the globe at any given moment. This point is called the **subsolar point**, and it receives the Sun's strongest rays. We call the latitude of the subsolar point the Sun's **declination**. We'll say more later about how the Sun's declination travels between the tropics of Cancer and Capricorn over the course of a year.

Light intensity at different angles • Figure 2.4

Light that shines vertically is more concentrated, whereas light that shines at an angle is dispersed over a wider area and is therefore less intense.

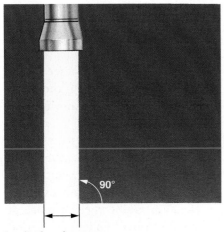

1 unit of surface area
One unit of light is concentrated over one unit of surface area.

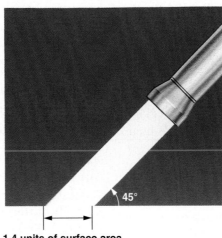

1.4 units of surface area
One unit of light is dispersed over 1.4 units of surface area.

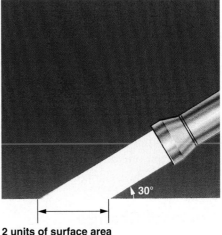

2 units of surface area
One unit of light is dispersed over 2 units of surface area.

The Earth takes 365.242 days to travel around the Sun in an elliptical orbit. The direction of travel is counterclockwise when viewed from above the Earth's North Pole. The point in its orbit at which the Earth is nearest to the Sun is called **perihelion**. The Earth is usually at perihelion on or about January 3. It is farthest away from the Sun, or at **aphelion**, on or about July 4. The distance between Sun and Earth varies by only about 3 % during one revolution, so it is not a significant factor in seasonal temperature differences.

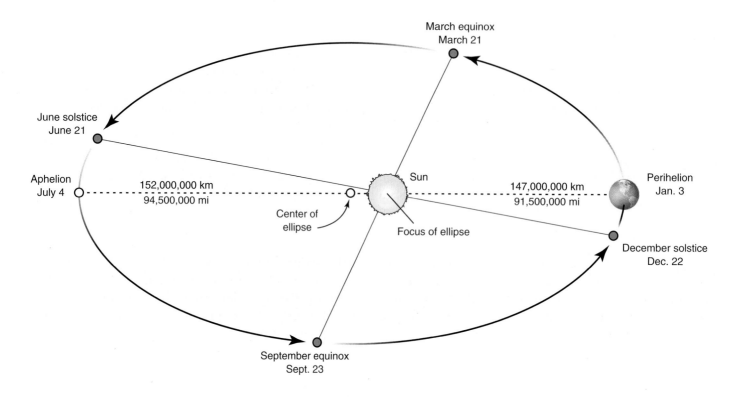

Day length The second factor that determines daily insolation is the amount of time that a given location is exposed to the Sun's rays. The Sun's rays always divide the Earth into two hemispheres—one that is bathed in light and one that is shrouded in darkness. The **circle of illumination** is the circle that separates the day hemisphere from the night hemisphere. The longer a point on the globe remains inside that circle of illumination in a 24-hour period, the longer its day, and the more solar energy it receives.

At the equator, night and day are 12 hours throughout the year. But at certain times of the year, the polar regions are bathed in 24-hour daylight or shadowed in 24-hour night. At other latitudes, the length of daylight changes slightly from one day to the next. Now let's find out why the length of days changes over the course of the year.

Seasonal Change

Daily patterns of insolation are related to the rotation of the Earth on its axis, as night turns to day and the

Sun rises in the sky, warming the Earth. But patterns of insolation also change over the course of the year, as days become longer and warmer during the summer and shorter and cooler in the winter. These seasonal changes are related to the **revolution** of the Earth around the Sun (**Figure 2.5**).

The tilt of the Earth's axis Seasonal changes occur because the Earth's axis is tilted compared to the plane of the Earth's orbit around the Sun, which is known as the **plane of the ecliptic** (**Figure 2.6**).

Because the tilt of the Earth's axis is nearly constant with respect to the plane of the ecliptic, the Earth tilts away from or toward the Sun at different points along its orbit (**Figure 2.7**). As the Earth moves along its orbit, different parts of the globe receive more of the Sun's direct rays and experience the longest period of daylight, which results in annual patterns of seasonal change.

Without the tilt of the Earth's axis, there would be no seasons. Annual insolation would always be very high at the equator because the Sun would always pass directly overhead at noon every day throughout the year. In contrast, annual insolation at the poles would always be near zero, because the Sun's rays would always skirt the horizon.

We can see that without a tilted axis, our planet would be a very different place. The tilt redistributes a very significant portion of the Earth's insolation from the equatorial regions toward the poles. So even though the pole does not receive direct sunlight for six months of the year, it still receives nearly half the amount of annual solar radiation as the equator. This distribution of solar insolation regulates the seasons.

Equinoxes and solstices
The two points along the Earth's orbit where the tilt is most extreme are the **solstices**. Midway between

> **solstice** The instant in time when the Earth's axis of rotation is fully tilted (23 1/2°) either toward or away from the Sun.

The tilt of the Earth's axis • Figure 2.6

As the Earth moves in its orbit, the axis of the Earth is tilted at an angle of $23\frac{1}{2}°$ away from a right angle to the plane of the ecliptic.

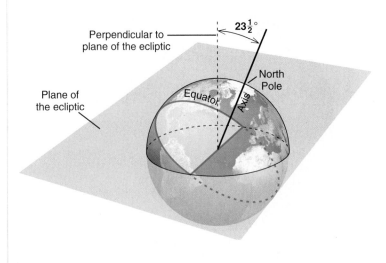

Earth–Sun relations through the year • Figure 2.7

As the Earth revolves around the Sun throughout a year, different parts of the globe are tilted toward the Sun, where they receive more of the Sun's rays at a higher angle and experience a longer day. While one hemisphere is experiencing the warmer conditions of summer, the opposite hemisphere is experiencing winter conditions, including sunlight at a lower angle and a shorter day.

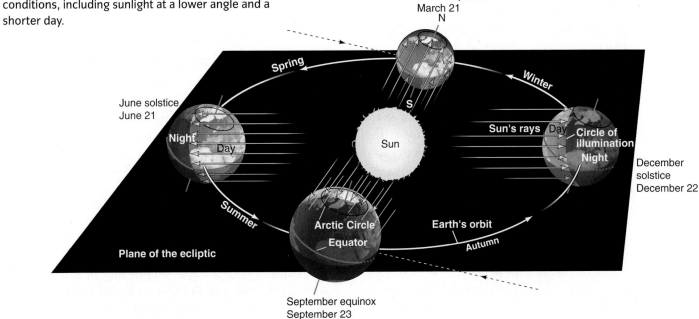

Put It Together

Review Figure 2.5, and complete this statement:
When the Earth is in the perihelion of its orbit, the North Pole is experiencing _____.

a. 24 hours of daylight
b. 24 hours of night
c. approximately 12 hours of both daylight and night
d. approximately 16 hours of daylight and 8 hours of night

Equinox conditions • Figure 2.8

At equinox, day and night are of equal length everywhere on the globe (ignoring twilight). The Sun's vertical rays fall on the equator, so that point receives the full force of solar illumination. The Sun's rays merely graze the surface at both poles, so the surfaces at the poles receive very little solar energy.

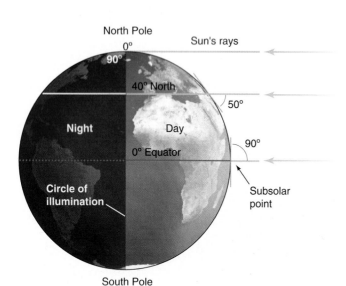

equinox The instant in time when the Sun is directly overhead at the equator and the circle of illumination passes through both poles.

the solstices are the **equinoxes,** when the Earth is neither tilted toward the Sun nor away from it. These four orbital positions and their corresponding calendar days traditionally mark the transitions between our seasons.

The March equinox occurs about March 21, and the September equinox occurs about September 23. The date of any solstice or equinox in a particular year may vary by a day or so, since the Earth's revolution period is not exactly 365 days. In the northern hemisphere, the March equinox is known as the *vernal equinox* because it represents the transition from winter to spring. Similarly, the September equinox is known as the *autumnal equinox* because it signals the transition to fall.

The conditions at the two equinoxes are identical as far as the Earth–Sun relationship is concerned. At equinox, the circle of illumination passes through the North and South Poles, as we see in **Figure 2.8,** and the Sun's vertical rays fall on the equator.

On June 21, the north polar end of the axis is tilted at 23 1/2° toward the Sun, while the South Pole and southern hemisphere are tilted away. This is the June solstice (also known as the *summer solstice* in the northern hemisphere). In the northern hemisphere, daily insolation will be greater at the June solstice than at the equinox because the Sun is in the sky longer and reaches a higher angle at noon.

On December 21 or 22, the north polar end of the Earth's axis leans at the maximum angle away from the Sun, 23 1/2°.

This event is the December solstice (also called the *winter solstice* in the northern hemisphere). In the midlatitudes of the northern hemisphere, the Sun's path is low in the sky and is visible for only about 9 hours. Sunrise is to the south of east, and sunset is to the south of west. Daily insolation reaching the surface at the December solstice will be less than at the equinox and much less than at the June solstice. At the same time, the southern hemisphere is tilted toward the Sun and enjoys strong solar heating (**Figure 2.9**).

Annual Insolation by Latitude

How does latitude affect annual insolation? Because it measures the flow of solar power available to heat the Earth's surface, insolation is the most important factor in determining air temperatures, and it is a major determinant of regional climates.

Because of the Earth's spherical shape, only one latitude can receive the Sun's most intense vertical rays at any given time. Although the declination of the Sun changes over the course of the year, throughout the year, the equator receives more direct rays from the Sun, whereas the poles receive more oblique rays (**Figure 2.10**).

Seasonal changes in day length vary by latitude. The equator is the only parallel that is divided exactly into two by the circle of illumination, which means that daylight and nighttime hours will be equal throughout the year. The farther north you go, the longer the days become at the June solstice. Once you move to lat. 66 1/2°, you will find that the day continues unbroken for 24 hours during the sum-

Solstice conditions • Figure 2.9

At the solstices, the north end of the Earth's axis of rotation is fully tilted either toward or away from the Sun. Because of the tilt, polar regions experience either 24-hour day or 24-hour night. At this same time, the subsolar point lies on one of the tropics at lat. $23\frac{1}{2}°$ N or S.

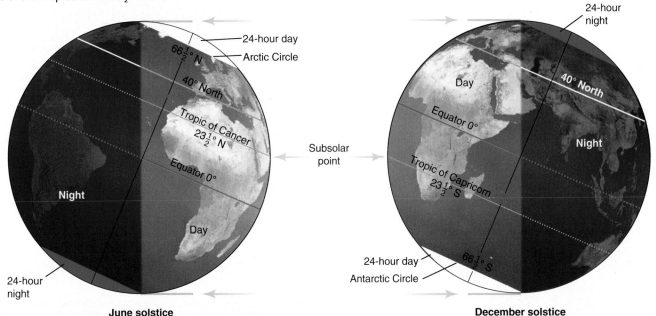

June solstice

December solstice

mer solstice on June 21. This parallel is known as the Arctic Circle. Even though the Earth rotates through a full cycle during a 24-hour period, the North Pole will remain in continuous daylight from March 21 to September 21. We can also see that the subsolar point is at a latitude of 23 1/2° N. This parallel is known as the Tropic of Cancer. Because the Sun is directly over the Tropic of Cancer at this solstice, solar energy is most intense along this parallel.

The conditions are reversed at the December solstice. Back at lat. 40° N, you will now find that the night is about 15 hours long, while sunlight lasts about 9 hours. All the area south of lat. 66 1/2° S lies under the Sun's rays, inundated with 24 hours of solar illumination. This parallel is known as the Antarctic Circle. The subsolar point has shifted to a point on the parallel at lat. 23 1/2° S, known as the Tropic of Capricorn.

Angle of sunlight by latitude • Figure 2.10

Because the angle of the solar beam striking the Earth varies with latitude, solar insolation is strongest near the equator and weakest near the poles.

Put It Together

Review Figures 2.4 and 2.8 and complete this statement:
At _____ latitude, the solar radiation would be spread out over twice as much area as at the equator.

a. 30°
b. 45°
c. 60°
d. 90°

A geographer's system of latitude zones is based on the seasonal patterns of daily insolation observed over the globe.

Subarctic zone
This zone has an extremely large yearly variation in day lengths, yielding enormous contrasts in insolation over the year. Much of the subarctic zone is covered by tundra or evergreen forest, seen here with a ground cover of snow in the Hudson Bay region of Canada.

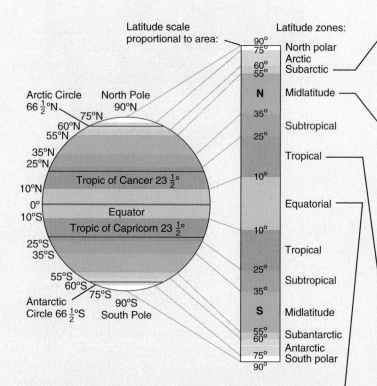

Latitude scale proportional to area:

Arctic Circle 66 ½°N   North Pole 90°N
75°N
60°N
55°N
35°N
25°N
Tropic of Cancer 23 ½°
10°N
0°
10°S
Equator
Tropic of Capricorn 23 ½°
25°S
35°S
55°S
60°S
75°S
Antarctic Circle 66 ½°S   90°S   South Pole

Latitude zones:
90°
75° North polar
60° Arctic
55° Subarctic
N Midlatitude
35°
25° Subtropical
Tropical
10°
Equatorial
10°
Tropical
25°
35° Subtropical
S Midlatitude
55°
60° Subantarctic
Antarctic
75° South polar
90°

Midlatitude zone
The length of a day can vary greatly from winter to summer here, and seasonal contrasts in insolation are quite strong. This summer landscape is from the Tuscany region of Italy.

Equatorial zone
The Sun provides intense insolation here, giving rise to lush plant life and great biodiversity, such as that found in this equatorial rainforest in the Gorge of the Rio Ole, Bioko Island, Equatorial Guinea.

Tropical zone
The tropical zone has high annual insolation and is the home of the world's driest deserts, such as this one in Rub' al Khali, Saudi Arabia, as well as fully vegetated regions.

Differences in annual insolation are a key factor in local climates, landscape, and life cycles. Near the equator, there is limited seasonal change, allowing for a year-round growing cycle for plants and ample food for animals. In contrast, toward the poles, seasonal changes are dramatic. The cold, dark winters force animals to hibernate or migrate southward in search of food. We can identify common climate patterns across loosely defined latitude zones around the globe (**Figure 2.11**).

CONCEPT CHECK

1. **Why** is sunlight most intense when the Sun is directly overhead?

2. **How** would conditions on the Earth be different if the Earth's axis were not tilted?

3. **How** do annual patterns of insolation affect regional climates?

Solar Energy and the Earth's Atmosphere

LEARNING OBJECTIVES

1. **Describe** the composition of the atmosphere.

2. **Explain** the fate of solar radiation as it passes through the atmosphere.

The solar radiation that actually reaches the Earth's surface is quite different from the solar radiation measured above the Earth's atmosphere. This is because the Earth's atmosphere interacts with solar radiation by absorbing, scattering, and reflecting incoming solar energy.

Composition of the Atmosphere

The atmosphere is a thin envelope of air surrounding the Earth, held in place by gravity. It consists of a mixture of various gases that reaches up to a height of many kilometers. Almost all the atmosphere (97%) lies within 30 km (19 mi) of the Earth's surface. Although there is no real boundary between the atmosphere and outer space, an altitude of 100 km (62 mi) is often regarded as the top of the atmosphere.

troposphere The lowest layer of the atmosphere, where human activity and most weather takes place.

Layers of the atmosphere

The atmosphere consists of several distinctive layers that can be characterized by their temperature, composition, and function. The **troposphere** is the lowest atmospheric layer. The troposphere is thickest in the equatorial and tropical regions, where it stretches from sea level to about 16 km (10 mi). It thins toward the poles, where it is only about 6 km (4 mi) thick. Approximately 50% of the atmosphere's volume is contained in the troposphere. Most of the Sun's harmful radiation is absorbed by the time it reaches the troposphere, as we will discuss later in the chapter. Temperatures in the troposphere normally decrease with altitude.

The troposphere ends at the boundary called the **tropopause**. The altitude of the tropopause varies somewhat with season, so the troposphere is not uniformly thick at any location. The **stratosphere** is the next layer, reaching up to roughly 50 km (about 30 mi) above the Earth's surface. It is the home of strong, persistent winds that blow from west to east and the protective ozone layer. Air does not readily mix between the troposphere and stratosphere, so the stratosphere normally holds very little water vapor or dust. Above the stratosphere we find the **mesosphere** and the **thermosphere**.

stratosphere The layer of atmosphere directly above the troposphere, where temperature slowly increases with height.

The gas composition of the lower part of the atmosphere is uniform (ignoring variations in water vapor from place to place). However, the air density decreases with

Component gases of the lower atmosphere • Figure 2.12

The atmosphere, seen here in profile, is an envelope of gases surrounding the Earth. The lower atmosphere, or homosphere, is 99% nitrogen and oxygen. The remaining 1% is mostly argon, an inactive gas of little importance in natural processes, with a very small, but increasing, amount of carbon dioxide.

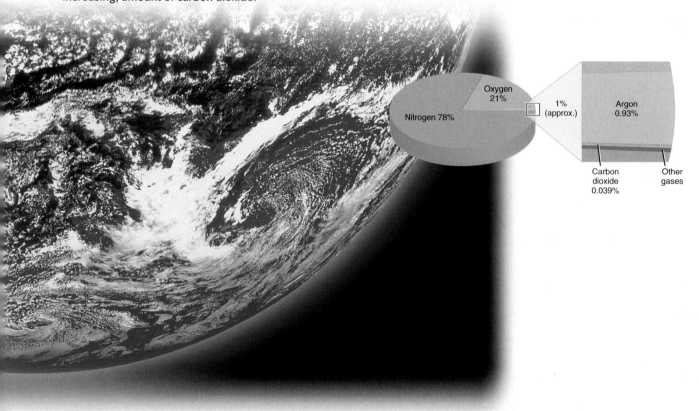

altitude for about the first 100 km of altitude, which includes the troposphere, stratosphere, mesosphere, and the lower portion of the thermosphere (**Figure 2.12**).

Considering the composition of the atmosphere, there are two primary layers. In the **homosphere**, gases are well mixed and, except for variations in water vapor from place to place, atmospheric composition is uniform. Above 100 km in altitude, gas molecules tend to be widely separated and sorted into layers by molecular weight and electric charge. This region is called the **heterosphere**.

Permanent gases

The two main component gases of the lower atmosphere, nitrogen and oxygen, are perfectly mixed, so pure, dry air behaves as if it is a single gas with very definite physical properties. These are considered to be permanent gases because their quantities are stable over human timescales.

Nitrogen gas (N_2) does not easily react with other substances, so we can think of it as a mainly neutral substance, adding inert bulk to the atmosphere. Its biological role is important, however, regulated in part by nitrogen-fixing soil bacteria.

By contrast, oxygen gas (O_2) is highly chemically reactive, combining readily with other elements in the process of oxidation. Fuel combustion is a rapid form of oxidation, while certain types of rock decay (weathering) are very slow forms of oxidation. Living tissues require oxygen to convert foods into energy.

Variable gases

The other gases that make up the atmosphere have variable rates, depending on environmental conditions. Although these variable gases make up less than 1% of the atmosphere, variations in the levels of some of these gases, such as carbon dioxide, water vapor, and ozone, can have a profound effect on weather and climate.

Carbon dioxide (CO_2) plays an important role in the Earth's climate because it absorbs radiation in specific wavelength regions, warming the lower atmosphere. We'll say more about how this process works later in the chapter.

Carbon dioxide is used by green plants, which convert CO_2 into its chemical compounds to build up their tissues, organs, and supporting structures during photosynthesis. But when the plants die, microorganisms digest the plant matter, returning the CO_2 to the atmosphere. CO_2 is also a byproduct of the burning of fossil fuels. Increased human combustion of fossil fuels over the past 150 years has led to an increase in the amount of CO_2 in our atmosphere.

The troposphere contains significant amounts of water vapor. When the water vapor content is high, vapor can condense into water droplets, forming low clouds and fog, or the vapor can be deposited as ice crystals, forming high clouds. Individual water molecules are mixed freely throughout the atmosphere, just like the other gases. But unlike the other component gases, water vapor can vary greatly in concentration. Water vapor usually makes up less than 1% of the atmosphere, but under very warm, moist conditions, as much as 2% of the air can be water vapor. Tropical atmosphere regions contain the highest percentage of water and polar regions the lowest. Because water vapor is a good absorber of longwave radiation, like carbon dioxide, it plays a major role in warming the lower atmosphere.

Another small but important constituent of the atmosphere is ozone (O_3). Ozone is mostly concentrated in a layer, sometimes called the **ozone layer**, that is contained in the stratosphere, beginning at an altitude of about 15 km (about 9 mi) and extending to about 35 km (about 22 mi) above the Earth.

Ozone serves the important function of absorbing ultraviolet radiation from the Sun, protecting the Earth's surface from this damaging form of radiation. The reactions are quite complicated, but the net effect is that ozone (O_3), molecular oxygen (O_2), and atomic oxygen (O) are constantly formed, destroyed, and reformed in the ozone layer, absorbing ultraviolet radiation with each transformation. Without the protective filter of the ozone layer, the full intensity of solar ultraviolet radiation on the Earth's surface would destroy bacteria and severely damage animal tissue. From an evolutionary perspective, life on the land surface could not evolve until the formation of the protective ozone layer, which in turn depended upon oxygen in the atmosphere.

Aerosols The lower atmosphere contains countless tiny particles called **aerosols** that are so small and light that

> **aerosols** Tiny particles present in the atmosphere that are so small and light that the slightest air movements keep them aloft.

the slightest movements of the air keep them aloft. These small particulates are swept into the air from dry desert plains, lakebeds, and beaches. In addition, volcanic eruptions produce significant amounts of particulates, particularly those associated with sulfur compounds. Oceans are also a source of aerosols. Strong winds blowing over the ocean lift droplets of spray into the air. These droplets of spray lose most of their moisture by evaporation, leaving tiny particles of watery salt that are carried high into the air. Forest fires and brushfires also generate particles of soot as smoke. And as meteors vaporize as they hit the atmosphere, they leave behind dust particles in the upper layers of air. Closer to the ground, industrial processes that incompletely burn coal or fuel oil, including automobile combustion, release measurable amounts of aerosols into the air as well.

Aerosols are important because water vapor can condense on them to form tiny droplets. When these droplets grow large and occur in high concentration, they are visible as clouds or fog. Aerosol particles scatter sunlight, brightening the whole sky while slightly reducing the intensity of the solar beam. As a result, efforts to reduce industrial aerosols may actually lead to increased global warming (see *Video Explorations*).

Video Explorations
The Dimming Sun

The reduction in industrial pollution in the United States and other countries has resulted in cleaner and clearer skies. But, as this video shows, this positive trend has also led to an increase in solar insolation reaching the Earth's surface. As a result, temperatures can be expected to increase, adding to the existing global warming trend.

The global energy balance • Figure 2.13

Energy flows continuously among the Earth's surface, atmosphere, and space. The relative size of each flow is based on an arbitrary "100 units" of solar energy reaching the top of the Earth's atmosphere. The difference between solar energy absorbed by the Earth system (100 units incoming – 31 units reflected = 69 units absorbed) and the energy absorbed at the surface (144 units) is the energy (75 units) that is recycled within the Earth system (144 – 69 = 75). The larger this number, the warmer the Earth system's climate.

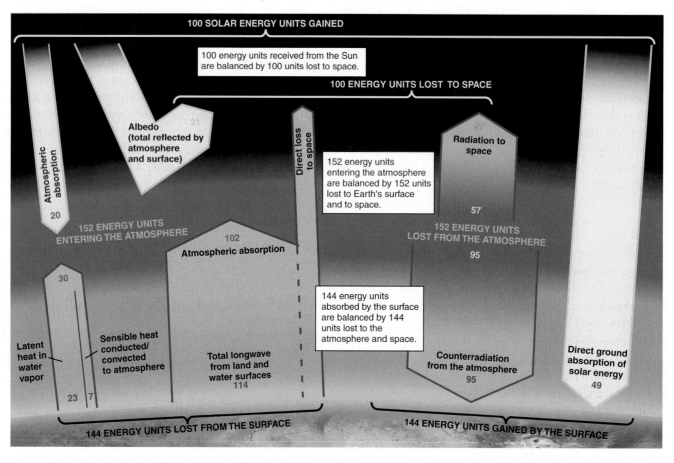

Absorption, Scattering, and Reflection of Insolation

The atmosphere acts as a filter for many wavelengths of the electromagnetic spectrum, as previously discussed. Some incoming solar radiation is absorbed, scattered, or reflected as it passes through the atmosphere and collides with gas molecules and particulates (**Figure 2.13**). The portion of solar radiation that reaches the surface of the Earth is either absorbed by or reflected from the surface.

Absorption About 16% of incoming solar radiation is absorbed in the atmosphere. This atmospheric **absorption** raises the temperature of the atmosphere. Carbon dioxide and water are the biggest absorbers. Because the water vapor content of air can vary greatly, absorption also varies from one global environment to another. For example, clouds absorb as much as 3 to 20% of radiation, so cloud cover significantly affects the amount of solar energy that reaches the Earth.

Absorption also plays an important role in filtering out some of the most damaging types of solar radiation, much as sunglasses filter bright rays from your eyes. Gamma rays and X-rays from the Sun are almost completely absorbed by the thin outer layers of the atmosphere, while much of the ultraviolet radiation is also absorbed, particularly by ozone, as we discussed earlier. Effective absorbers, such as carbon dioxide and water vapor, also absorb outgoing radiation, as we will see later in this chapter.

> **absorption** The process in which electromagnetic energy is absorbed when radiation strikes the molecules or particles of a gas, liquid, or solid, raising its energy content.

Scattering Solar rays can also be **scattered** in different directions when they collide with molecules or particles in the atmosphere. Rays can be diverted back up into space or down toward the surface. Scattered radiation is redirected, but its wavelength is unchanged.

Scattered radiation moving in all directions through the atmosphere is known as *diffuse radiation*. Some scattered radiation flows down to the Earth's surface, while some flows upward. This upward flow of diffuse radiation escaping back to space amounts to about 3% of incoming solar radiation.

Reflection Some incoming solar radiation bounces directly back into space, reflected by the atmosphere or by surfaces on the Earth. The proportion of shortwave radiant energy reflected upward by a surface is called its **albedo** and is measured on a scale of 0 to 1, with black surfaces near 0 and white surfaces near 1. A surface that reflects 40% of incoming shortwave radiation has an albedo of 0.40.

Because the albedos of different surfaces vary, the amount of sunlight reflected back into space varies depending on the time of year and surface composition (**Figure 2.14**). For example, snow-covered ground has high reflectance and a high albedo, whereas fields, forests, and bare ground have intermediate albedos with lower reflected radiation. The energy absorbed by a surface warms the air immediately above it by conduction and convection, so surface temperatures are warmer over low-albedo surfaces than over high-albedo surfaces.

The albedo of water is usually low, but it depends on the depth, clarity, and sediment load of the water column. Water albedo can be higher when the Sun's rays strike the water at a low angle, creating sunglint that raises the albedo value.

> **scattering** The process by which particles and molecules deflect incoming solar radiation in different directions on collision; atmospheric scattering can redirect solar radiation back to space.

> **albedo** The proportion of solar radiation reflected upward from a surface.

Albedo contrasts • Figure 2.14

a. Snow and ice have high albedos (0.45 to 0.85), reflecting much of the solar radiation that hits them and absorbing only a small amount. ▼

▲ **b.** Black pavement, which has a low albedo (0.03), absorbs nearly all the incoming solar energy and so can reach high temperatures on sunny days.

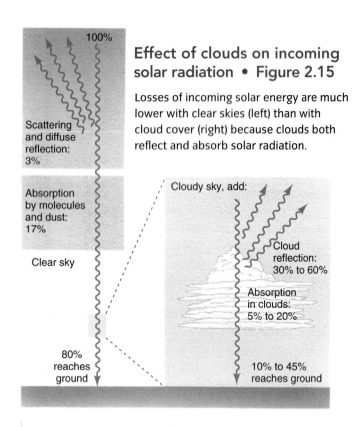

Effect of clouds on incoming solar radiation • Figure 2.15

Losses of incoming solar energy are much lower with clear skies (left) than with cloud cover (right) because clouds both reflect and absorb solar radiation.

Clear sky (left):
- 100%
- Scattering and diffuse reflection: 3%
- Absorption by molecules and dust: 17%
- Clear sky
- 80% reaches ground

Cloudy sky, add:
- Cloud reflection: 30% to 60%
- Absorption in clouds: 5% to 20%
- 10% to 45% reaches ground

Certain orbiting satellites carry instruments that can measure shortwave and longwave radiation at the top of the atmosphere, helping us estimate the Earth's average albedo. The albedo values obtained in this way vary between 0.29 and 0.34. This means that the Earth–atmosphere system reflects slightly less than one-third of the solar radiation it receives back into space.

Clouds can greatly increase the amount of incoming solar radiation reflected back to space. Reflection from the bright white surfaces of thick low clouds returns about 30 to 60% of incoming radiation back into space (**Figure 2.15**).

CONCEPT CHECK STOP

1. **What** important role is played by ozone in the atmosphere?

2. **How** does cloud cover affect the amount of solar radiation reaching the Earth's surface?

The Global Energy System

LEARNING OBJECTIVES

1. **Explain** how the greenhouse effect arises.
2. **Define** net radiation.
3. **Identify** key areas where human activities influence the global energy balance.

The Sun continually radiates shortwave energy in all directions. The Earth intercepts a tiny portion of this radiation, absorbing it and in turn emitting longwave radiation. Over time, this gain and loss of radiant energy is balanced, so the Earth's solar-heated surface remains at a constant average temperature, currently about 12.8°C (55°F). The energy flows we have been looking at between the Sun and the Earth's atmosphere and surface must balance over the long term in order to maintain this constant average temperature.

The Earth's Energy Output

So far we have focused on the flow of energy through the Earth's atmosphere from the Sun. Meanwhile, the Earth's surface emits longwave radiation upward.

> **greenhouse gases**
> The collection of gases in the atmosphere that absorb the Earth's emitted longwave radiation, raising temperatures in the Earth's lower atmosphere.

Remember that an object's temperature controls both the range of wavelengths and the intensity of radiation it emits. The Earth's surface and atmosphere are much colder than the Sun's surface, so our planet radiates less energy than the Sun, and that energy is emitted at longer wavelengths. Radiation emitted by the Earth peaks at about 10 μm in the thermal infrared region.

The flow of outgoing energy from the Earth to space mostly occurs in three wavelength ranges—4 to 6 μm, 8 to 14 μm, and 17 to 21 μm. The atmosphere absorbs other outgoing radiation wavelengths.

The greenhouse effect We have already established that the atmosphere plays an important role in absorbing incoming solar radiation, protecting the Earth from harmful solar radiation wavelengths and warming the atmosphere. The atmosphere also plays an important role in absorbing outgoing radiation.

Much of the longwave radiation emitted upward from the Earth's surface is absorbed by carbon dioxide, water vapor, and a few other gases found in the atmosphere, commonly called **greenhouse gases**. Some greenhouse gases, such as methane and sulfur hexafluoride,

Greenhouse gases characterization   Table 2.1

Greenhouse gases	Chemical formula	Pre-industrial concentration ppb*	2010 concentration	Anthropogenic sources	Global warming potential**
Carbon dioxide	CO_2	278,000	389,000	Fossil fuel, combustion, land use conversion, cement production	1
Methane	CH_4	700	1866	Fossil fuels, rice paddies, waste dumps, livestock	21
Nitrous oxide	N_2O	275	323	Fertilizer, industrial processes, combustion	310
CFC-12	CCl_2F_2	0	0.537	Liquid coolants, foams	7000
HCFC-22	$CHClF_2$	0	0.210	Liquid coolants	1350
Sulfur hexafluoride	SF_6	0	0.684	Dielectric fluid	23,900

*Parts per billion per volume of air in the atmosphere.

**Global warming potential for 100-year time frame, based on CO_2 value of 1.

occur in smaller concentrations than CO_2 but have a much greater global warming potential. Many greenhouse gases are byproducts of industrial processes and remain in the atmosphere for hundreds of years (**Table 2.1**). The energy absorbed by these gases raises the temperature of the atmosphere, causing it to emit energy back to the Earth in the form of **counterradiation**.

Cloud layers, which are composed of tiny water droplets, are even more important than carbon dioxide and water vapor in producing counterradiation because liquid water is also a strong absorber of longwave radiation.

Absorption of outgoing longwave radiation and re-emission of counterradiation back toward the Earth's surface is known as the **greenhouse effect** (**Figure 2.16**). Unfortunately, the term *greenhouse*

greenhouse effect Absorption of outgoing longwave radiation by components of the atmosphere and reradiation back to the surface, which raises surface temperatures.

The greenhouse effect • Figure 2.16

The Earth's atmosphere creates the greenhouse effect by absorbing a portion of outgoing longwave radiation and reradiating it back toward Earth as counterradiation.

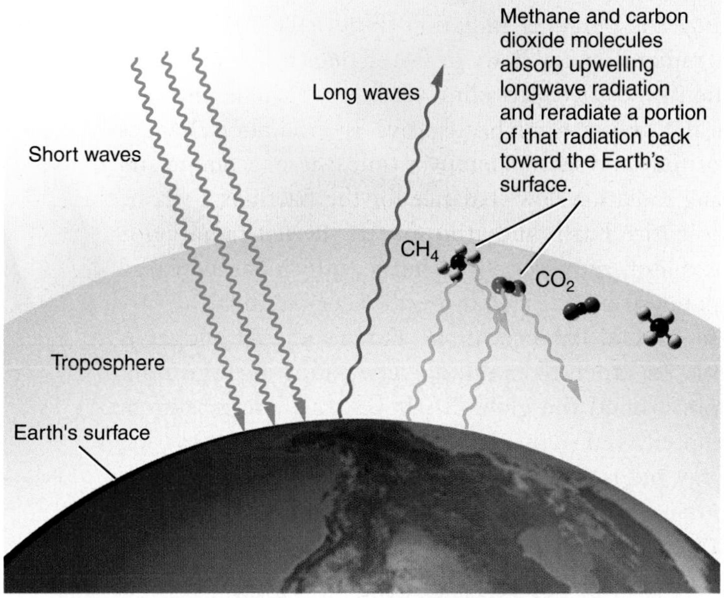

Methane and carbon dioxide molecules absorb upwelling longwave radiation and reradiate a portion of that radiation back toward the Earth's surface.

Long waves

Short waves

CH_4

CO_2

Troposphere

Earth's surface

Think Critically

Why might the greenhouse effect be stronger in humid regions, such as the Amazon Basin, than in dry regions?

PROCESS DIAGRAM

is not accurate as applied to the atmosphere. In a real greenhouse, glass windows permit entry of shortwave energy, as does our atmosphere. However, the primary reason that temperatures are warmer inside a greenhouse is because the warm inside air is not able to mix with cooler outside air. In contrast, the greenhouse effect in the atmosphere refers to the process in which longwave radiation that would have left the top of the atmosphere is absorbed and counterradiated back to the surface. This counterradiation from the atmosphere to the Earth's surface helps warm the surface through longwave heating, not by shortwave heating as in a real greenhouse.

Without the natural warming process of the greenhouse effect, the Earth would be too cold for human habitation. During the past 10,000 years, the Earth's energy balance has been relatively stable with respect to the conditions of this greenhouse system of thermal regulation. However, as levels of greenhouse gases in the atmosphere increase as a byproduct of human activities, the greenhouse effect warms the Earth more, disrupting our climate. We will come back to this point later in this chapter and in Chapter 3.

Net Radiation and the Global Energy Budget

The heat level of our planet rises as the Earth absorbs solar energy. But at the same time, it radiates energy into outer space in the form of longwave radiation, cooling itself. The global energy budget takes into account all the important energy flows to and from the Earth. Analysis of these energy inputs and outputs can help us understand how any changes to energy flow might affect the Earth's climate.

Net radiation is a measure of the difference between incoming and outgoing radiation (**Figure 2.17**). In places where radiant energy flows in faster than it flows out, net radiation is positive, providing an energy surplus. In other places, net radiation can be negative. Net radiation changes with latitude and season, but over time, these incoming and outgoing radiation flows balance for the Earth as a whole. For the entire Earth and atmosphere, the net radiation is approximately zero over a year, but recently it has increased enough to warm the Earth by over 1°C per century.

The annual balance in the Earth's energy budget is essential in order to maintain a constant average temperature around the globe. If the Earth's energy system is disrupted even slightly, with the addition or subtraction of energy, net radiation might be positive or negative over the course of a year or decade, substantially disrupting our climate. We will explore the risks and consequences of climate change in greater detail in upcoming chapters.

Human Impacts on the Global Energy Balance

Human activity around the globe has changed the planet's land surface cover and added gases and particulates to the atmosphere. Have we irrevocably shifted the balance of energy flows? Is the Earth absorbing more solar energy and becoming warmer? Or is it absorbing less and becoming cooler? If we want to understand human impact on the Earth–atmosphere system, we need to examine the global energy balance in detail.

Rising concentrations of greenhouse gases
Recall that although carbon dioxide, methane, and other greenhouse gases make up only a small portion of our atmosphere, they are a significant byproduct of human industry, and they can have a big impact on Earth's climate. Only in recent decades have scientists begun to establish the long-term trends in the atmospheric levels of these gases.

Climate physicist Charles Keeling was the first to track atmospheric levels of CO_2 with engineering precision to demonstrate trends in CO_2 levels, both over

Global net radiation • Figure 2.17

Net radiation measures the balance between incoming and outgoing radiation. In the sunlit part of the Earth, incoming shortwave radiation exceeds outgoing longwave radiation, and net radiation is positive. In the part of the Earth where it is night, there is no incoming shortwave radiation, yet the Earth's surface and atmosphere still emit outgoing longwave radiation. This causes a negative net radiation for those nighttime hours. Over the long term, net radiation balances out to zero.

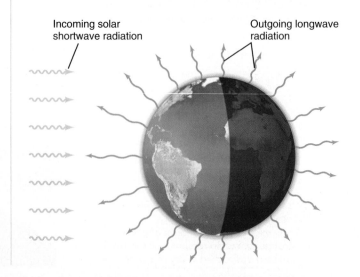

Incoming solar shortwave radiation

Outgoing longwave radiation

Rising atmospheric CO₂ levels • Figure 2.18

Carbon dioxide levels were 280 parts per million of the atmosphere up until the Industrial Revolution. Since 1958, carbon dioxide levels have been measured at Mauna Loa, Hawaii. These levels show both the annual cycle related to vegetation growth activity (inset) and the rising concentration of CO_2 in the atmosphere.

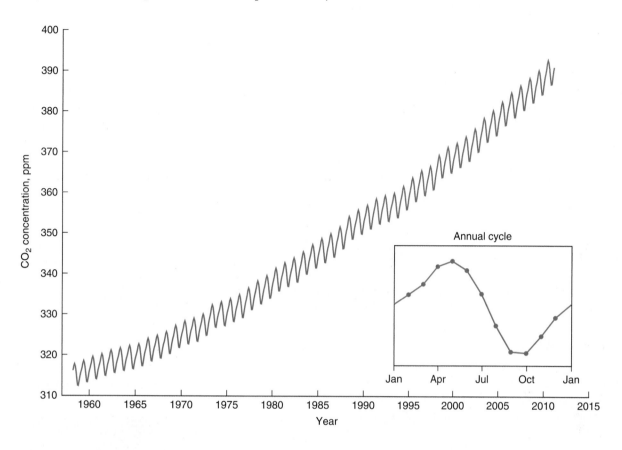

the course of a year and over the course of decades. He began making measurements of atmospheric CO_2 concentrations first from the South Pole and then, beginning in 1958, from the top of a volcano in Hawaii. The U.S. government has taken responsibility for continuing these measurements since his death so that the data are current to the present day.

When plotted on a graph, commonly called the Keeling curve, these measurements show the annual shift in atmospheric CO_2 as a repeating pattern of wiggles on the graph (**Figure 2.18**). When it is spring in the northern hemisphere, which contains the great majority of the Earth's land mass, vegetation grows and absorbs more CO_2. This is the downward turn seen each year in the curve's wiggling line. In the fall, the loss of leaves results in the increase of CO_2 as leaves decay and plant CO_2 uptake goes dormant through the winter.

Plotted over many years, the Keeling curve shows a distinct long-term upward trend, about 3% each year.

Keeling's research, using high-precision and continuous air-sampling measurements was the first to connect the rising of CO_2 levels with the excess CO_2 emissions coming from burning fossil fuels. Since then, governments around the world have continued to closely monitor the levels of CO_2 and other greenhouse gases in the atmosphere.

Methane (CH_4) is naturally released from wetlands when organic matter decays. But human activity generates about double that amount in rice cultivation, farm animal wastes, bacterial decay in sewage and landfills, fossil fuel extraction and transportation, and biomass burning.

Air pollution also increases the amount of ozone in the lower troposphere, where it can affect human health. Ozone in urban areas is distinct from the stratospheric ozone previously mentioned. Nitrous oxide (N_2O) is released by bacteria acting on nitrogen fertilizers in soils and runoff water. Motor vehicles also emit significant amounts of N_2O.

Scientists are investigating the wide range of effects that rising concentrations of greenhouse gases will have on our climate and our ecosystems. There is already evidence that an increase in the concentration of all these greenhouse gases in our atmosphere increases the greenhouse effect, warming the surface of the Earth more. We will take a closer look at the impact these warming trends have on the Earth's climate in Chapter 3.

Changes to surface albedo Clearing forests for agriculture and turning agricultural lands into urban and suburban areas normally decreases surface albedo. In that case, more energy is absorbed by the ground, raising its temperature. That, in turn, increases the flow of surface longwave radiation to the atmosphere, where it is absorbed and reradiated toward the Earth as counter-radiation. In Chapter 3, you will learn more about how urbanization changes the Earth's energy balance.

Air pollution Air pollution changes the flow of energy to and from the Earth. Air pollution increases the formation of low, thick clouds. Because low clouds increase short-wave reflection back to space, greater cloud cover causes the Earth's surface and atmosphere to cool.

Burning fossil fuels produces tropospheric aerosols. These aerosols are a potent form of air pollution that includes sulfate particles, fine soot, and organic compounds. Aerosols scatter solar radiation back to space, reducing the flow of solar energy to the Earth's surface. They also enhance the formation of low, bright clouds that reflect solar

WHAT A GEOGRAPHER SEES

Hazy Asian Atmosphere

You might recognize India's east coast on the left, the Bay of Bengal centered and to the south of Bangladesh in the middle, and Burma (Myanmar) to the right, with the arch of the Himalayan Mountains along the top of this MODIS-Terra satellite image (Photo). You might also notice the blurry haze that obscures most of the Indian and Bangladesh landscape as it appears to drift offshore. A geographer would identify this hazy regional air pollution as the Asian Brown Cloud (ABC), a pattern of haze that typically follows the drainage patterns of the Ganges and Brahmaputra rivers, over some of the most densely human-populated areas on the globe. This haze forms from aerosols created from farming and cooking fires, vehicle exhaust, and industrial emissions.

The UNEP (United Nations Environment Programme) Indonesian Observation Experiment (Graph) measured a 10% seasonal reduction (November to April) in solar irradiation reaching the land surface as a result of the haze. (Heavy rains clean the air of pollutants during other times of the year.) The reduced insolation can have a dramatic effect on regional climate, including changing rainfall patterns. The UNEP experiment documented a 10% reduction in crop yields as a result of the decrease in sunlight reaching the Earth's surface. The researchers also found that dark soot particles had lowered the albedo of snowpack in the region, causing the snow to absorb more radiation and melt sooner, leading to the shrinkage of Himalayan glaciers.

Solar radiation reduction from air pollutants

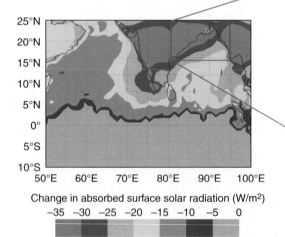

Change in absorbed surface solar radiation (W/m²)
−35 −30 −25 −20 −15 −10 −5 0

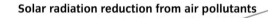

radiation back to space. These and other changes in cloud cover caused directly or indirectly by human activity lead to a significant cooling effect (see *What a Geographer Sees*).

In contrast, condensation trails (contrails) from jet aircraft create more high, thin clouds, which absorb more longwave energy than they reflect shortwave energy. This should make the atmosphere warmer by boosting counter-radiation and increasing the greenhouse effect. Scientific data remain inconclusive, however, on how contrails might contribute to global warming.

Thinning of the ozone layer

Changes in the Earth's energy system are also related to a decrease in the filtering provided by the atmosphere. In the mid-1980s, scientists discovered thinning in the ozone layer over the continent of Antarctica, sometimes called the ozone hole (see **Figure 2.19** on the next page).

Since 1978, scientific measurements have demonstrated that the amount of ultraviolet radiation reaching the Earth's surface has been increasing, apparently as a result of thinning of the ozone layer. Over most of North America, the increase has been about 4% per decade. Crop yields and some forms of aquatic life may suffer as a result. Increased ultraviolet radiation also puts humans at increased risk of skin cancer. In Australia and New Zealand, where the ozone layer is thinner, children commonly wear hats and stay covered for all outdoor activities.

Although numerous factors influence the rate of ozone in the atmosphere, scientific research and monitoring indicate that the thinning is related to certain forms of air pollution, especially chlorofluorocarbons, or CFCs, which

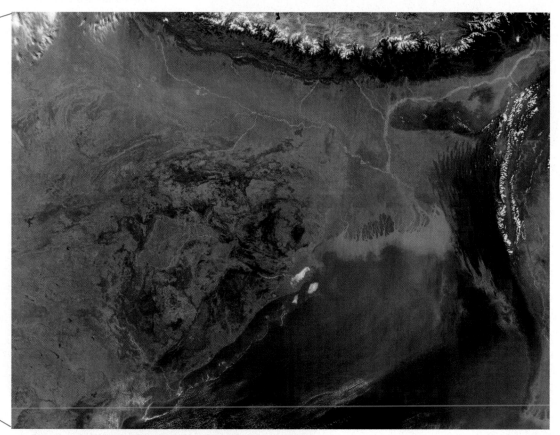

Satellite view of Asian Brown Cloud over Bay of Bengal region

Ask Yourself

If tomorrow an international treaty could stop air pollution from forming the ABC, what would happen to the amount of solar radiation reaching the Earth's surface?

a. It would increase.
b. It would decrease.
c. It would stay the same.
d. The answer is impossible to determine.

Calibrating sensors for atmospheric ozone monitoring • Figure 2.19

Prior to 1986, NASA's satellites failed to record a growing hole in the ozone layer over the South Pole. Fieldwork was needed to ensure that measurements were accurate and meaningful.

a. In 1986, NOAA's research teams launched balloon-mounted ozonesondes (chemistry sampling sensors) that take air samples along the air column as they rise. These calibrated field readings of ozone concentrations were then used to adjust the satellite sensors to ensure continuous monitoring of the regional ozone levels.

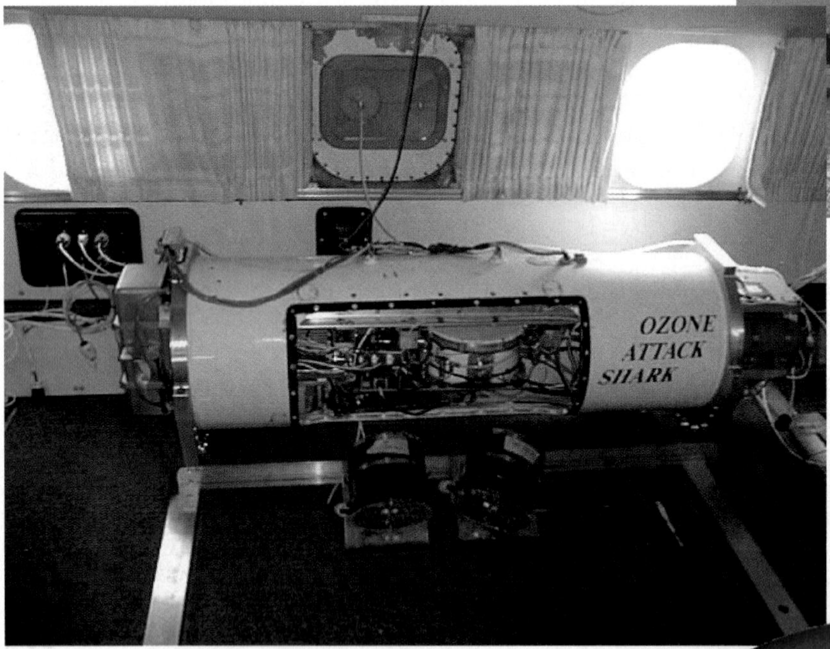

b. Aircraft sampling of air chemistry helped to increase the precision of the satellite remote-sensing measurements and helped scientists better understand the complex chemistry involved in the creation and destruction of atmospheric ozone molecules.

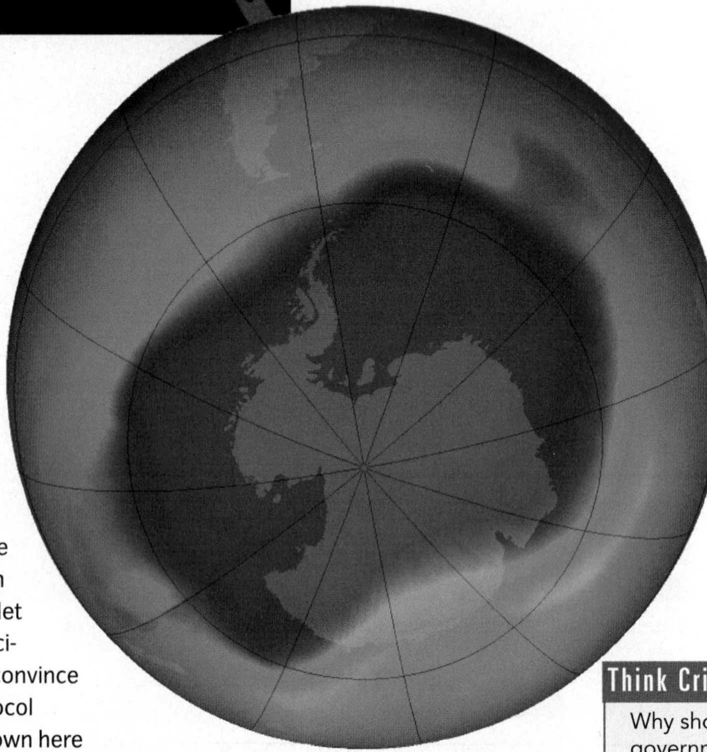

c. Once NASA's satellites were recalibrated based on the sampling data, they were able to detect the emergence of the ozone hole. Their findings confirmed that chlorine molecules originating in refrigerants and industrial processes were interfering with the Earth's protective ozone layer and creating a hole over the southern hemisphere, which was allowing harmful ultraviolet wavelengths to penetrate the atmosphere. The science team's findings led to the political effort to convince industrial nations to sign the 1987 Montreal Protocol international treaty banning CFC refrigerants. Shown here is an ozone hole observed September 24, 2006—one of the largest holes yet observed.

Think Critically

Why should the U.S. government fund both satellites and field work to conduct climate change research?

Where Geographers CLICK

The Ozone Layer

http://ozonewatch.gsfc.nasa.gov

The Antarctic ozone hole is constantly monitored by instruments on board a series of NASA and NOAA (National Oceanic and Atmospheric Administration) polar orbiting satellites as well as ground-based monitoring stations, aircraft, and balloons. You can investigate still images, animations, and a variety of science visualization tools to follow the history and current status of the ozone hole.

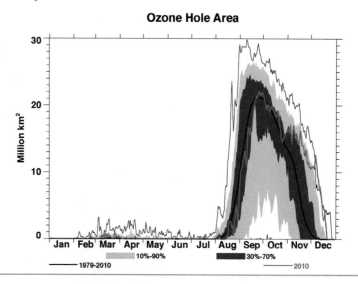

substantially reduce ozone concentrations. CFCs are synthetic industrial chemical compounds that contain chlorine, fluorine, and carbon atoms. Although CFCs were banned in aerosol sprays in the United States in 1976, they are still used as cooling fluids in some refrigeration systems and on the illegal black market in developing nations. When appliances containing CFCs leak or are discarded, their CFCs are released into the air.

CFC molecules move up to the ozone layer, where they decompose into chlorine oxide (ClO), which attacks ozone, converting it to ordinary oxygen through a chain reaction. With less ozone, there is less absorption of ultraviolet radiation.

In response to the global threat of ozone depletion, 23 nations signed the Montreal Protocol treaty in 1987 to cut global CFC consumption by 50% by 1999. The treaty was effective. By late 1999, scientists confirmed that stratospheric chlorine concentrations had topped out in 1997 and were beginning to fall. Despite the reduction in CFCs, however, the ozone layer is not expected to be completely restored until the middle of the twenty-first century. Scientists continue to monitor changes in the concentration of ozone throughout each year (see *Where Geographers Click*).

CONCEPT CHECK

1. **How** might clouds influence the greenhouse effect?
2. **How** does net radiation change at night?
3. **What** is the impact of air pollution on the flow of global energy?

THE PLANNER ✓

Summary

1 Electromagnetic Radiation 38

- All objects give off **electromagnetic radiation** at different wavelengths, as shown here.

- Hotter objects emit greater amounts of electromagnetic radiation. They also emit radiation at shorter wavelengths than cooler objects.

- The Sun emits mostly **shortwave radiation**, and the Earth gives off **longwave radiation**.

The electromagnetic spectrum • Figure 2.1

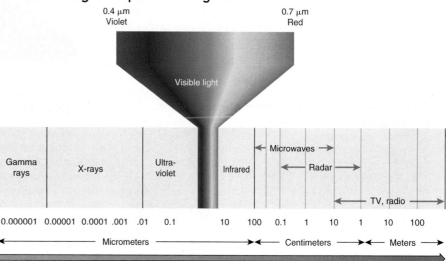

2 Insolation over the Globe 41

- **Insolation** is the rate of solar radiation flow at a location at a given time.

- Daily insolation is greatest when there is more daylight and the Sun is higher in the sky.

- Annual insolation varies due to changes in the angle of the Sun and day length. It is greatest near the equator and least at the poles, as shown here.

Angle of sunlight by latitude • Figure 2.10

- The seasons arise from the Earth's revolution around the Sun and the tilt of the Earth's axis.

- At the June **solstice**, the northern hemisphere is tilted toward the Sun. At the December solstice, the southern hemisphere is tilted toward the Sun. At the **equinoxes**, day and night are of approximately equal lengths.

- We can divide the globe into latitude zones according to how much insolation different latitudes receive.

3 Solar Energy and the Earth's Atmosphere 47

- The **troposphere** is the lowest layer of the atmosphere. The next layer is the **stratosphere**.

- The Earth's atmosphere is dominated by nitrogen and oxygen, as shown here. Carbon dioxide, water vapor, and ozone make up only a small part of the atmosphere but play an important role in climate.

Component gases of the lower atmosphere • Figure 2.12

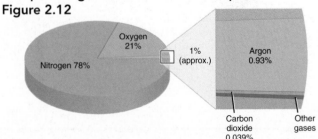

- Incoming solar radiation can be **absorbed**, reflected, or **scattered** as it travels through the atmosphere.

- The lower atmosphere contains small particles called aerosols that reflect or scatter solar radiation.

- The **ozone layer**, which helps shelter the Earth from ultraviolet rays, has been depleted as a result of chemicals such as CFCs being released into the air by human activities.

- Incoming solar radiation is partly absorbed or scattered by molecules in the atmosphere.

- The **albedo** of a surface is the proportion of solar radiation it reflects.

4 The Global Energy System 52

- The atmosphere absorbs longwave energy from the Earth and counterradiates some of it back to the Earth, creating the **greenhouse effect**, as shown in the diagram.

The greenhouse effect • Figure 2.16

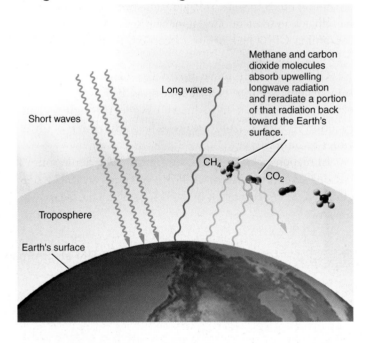

- Net radiation describes the balance between incoming and outgoing radiation.

- Human activities, such as the burning of fossil fuels and air pollution from industry, affect the global energy budget in complex ways.

Key Terms

- absorption 50
- aerosol 49
- albedo 51
- aphelion 42
- circle of illumination 42
- counterradiation 53
- declination 41
- electromagnetic radiation 38

- electromagnetic spectrum 38
- equinox 44
- greenhouse effect 53
- greenhouse gases 52
- heterosphere 48
- homosphere 48
- insolation 41
- longwave radiation 39

- mesosphere 47
- ozone layer 49
- perihelion 42
- plane of the ecliptic 42
- revolution 42
- scattering 51
- shortwave radiation 39
- solar constant 40

- solstice 43
- stratosphere 47
- subsolar point 41
- thermosphere 47
- tropopause 47
- troposphere 47

Critical and Creative Thinking Questions

1. Suppose that the Earth's axis were tilted at 40° to the plane of the ecliptic, instead of 23 1/2°. How would the seasons change at your location? What would be the global effects of the change?

2. What factors influence the amount of dangerous ultraviolet radiation that reaches the Earth? In what parts of the world are people most likely to be affected by this problem? Why?

3. The terms *greenhouse effect* and *global warming* are often used interchangeably. Are these terms synonyms? Explain your answer.

4. Because the Sun is much hotter than the Earth and therefore emits a lot more radiation than the Earth, how can the amount of outgoing infrared radiation balance the amount of solar radiation that the Earth intercepts?

5. The fossil fuel combustion that releases carbon dioxide into the atmosphere also consumes oxygen. The ratio is approximately one to one. If fossil fuel combustion consumes oxygen from the Earth's atmosphere, why are scientists not worried about depleting the atmospheric supply of oxygen?

6. Volcanic aerosols are described as "white"—that is, they have a high albedo. With this in mind, what effect do you expect a volcano to have on the Earth's surface temperature?

What is happening in this picture?

Thermal infrared radiation can be seen using a special imaging camera. Here we can see a suburban street on a cool night. Different temperature regions are represented by different colors in the image, ranging from black and violet (colder) to yellow and red (warmer).

Think Critically
1. Which surfaces are the warmest and which are the coolest in this scene?
2. What would this picture look like if it were taken at noon?

Self-Test

(Check your answers in the Appendix.)

1. In the figure, label the regions of the electromagnetic spectrum that correspond to (a) visible light, (b) gamma rays, (c) X-rays, and (d) ultraviolet light.

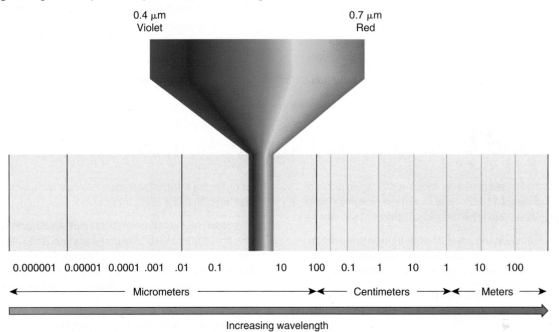

0.4 μm
Violet

0.7 μm
Red

0.000001　0.00001　0.0001　.001　.01　0.1　　10　100　0.1　1　10　1　10　100

◄────────────── Micrometers ──────────────►◄── Centimeters ──►◄── Meters ──►

Increasing wavelength

2. In the case of electromagnetic energy, warmer objects radiate _____ energy at _____ wavelengths than cooler objects.

a. more, shorter

b. more, longer

c. less, shorter

d. less, longer

3. The highest-energy, shortest-wavelength form of electromagnetic radiation that the Sun emits is _____.

a. shortwave infrared

b. visible light

c. X-rays

d. ultraviolet radiation

4. The Earth, maintaining a significantly cooler surface temperature than the Sun, emits _____.

a. ultraviolet radiation

b. shortwave infrared radiation

c. longwave infrared radiation

d. visible light

5. The amount of incoming solar radiation, or *insolation*, that a point on the surface of the Earth receives, is most dependent upon the _____.

a. angle at which the insolation is received

b. variations in the total solar radiation output of the Sun

c. amount of glacial coverage in a particular area

d. amount of ocean surface in a particular region

6. The Earth completes a(n) _____ revolution around the Sun in _____ days.

a. circular, 366

b. elliptical, 365

c. elliptical, 365 1/4

d. circular, 365

7. Which of the following is not a greenhouse gas?

a. carbon dioxide

b. water vapor

c. oxygen

d. methane

8. What time of year is represented by this diagram? Label the Tropic of Capricorn and explain its significance at this time.

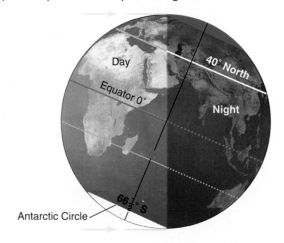

9. The _____ regions of the Earth receive the greatest amount of insolation.

a. polar

b. equatorial

c. midlatitude

d. subtropical

10. The Earth's _____ regions have the greatest insolation variation through the year.

a. polar

b. equatorial

c. midlatitude

d. tropical

11. The diagram shows the proportion of gases that make up the Earth's atmosphere. Label the following gases on the figure: (a) argon, (b) carbon dioxide, (c) nitrogen, and (d) oxygen.

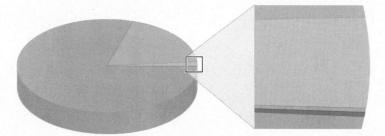

12. The seasons are caused by _____.

a. the varying distance between the Earth and the Sun due to the Earth's elliptical orbit

b. changes in the solar constant of the Sun

c. the tilt of the Earth's axis of rotation relative to the plane of the ecliptic

d. changes in the speed of the Earth's rotation on its axis

13. Of the following gases, _____ contributes the most to destruction of ozone.

a. nitrogen

b. carbon dioxide

c. chlorofluorocarbons (CFCs)

d. argon

14. Scattered radiation moving in all directions through the atmosphere is known as _____.

a. diffuse radiation

b. diffuse reflection

c. direct radiation

d. refracted radiation

15. The percentage of radiant energy scattered upward by a surface is called its _____.

a. outflow

b. output

c. albedo

d. reflection

THE PLANNER ✓

Review your Chapter Planner on the chapter opener and check off your completed work.

Air Temperature

In 1991, a sleeping giant awoke with a roar. Mt. Pinatubo, a volcano in the Philippines that had lain dormant for more than 500 years, erupted. The eruption blew off the top of the mountain and ejected between 8 and 10 square kilometers of material. Floods of lava spewed out, and hundreds of cubic meters of sand- and gravel-sized debris rained down on the mountain's upper slopes.

Before the eruption, many people lived on the mountain's forested lower slopes. Although more than 50,000 people were evacuated, several hundred died as roofs collapsed under the weight of falling ash, and tens of thousands suffered the effects of mudslides, ash-clogged rivers, and crop devastation.

Reverberations were felt globally. The volcano sent a vast plume of sulfur dioxide gas and dust into the atmosphere—some 15 to 20 million tons. A haze of sulfuric acid droplets spread throughout the stratosphere over the following year. These tiny particles reduced the sunlight reaching the Earth's surface by 2 to 3% for the year, and average global temperatures fell by about 0.3°C (0.5°F).

Our planet's atmosphere is a dynamic system, and we must keep a close watch on the natural and artificial phenomena that can perturb it.

Mt. Pinatubo erupts in a deadly cloud of ash and pumice, July 8, 1991.

CHAPTER PLANNER ✓

- ☐ Study the picture and read the opening story.
- ☐ Scan the Learning Objectives in each section:
  p. 66 ☐   p. 69 ☐   p. 76 ☐   p. 80 ☐   p. 83 ☐
- ☐ Read the text and study all figures and visuals. Answer any questions.

Analyze key features

- ☐ Where Geographers Click p. 71 ☐   p. 85 ☐
- ☐ Process Diagram, p. 72
- ☐ Geography InSight p. 74 ☐   p. 88 ☐
- ☐ What a Geographer Sees, p. 79
- ☐ Video Explorations, p. 87
- ☐ Stop: Answer the Concept Checks before you go on.
  p. 69 ☐   p. 75 ☐   p. 79 ☐   p. 82 ☐   p. 90 ☐

End of chapter

- ☐ Review the Summary and Key Terms.
- ☐ Answer the Critical and Creative Thinking Questions.
- ☐ Answer What is happening in this picture?
- ☐ Complete the Self-Test and check your answers.

Temperature and Heat Flow Processes

LEARNING OBJECTIVES

1. **Explain** how air temperatures close to the ground are related to surface temperatures.

2. **Explain** three important processes of heat energy transfer.

3. **Define** latent heat and explain how it is related to the state (solid, liquid, or gas) of matter.

I n Chapter 2, you saw how electromagnetic radiation from the Sun reaches the Earth and is either reflected or absorbed. Now let's take a closer look at how that heat energy is related to the temperatures we experience at the Earth's surface.

Measuring Temperature

Temperature is a very familiar concept. But how exactly is temperature defined? You may recall from earlier study of physics or chemistry that all matter possesses internal energy, which is in the form of the internal motion of atoms and molecules that make up the matter. The level of that internal energy is measured by the **temperature** of the matter.

In measuring temperature, we use a standard for comparison. For example, in using the Celsius temperature scale, we regard the internal energy level of a mixture of ice and water under specific conditions as 0°C and the internal energy level of boiling water under specific conditions as 100°C.

In the United States, temperature is still widely measured and reported using the Fahrenheit scale. The freezing point of water on the Fahrenheit scale is 32°F, and the boiling point is 212°F. In this book, we follow scientific convention and use the Celsius temperature scale, which is the international standard (**Figure 3.1**). A third temperature scale to recognize is the Kelvin (K) scale, used in the physical sciences, which begins at absolute zero (–273°C [–460°F])—a theoretical temperature that would be achieved by an object that cooled forever.

Surface temperature Thanks to gravity, we spend most of our time on or near the boundary of the atmosphere and the Earth's surface. When we think of the temperature of our environment, we are usually thinking of the layer of air within a few meters of the ground. But the surface of the ground is often at a different temperature.

Temperature scales • Figure 3.1

At sea level, the freezing point of water is at Celsius (C) temperature 0°, while it is at 32° on the Fahrenheit (F) scale. Boiling occurs at 100°C, or 212°F. The Kelvin scale provides an absolute measure of temperature and is often used by physicists.

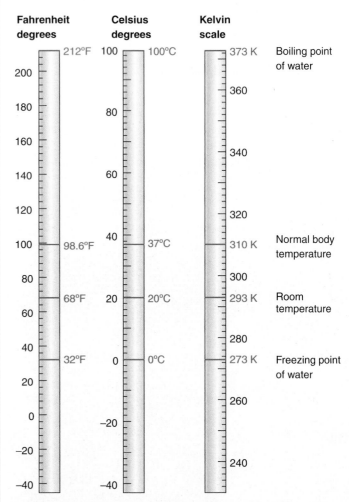

When you walk across a blacktop parking lot on a clear summer day in sandals, you will notice that the pavement, which is heated directly by the Sun, is a lot hotter than the air on the upper part of your body. At night, the ground surface is often cooler than the air above it because the surface loses heat by radiation. This means that air temperatures measured at the standard height of 1.2 m (4.0 ft) above a surface reflect the same day-to-night trends as ground surface temperatures, but ground temperatures are likely to be more extreme.

Wind chill and heat index The way we experience temperature is more complex than what we read on a thermometer because of additional factors, such as winds

Wind chill index and heat index • Figure 3.2

Dry, windy air makes us feel colder than the measured temperature, whereas humid, still air makes us feel warmer.

a. Wind chill index is an apparent temperature that takes wind speed into account. The higher the wind speed, the faster the rate at which heat flows from the skin, and the colder we feel.

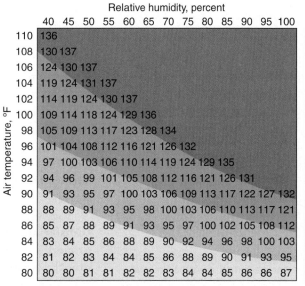

Temperature, °F

Wind speed, mph	40	35	30	25	20	15	10	5	0	−5	−10	−15	−20	−25	−30	−35	−40	−45
5	36	31	25	19	13	7	1	−5	−11	−16	−22	−28	−34	−40	−46	−52	−57	−63
10	34	27	21	15	9	3	−4	−10	−16	−22	−28	−35	−41	−47	−53	−59	−66	−72
15	32	25	19	13	6	0	−7	−13	−19	−26	−32	−39	−45	−51	−58	−64	−71	−77
20	30	24	17	11	4	−2	−9	−15	−22	−29	−35	−42	−48	−55	−61	−68	−74	−81
25	29	23	16	9	3	−4	−11	−17	−24	−31	−37	−44	−51	−58	−64	−71	−78	−84
30	28	22	15	8	1	−5	−12	−19	−26	−33	−39	−46	−53	−60	−67	−73	−80	−87
35	28	21	14	7	0	−7	−14	−21	−27	−34	−41	−48	−55	−62	−69	−76	−82	−89
40	27	20	13	6	−1	−8	−15	−22	−29	−36	−43	−50	−57	−64	−71	−78	−84	−91
45	26	19	12	5	−2	−9	−16	−23	−30	−37	−44	−51	−58	−65	−72	−79	−86	−93
50	26	19	12	4	−3	−10	−17	−24	−31	−38	−45	−52	−60	−67	−74	−81	−88	−95
55	25	18	11	4	−3	−11	−18	−25	−32	−39	−46	−54	−61	−68	−75	−82	−89	−97
60	25	17	10	3	−4	−11	−19	−26	−33	−40	−48	−55	−62	−69	−76	−84	−91	−98

Frostbite times   ☐ 30 minutes   ▨ 10 minutes   ▨ 5 minutes

To convert the wind chill index from °F to °C, use the following equation: $°C = (°F − 32) \times \frac{5}{9}$

b. The heat index is an apparent temperature accounting for the fact that higher levels of humidity make us feel hotter. When it is humid, the cooling effect of evaporation of sweat from the skin is inhibited.

Relative humidity, percent

Air temperature, °F	40	45	50	55	60	65	70	75	80	85	90	95	100
110	136												
108	130	137											
106	124	130	137										
104	119	124	131	137									
102	114	119	124	130	137								
100	109	114	118	124	129	136							
98	105	109	113	117	123	128	134						
96	101	104	108	112	116	121	126	132					
94	97	100	103	106	110	114	119	124	129	135			
92	94	96	99	101	105	108	112	116	121	126	131		
90	91	93	95	97	100	103	106	109	113	117	122	127	132
88	88	89	91	93	95	98	100	103	106	110	113	117	121
86	85	87	88	89	91	93	95	97	100	102	105	108	112
84	83	84	85	86	88	89	90	92	94	96	98	100	103
82	81	82	83	84	84	85	86	88	89	90	91	93	95
80	80	80	81	81	82	82	83	84	84	85	86	86	87

With prolonged exposure and/or physical activity

Extreme danger	Heat stroke or sunstroke highly likely
Danger	Sunstroke, muscle cramps, and/or heat exhaustion likely
Extreme caution	Sunstroke, muscle cramps, and/or heat exhaustion possible
Caution	Fatigue possible

To convert the heat index from °F to °C, use the following equation: $°C = (°F − 32) \times \frac{5}{9}$

Ask Yourself

If the relative humidity is 85%, at what temperature should someone exercising outdoors use "extreme caution"?

and humidity. The **wind chill index** and **heat index** reflect our sense of how hot or cold it is, which may differ from the actual temperature of the air molecules (**Figure 3.2**). These indexes often give us a better idea of how comfortable we might expect to be than a simple measure of air temperature. Public health officials have been using the heat index more visibly over the past few years due to the increased incidence of heat-related fatalities during summer months.

Energy Transfer

Physicists define the term **heat** as internal energy transferred from one substance to another as a result of their differences in temperature. Internal energy can flow in various ways. As we saw in Chapter 2, all objects lose internal energy to their surroundings by radiating electromagnetic energy. This radiant energy flow is thus a form of heat. Heat flow also occurs by conduction, convection, and advection.

Energy flow by convection and conduction • Figure 3.3

Conduction
Gas molecules of air in contact with a warm ground surface absorb internal energy from the ground, raising their temperature. This causes the surface layer of air to expand and become less dense.

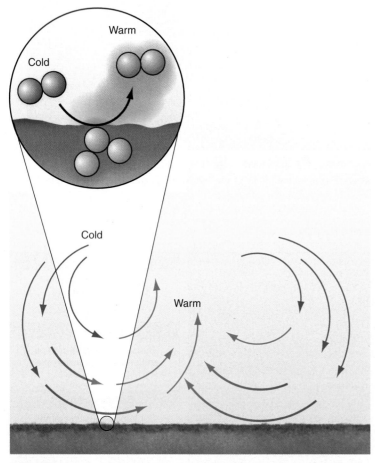

Convection
The warmer air, now less dense than the surrounding air, is buoyed upward. The warmer gas molecules, moving together in a current, carry their internal energy upward in an energy flow called convection. Cooler, denser air replaces the warmer air at the surface.

Conduction Internal energy flows from warm objects to colder ones when they are placed in contact. This transfer of internal energy from adjacent particles (molecules or atoms) is called **conduction** (**Figure 3.3**). Neighboring particles exchange energy until all particles have the same level of internal energy. In other words, the cooler object becomes warmer, and the warmer object becomes cooler.

Convection **Convection** is a flow of internal energy that occurs when matter moves from one place to another. This motion happens spontaneously in fluids—liquids or gases—when parts or regions of the fluid are at different temperatures and therefore have different densities.

For example, if you place your hand above a warm burner of an electric stove, you will find that a rising current of warm air warms your hand. Air in contact with the warm burner absorbs internal energy by conduction. This causes the gas molecules of the air to move more rapidly and collide more frequently. As they push at one another more strongly, the air expands and becomes less dense. It is then buoyed upward by surrounding cooler, denser air. As the warm air moves, it carries with it the internal energy absorbed from the surface, which warms the skin of your hand by conduction. This process of heat flow by fluid currents is convection.

As we will see later in this chapter, convection helps drive the circulation of the Earth's atmosphere and oceans and redirects much of the solar heat flow received between the tropics to higher latitudes.

Advection **Advection** occurs when a mass of air moves to a new location, bringing its properties, such as temperature or moisture content, along with it. For example, a layer of warm, moist air may flow into a region in summer, displacing cooler, drier air. If the air is warmer than the ground surface, heat will flow from the air to the ground, warming the surface. In winter, advection of cold air into a warmer region can cause heat to flow from the ground to the air, causing surface temperature to drop while the cold air layer warms.

Let's apply what we've learned about energy transfer to what happens when sunlight hits the ground surface. The solar shortwave radiation is quickly absorbed by a very thin layer of soil, which raises its temperature. As a result, the surface layer radiates a portion of the absorbed solar energy into the surrounding atmosphere. Because the surface layer is warmer than soil layers below, another portion of the absorbed energy flows into the ground by conduction. If the surface layer is warmer than the air above it, energy flows into the air by conduction and convection.

During the night, the soil surface no longer receives solar energy, but it continues to radiate heat to the sky and becomes cooler than the underlying soil layers. Heat then flows from the lower layers to the soil surface. Heat also flows to the surface from the warmer air above, creating a layer of cooler, denser air. If the surface is on a slope, the cool air may flow downhill, causing advection to occur.

Latent Heat

The flow of heat between two objects resulting in a temperature change is referred to as **sensible heat**. But a flow of heat to an object can also occur without a change in temperature, when the energy is absorbed in a change of state (solid, liquid, or gas)—for example, when liquid water evaporates to become water vapor. An energy flow that changes the state of a substance is known as **latent heat**. The latent heat changes the potential energy that is held in the positioning of atoms or molecules within a substance. Atoms or molecules moving freely in a gas have the highest potential energy, while atoms or molecules in a liquid state have lower potential energy. In a solid, the atoms or molecules have the lowest potential energy.

Latent heat taken up in a change of state is completely released when the change of state is reversed. For example, the amount of energy required to evaporate 1 kilogram of water liquid at 25°C to water vapor is 2441 kilojoules (kJ). If that water vapor is then condensed to liquid water at the same temperature, 2441 kJ of energy is released to the surroundings. For freezing and thawing, the latent heat flow required to melt 1 kg of ice is 334 kJ.

When the water refreezes, it will provide a latent heat flow to the surroundings of 334 kJ.

The flow of heat absorbed in the evaporation of water liquid to water vapor and its release in the condensation of water vapor to water liquid is a very important mechanism for moving energy from one region of the Earth to another. When tropical oceans are warmed by solar radiation, some of the Sun's energy is taken up in the evaporation of water liquid to water vapor. As we will see in later chapters, global winds move much of the water vapor poleward. At higher latitudes, the air cools and the water vapor changes state, becoming clouds of water droplets or ice crystals that can eventually form precipitation. Latent heat flows into the surrounding air, making it warmer. In this way, solar energy is exported from the tropics to mid- and high-latitude regions, making their climates significantly warmer.

CONCEPT CHECK

1. **How** is the air temperature directly above a ground surface related to the temperature of the ground surface?
2. **What** is one example of heat transfer by convection?
3. **How** does water vapor move solar energy from the tropics poleward?

Daily and Annual Cycles of Air Temperature

LEARNING OBJECTIVES

1. **Describe** the origin of daily cycles of air temperature.
2. **Describe** the causes of annual cycles of air temperature.
3. **Explain** why temperature patterns in maritime and continental regions differ.
4. **Explain** how latitude affects annual air temperature cycles.

We have already touched on some of the issues related to climate change, and we will return to them again later in this chapter. But first we need to understand the interplay between the Earth's surface and air temperatures and examine how and why air temperature changes from day to day, month to month, and year to year in different regions of the globe. Four important controls of air temperature are time of day, season, surface type (continental or maritime), and latitude.

The Daily Cycle of Air Temperature

Net radiation at a given place is normally positive during the day, as the surface gains heat from the Sun's rays. Over

Daily cycles of air temperature • Figure 3.4

These graphs compare idealized daily cycles of net radiation and air temperature for a midlatitude station at a continental interior location at the September equinox.

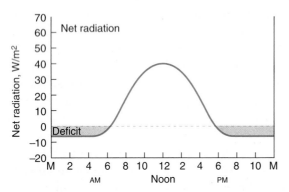

a. Net radiation

At midnight, net radiation is negative. Shortly after sunrise, it becomes positive, rising sharply to a peak at noon. In the afternoon, net radiation decreases as insolation decreases. Shortly before sunset, net radiation is zero: Incoming and outgoing radiation are balanced. Net radiation then becomes negative.

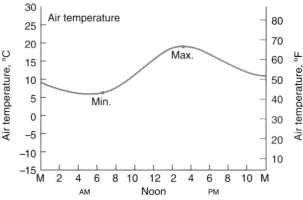

b. Air temperature

Because net radiation has been negative during the night, heat has flowed from the ground surface, and the ground has cooled the surface air layer to its lowest temperature about a half hour after sunrise. As net radiation becomes positive, the surface warms quickly, and heat flows into the air above. Air temperature rises sharply in the morning hours and continues to rise long after the noon peak of net radiation.

the course of the day, insolation, and therefore net radiation, increases as the angle of the Sun approaches its highest position in the sky at solar noon, then decreases later. At night, the flow of incoming shortwave radiation stops, but the Earth's surface and atmosphere continue to radiate longwave radiation, so net radiation becomes negative. Because the air next to the surface is warmed or cooled as well, air temperature changes through the day as net radiation changes (**Figure 3.4**).

Why does the temperature peak in the midafternoon? We might expect it to continue rising as long as the net radiation is positive. On sunny days in the early afternoon, large convection currents develop within several hundred meters of the surface, complicating the pattern. They carry hot air near the surface upward, and they bring cooler air downward. Temperatures typically peak between 2 and 4 PM. By sunset, air temperatures are falling. Temperature continues to fall more slowly throughout the night.

The daily air temperature variation is greatest just above the surface. The air temperature at standard height is far less variable (**Figure 3.5**). In the soil, the daily cycle becomes gradually less pronounced with

Daily temperature profiles close to the ground • Figure 3.5

The red curves show a set of temperature profiles for a bare, dry soil surface from about 30 cm (12 in.) below the surface to 1.5 m (4.9 ft) above it at five times of day.

At 8 AM, the temperature of air and soil is the same, producing a vertical line on the graph.

By noon, the surface is considerably warmer than the air at standard height, and the soil below the surface has been warmed as well.

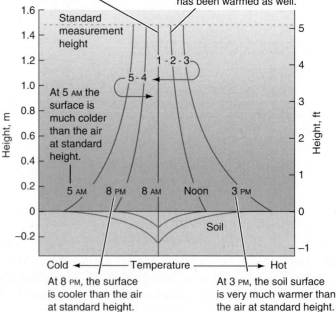

At 5 AM the surface is much colder than the air at standard height.

At 8 PM, the surface is cooler than the air at standard height.

At 3 PM, the soil surface is very much warmer than the air at standard height.

Annual cycles of insolation and air temperature • Figure 3.6

These graphs show idealized daily cycles of insolation and air temperature for a midlatitude station at a continental interior location at different times of the year.

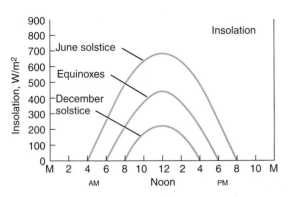

a. Insolation

At the equinox (middle curve), insolation begins at about sunrise (6 AM), peaks at noon, and falls to zero at sunset (6 PM). The June solstice peak is much greater than at equinox, and there is much more total insolation. The daily total insolation is greatly reduced in December.

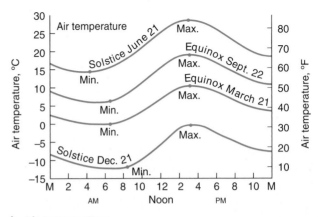

b. Air temperature

The height of the temperature curves varies with the seasons. In the summer, temperatures are warm, and the daily curve is high. In winter, the temperatures are colder. The September equinox is considerably warmer than the March equinox.

depth, until we reach a point where daily temperature variations on the surface cause no change at all. People have taken advantage of these thermal properties for millennia for storing perishable foods in deeply dug soil cellars, where the temperatures remains a constant 12.5°C (55°F). Modern geothermal heat pumps also utilize the consistent temperature conditions of deep soil or rock.

Annual Cycles of Air Temperature

Daily insolation varies over the seasons of the year, due to the Earth's motion around the Sun and the tilt of the Earth's axis. Recall from Chapter 2 that insolation changes over the course of the year, with changes in the day length and the path of the Sun through the sky. That rhythm produces a net radiation cycle that, in turn, causes an annual cycle in mean monthly air temperatures (**Figure 3.6**). NASA scientists provide regular reports and maps that depict the annual cycles of surface air temperatures (see *Where Geographers Click*).

Where Geographers CLICK

Goddard Institute for Space Studies Surface Temperature Analysis

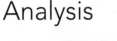

http://data.giss.nasa.gov/gistemp/

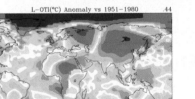

Calculating the Earth's surface temperature from ground, air, and satellite measurements is extremely complex. Among many other sources of data, geographers refer to the Goddard Institute for Space Studies for authoritative and current information about the Earth's temperature trends.

PROCESS DIAGRAM

Warming a water body with solar radiation • Figure 3.7

Coastal climates are affected by the fact that solar radiation raises the temperature of water bodies less than it raises the temperature of land surfaces.

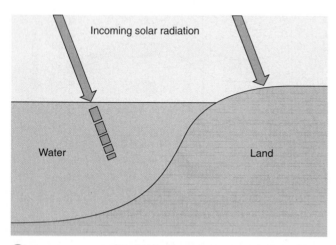

1 **Solar radiation penetrates the water.**
Solar radiation penetrates the surface of the water. As the solar beam is absorbed, it warms the upper layers of water. On land, the radiation cannot penetrate the soil or rock, so absorption is concentrated at the surface.

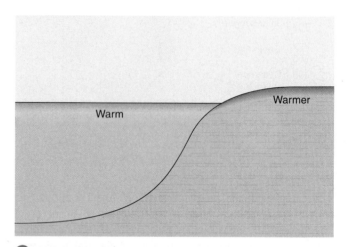

2 **Water temperature rises in the upper layers.**
The energy from the solar radiation warms the water, with the most warming near the surface. However, water is a difficult substance to heat. It can take as much as five times as much energy to raise water temperature one degree as it takes to raise the temperature of the same volume of rock one degree. So the land gets warmer from the same flow of solar energy.

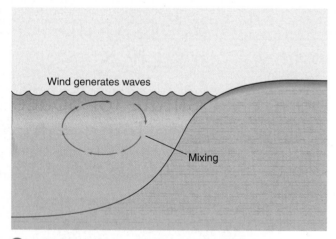

3 **Mixing distributes the absorbed energy to lower layers.**
Wind-generated waves and currents mix the water, distributing the absorbed energy more deeply. This also helps keep the increase in water temperature low. In rock and soil, mixing does not occur, and the soil temperature remains higher.

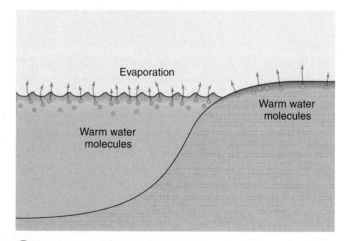

4 **Evaporation leads to cooling.**
As water molecules evaporate, they draw energy from the surface, lowering the water temperature. Land surfaces also can be cooled by evaporation if water exists near the soil surface, but once the land surface dries, evaporation stops, and so does cooling.

Land and Water Contrasts

Daily and annual air temperature cycles are more moderate at coastal, or maritime, locations than at interior locations due to four primary factors in the way heat is transferred differently on land and in water: transmission of light, specific heat capacity, mixing, and evaporation (**Figure 3.7**).

Specific heat is the amount of heat required to raise the temperature of a unit mass of a substance by 1 °C. Rock and soil have low specific heat values, which means it requires less energy to raise the temperature of the land surface. For the same reason, rock and soil cool more quickly. Water has a much higher specific heat capacity than rock and soil, so the surface layer of any extensive, deep body of water heats more slowly and cools more slowly than the surface layer of a large body of land when both are subjected to the same intensity of insolation. As a result, air temperatures above water are less variable than those over land.

As a result of these effects, a coastal city has more moderate, uniform temperatures, while an inland city has more extremes. In a seacoast city, summer heat is moderated by the presence of water, and so cooler temperatures prevail. Similarly, winter chill is moderated in a seacoast city by the water and so is warmer.

The same differences as those experienced seasonally are reflected in temperature differences between day and night. Places located well inland and far from oceans tend to have stronger temperature contrasts between night and day (**Figure 3.8**).

Temperature by Latitude

Recall from Chapter 2 that the amount of daily and annual insolation varies significantly by latitude. Locations near the equator receive the highest and steadiest rate of insolation throughout the year, and locations closer to the poles receive a variable rate of insolation, depending on the time of year. **Figure 3.9** (on the next page) applies what we've learned about daily and annual cycles of air temperature to demonstrate how mean monthly air temperature varies by latitude.

Daily temperature cycles in maritime and continental locations • Figure 3.8

Inland locations have greater variation in temperature from day to night. A recording thermometer made continuous records of the rise and fall of air temperature for a week in summer at San Francisco, California, and at Yuma, Arizona.

a. Temperatures in San Francisco, on the tip of a peninsula between the Pacific Ocean and the San Francisco Bay, hover around 13 °C (55 °F) and change only a little from day to night.

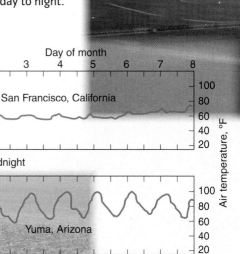

b. Air temperatures in Yuma, Arizona, which is located in the Sonoran Desert, are much warmer, on average—about 28 °C (82 °F). The daily range is also much greater. The hot desert drops by nearly 20 °C (36 °F) from its daytime temperature, producing cool desert nights.

Geography InSight

Net radiation and air temperature by latitude • Figure 3.9

The monthly mean net radiation and air temperature at latitudes near the equator are higher and steadier throughout the year. Moving closer to the poles, the annual average net radiation is lower, and the seasonal variations are more extreme.

Monthly average net radiation

Place	Lat	Annual avg W/m²
Yakutsk	62°N	42
Hamburg	54°N	47
Aswan	24°N	87
Manaus	3°S	98

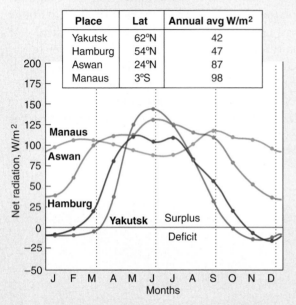

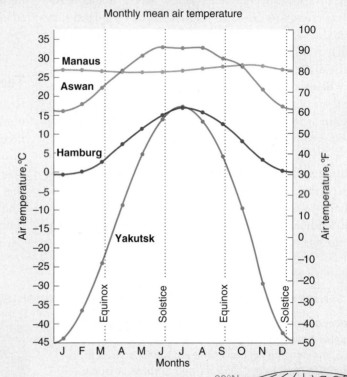

Monthly mean air temperature

a. Equatorial latitude

At Manaus, Brazil (03°09′N/57°59′W), the average net radiation rate is strongly positive every month and relatively constant. As a result, Manaus has uniform air temperatures, averaging about 27°C (81°F), and no discernable warm or cool seasons, with only a small difference of 1.7°C (3°F) between the highest and lowest mean monthly temperature.

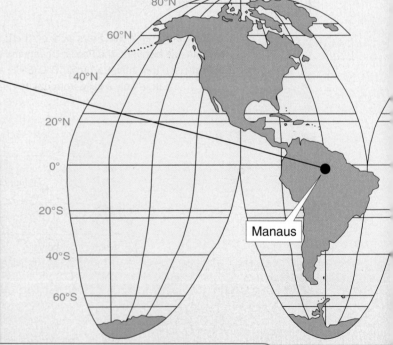

Manaus

Put It Together

Review Figure 3.7 and answer this question:.
Although Yakutsk is only about 9.5° farther north than Hamburg, the annual temperature cycles of these two cities are drastically different. While summer temperatures are similar, winter temperatures at Yakutsk average −45°C (−49°F) compared to just about freezing at Hamburg.

Which is the best explanation for this observation?
a. The elevation is much higher in Yakutsk.
b. The elevation is much higher in Hamburg.
c. Winds bring air from the Arctic Ocean to Yakutsk.
d. Winds bring air from the Atlantic Ocean to Hamburg.

b. Midlatitude

In Hamburg, Germany (53°38′N/10°00′E), there is a radiation surplus for nine months and a deficit for three winter months. The temperature cycle for Hamburg reflects the reduced total insolation at this latitude. Summer months reach an average maximum of just over 16°C (61°F), while winter months reach an average minimum of just about freezing (0°C [32°F]). The annual range is about 17°C (31°F), the same as at Aswan.

c. Arctic latitude

During the long, dark winters in Yakutsk, Siberia (62°05′N/129°45′E), there is a radiation deficit that lasts about six months when temperatures are between −35 and −45°C (about −30 and −49°F). In summer, when daylight lasts most of a 24-hour day, the net radiation rate rises to a strong peak, and air temperatures rise to over 13°C (55°F). Because of Yakutsk's high latitude and continental interior location, its annual temperature range is enormous—over 60°C (108°F).

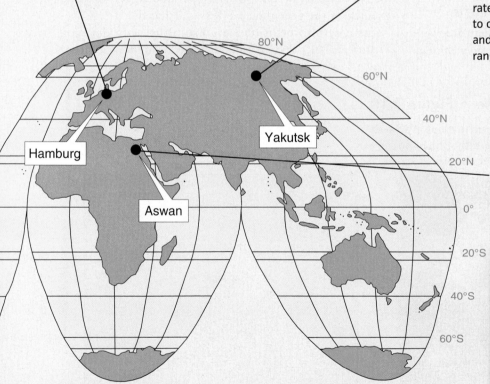

d. Tropical latitude

Aswan, Egypt (23°58′N/32°47′E), shows a large surplus of positive net radiation every month. The net radiation rate curve has a strong annual cycle; values for June and July are triple those of December and January. As a result, Aswan has an annual temperature range of about 17°C (31°F). June, July, and August are terribly hot, averaging over 32°C (90°F).

CONCEPT CHECK

1. **When** during the day would you expect air temperature to be at its minimum and maximum values? Why?

2. **Why** is insolation greater at noon during the summer than in the winter?

3. **Why** do San Francisco, California, and Yuma, Arizona, have such different daily temperature cycles?

4. **Why** is the air temperature in Manaus, Brazil, largely uniform across the year, while Yakutsk, Siberia, displays a wider temperature range?

Local Effects on Air Temperature

LEARNING OBJECTIVES

1. **Explain** how and why atmospheric conditions change at high elevation.

2. **Explain** how surface type affects urban and rural temperatures.

In addition to temperature effects related to season, time of day, position relative to the coast, and latitude, local areas also experience temperature differences related to elevation, land use, and urbanization. Ocean currents also play a major role in distributing heat around the globe, as you will see in greater detail in Chapter 5. Local effects can also produce **microclimates**—local atmospheric zones where the climate differs from surrounding areas.

Effect of Elevation on Temperature

In general, the air is cooler at higher altitudes than at lower altitudes. Remember from Chapter 2 that most incoming solar radiation passes through the atmosphere and is absorbed by the Earth's surface. The atmosphere is then warmed at the surface by latent and sensible heat flows. It is logical to consider that, in general, air farther from the Earth's surface is cooler.

Air pressure also changes with elevation. For example, as you climb higher on a mountain, you may become short of breath, and you might notice that you sunburn more easily. You may also feel the temperature drop as you ascend. If you camp out, the nighttime temperature will get lower than you might expect, even given that temperatures are generally cooler the farther up you go.

What causes these effects? At high elevations, there is significantly less air above you, so air pressure is low. It becomes more difficult to breathe simply because of the reduced oxygen pressure in your lungs. And with fewer molecules to scatter and absorb the Sun's light, the Sun's rays will be stronger as they beat down on you. There is less carbon dioxide and water vapor, and so the thermal response of the greenhouse effect is reduced.

As a result, mean temperatures drop with elevation (**Figure 3.10**). We call the decrease in air temperature

Environmental temperature lapse rate • Figure 3.10

On a typical summer day in the midlatitudes, temperature drops at an average rate of 6.49°C/1000 m (3.56°F/1000 ft). **a.** Looking at the graph, we see that the air temperature at sea level is a pleasant 20°C (68°F).

The aster daisies along the Kangshung Valley, Tibet, at 3650 m (11,975 ft) will experience an air temperature of around 5°C (41°F), while the air at the top of Mt. Chomolonzo, seen in the background of the valley at an altitude of 7800 m (25,590 ft), will be a bone-chilling −25°C (−10°F).

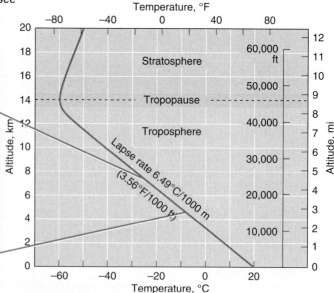

Ask Yourself

If the temperature at the ground were 0°C instead of 20°C, what would be the approximate temperature at 6096 m (20,000 ft) elevation, assuming the same lapse rate?

a. −40°C
b. −25°C
c. −15°C
d. −5°C
e. 10°C

Temperature structure of the upper atmosphere • Figure 3.11

Temperature trends at different layers of the atmosphere depend on the composition of gas molecules at each layer.

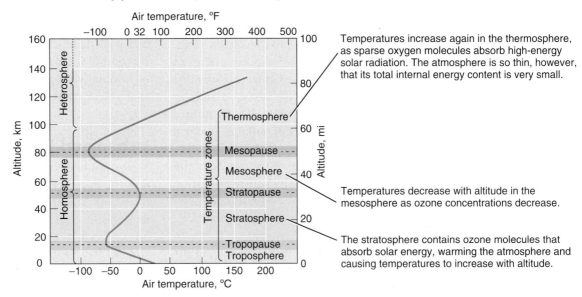

Temperatures increase again in the thermosphere, as sparse oxygen molecules absorb high-energy solar radiation. The atmosphere is so thin, however, that its total internal energy content is very small.

Temperatures decrease with altitude in the mesosphere as ozone concentrations decrease.

The stratosphere contains ozone molecules that absorb solar energy, warming the atmosphere and causing temperatures to increase with altitude.

with increasing altitude the **environmental temperature lapse rate**, or simply the environmental lapse rate. We measure the temperature drop in degrees Celsius per 1000 m (or degrees Fahrenheit per 1000 ft). Keep in mind that the environmental temperature lapse rate is an average value and that on any given day, the observed lapse rate might be quite different, as atmospheric pressure and temperature change.

Temperatures continue to decrease with elevation up through the troposphere. Above the tropopause, however,

> **environmental temperature lapse rate** The average rate at which air temperature drops with increasing altitude.

temperatures increase or decrease, depending on the gas composition of each atmospheric layer (**Figure 3.11**).

Effect of elevation on temperature variation Higher elevations experience more variation in day and night temperatures (**Figure 3.12**). At higher elevations, there is less carbon dioxide and water vapor, and so the greenhouse effect is reduced. With less atmospheric counterradiation returning to the Earth, surface temperatures tend to drop even lower at night.

The effect of elevation on air temperature cycles • Figure 3.12

The graph shows daily temperature for five weather stations at different altitudes in the Andes Mountains. With higher elevations, mean temperatures decrease, and the daily range between maximum and minimum temperatures increases.

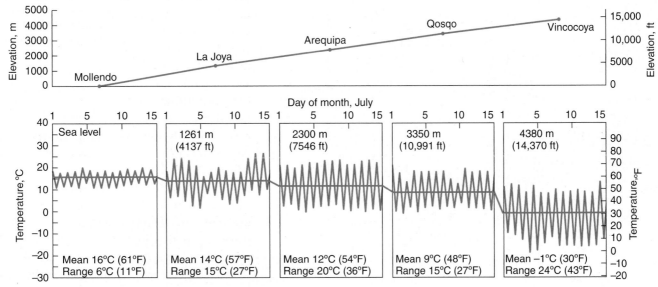

Temperature inversion Occasionally, weather conditions in the troposphere lead to an exception to the rule that temperature decreases with elevation, creating a temporary state called a **temperature inversion**. On a clear, calm night, the ground surface radiates longwave energy to the sky, and net radiation becomes negative. As the surface cools, air near the surface also cools by conduction and convection. If the surface stays cold, a layer of cooler air above the ground will build up under a layer of warmer air, as shown in **Figure 3.13**.

> **temperature inversion** A state of the atmosphere in which air temperature increases with elevation.

In a temperature inversion, the temperature of the air near the ground can fall below the freezing point. This temperature condition can cause a killing frost—even though actual frost may not form—because of its effect on sensitive plants during the growing season.

Growers of fruit trees and other crops use several methods to break up an inversion. Large fans can be used to mix the cool air at the surface with the warmer air above, and oil-burning heaters are sometimes used to warm the surface air layer and increase air mixing. When temperature inversions occur over cities, such as Los Angeles or Denver, harmful pollutants can become trapped near the ground, creating a temporary toxic air mixture and sparking public health alerts.

Surface inversions forming on cold nights create a layer of cold, dense air that can flow downhill. The cold air typically follows slopes and watercourses to lower elevations, where it accumulates in valleys. This can place crops in lowland valleys at a greater risk of killing frosts than crops on nearby hill slopes.

Urban and Rural Environments

When you walk across a parking lot on a hot day, the pavement can become hot enough to burn your bare feet. It is different when you step across a grassy field: The ground does not feel as hot, even if the air temperature is similar. Remember from Chapter 2 that different surfaces have different albedos, and black pavement absorbs more radiant energy than green grass. Another factor is that soil and vegetation contain more moisture, which cools the surface as it evaporates.

Because different land surface types have different properties, land use can affect local climate. Where forests are cleared to provide cropland, air temperatures generally increase. Forests are cooler because trees take up and transpire soil water from the root zone, providing a constant stream of evaporative cooling. Cropland soils dry much more quickly under the Sun and heat the air above them more readily.

Where urban development has replaced farmland, there are also temperature differences. Because urban surfaces tend to be drier (less evaporation) and darker (with lower albedo) than rural ones, city centers tend to be several degrees warmer than the surrounding farms, suburbs, and countryside. We call this thermal center an **urban heat island** because it has a significantly elevated temperature as the city both generates and traps heat (see *What a Geographer Sees*).

> **urban heat island** An area at the center of a city that has a higher temperature than surrounding regions.

The urban heat island effect has important economic and health consequences. We use more air conditioning and more electric power to combat higher temperatures in the city center. Smog is also more likely to form in warm temperatures, creating conditions that

Temperature inversion • Figure 3.13

Air temperature normally decreases with altitude (dashed line), but in an inversion, temperature increases with altitude. In this example, the surface temperature (A) is the same as the temperature at 760 m (2500 ft) off the ground (B).

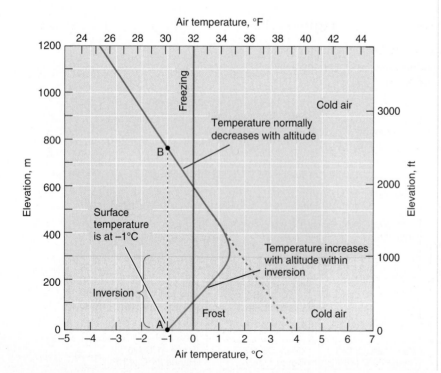

WHAT A GEOGRAPHER SEES

Urban Heat Islands

Residents of Atlanta, Georgia, are well aware that their city gets very hot in the summer. They might have also noticed that the air feels cooler as they leave the city center and move into the more rural surroundings. A geographer could demonstrate this temperature difference by using airborne or satellite sensors to generate a color-coded map showing local temperature patterns. The Landsat satellite image pair, captured September 28, 2000, provides two perspectives of Atlanta's urban metropolitan area.

Figure a shows an aerial view of the city that we would see with our own eyes. Trees and other vegetation are green, roads and dense development appear cement-gray, and bare ground appears tan or brown. The thermal infrared image (Figure b) shows surface temperature, where cooler temperatures are yellow and hotter temperatures are red. The highest temperatures are concentrated in the city center, while lower temperatures occur in the suburbs, and the lowest temperatures are detected in the outlying rural areas.

These temperature differences occur in part because many city surfaces are dark and strongly absorb solar energy. In fact,

asphalt paving absorbs more than twice as much solar energy as vegetation. Notice how roads and highways stand out on the image as warmer orange or red surfaces. Vegetated surfaces are cooled by evaporation of water from green leaves.

High electricity use during the summer also releases waste heat that raises temperatures. Air conditioners cool buildings inside by pumping excess heat outdoors. The buildings of the city also provide many vertical surfaces that reflect radiation between surfaces. This traps solar energy, which bounces back and forth until most of it is absorbed.

b.

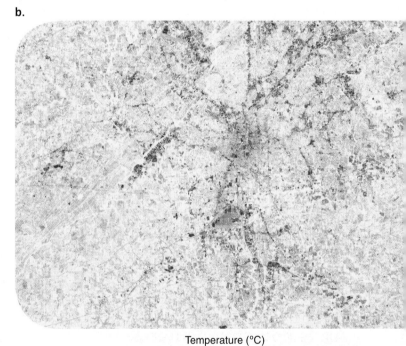

a.

Temperature (°C)

18   24   30

Think Critically Imagine a green, irrigated city filled with parks and tree-lined boulevards surrounded by a dry desert of sand and rocks.
1. Would the city be warmer or cooler than the surrounding desert?
2. Why?

increase mortality and heat-related illness. Heat-related deaths have increased significantly over the past couple of decades. Many cities are now planting more vegetation, including roof gardens, in an attempt to counteract these problems. In addition, painting rooftops with white, or high-albedo, paint is part of the effort to reduce heat island conditions in many cities.

CONCEPT CHECK

1. **What** atmospheric conditions are necessary for a temperature inversion?

2. **What** are urban heat islands, and how do they form?

World Patterns of Air Temperature

LEARNING OBJECTIVES

1. **Interpret** air temperature maps.
2. **Explain** world temperature patterns.

Surface type (urban or rural), elevation, latitude, daily and annual insolation cycles, and location (maritime or continental) can all influence air temperatures. Now let's put all these factors together and see how they affect world air temperature patterns.

Air Temperature Maps

Before we begin our discussion of world air temperature patterns, we need a quick explanation of air temperature maps. **Figure 3.14** shows a set of **isotherms**—lines connecting locations that have the same temperature. Usually, we choose isotherms that are separated by 5 or 10 degrees, but they can be drawn at any convenient temperature interval.

Isothermal maps clearly show centers of high and low temperatures. They also illustrate the rate at which temperature changes with distance, known as a **temperature gradient**.

> **isotherm** A line on a map drawn through all points with equal temperatures.
>
> **temperature gradient** The rate of temperature change along a selected line or direction.

Air Temperature Patterns Around the Globe

You have been exposed to the three main factors that explain world temperature patterns. The first is latitude. As latitude increases, average annual insolation decreases, and temperatures decrease as well, making the poles colder than the equator. Latitude also affects seasonal temperature variation. For example, the poles receive more solar energy at the summer solstice than does the equator due to the 24-hour exposure to what is sometimes called the midnight sun. So we must remember to note the time of year and the latitude when looking at temperature maps.

The second factor is the maritime–continental contrast. As we have noted, coastal weather stations record more uniform temperatures and are cooler in summer and warmer in winter compared to interior landscapes. Interior weather stations, on the other hand, record much larger annual temperature variations. Ocean currents can also have an effect because they can keep coastal waters warmer or cooler than you might expect. The eastern seaboard of the United States is a prime example of maritime warming from the warm Gulf current.

Elevation is the third important factor. At higher elevations, temperatures will be cooler, so we expect to see lower temperatures in mountainous environments.

The combination of these three factors dictates the range of temperature extremes experienced on our planet. The warmest recorded temperature of 58°C (136°F) at El Azizia, Libya, and the coldest recorded temperature of –89°C (–128°F) at Vostok Station, Antarctica, set a formidable range for human habitability. Within North America, temperatures range from 57°C (134°F) in Death Valley, California, to –63°C (–81°F) in Snag, Yukon, Canada. Human adaptability has enabled sparse populations to eke out a living and survive for millennia in these extreme climates. With the population expected to grow to 9 to 10 billion people by the middle of the 21st century and more land needed for human habitation and agriculture, however, governments and scientists will need to recognize these climatic boundaries.

Isotherms • Figure 3.14

Isotherms are used to make air temperature maps. Each line connects points having the same temperature. Each point denotes a temperature measuring station. Where temperature changes along one direction, a temperature gradient exists. Where isotherms close in a tight circle, a center exists. This example shows a center of low temperature.

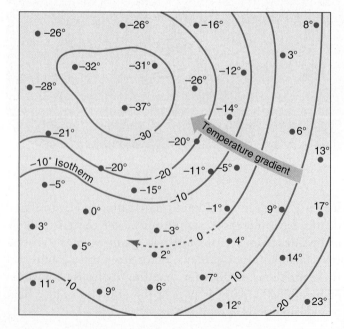

Air temperature at the equator and midlatitudes Maps of mean monthly air temperatures around the world show how air temperature is related to latitude, elevation, and position relative to the ocean (**Figure 3.15**). For example, highlands are always colder than surrounding lowlands because temperatures decrease with elevation.

By comparing maps from January and July, we can recognize the degree of seasonal variation in temperature. Isotherms make a large north–south shift from January to July over continents in the midlatitude and subarctic zones. Meanwhile, temperatures in equatorial regions change little from January to July because insolation at the equator does not vary greatly with the seasons.

Mean monthly air temperatures for January and July • Figure 3.15

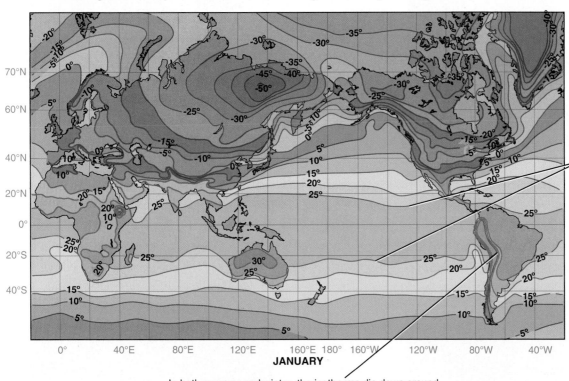

JANUARY

In the coastal equatorial region, the average temperature is usually greater than 25°C (77°F) but less than 30°C (86°F). Temperatures vary depending on how far inland from the coast you are located and how high the elevation is. Although the two isotherms move a bit from winter to summer, the equator always falls between them, showing how uniform the temperature is over the year.

In both summer and winter, the isotherms dip down around the mountains because of the high elevation.

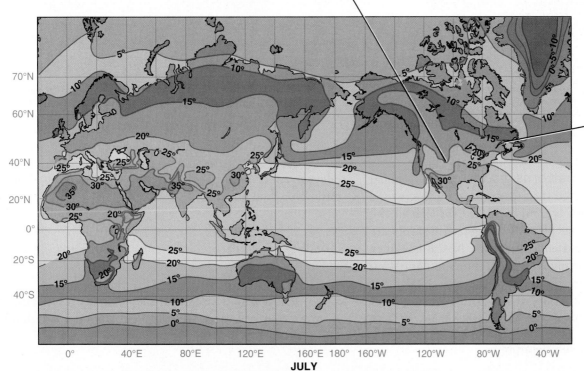

JULY

15°C (59°F) isotherm lies over central Florida in January, but by July it has moved far north, cutting the southern shore of the Hudson Bay and then looping far up into northwestern Canada and Alaska. In contrast, isotherms over oceans shift much less. This is because continents heat and cool more rapidly than oceans.

The general pattern of the isotherms as nested circles shows the coldest temperatures at the center, near the poles, in both January and July.

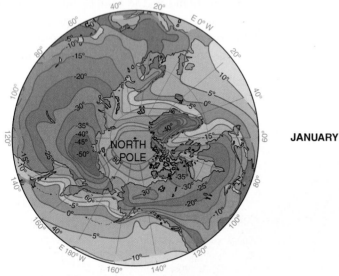

JANUARY

In January, the North Pole is experiencing its winter, with its minimum annual insolation. The cold center in Siberia, reaches –50°C (–58°F), while northern Canada is also quite cold –35°C (–32°F).

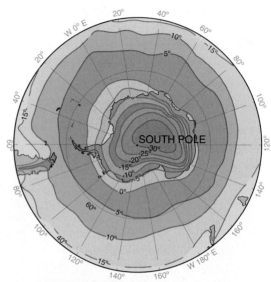

Meanwhile, the South Pole is experiencing its summer. Despite peak levels of insolation, the temperature there is still extremely cold as a result of high elevation and year-round snow cover.

JULY

In July, the North Pole is experiencing its summer. Nonetheless, temperatures in Greenland remain extremely cold as a result of its high elevation, snow cover, and continental land mass.

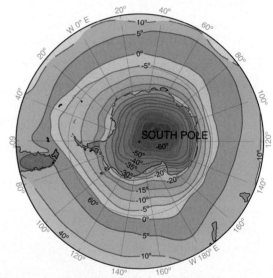

Meanwhile, the South Pole is experiencing its winter. With virtually no daylight, temperatures there reach as low as –70°C (–94°F).

Air temperature patterns at the poles Annual insolation decreases from the equator to the poles, so the North and South poles are intensely cold throughout the year (**Figure 3.16**). Our planet's two great ice sheets are contained in Greenland and Antarctica. Notice how they stand out on the polar maps as cold centers in both January and July. They are cold for three reasons. First, their surfaces are high in elevation, rising to over 3000 m (about 10,000 ft) in their centers. Second, the white snow surfaces reflect much of the insolation. Third, they are located on continental land masses.

CONCEPT CHECK STOP

1. **What** is an isotherm, and why is it useful?

2. **How** and **why** do isotherms shift on monthly temperature maps of the world between January and July?

The Temperature Record and Global Warming

LEARNING OBJECTIVES

1. **Describe** how temperature records can be established for periods before recorded temperatures are available.

2. **Explain** the causes of the current trend in global warming.

3. **Describe** the possible impact of global warming in the future.

For the past 10,000 years or so, the Earth has experienced a period of moderate climate in which human civilization has developed and thrived. We understand how temperature varies based on the time of day or year and that our planet can undergo short- and longer-term changes in temperature patterns as a result of natural events. But issues of global, human-induced climate change are new to our social, historical, and scientific understanding. Today, the science community is just beginning to understand the effects of human activities on our climate and to help illuminate preventable causes of global warming trends.

The Temperature Record

Today, satellite technology allows scientists to monitor the surface temperature of the Earth on a daily basis. Establishing a baseline for the Earth's temperature in different regions and monitoring changes to temperature patterns requires taking the sea surface temperature and the land surface temperature from a network of sea-based and ground-based meteorological stations, in combination with remote-sensing thermal temperature sensors on satellites and airplanes (**Figure 3.17**). These quantitative measurements are used for a range of applications, from ocean fisheries to crop forecasting to climate change assessments.

SCIENCE TOOLS

Measuring the Earth's temperature • Figure 3.17

NASA and NOAA employ a constellation of weather and special instrument satellites to provide daily temperature profiles of the Earth's surface, the air mass directly above, and the profiles along the atmospheric layers.

a. Sea surface temperatures averaged for the month of January 2011 are measured for the globe as recorded from Moderate Resolution Imaging Spectroradiometer (MODIS) thermal sensors aboard the Aqua satellite.

Think Critically Besides the influence of urban heat islands, what natural geographic feature is influencing temperature patterns in Figure 3.17b? (Ignore the black, which denotes bodies of water.)

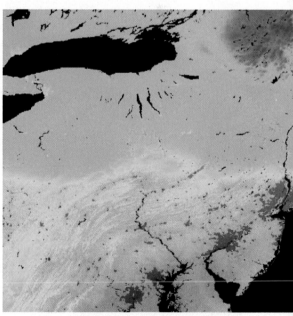

b. Weather satellites provide data for weather prediction, here forecasting land surface temperatures for urban heat islands affecting the east coast from Virginia to New York. This image shows a temperature prediction for June 11, 2001.

We have direct records of air temperature measurements dating to the middle of the 19th century. If we want to know about temperatures at earlier times, we need to use indirect methods, often called proxies. In climates with distinct seasons, many tree species grow annual rings. For trees along the timberline in North America, the ring width is related to temperature: The trees grow better when temperatures are warmer. Because only one ring is formed each year, we can work out the date of each ring by counting backward from the present. Because these trees live a long time, we can extend the temperature record and correlate the tree ring measurements with other proxy measures going back several centuries. Additional climate-monitoring proxies include ancient sediment and coral reef coring. These proxies provide growth or layering clues to past climate variability that also presents data on localized differences in past climate conditions.

Ice cores from glaciers and polar regions can show annual layers that are related to annual precipitation cycles. The crystalline structure of the ice, the concentrations of salts and acids, dust and pollen trapped in the layers, and amounts of trapped atmospheric gases, such as oxygen, carbon dioxide, and methane, all reveal information about long-term climate change patterns (**Figure 3.18**).

Reconstruction of past temperature records shows cycles of higher and lower temperatures. There have also been wide swings in the mean annual surface temperature. Some of this variation is caused by volcanic eruptions. Volcanic activity propels particles and gases—especially sulfur dioxide (SO_2)—into the stratosphere, forming stratospheric aerosols. Strong winds spread the aerosols quickly throughout the entire layer, where they reflect incoming solar radiation and have a cooling effect.

SCIENCE TOOLS

Reconstructing temperature records through ice core analysis • Figure 3.18

Ice core analysis has helped establish a record of long-term temperature records on the Earth that is now well accepted by the scientific community.

a. Scientists drill down into deep ice to extract a column of ice, known as an ice core. Annual cycles of precipitation form distinct layers in the ice core that allow scientists to determine the age of different portions of the sample. Tiny air bubbles in the ice preserve samples of the air from the period in which it froze. Shown here is a scientist extracting an ice core from a glacier in Kluane National Park, Canada.

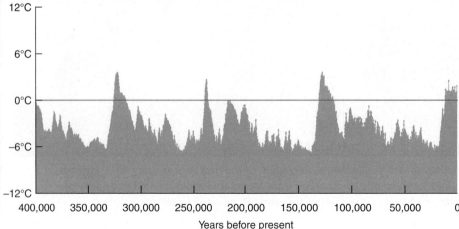

b. In ice cores from the Russian antarctic Vostok research station, oxygen isotope ratios have been used to calculate global temperatures for the past 400,000 years, using known laws of physics and water evaporation rates.

Mean annual surface temperature of the Earth, 1880–2010 • Figure 3.19

The years 2005 and 2010 are tied for the hottest years on record, and the first decade of the 21st century is now the warmest decade on record. Of the top 11 warmest years so far, 10 occurred between 2000 and 2010. The yellow line shows the mean for each year. The red line shows a running five-year average. Note the effect of the eruption of Mt. Pinatubo, mentioned in the chapter opener.

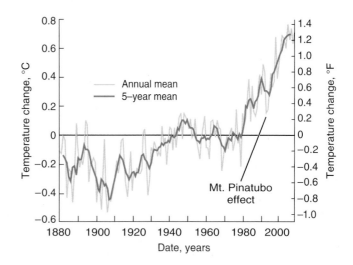

The eruption of Mt. Pinatubo in the Philippines lofted 15 to 20 million tons of sulfuric acid aerosols into the stratosphere in the spring of 1991. The aerosol layer produced by the eruption reduced solar radiation reaching the Earth's surface between 2 and 3% for the year or so following the blast. In response, global temperatures fell about 0.3°C (0.5°F) in 1992 and 1993.

Global Warming

The Earth has been getting warmer, especially in the past 50 years (**Figure 3.19**). In February 2007, the Intergovernmental Panel on Climate Change (IPCC), a United Nations–sponsored group of more than 2000 scientists, issued a consensus report stating that global warming is "unequivocal" (see *Where Geographers Click*). The National Academy of Sciences, the premier science advisors for the U.S. government, has recently amplified the IPCC findings by stating, "Climate change is occurring, is caused largely by human activities, and poses significant risks for—and in many cases is already affecting—a broad range of human and natural systems."

Where we live in the world has a lot to do with our view of climate change. Small island nations have expressed alarm at the rising sea levels they witness. People living in the Arctic are witnessing a greater rise in temperature

Where Geographers CLICK

Intergovernmental Panel on Climate Change

www.ipcc.ch

In 2007, the Intergovernmental Panel on Climate Change, along with former Vice President Al Gore, was awarded the Nobel Peace Prize, shown here. Visit the IPCC Website to keep close watch on the topic of climate change. The international science community depends on this resource to follow the scientific method in monitoring the Earth's climate system.

Temperature trends by latitude • Figure 3.20

The Earth's recent temperatures, measured from all latitudes and compared with the temperature averages over the 1951–1980 time period, demonstrate notable warming differences at the highest latitudes, both north and south of the equator.

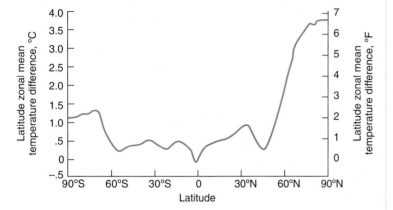

than people living in lower latitudes (**Figure 3.20**). Latitude therefore plays a significant role in where the effects of global warning are first occurring. Currently, the Arctic region is warming at a rate of 2.5 times the global average.

Causes of global warming In 1995, the IPCC concluded that human activity has probably caused climatic warming by increasing the concentration of greenhouse gases in the atmosphere. Recall from Chapter 2 that adding greenhouse gases to the atmosphere enhances the greenhouse effect and increases warming. This judgment, which was upgraded to "likely" in 2001 and "very likely" in 2007, was based largely on computer simulations showing that the release of CO_2, CH_4, and other gases from fossil fuel burning and other human activity over the past century has accounted for the pattern of warming recently documented.

Figure 3.21 shows how a number of important factors have influenced global warming since about 1850. Although human activity has both warming and cooling effects on the atmosphere, when we add all these factors together, the warming effects of the greenhouse gases outweigh the cooling effects. The result is a net warming effect of about 1.6 to 2.4 W/m^2, which is about 1% of the total solar energy flow absorbed by the Earth and atmosphere.

The years 2005 and 2010 are tied as the warmest years on record since the middle of the 19th century, according to data compiled by NASA scientists at New York's Goddard Institute of Space Science. These were also the warmest years of the past millennium, according to reconstructions of past temperatures using tree rings and glacial ice cores by University of Massachusetts scientists. In fact, the data from land and ocean temperatures collected from around the globe document that the first 10 years of the 21st century are the warmest decade on record since 1400. In the past 30 years, the Earth has warmed by 0.6°C (1.1°F). In the past century, it has warmed by 0.8°C (1.4°F).

Factors affecting global warming and cooling • Figure 3.21

Greenhouse gases act primarily to enhance global warming, while aerosols, cloud changes, and land-cover alterations caused by human activity act to retard global warming. Natural and human-induced factors may be either warming (red) or cooling (blue) or both.

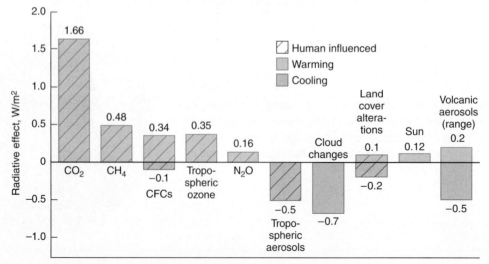

Computer projections for global temperature increases in the 21st century • Figure 3.22

The IPCC considers numerous computer models in its projections for the next century, based on continued human releases of CO_2, CH_4, and other industry-generated gases. Note that increasing aerosols are predicted to counteract some of the warming trend by reducing surface insolation. (Newest predictions are slightly different from this year 2000 graph.)

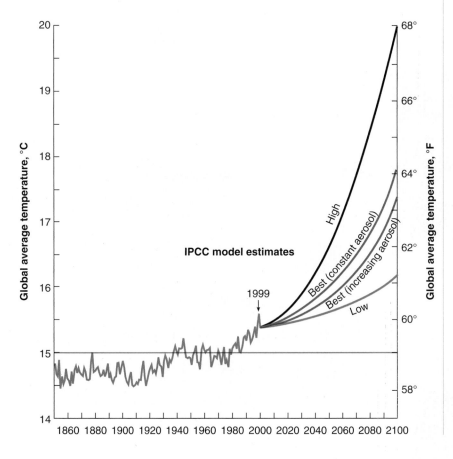

Consequences of Global Warming

What will be the future effect of anthropogenic greenhouse warming? Using computer climate models, the scientists of the IPCC have projected that global temperatures will warm between 1.8°C (3.2°F) and 4.0°C (7.2°F) by the year 2100 (**Figure 3.22**). The range of temperatures predicted by the science models results in part from differences in the models' technical performance and in use of assumptions regarding future states of atmospheric conditions. Aerosols from dirty air, for example, could have a cooling effect, as noted in the range of predictions. It should also be noted that the IPCC uses scientific consensus from thousands of scientists to publish these predictions, which means that at this time, the higher temperatures predicted are just as likely as the best, and lower, estimates.

A temperature increase of a few degrees may not sound like very much, but consider the fact that the increase of just a few degrees above 37°C (98.6°F) for human body temperature can have serious medical consequences, including brain damage and even death. We are only recently learning about the Earth's temperature norms for a much larger system and the effects of small temperature change. The problem is compounded by the realization that many other changes, including melting sea ice and rising sea levels, may accompany a rise in temperature (see *Video Explorations* and **Figure 3.23** on the next page).

Video Explorations

Global Warming Update

Recent studies by NASA show that the Greenland ice sheet is melting more rapidly at its margins and appears to be steadily losing mass. This video demonstrates that the vast Antarctic ice sheet may be starting to lose mass as well, according to gravity-measuring satellites. Arctic sea ice is also diminishing, leaving more open water for more of the year. Because water absorbs most solar energy while ice reflects most solar energy, reduced ice cover will lead to warmer waters, which in turn will lead to more loss of ice. This heating cycle may bring massive dangerous effects to our world through sea-level rise and disruptions in the hydrologic cycle that will affect the whole world.

Global warming is changing habitats and ecosystems around the globe. The risk of habitat loss is greatest at transitional areas, such as coastlines and the polar regions in North America and Eurasia.

a. Arctic thawing
Longer summers and warmer winters are disrupting native people's way of life and eroding the coastal communities as melting permafrost undermines community infrastructure and homes. Inuit villages are being relocated farther inland, at a cost of millions of dollars.

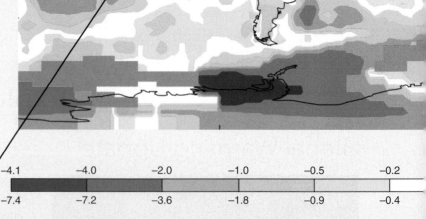

−4.1	−4.0	−2.0	−1.0	−0.5	−0.2
−7.4	−7.2	−3.6	−1.8	−0.9	−0.4

b. Sea-level rise
As glaciers melt and ocean water expands in response to the warming, sea levels are predicted to rise 28 to 43 cm (11.0 to 16.9 in.) by the year 2100, overwhelming the ability of both humans and native biodiversity to adapt and placing as many as 92 million people within the risk of annual flooding. The combination of sea-level rise and increasingly frequent and more intense storms threatens coastal communities.

c. Polar sea ice melting

Careful monitoring of the arctic sea ice pack has shown a decrease in September sea ice of about 10% per decade since 1979. Accelerated warming is predicted as the high albedo of the ice is replaced by the lower albedo of water that will absorb increased radiation, thereby raising water temperatures. If these trends continue, the Arctic Ocean will eventually become free of summer ice.

Mar | Sep

Temperature difference from long-term monthly mean (1951-1980) October, 2010

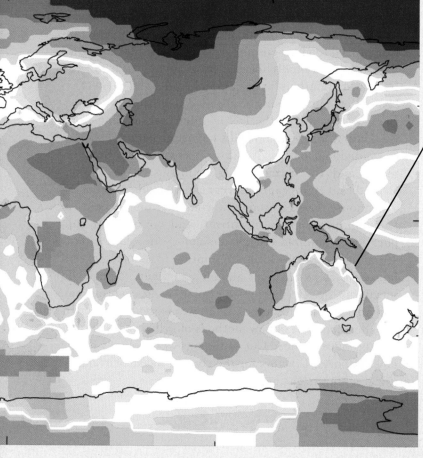

d. Habitat loss

As a result of rapidly rising sea levels, low-lying islands with their surrounding coral reef ecosystems are subject to increased exposure from storm surges.

Temperature difference, °C

0.2	0.5	1.0	2.0	4.0	7.1
0.4	0.9	1.8	3.6	7.2	12.8

Temperature difference, °F

Put It Together

Review Figure 3.20 and answer this question:
Where do you expect to find the largest observed changes in global temperature as a function of latitude?

a. Equatorial regions
b. Tropical regions
c. Subtropical regions
d. Midlatitude regions
e. Polar regions

The Temperature Record and Global Warming 89

Climate change Global warming is triggering a variety of complex changes in our global climate, which may even include patterns of cooler weather in some regions. Weather seems to be becoming more variable and more extreme in many locations. The scientific community is monitoring a variety of ongoing changes in our current climate regime, such as the early onset of spring and the delay in fall. These climate pattern shifts are affecting ecosystems from pine beetle infestations in the Pacific Northwest to bleaching of coral reefs in the Indian Ocean. Climate change could also promote the spread of insect-borne diseases such as malaria. Climate range boundaries may shift positions, making some regions wetter and others drier and adding new extremes to the new climate settings. Shifts in agricultural patterns, including desert expansion, could displace large human populations as well as natural ecosystems.

Global warming is predicted to increase the variability of our climate and increase the amplitude of storms and droughts. Very high 24-hour precipitation—extreme snowstorms, rainstorms, sleet, and ice storms—have become more frequent since 1980. More frequent and more intense spells of hot and cold weather also may be related to climate warming.

International response to global warming
The combination of rising population numbers, increased living standards, conversion of forests to farmlands, and the ever-increasing consumption of fossil fuel energy are creating a complex turbulence of human affairs affecting our future climate. The world has become widely aware of the problem of global warming, and, at the Rio de Janeiro Earth Summit in 1992, nearly 150 nations signed a treaty limiting emissions of greenhouse gases. The IPCC, formed in 1988, prepared the First Assessment Report in 1990, which set into motion the creation of the United Nations Framework Convention on Climate Change (UNFCCC) by concluding that:

- Anthropogenic climate change will persist for many centuries.

- Further action is required to address remaining gaps in information and understanding.

- There is a continuing imperative to communicate research advances in terms that are relevant to decision making.

Despite the bold start under the UNFCCC, international consensus has not been reached. Notably, the United States has not been as supportive of international efforts to address climate change as many other nations.

The Kyoto Protocol, initiated in 1997, was the first attempt to codify an agreement whereby industrial nations would reduce their greenhouse gas emission to about 5% below 1990 levels and would assist developing nations with non-fossil fuel technologies. Although many lessons were learned in the ensuing years regarding carbon monitoring and carbon market trading, implementation proved too burdensome, leaving individual countries to develop nationally focused programs outside the UN framework. Carbon taxes, cap-and-trade, and various hybrid policy mechanisms have failed to garner the collective political will of the UN member nations, including the United States, the world's largest economy.

Individual nations, however, especially China, Germany, and the Scandinavian countries, have demonstrated that significant progress can be made in reducing greenhouse gases through a combination of political leadership, economic incentives, and government support for renewable energy technology development. It seems clear that the ultimate solution to reducing greenhouse gas concentrations and global climate change will inevitably involve harnessing solar, wind, geothermal, and nuclear energy sources. Energy conservation and parallel development of new methods of utilizing fossil fuels that produce reduced CO_2 emissions will probably be the most important factors for the coming decades in reducing greenhouse gas emissions.

CONCEPT CHECK STOP

1. **How** do ice cores provide a record of past temperatures?

2. **What** is the relationship between the increase in greenhouse gases and current warming trends?

3. **What** areas are at greatest risk if global warming continues to escalate? Why?

Summary

1 Temperature and Heat Flow Processes 66

• **Temperature** is commonly measured in the Fahrenheit, Celsius, and Kelvin temperature scales. **Sensible heat** is a flow of energy that changes an object's temperature.

• Energy is transferred through the ground and to and from the air through radiation, **conduction**, latent heat transfer, **convection**, and **advection**, as shown in the diagram.

Energy flow by convection and conduction • Figure 3.3

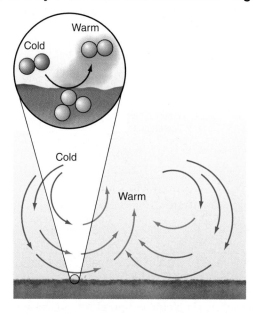

• Latent heat is a flow of energy taken up or released when substances change between solid, liquid, and gaseous states.

2 Daily and Annual Cycles of Air Temperature 69

• Daily and annual air temperature cycles are produced by the Earth's rotation and revolution, which create insolation and net radiation cycles.

• Annual air temperature cycles are influenced by annual net radiation patterns, which depend on latitude.

• Maritime locations show smaller ranges of both daily and annual temperature than do continental regions. This is because, as shown, water heats more slowly, absorbs energy throughout a surface layer, and can mix and evaporate freely.

Warming a water body with solar radiation • Figure 3.7

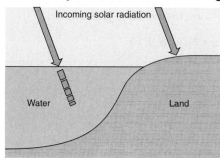

3 Local Effects on Air Temperature 76

• Air temperatures generally decrease with elevation. The average rate of decrease is known as the **environmental temperature lapse rate**, shown throughout the atmosphere in this graph.

Environmental temperature lapse rate • Figure 3.10

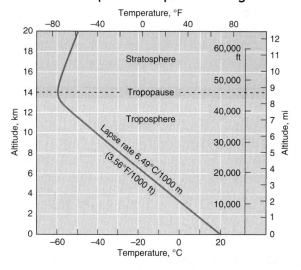

• In the stratosphere, temperature increases slightly with altitude, as ozone and other gases intercept and absorb incoming solar radiation.

• Air temperature increasing with elevation is known as a **temperature inversion**.

• Rural surfaces heat and cool slowly. Urban surfaces absorb and store solar energy more effectively, resulting in **urban heat islands**.

4 World Patterns of Air Temperature 80

• Temperatures at the equator vary little from season to season.

• **Isotherms** display temperature profiles by connecting points of the same temperatures with a line, as in this isotherm map. The rate at which temperature changes with distance is the temperature gradient.

Isotherms • Figure 3.14

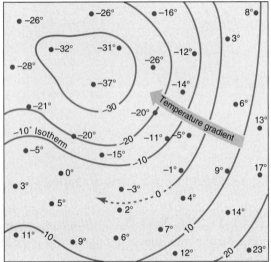

• Poleward, temperatures decrease with latitude, and continental surfaces at high latitudes can become very cold in winter.

• The annual range in temperature increases with latitude and is greatest in northern hemisphere continental interiors.

5 The Temperature Record and Global Warming 83

• A record of past temperatures can be established for the past 450,000 years by analyzing tree rings, coral growth, ice cores, and other data.

• Global temperatures have increased over the past few decades, as shown in the graph. CO_2 released by fossil fuel burning and other greenhouse gases plays an important role in causing warming.

Mean annual surface temperature of the Earth, 1880–2010 • Figure 3.19

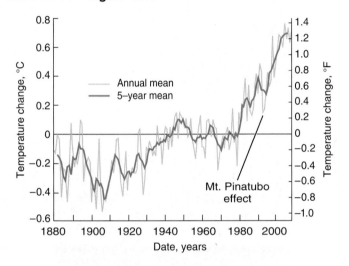

• A significant rise in global temperatures will be accompanied by increasing sea level.

Key Terms

- advection 68
- conduction 68
- convection 68
- environmental temperature lapse rate 77

- heat 67
- heat index 67
- isotherm 80
- latent heat 69

- microclimate 76
- sensible heat 69
- specific heat 73
- temperature 66

- temperature gradient 80
- temperature inversion 78
- urban heat island 78
- wind chill index 67

Critical and Creative Thinking Questions

1. Describe the changing conditions that you might notice as you climb higher up a mountain. Explain what causes each of these changes.

2. Explain how latitude affects the annual cycle of air temperature through net radiation by comparing Key West, Las Vegas, New York, and Anchorage.

3. Explain the three temperature controls that affect San Francisco, California, and Yuma, Arizona.

4. Describe how global air temperatures have changed in the recent past. If you were asked to make a model to predict how global climates will warm or cool 50 years in the future, what factors would you consider?

5. In the graph, how many times has the Earth reached temperatures similar to today's temperatures? About how much time is spent between these warm periods? Finally, what happens to ice volume every time the Earth is in a warm period?

6. Your *carbon footprint* is used to describe the impact you have on emissions of carbon dioxide into the atmosphere as a result of daily activities. Work out your personal carbon footprint over the past month. (There are a number of carbon footprint calculators on the Web.) What measures are being taken to help individuals and organizations in your area reduce their carbon footprint?

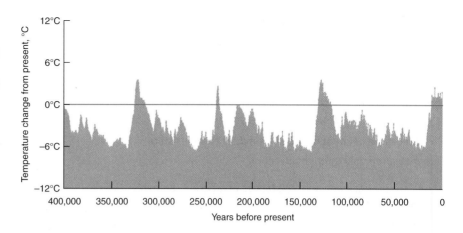

What is happening in these pictures?

These remarkable photos of the confluence of the Muir and Riggs glaciers, in Alaska's Glacier Bay, show a major recession of the glaciers between 1941 and 2004. In the earlier photo, the two glaciers are easily visible as they meet in the middle of the image and collectively flow into the foreground. By 2004, these glaciers had retreated into the far background.

1941

2004

Think Critically

1. Is the retreat of the Muir and Riggs glaciers an effect of global warming?

2. What approach would you take to monitor the glaciers related to other latitudinal temperature shifts?

What is happening in these pictures? 93

Self-Test

(Check your answers in the Appendix.)

1. Temperature is _____.

    a. a measure of the internal energy level of matter

    b. measured only with a thermometer

    c. a measure of latent heat flow as matter changes state

    d. measured only through advection

2. _____ occurs when two substances at different temperatures are in contact.

    a. Advection

    b. Particle acceleration

    c. Convection

    d. Conduction

3. Correctly add the following labels to the figure: "June solstice," "Equinoxes," and "December solstice." Make clear which line corresponds to each of the labels.

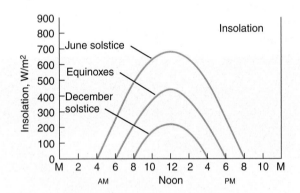

4. Urban surface temperatures tend to be warmer than rural temperatures during the day because _____.

    a. drier surfaces are cooler than wet soils

    b. drier surfaces have less water to evaporate than do moist soils

    c. paved surfaces reflect most solar radiation back to the atmosphere

    d. paved surfaces absorb little solar insolation

5. Sketch a line on the graph to show the relationship between air temperature and elevation in a low-level temperature inversion. What unwanted consequences can such an inversion have? How do people attempt to combat these effects?

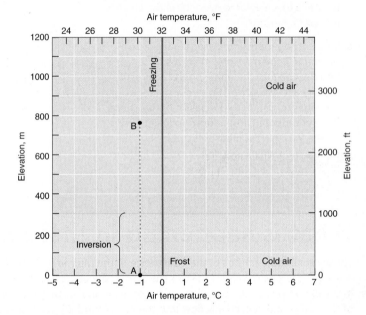

6. Positive net radiation values usually begin _____.

    a. at noon

    b. shortly after sunrise

    c. in the afternoon

    d. at night

7. Minimum daily temperatures usually occur _____.

    a. just before sunrise

    b. at midnight

    c. one hour before sunrise

    d. about one-half hour after sunrise

8. Continental locations generally experience _____ seasonal temperature variations than do ocean-adjacent locations.

   a. weaker

   b. stronger

   c. more stable

   d. more unstable

9. Because large bodies of water heat and cool more _____ compared to land surfaces, monthly temperature maximums and minimums tend to be delayed at coastal stations.

   a. slowly

   b. rapidly

   c. constantly

   d. randomly

10. _____ are lines of equal temperature drawn on a weather map.

    a. Isohyets

    b. Isobars

    c. Isopachs

    d. Isotherms

11. Air temperature changes with altitude in the atmosphere. On the graph, sketch a line showing how air temperature varies as you travel from the troposphere up to the thermosphere.

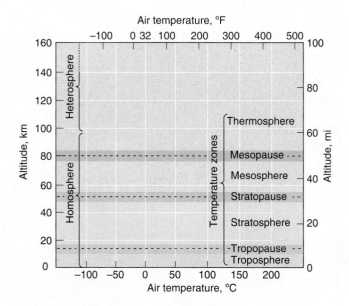

12. Air temperature changes with latitude on the planet because of _____ .

    a. lack of vegetative surfaces near the poles

    b. the zone defined by the lapse rate

    c. the decrease of annual insolation with latitude

    d. maritime influence around the poles

13. Assuming the "Best" estimate scenario with constant aerosols, what will be the increase in the average temperature (°C) of the Earth between the 2000 and 2050?

    a. 0.2   c. 0.6

    b. 0.4   d. 0.8

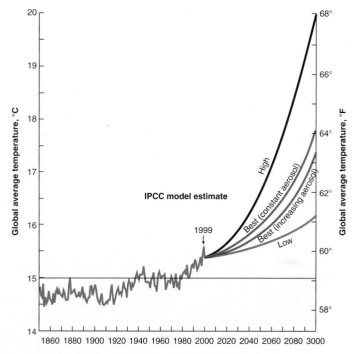

14. According to NASA scientists using a network of meteorology stations, the hottest decade on record is _____.

    a. 1770–1779

    b. 1960–1969

    c. 1840–1849

    d. 2000–2009

15. Potential consequences of global warming include _____.

    a. severe ocean evaporation and lowering sea level

    b. increased glacier formation

    c. increases in insect-borne diseases

    d. disruption of lapse rates in the thermosphere

THE PLANNER ✔

Review your Chapter Planner on the chapter opener and check off your completed work.

Atmospheric Moisture and Precipitation

Water is the principal life-support ingredient for the planet. All human development and all living resources depend upon this critical element. The lifeblood of our biosphere, water moves through the hydrosphere, in all three physical states, driven by the Sun's energy and by gravity as it shapes the lithosphere and nourishes our world. So important is the hydrosphere to our understanding of physical geography that we have dedicated Chapters 10 through 14 to exploring in detail how water shapes our planet.

The hydrosphere exhibits tremendous variability in how water is delivered from moisture in the atmosphere to the surface of the Earth. Geographers study these differences to explain how variations in rainfall affect various regions. Precipitation on Mt. Waialeale on the Hawaiian island of Kauai is reported by U.S. National Climate Data Center to be the highest on the Earth, with an annual average of 11,684 millimeters (460 inches). At the other end of the rainfall spectrum is the Atacama Desert in Chile, famous for having no record of precipitation.

Precipitation can arrive as a persistent and gentle rain or as a devastating deluge, causing catastrophic damage to crops and human settlements. In 2010, Pakistan experienced exceptionally heavy monsoon rains that flooded one-fifth of the nation and affected 20 million people. Climate scientists predict more extreme flooding events and droughts in the coming decades as a result of climate change.

Heavy monsoon rains in Pakistan in 2010 wreaked havoc.

CHAPTER OUTLINE

CHAPTER PLANNER ✓

- ❑ Study the picture and read the opening story.
- ❑ Scan the Learning Objectives in each section:
  p. 98 ❑   p. 102 ❑   p. 105 ❑   p. 108 ❑   p. 111 ❑   p. 119 ❑
- ❑ Read the text and study all figures and visuals. Answer any questions.

Analyze key features

- ❑ What a Geographer Sees, p. 103
- ❑ Process Diagram p. 107 ❑   p. 114 ❑   p. 115 ❑   p. 116 ❑   p. 118 ❑
- ❑ Video Explorations, p. 111
- ❑ Geography InSight, p. 112
- ❑ Where Geographers Click, p. 121
- ❑ Stop: Answer the Concept Checks before you go on.
  p. 101 ❑   p. 105 ❑   p. 106 ❑   p. 111 ❑   p. 118 ❑   p. 121 ❑

End of chapter

- ❑ Review the Summary and Key Terms.
- ❑ Answer the Critical and Creative Thinking Questions.
- ❑ Answer What is happening in this picture?
- ❑ Complete the Self-Test and check your answers.

Water and the Hydrosphere

LEARNING OBJECTIVES

1. **Explain** how water changes state.
2. **Identify** the primary water reservoirs in the hydrosphere.
3. **Describe** the hydrologic cycle.

Water is often described as both the universal solvent and life's elixir, due to its capacity to dissolve the majority of the Earth's chemical components and its vital role in supporting living systems. Water is therefore one of the principal elements for our geographic understanding of how our world was shaped and how the future will likely unfold.

The Three States of Water

Water can exist in three states—as a solid (ice), a liquid (water), or an invisible gas (water vapor). Recall from Chapter 3 that energy is required to change the state of water from solid to liquid, liquid to gas, or solid to gas. This flow of energy, called latent heat, changes the potential energy of the water molecules, allowing them to move more freely. When the change goes the other way, from liquid to solid, gas to liquid, or gas to solid, this potential energy is released (**Figure 4.1**).

Several of the state changes of water should be very familiar to you. Liquid water can freeze to become solid ice or **evaporate** to become vapor. Solid ice melts when heated, and vapor **condenses** on a surface as it cools, such as when car windows fog up on a cold day. **Sublimation** is the direct transition from solid to vapor. For example, ice cubes left in the freezer eventually shrink away from the sides of the ice cube tray and get smaller. They shrink through sublimation; they never melt but lose bulk directly as vapor. In this book, we use the term **deposition** to describe the reverse process, in which water vapor crystallizes directly as ice. Frost forming on a cold winter night is a common example of deposition.

The Hydrosphere

The realm of water in all its forms, including the flows of water among ocean, land, and atmosphere, is known as the **hydrosphere**. Nearly all the Earth's water is contained in the world ocean. Ice-bound water, fresh surface water, and soil water make up only a small fraction of the total volume of global water (**Figure 4.2**).

> **hydrosphere** The total water realm of the Earth's surface, including the oceans, surface waters of the lands, ground water, and water held in the atmosphere.

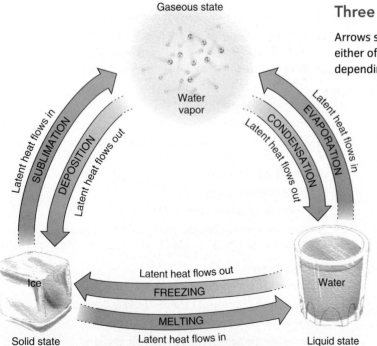

Gaseous state

Water vapor

SUBLIMATION — Latent heat flows in
DEPOSITION — Latent heat flows out

CONDENSATION — Latent heat flows out
EVAPORATION — Latent heat flows in

Ice — Solid state

Latent heat flows out
FREEZING

MELTING
Latent heat flows in

Water — Liquid state

Three states of water • Figure 4.1

Arrows show the ways that any one state of water can change into either of the other two states. Energy is absorbed or released, depending on the direction of change.

Ask Yourself

On a cold, dry day, snow covering a sidewalk slowly disappears, and there is no visible melting. Which process is at work?

a. sublimation
b. deposition
c. condensation
d. evaporation

The hydrosphere includes all of the Earth's water, in oceans, on land, and in the atmosphere.

a. Oceans

About 97.5% of the hydrosphere consists of ocean salt water. Here, a southern stingray swims along a shallow ocean bottom near Grand Cayman Island.

Total water

Salt water 97.5%

Fresh water 2.5%

Fresh water

Glaciers 68.7%

Permafrost 0.8%

Ground water 30.1%

Surface and atmospheric water 0.4%

Surface and atmospheric water

Freshwater lakes 67.4%

Soil moisture 12.2%

Plants and animals 0.8%

Rivers 1.6%

Wetlands 8.5%

Atmosphere 9.5%

b. Ice

Most of the world's fresh water is stored as ice in the world's ice sheets and mountain glaciers, accounting for 1.7% of total global water. Although glaciers are too cold and forbidding for most forms of animal life, Antarctic ice sheets provide a habitat for these seals.

c. Surface water

Surface water, including lakes, is only a very tiny fraction of Earth's water volume. Here a moose wades along the margin of a lake, in search of tasty aquatic plants.

Most of the world's fresh water is stored as ice in ice sheets and glaciers. The rapid melting of the ice-bound fresh water as a result of global warming is predicted to accelerate rising sea levels, as glaciers continue to lose their mass. Fresh liquid water from glaciers, snow, and rain is found above and below the Earth's land surfaces. Subsurface water, accounting for 0.75% of the hydrosphere, occupies cavities in the soil and rock. Most of the ground water is held in deep storage, beyond the reach of plant roots.

The small remaining proportion of the Earth's water includes the water available for plants, animals, and human use. Soil water near the surface is within the reach of plant roots and is heavily influenced by the vegetative cover. Surface water is held in streams, lakes, marshes, and swamps. Most of this surface water is in lakes, which are about evenly divided between freshwater lakes and saline (salty) lakes. An extremely small proportion of surface water makes up the streams and rivers, which flow with gravity toward the sea or inland lakes.

Only a very small quantity of water is held as vapor and cloud water droplets in the atmosphere—just 0.001% of the hydrosphere. But this small reservoir of water is enormously energetic and important. As you will see later in the chapter, air moisture supplies water and ice to replenish all fresh water stocks on land. Radiant solar energy transferred to water molecules and latent heat energy transfers associated with changing water states fuel nature's water recycling system. In the next chapter, you will see that the flow of water vapor—with its internal and potential energy from warm tropical oceans to cooler regions—provides a global heat exchange from low to high latitudes.

Approximately three-quarters of the Earth's surface is covered by water, but this ocean layer is actually very thin relative to the Earth's size. Furthermore, the total volume of all water in the hydrosphere represents a finite resource. This finite volume of water is in constant motion as it shifts from ocean to atmosphere to the land surface and into all living systems (**Figure 4.3**). Human policies and behaviors will need to change in order to decrease water pollution, sustain fishing, and avoid wasting this precious finite resource.

The Hydrologic Cycle

The **hydrologic cycle**, or water cycle, moves water from land and ocean to the atmosphere and then returns the water as precipitation, replenishing the land and ocean (**Figure 4.4**). Portions of this cycle move very rapidly, such as a swift mountain stream or a hurricane. Deep-ocean currents are much slower, moving huge volumes of water at the speed of centuries from one ocean to the next. Water in other portions of the hydrologic cycle, such as glaciers and ancient underground aquifers, can remain essentially dormant for decades or centuries.

Water from the ocean and from land surfaces evaporates, changing state from liquid to vapor using solar energy and entering the atmosphere. Total evaporation is about six times greater over oceans than over land because oceans cover most of the planet and because land surfaces are not always wet enough to yield much water. However, land surfaces do contribute large amounts of water vapor back into the atmosphere as plants lose water by evaporation through leaf pores, a process called **transpiration**. The combined water flow to the atmosphere by evaporation from soil and transpiration from plants is called **evapotranspiration**.

Once in the atmosphere, water vapor can condense or deposit to form clouds and **precipitation**, which falls to Earth as rain, snow, or hail. There is nearly four times as much precipitation over oceans as over land.

precipitation Particles of liquid water or ice that fall from the atmosphere and reach the ground.

Water and the Earth's hydrosphere • Figure 4.3

The small blue sphere represents the planet's total water volume in proportion to the Earth's size, demonstrating the limits on this critical resource.

The hydrologic cycle • Figure 4.4

In the hydrologic cycle, water moves among the ocean, the atmosphere, the land, and back to the ocean in a continuous process.

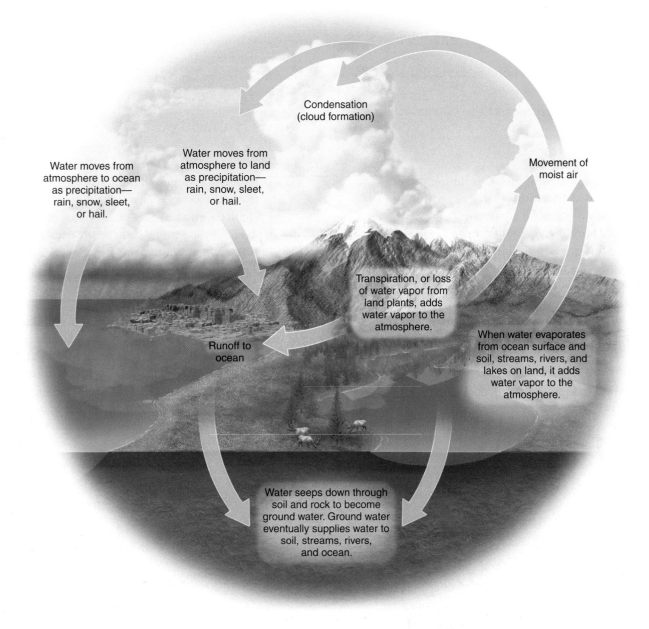

Condensation (cloud formation)

Water moves from atmosphere to ocean as precipitation— rain, snow, sleet, or hail.

Water moves from atmosphere to land as precipitation— rain, snow, sleet, or hail.

Movement of moist air

Transpiration, or loss of water vapor from land plants, adds water vapor to the atmosphere.

When water evaporates from ocean surface and soil, streams, rivers, and lakes on land, it adds water vapor to the atmosphere.

Runoff to ocean

Water seeps down through soil and rock to become ground water. Ground water eventually supplies water to soil, streams, rivers, and ocean.

When it hits land, precipitation has three fates. First, it can evaporate and return to the atmosphere as water vapor. Second, it can sink into the soil and then into the surface rock layers below. As you will see in later chapters, portions of this subsurface water emerges from below to feed rivers, lakes, and even ocean margins. Third, precipitation can run off the land, concentrating in streams and rivers that eventually carry it to the ocean or to lakes. This flow is known as runoff.

CONCEPT CHECK   STOP

1. **What** are the six processes through which water changes state, and **how** are they related to the exchange of latent heat?

2. **How** is water stored around the Earth and its atmosphere?

3. **What** happens to precipitation when it reaches the Earth's surface?

Humidity

LEARNING OBJECTIVES

1. **Describe** the relationship between temperature and humidity.

2. **Distinguish** between specific humidity and relative humidity.

3. **Explain** the importance of the dew-point temperature.

The amount of moisture in the air, called **humidity**, varies widely from place to place and time to time. In the cold, dry air of polar regions in winter, the humidity is almost zero, and it can reach up to as much as 4 or 5% of a given volume of air in the warm, wet regions near the equator. This is because the maximum quantity of moisture that can be held at any time in the air depends on the temperature of the air, with warmer air holding more water vapor (**Figure 4.5**).

> **humidity** The amount of water vapor in the air.

Humidity has a strong influence on how we experience heat (see *What a Geographer Sees*) and on daily temperature variations. In the humid regions of the southern and continental United States, for example, the difference between daytime and nighttime temperatures can be as little as 3 to 5°C (5 to 10°F). In contrast, at the same time of year in a dry desert or an arid mountain area with low humidity, air temperature can drop to near freezing at night from daytime highs in the 30s °C (90s °F).

Relative Humidity

When radio or television weather forecasters speak of humidity, they are referring to **relative humidity**. This measure compares the amount of water vapor present to the maximum amount that the air can hold at that temperature. It is expressed as a percentage. So, for example, if the air currently holds half the moisture possible at the present temperature, then the relative humidity is 50%. When the humidity is 100%, the air holds the maximum amount of water vapor possible. We say that it is saturated.

The relative humidity of the atmosphere can change in one of two ways. First, the atmosphere can directly gain or lose water vapor. For example, additional water vapor can enter the air from an exposed water surface, from a moist soil layer, or by evaporation from the leaves of vegetation. This is a slow process because the water vapor molecules diffuse upward from the surface into the air layer above.

Humidity and temperature • Figure 4.5

The maximum volume of water vapor, or humidity, of a mass of air increases sharply with rising temperature. Air at room temperature (20°C [68°F]) can hold about three times as much water vapor as freezing air (0°C [32°F]).

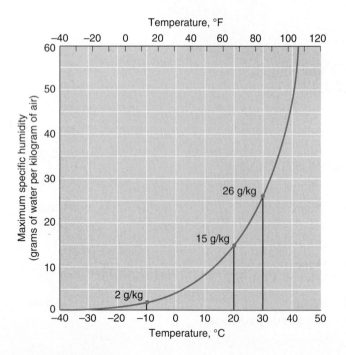

WHAT A GEOGRAPHER SEES

Temperature and Humidity

Anyone can see that when the thermometer reaches 90°F (32°C), it is a hot day. But temperature alone can be misleading. A geographer would know that humidity has an important effect on how we experience heat. The low humidity of Death Valley in California (Figure a) creates a warm but comfortable day for a trek across the sand. However, in the high-humidity conditions of a Florida wetland (Figure b), the same temperature reading can be unbearably hot and have possible serious health-related effects on the unprepared.

High humidity slows the evaporation of sweat from your body, reducing its cooling effects and increasing the health risks of high temperatures. Blistering summer heat waves can be deadly, with the elderly and the ill at greatest risk. But even healthy young people need to be careful of the heat, especially in hot, humid weather.

a.
Low humidity and heat

b.
High humidity and heat

Pole-to-pole profiles of specific humidity (left) and temperature (right) show similar trends because the ability of air to hold water vapor (measured by specific humidity) is limited by temperature.

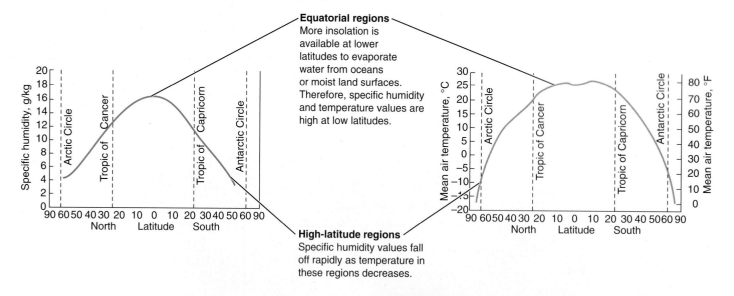

Equatorial regions
More insolation is available at lower latitudes to evaporate water from oceans or moist land surfaces. Therefore, specific humidity and temperature values are high at low latitudes.

High-latitude regions
Specific humidity values fall off rapidly as temperature in these regions decreases.

Put It Together *Review Figure 4.5 and answer this question.*
What is the average relative humidity at 40° N latitude?

The second way is through a change of temperature. When the temperature is lowered, the relative humidity rises, even if no water vapor is added. This is because the capacity of air to hold water vapor depends on temperature. When the air is cooled, this capacity is reduced. The existing amount of water vapor will then represent a higher percentage of the total water-holding capacity.

Specific Humidity

The actual quantity of water vapor held by a parcel of air is known as its **specific humidity** and is expressed as grams of water vapor per kilogram of air (g/kg).

Climatologists often use specific humidity to describe the moisture in a large mass, cell, or parcel of air. Specific humidity is largest at the equatorial zones and falls off rapidly toward the poles, as shown in **Figure 4.6**. Extremely cold, dry air over arctic regions in winter may have a specific humidity as low as 0.2 grams/kilogram, while the extremely warm, moist air of equatorial regions often holds as much as 18 g/kg. More insolation is available at lower latitudes to evaporate water in oceans or on moist land surfaces, so specific humidity values tend to be higher at low latitudes than at high latitudes.

We can see that the global profile of mean (average) surface air temperature, also shown in Figure 4.6, has a shape similar to that of the specific humidity profile. This is because air temperature and maximum specific humidity are positively correlated and vary together.

Specific humidity is also a geographer's yardstick for a basic natural resource: water. It is a measure of the quantity of water in the atmosphere that can be extracted as precipitation. Whereas cold, moist air can supply only a small quantity of rain or snow, warm, moist air is capable of supplying large amounts.

Dew Point

Another way of describing the water vapor content of air is by its **dew-point temperature** (also called *dew point*). If air is slowly chilled, it will eventually reach saturation. At this temperature, the air holds the maximum amount of water vapor possible. If the air is cooled further, condensation will begin, and dew will start to form. This condensation can be seen on the ground surface as drops of dew on vegetation

> **dew-point temperature** The temperature at which air with a given humidity will reach saturation when cooled without changing its pressure.

that form during the night when temperatures fall. Moist air has a higher dew-point temperature than dry air.

The dew-point temperature depends to some degree on atmospheric pressure. At higher elevations, where pressure is lower, the air must cool somewhat more before condensation will form. This means that the dew-point temperature of air with a given moisture content decreases somewhat with elevation.

1. **Where** would you expect the air to be more humid: near the equator or near the poles?
2. **What** atmospheric changes would cause the measured specific humidity and relative humidity to change?
3. **What** happens as air is cooled below the dew-point temperature?

Adiabatic Processes

LEARNING OBJECTIVES

1. **Describe** the relationship between air pressure and air temperature.
2. **Explain** the role of adiabatic processes in cloud formation.

Precipitation forms only when a substantial mass of air experiences a steady drop in temperature below the dew point. Because atmospheric pressure decreases as altitude increases, this cooling process happens as an air parcel is lifted to a higher level in the atmosphere. So, as a parcel of air is uplifted, atmospheric pressure on the parcel becomes lower, and the parcel expands and cools. According to the same principle, as a parcel of air descends, atmospheric pressure becomes higher, and the air is compressed and warmed (**Figure 4.7**). These are called **adiabatic processes** because the change in temperature is not caused by heat flow to or away from the air parcel but only by a change in pressure within the air mass.

> **adiabatic process**
> A process in which the temperature of a parcel of air changes in response to a change in atmospheric pressure.

Adiabatic cooling and heating • Figure 4.7

When air is forced to rise, it expands and its temperature decreases. When air is forced to descend, it is compressed and its temperature increases.

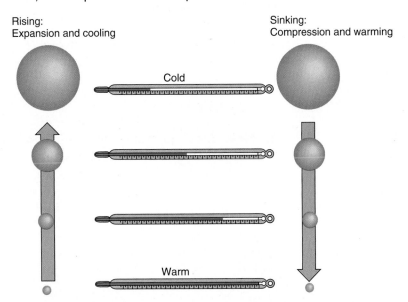
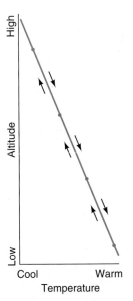

The Dry Adiabatic Lapse Rate

dry adiabatic lapse rate The rate at which rising air cools or descending air warms when no condensation is occurring; 10°C per 1000 m (or 5.5°F per 1000 ft).

For a rising or descending parcel of air that does not contain liquid or solid water, the temperature changes at the **dry adiabatic lapse rate**. This is called the "dry" rate because there is no condensation or evaporation as a result of the change in temperature or pressure. This rate has a value of about 10°C per 1000 meters (5.5°F per 1000 feet) of altitude change.

For example, when a parcel of air rises 1 km, it expands and its temperature drops by 10°C. Or, in English units, if a parcel of air rises 1000 ft, its temperature will drop by 5.5°F. Descending air masses are warmed by the reverse process. As an air parcel descends in elevation, the increase in temperature lowers the relative humidity of the parcel, creating a drier volume of air that can readily absorb any moisture it may encounter.

In the previous chapter, we discussed the environmental temperature lapse rate. There's an important difference to note between the environmental temperature lapse rate and the dry adiabatic lapse rate. The temperature lapse rate expresses how the temperature of still air varies with altitude, as we would sense if we hiked up a mountainside. This rate varies from time to time and from place to place. In contrast, the dry adiabatic rate applies to a mass of air in motion and does not depend on local conditions.

The Moist Adiabatic Lapse Rate

If an air mass rises high enough, to the **lifting condensation level**, it cools to the dew point, and condensation begins (**Figure 4.8**). Think about extracting water from a moist sponge. To release the water, you have to squeeze the sponge—that is, reduce its ability to hold water. In the atmosphere, chilling the air beyond the dew point is like squeezing the sponge: It reduces the air's ability to hold water, forcing some water vapor molecules to change state to form water droplets or ice crystals, forming a cloud. The lifting condensation level occurs where you observe the flat bottom surface of clouds.

Once an air parcel reaches saturation, it cools at a different rate, known as the **moist adiabatic lapse rate**. Unlike the dry adiabatic lapse rate, which remains constant, the moist adiabatic lapse rate is variable because it depends on the temperature and pressure of the air and its moisture content. But for most situations, we can use a value of 5°C/1000 m (2.7°F/1000 ft).

moist adiabatic lapse rate The rate at which rising air is cooled by expansion when condensation is occurring; ranges from 4 to 9°C per 1000 m (2.2 to 4.9°F per 1000 ft).

As the air parcel is lifted higher, water vapor condenses into droplets or deposits as ice particles, providing a flow of latent heat to warm the surrounding atmosphere. This heat flow can keep a rising parcel of air warmer than the surrounding air, fueling the convection process and driving the parcel ever higher. When the parcel reaches a high altitude, most of its water will have condensed. As adiabatic cooling continues, the latent heat flow will weaken, and so will the uplift. Eventually, uplift stops, since the energy source is gone. The cell dissipates into the surrounding air. The continuous cycle of rising moist air, precipitation, air parcel dissipation, and evaporation of the precipitation to form moist air describes a major loop in the hydrologic cycle. Atmospheric scientists study the pace and spatial distribution of this loop, along with factors causing its disruption.

CONCEPT CHECK **STOP**

1. **How** does the dry adiabatic lapse rate differ from the environmental temperature lapse rate?

2. **How** does the rate of adiabatic cooling change at the lifting condensation level?

Cloud formation and the adiabatic process • Figure 4.8

✓ THE PLANNER

Adiabatic decrease in temperature in a rising parcel of air leads to condensation of water vapor into water droplets and the formation of a cloud.

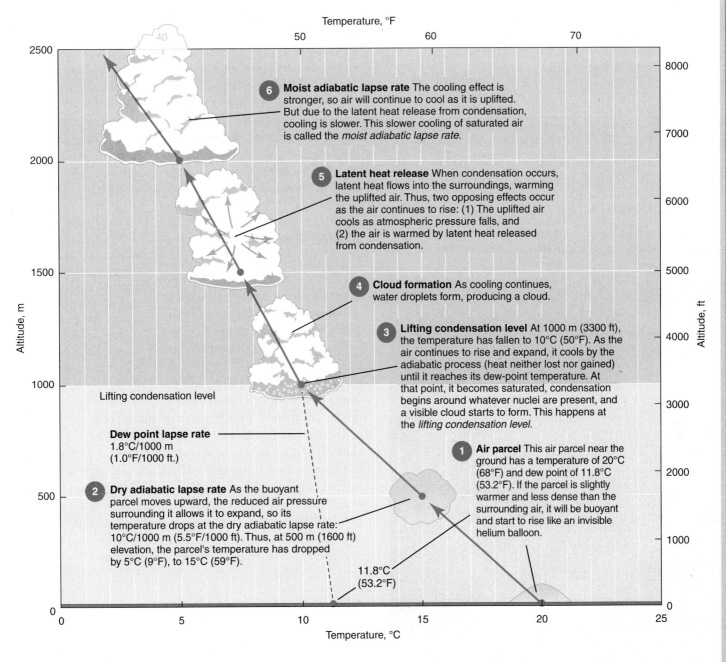

6 **Moist adiabatic lapse rate** The cooling effect is stronger, so air will continue to cool as it is uplifted. But due to the latent heat release from condensation, cooling is slower. This slower cooling of saturated air is called the *moist adiabatic lapse rate*.

5 **Latent heat release** When condensation occurs, latent heat flows into the surroundings, warming the uplifted air. Thus, two opposing effects occur as the air continues to rise: (1) The uplifted air cools as atmospheric pressure falls, and (2) the air is warmed by latent heat released from condensation.

4 **Cloud formation** As cooling continues, water droplets form, producing a cloud.

3 **Lifting condensation level** At 1000 m (3300 ft), the temperature has fallen to 10°C (50°F). As the air continues to rise and expand, it cools by the adiabatic process (heat neither lost nor gained) until it reaches its dew-point temperature. At that point, it becomes saturated, condensation begins around whatever nuclei are present, and a visible cloud starts to form. This happens at the *lifting condensation level*.

Lifting condensation level

Dew point lapse rate
1.8°C/1000 m
(1.0°F/1000 ft.)

2 **Dry adiabatic lapse rate** As the buoyant parcel moves upward, the reduced air pressure surrounding it allows it to expand, so its temperature drops at the dry adiabatic lapse rate: 10°C/1000 m (5.5°F/1000 ft). Thus, at 500 m (1600 ft) elevation, the parcel's temperature has dropped by 5°C (9°F), to 15°C (59°F).

1 **Air parcel** This air parcel near the ground has a temperature of 20°C (68°F) and dew point of 11.8°C (53.2°F). If the parcel is slightly warmer and less dense than the surrounding air, it will be buoyant and start to rise like an invisible helium balloon.

11.8°C
(53.2°F)

Temperature, °F
Temperature, °C
Altitude, m
Altitude, ft

Think Critically 1. Suppose the air parcel shown in the diagram contained more water vapor. How would that affect the lifting condensation level?
2. What would be the effect if there were less water vapor in the air parcel?

Clouds and Fog

LEARNING OBJECTIVES

1. **Explain** the role of condensation nuclei in the formation of clouds.

2. **Describe** how clouds are classified.

3. **Explain** how fog forms.

I mages of the Earth from space show that about half of our planet is blanketed in clouds. Clouds play a complicated temperature role—both cooling and warming the Earth and atmosphere. In this chapter, we will look at one of the most familiar roles of clouds: producing precipitation.

Clouds are made up of condensed water droplets, ice particles, or a mixture of both, suspended in air. Liquid water turns to ice at 0°C (32°F), but when water is dispersed as tiny droplets in clouds, it can remain in the liquid state at temperatures far below freezing. In fact, clouds consist entirely of supercooled water droplets at temperatures down to about –12°C (10°F). As cloud temperatures grow colder, ice crystals begin to appear. The coldest clouds, with temperatures below –40°C (–40°F), occur at altitudes of 6 to 12 km (20,000 to 40,000 ft) and are made up entirely of ice particles.

Ocean aerosols • Figure 4.9

Breaking or spilling waves in the open ocean, shown here from the deck of a ship, are an important source of cloud-forming salt particle aerosols.

Cloud Formation

Cloud particles grow around a tiny center of solid matter. This dust speck of matter, called a **condensation nucleus**, typically has a diameter between 0.1 and 1 micrometers (0.000004 and 0.00004 in.).

> **condensation nucleus** A tiny bit of solid matter, or dust particle, in the atmosphere on which water vapor condenses to form a water droplet.

The surface of the sea is an important source of condensation nuclei (**Figure 4.9**). Droplets of spray from the crests of the waves are carried upward by turbulent air. When these droplets evaporate, they leave behind a tiny residue of concentrated salt, suspended in the air, that strongly attracts liquid water molecules, stimulating cloud formation. Nuclei are also thrown into the atmosphere from polluted air over cities, where particulate matter and soot aid condensation and the formation of clouds and fog, therefore increasing rates of precipitation.

Cloud Forms

Cloud forms range from the small, white puffy billows often seen in summer to the gray hazy layers that produce a good, old-fashioned rainy day. Meteorologists classify clouds into four families, arranged by their height in the sky (**Figure 4.10**), and assign them compound names depending on their rain or ice potential.

Cloud families and types • Figure 4.10

Clouds are grouped into families on the basis of height: high clouds, middle clouds, low clouds, and clouds with vertical development. Individual cloud types are given compound names that indicate their form and whether they produce precipitation.

a. Cirriform clouds, at the top of the troposphere, are high, thin, wispy clouds drawn out into streaks. They are composed of ice crystals and form when moisture is present high in the air.

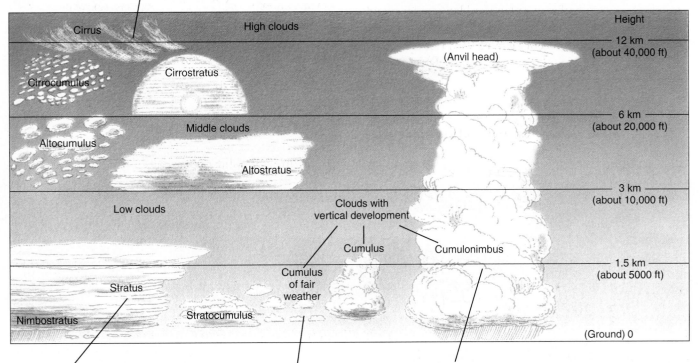

b. Stratiform clouds are blanket-like layers that cover large areas. A common type is stratus clouds, which often cover the entire sky. In this photo, high cumulus clouds (left) grade into a high stratus layer (right).

d. Nimbus clouds are clouds of any of type that produce precipitation. An isolated cumulonimbus cell discharges its precipitation as heavy rain in this photo.

c. Cumuliform clouds are globular masses of cloud that are associated with small to large parcels of moist rising air. In this photo, puffy, fair-weather cumulus clouds drift over a lake.

Fog

Fog is simply a cloud layer at or very close to the Earth's surface (**Figure 4.11**). One type of fog, known as **radiation fog**, usually forms at night when the temperature of the ground surface drops rapidly due to cooling by radiation. This chills the air layer at the ground level below the dew point. Another fog type, **advection fog**, results when a warm, moist air layer moves across a cold surface. As the warm air layer transfers heat energy to the surface, its temperature drops below the dew point, and condensation sets in. Advection fog commonly occurs in mountains and valleys or over oceans where warm and cold currents occur side by side. This sea fog is frequently found along the California coast; for example, Point Reyes, just north of San Francisco, records 200 fog days each year. It forms from a cool marine air layer, which makes direct contact with the colder water of the California current.

Fog formations • Figure 4.11

Fog forms when warm moist air encounters a cooler surface.

a. Moist air above lakes, rivers, irrigated farmland, and, as pictured, the wet Murnauer Moor in Germany, can lose heat at night, leading to condensation that produces a fog layer.

b. Mountains supply gravity-fed cool air that moves down drainage paths, creating valley fog, a type of advection fog, as seen here in Borneo, Malaysia.

c. Coastal regions, such as Ketchikan, Alaska, often experience advection fog as moist maritime air moves over colder water currents.

Video Explorations

Rainy Days and Weekdays

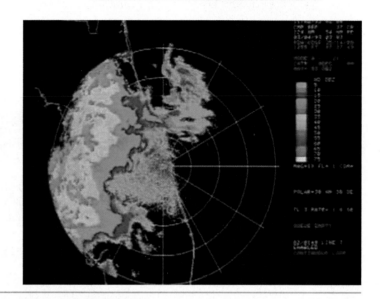

It seems that even downpours like to take a break on the weekends. This video shows that atmospheric scientists have recently found that weekdays bring heavier rainfall than weekends. The researchers examined data collected by NASA's Tropical Rainfall Measuring Mission satellite during summertime storms in the southern United States from 1998 to 2005. They found that it rains, on average, more between Tuesday and Thursday than from Saturday through Monday. The driest day of the week was Saturday; the wettest day, Tuesday afternoon, had nearly twice the rainfall as Saturday.

What could be the cause? Comparing the rainfall with corresponding air-pollution records from the U.S. Environmental Protection Agency showed that higher levels of air pollution created by traffic and businesses during the week could be a factor.

In our industrialized world, fog can be a major environmental hazard. For centuries, fog at sea has increased the danger of ship collisions and groundings. Dense fog on high-speed highways can cause chain-reaction accidents, sometimes involving dozens of vehicles.

Most travel occurs during the work week, and it appears that most inclement weather also coincides with these travel times, increasing hazardous driving conditions (see *Video Explorations*).

CONCEPT CHECK

1. **Where** do condensation nuclei come from?
2. **How** do stratiform clouds differ from cumuliform clouds?
3. **How** is sea fog formed?

Precipitation

LEARNING OBJECTIVES

1. **Describe** the two cloud processes that lead to the formation of precipitation.
2. **Distinguish** among the four types of atmospheric lifting that lead to precipitation.

Precipitation provides the fresh water essential for terrestrial life forms. Precipitation includes not only rain but also snow, sleet, freezing rain, and ice. Annual rates of precipitation vary greatly around the world, as shown in **Figure** 4.12 on the next page. Depending on the circumstances, precipitation can be a welcome relief, a minor annoyance, or a life-threatening hazard.

We talk about precipitation in terms of the amount that falls during a certain period of time. So, for example, we often speak of centimeters or inches per hour or per day. One centimeter (or inch) of rainfall would cover the ground to a depth of 1 cm (or 1 in.) if the water did not run off or sink into the soil. To convert rainfall units, we simply multiply the number of inches by 2.54 to obtain centimeters or multiply centimeters by 0.394 to yield inches.

Geography InSight

Global precipitation • Figure 4.12

a. Snow

When air temperatures are low, precipitation falls as snow. These bison in Yellowstone National Park in Montana are well prepared for the region's cold weather, deriving moisture from snow or rain.

b. Mean annual precipitation

On a global scale, rainfall regimes range from tropical to desert to polar. These differences profoundly influence landforms, the biosphere, and human activities. Most species have evolved to survive within narrow ranges of annual precipitation.

Mean annual precipitation, cm

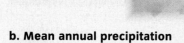

| More than 300 | 200 – 300 | 150 – 200 | 100 – 150 | 60 – 100 | 40 – 60 | 20 – 40 | 10 – 20 | Less than 10 | Non-land area | No data available |

| More than 118 | 79 – 118 | 59 – 79 | 39 – 59 | 24 – 39 | 16 – 24 | 8 – 16 | 4 – 8 | Less than 4 | Non-land area | No data available |

Mean annual precipitation, in.

c. Equatorial regions

Rainfall is greatest in the equatorial regions, as in Matara, Peru, where abundant solar energy drives evaporation.

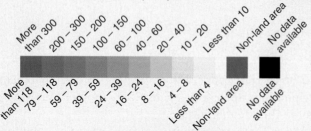

112 CHAPTER 4 Atmospheric InSight and Precipitation

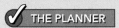

d. Tropical regions
Abundant rainfall occurs in a strong seasonal cycle in some tropical regions. Here, heavy seasonal rains drench flooded fields in Bangladesh. Rice farming traditions closely follow the annual rainfall patterns.

e. Wet-dry regions
African savannas support large herds of mammals adapted to short rainy seasons followed by extended dry seasons.

f. Deserts
The world's deserts occur where moisture sources are far away or where circulation patterns retard precipitation formation. In this photo, a herd of camels crosses Australia's Simpson Desert.

Precipitation **113**

Snowfall is measured by melting a sample column of snow and reducing it to an equivalent in rainfall. In this way, we can combine rainfall and snowfall into a single record of precipitation. Ordinarily, a 100-cm (or 10-in.) layer of snow is assumed to be equivalent to 10 cm (or 1 in.) of rainfall, but this ratio may range from 30 to 1 in very loose snow to 2 to 1 in old, partly melted snow.

Formation of Precipitation

Precipitation forms in clouds by two different processes, depending on the temperature of the clouds. In warm clouds, fine water droplets condense, collide, and coalesce into larger and larger droplets that can fall as rain. This type of precipitation formation occurs in clouds typical of the equatorial and tropical zones (**Figure 4.13**).

When saturated air rises rapidly, the updraft lifts tiny suspended cloud droplets that grow in volume by collisions with other droplets. Droplets are kept aloft by the force of the updraft. Eventually, they grow large enough to fall, sweeping up more tiny droplets as they descend. On their way to Earth, they can experience evaporation and decrease in size or even disappear before reaching the surface, as is often the case over desert regions.

In colder clouds, a second process occurs when ice crystals form and grow in a cloud that contains a mixture of both ice crystals and supercooled water droplets (**Figure 4.14**). The ice crystals take up water vapor and grow by deposition. At the same time, the supercooled water droplets lose water vapor by evaporation and shrink. Eventually, the ice crystals may become large enough to fall from the cloud as snowflakes. Ice crystals can also grow by collision with droplets of supercooled water, which freeze on contact, or by collision with other ice particles. Depending on air temperatures closer to the ground, the snow particles may melt and arrive as rain.

Types of Precipitation

Precipitation can take a variety of forms, depending on the temperature where it is formed and where it falls and on the conditions in which it is produced. The most common types of precipitation are rain, snow, freezing rain, and hail. The U.S. Weather Service uses radar to track the volume and type of precipitation (**Figure 4.15**)

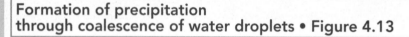

PROCESS DIAGRAM

✓ THE PLANNER

Formation of precipitation through coalescence of water droplets • Figure 4.13

The type of precipitation formation shown here occurs in warm clouds typical of the equatorial and tropical zones.

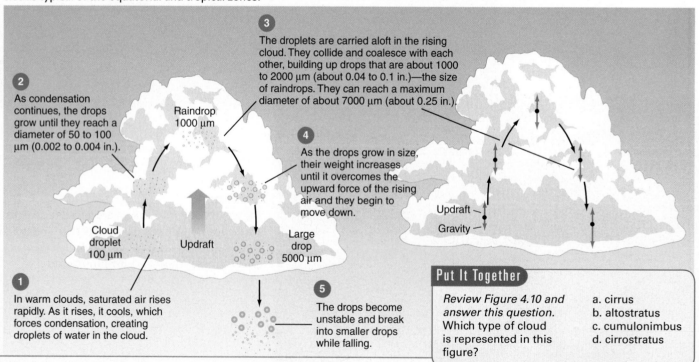

3 The droplets are carried aloft in the rising cloud. They collide and coalesce with each other, building up drops that are about 1000 to 2000 μm (about 0.04 to 0.1 in.)—the size of raindrops. They can reach a maximum diameter of about 7000 μm (about 0.25 in.).

2 As condensation continues, the drops grow until they reach a diameter of 50 to 100 μm (0.002 to 0.004 in.).

Raindrop 1000 μm

4 As the drops grow in size, their weight increases until it overcomes the upward force of the rising air and they begin to move down.

Cloud droplet 100 μm

Updraft

Large drop 5000 μm

Updraft

Gravity

1 In warm clouds, saturated air rises rapidly. As it rises, it cools, which forces condensation, creating droplets of water in the cloud.

5 The drops become unstable and break into smaller drops while falling.

Put It Together

Review Figure 4.10 and answer this question. Which type of cloud is represented in this figure?

a. cirrus
b. altostratus
c. cumulonimbus
d. cirrostratus

Formation of precipitation by ice crystallization • Figure 4.14

THE PLANNER

In cold clouds, precipitation forms as water vapor evaporates from supercooled liquid cloud drops. The water vapor is then deposited on ice crystals, forming snowflakes.

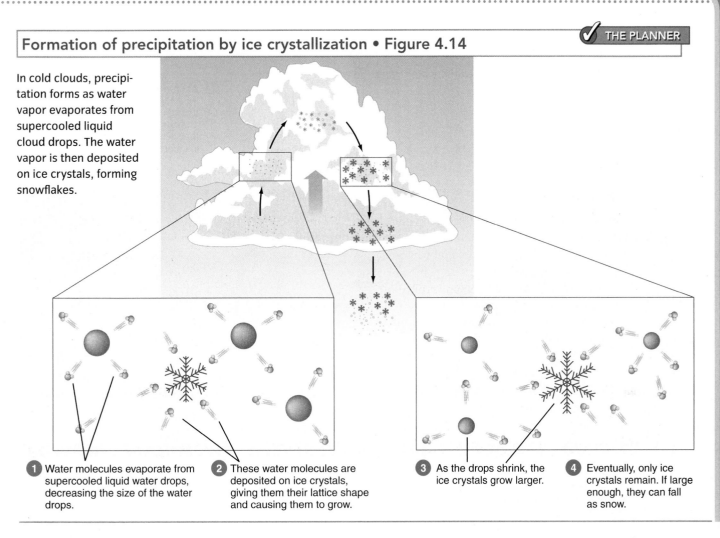

1. Water molecules evaporate from supercooled liquid water drops, decreasing the size of the water drops.

2. These water molecules are deposited on ice crystals, giving them their lattice shape and causing them to grow.

3. As the drops shrink, the ice crystals grow larger.

4. Eventually, only ice crystals remain. If large enough, they can fall as snow.

Rain and snow In equatorial and tropical regions, rain is formed through the coalescence of water droplets in warm clouds. In cooler regions, precipitation forms in clouds as snow and then turns to rain as it descends through warmer temperatures closer to the ground. When temperatures close to the ground are cold enough, the precipitation reaches the ground as snow.

Freezing rain Freezing rain occurs when the ground is frozen and the lowest air layer is also below freezing. Rain falling through the cold air layer is chilled and freezes onto ground surfaces as a clear, slippery glaze, making roads and sidewalks extremely hazardous. Ice storms cause great damage, especially to telephone and power lines and to tree limbs pulled down by the weight of the ice.

SCIENCE TOOLS

Measuring precipitation • Figure 4.15

The U.S. National Weather Service's NEXRAD (**Next** Generation **Rad**ar) employs a network of 159 stations across the nation, each capable of measuring local rainfall volume and intensity. Upgraded radars are also capable of distinguishing between rain, hail, and snow.

b. The concentration of raindrops or snowflakes is measured by the strength of the radar return in decibels (DBZ). This map shows stormy weather passing though southern New England.

a. Weather Surveillance Radars like the one shown in the photo measure the motion and concentration of particles in the air, including dust, bugs, birds, and precipitation.

Hail formation • Figure 4.16

Hail can form during thunderstorms when there are strong updrafts.

2 Growth
Kept aloft by updrafts, the ice pellets continue to grow, forming hailstones.

1 Accretion
As small ice pellets move through below-freezing portions of the cloud, they accumulate layers of ice around them.

Updrafts

0˚ C

3 Precipitation
Eventually, the hailstones become too big and heavy to be kept aloft by updrafts and hail falls to the Earth.

Hail Hail is another common type of precipitation that consists of pea-sized to grapefruit-sized lumps of ice—that is, with a diameter of 5 mm (0.2 in.) or larger. Hailstones are formed when layers of ice build up on ice pellets that are suspended in the strong updrafts of a thunderstorm (**Figure 4.16**). They can reach diameters of 3 to 5 cm (1.2 to 2.0 in.) or larger. When they become too heavy for the updraft to support, they fall to Earth. Crop and property destruction caused by hailstorms can add up to losses of several hundred million dollars. Damage to wheat and corn crops is particularly severe in the Great Plains, running through Nebraska, Kansas, Missouri, Oklahoma, and northern Texas.

Atmospheric Lifting

So far, you have seen how precipitation occurs when air that is moving upward is chilled by the adiabatic process. But one key piece of the precipitation puzzle has been missing: What causes air to move upward in the first place?

Air can move upward in four ways: through orographic, convective, convergent, or frontal lifting (**Figure 4.17**). In this chapter we examine the precipitation resulting from orographic and convective lifting. We will save our discussion of frontal and convergent lifting until Chapter 6 because these processes involve air mass movement by larger weather systems, whereas orographic and convective are not linked to such weather systems.

Orographic lifting Orographic precipitation occurs when a current of moist air is forced upward over a mountain range (**Figure 4.17a**). As the moist air is lifted, it is cooled, causing condensation and rainfall to occur. As it passes over the mountain summit, the moisture-depleted air mass descends the leeward slopes of the range, where it is compressed and warmed. At the base of the mountain on the leeward, or downwind side, the air is warmer and drier.

orographic precipitation Precipitation that is induced when moist air is forced to rise over a mountain barrier.

convective precipitation

Precipitation that is induced when warm, moist air is heated at the ground surface, rises, cools, and condenses to form water droplets, raindrops, and eventually rainfall.

Moisture falls as precipitation on the windward slope, creating a rain shadow effect on the far side of the mountain—a belt of dry climate that extends down the leeward slope and beyond.

Convective lifting Air can also be forced upward through convection, leading to **convective precipitation** (**Figure 4.17b**).

The convection process begins when surfaces are heated unequally. Think of an agricultural field surrounded by a forest, for example. The field surface is largely bare soil, with only a thin layer of vegetation. Therefore, with steady sunshine, the field will be warmer than the adjacent forest.

The density of air depends on its temperature: Warm air is less dense than cooler air. Because it is less dense, warm air rises above cold air. This process is convection, and hot-air balloons operate on this principle. In the case of our field, a bubble of air will form over the field, rise,

Four types of atmospheric lifting • Figure 4.17

Precipitation occurs when moist air parcels or masses are lifted and cooled.

a. Orographic lifting occurs when air is cooled as it passes over a mountain summit.

b. Convective lifting occurs when air is buoyed upward by convection.

c. Convergent lifting occurs when air masses converge together at a location from different directions. Air "piles up" and is forced upward, creating a low-pressure cell.

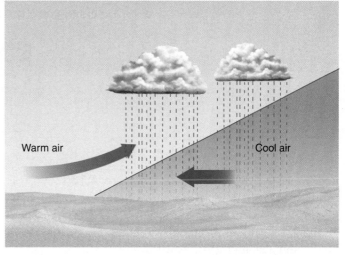

Warm air   Cool air

d. Frontal lifting occurs when a mass of warm air is forced to rise over a dense mass of cooler air.

Convective lifting • Figure 4.18

✓ THE PLANNER

Under the right conditions, a parcel of air that is heated enough to rise will continue to rise to great heights.

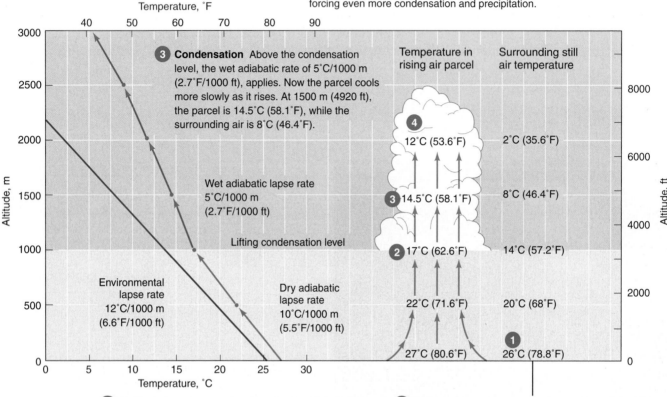

4 **Rising parcel** The parcel is still warmer than the surrounding air, so it continues to rise. Notice that the difference in temperature between the rising parcel and the surrounding air now actually increases with altitude. This means that the parcel will be buoyed upward ever more strongly, forcing even more condensation and precipitation.

3 **Condensation** Above the condensation level, the wet adiabatic rate of 5°C/1000 m (2.7°F/1000 ft), applies. Now the parcel cools more slowly as it rises. At 1500 m (4920 ft), the parcel is 14.5°C (58.1°F), while the surrounding air is 8°C (46.4°F).

Wet adiabatic lapse rate
5°C/1000 m
(2.7°F/1000 ft)

Lifting condensation level

Environmental lapse rate
12°C/1000 m
(6.6°F/1000 ft)

Dry adiabatic lapse rate
10°C/1000 m
(5.5°F/1000 ft)

Temperature in rising air parcel / Surrounding still air temperature

12°C (53.6°F) — 2°C (35.6°F)
3 14.5°C (58.1°F) — 8°C (46.4°F)
2 17°C (62.6°F) — 14°C (57.2°F)
22°C (71.6°F) — 20°C (68°F)
27°C (80.6°F) — 26°C (78.8°F)

2 **Uplift** At first, the parcel cools at the dry adiabatic rate. At 500 m (1640 ft), the parcel is at 22°C (71.6°F), while the surrounding air is at 20°C (68°F). Since it is still warmer than the surrounding air, it continues to rise. At 1000 m (3280 ft), it reaches the lifting condensation level. The temperature of the parcel is 17°C (62.6°F).

1 **Initial heating** At ground level, the surrounding air is at 26°C (78.8°F) and has a lapse rate of 12°C/1000 m (6.6°F/1000 ft). The air parcel is heated by 1°C (1.8°F) to 27°C (80.6°F) and—because it is less dense than the surrounding air—it begins to rise.

Ask Yourself

If the environmental lapse rate increased (that is, if it were cooler at higher altitudes), how would the lifting condensation level change?

a. It would be higher.
b. It would be lower.
c. It would stay the same.
d. It is impossible to determine.

and break free from the surface. **Figure 4.18** shows this process. As the bubble rises, it is cooled at the dry adiabatic rate. At the lifting condensation level, it becomes cool enough for condensation and water droplet formation to begin, and a cloud forms. Now the cloud cools less rapidly, at the moist adiabatic rate, and uplift and condensation continue. Eventually the cloud mixes with the surrounding air and dissolves.

CONCEPT CHECK STOP

1. **How** do the atmospheric conditions for rain formation and snow formation differ?

2. **What** atmospheric and topographic conditions lead to orographic precipitation?

Human Impacts on Clouds and Precipitation

LEARNING OBJECTIVES

1. **Describe** acid rain and its causes and effects.

2. **Explain** the different roles that water vapor could play in affecting climate change over the long term.

As you have already seen, human activity has the potential to disrupt the complex balance of the Earth's systems. The introduction of pollutants into the air as a byproduct of human activity can have damaging effects in the form of acid rain and global climate change.

Acid Rain

Acid rain, also called acid deposition, is made up of raindrops that have been chemically acidified by industrial air pollutants. Acids have a low pH value, less than that of distilled water (pH = 7). The lower the pH value, the more acidic the liquid.

Fossil fuel burning releases sulfur dioxide (SO_2) and nitric oxide (NO_2) into the air where they readily combine with oxygen and water in the presence of sunlight and dust particles to form condensation nuclei of sulfuric acid and nitric acid aerosols. The tiny water droplets created around these nuclei are acidic, and when the droplets coalesce in precipitation, the resulting raindrops or ice crystals are also acidic. Sulfuric and nitric acids can also be formed on dust particles, creating dry acid particles. Dry acid deposition can be as damaging to plants, soils, and aquatic life as liquid acid rain. Acid rain in the United States is most significant in regions downwind from where polluting industries are concentrated (**Figure 4.19**).

Acidity of rainwater and major industrial source regions for the United States in 2005 • Figure 4.19

Two major source areas for sulfur dioxide and nitric oxide, pollutants that cause acid rain, are the Texas–Louisiana oil and gas industrial region and the Ohio Valley manufacturing industrial region. While much of the acid precipitation remains local, downwind regions are also affected as the prevailing winds transport the acid particulates.

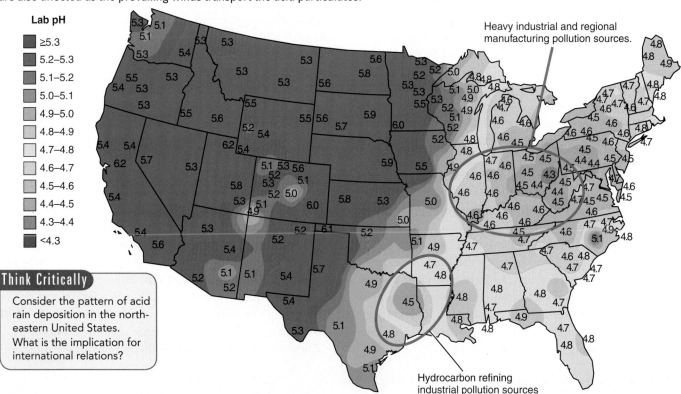

Think Critically

Consider the pattern of acid rain deposition in the northeastern United States. What is the implication for international relations?

a. Acid fallout from a nearby nickel smelter killed this previously lush forest in Monchegorsk, Russia, which then burned.

b. Acid rain has eroded the face of this stone angel in London, England, and it has ruined many historic stone buildings throughout Europe.

Effects of acid rain • Figure 4.20

What are the effects of acid rain? Acid deposition in Europe and North America has had a severe impact on some ecosystems (**Figure 4.20**). Scientists have long known that most animals, especially aquatic fish and invertebrates, are extremely sensitive to slight changes in the pH of their aquatic ecosystems. In Norway, acidification of stream water has virtually eliminated many salmon runs by inhibiting salmon egg development. The U.S. Environmental Protection Agency's National Surface Water Survey sampled more than 1000 lakes and determined that 75% of the lakes in the northeastern United States are acidic due to acid rain. Forests, too, have been damaged by acid deposition. In western Germany, the impact has been especially severe in the Harz Mountains and the Black Forest.

During the 1990s, the United States, using the Clean Air Act, significantly reduced the release of sulfur oxides, nitrogen oxides, and volatile organic compounds, largely by strengthening controls on industrial emissions. But coal burning for electricity and heating is an important source of acid deposition in many parts of the world because of the high sulfur content of coal. In Eastern Europe and the states of the former Soviet Union, air pollution controls have been virtually nonexistent for decades. Reducing pollution levels and cleaning up polluted areas will be a major task for these nations over the next decades.

Cloud Cover, Precipitation, and Global Warming

Clouds and precipitation represent the most visible and dynamic elements of the hydrologic cycle. Climate change scientists are confronted by the complexity of sophistication of their climate models as they address the question of how cloud cover and precipitation patterns will interact with global warming. Recall that global temperatures have been rising over the past few decades, along with human-induced increased atmospheric levels of CO_2. Satellite data show a rise in temperature of the global ocean surface of about 1°C (1.8°F) over the past century. Any rise in sea-surface temperature will increase the rate of evaporation, and an increase in evaporation will raise the average atmospheric content of water vapor. What effect will this have on climate?

Water plays several roles in global climate systems. First, in its vapor state, it is one of the greenhouse gases that absorbs and emits longwave radiation, enhancing the warming effect of the atmosphere above the Earth's surface. In fact, water is a more powerful greenhouse gas than CO_2. So, we can predict that an increase in water vapor in the atmosphere should enhance warming.

Second, increased water vapor forms clouds. Will more clouds increase or decrease global temperatures? Clouds can have different effects on the surface radiation balance, depending on the type of cloud, its thickness, and its moisture content. Large areas of low, white clouds, with their high albedo, can reflect a significant proportion of incoming shortwave radiation back to space, thus acting to lower global temperatures. But cloud droplets and ice particles also absorb longwave radiation from the ground, returning some of that energy as counterradiation. This absorption is an important part of the greenhouse effect, and it is much stronger for water as cloud droplets or ice particles than as water vapor. So, clouds also act to warm global temperatures by enhancing longwave reradiation from the atmosphere to the surface.

Which effect—longwave warming or shortwave cooling—will dominate? This question is at the forefront of climate science. At present, shortwave cooling is stronger than longwave warming, but global climate models predict that shortwave cooling will be reduced in the future, leading to even warmer temperatures.

Precipitation is also linked to global climate. With more water vapor and more clouds in the air on a warming Earth, more precipitation should result. Think, however, about what might happen if precipitation increases in arctic and subarctic zones. In this case, more of the Earth's surface could be covered by snow and ice. Because snow is a good reflector of solar energy, this would increase the Earth's albedo, thus tending to reduce global temperatures. Another effect might be to increase the depth of snow, tying up more water in snow packs and reducing runoff to the oceans. Reducing runoff would reduce the rate at which sea level rises.

The effects of climate change are complex. At this time, global climate models predict that the global climate system will be 1.8°C (3.2°F) to 4.0°C (7.2°F) warmer by the year 2100. This warming will be induced primarily by the CO_2 increases predicted for the 21st century. But model predictions always include some uncertainty. NASA relies on data collection and modeling to help scientists and policy-makers foresee the consequences of climate change and make informed decisions. *Where Geographers Click* describes efforts to examine the situation. As time goes by, our understanding of global climate and our ability to predict its changes are certain to increase.

Where Geographers CLICK

Clouds and Global Climate

http://isccp.giss.nasa.gov/index.html

Visit this Web site to access information and resources related to NASA's International Satellite Cloud Climatology Project, which was set up in 1982 to study the global distribution of clouds and their properties. The project takes data collected by weather satellites to help improve our understanding of how clouds affect the global radiation balance.

The project analyzes data collected from a suite of weather satellites operated by several nations and processed by several groups in government agencies, laboratories, and universities. It monitors the amount, temperature, thickness, and surface reflectance of clouds and how these vary between regions.

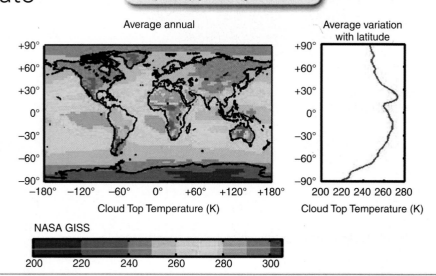

CONCEPT CHECK STOP

1. **What** are the effects of acid rain, and how can acid rain be reduced?

2. **How** and **why** will the amount of cloud cover change if global temperatures rise?

Summary

1 Water and the Hydrosphere 98

• Water can change state by **evaporation**, **condensation**, melting, freezing, **sublimation**, and **deposition**.

• The **hydrosphere** is the realm of water in all its forms. The fresh water in the atmosphere and on land in lakes, streams, rivers, and ground water is only a very small portion of the total water in the hydrosphere.

• The **hydrologic cycle**, depicted in this diagram, moves water from land and ocean to the atmosphere and then returns the water as precipitation.

The hydrologic cycle • Figure 4.4

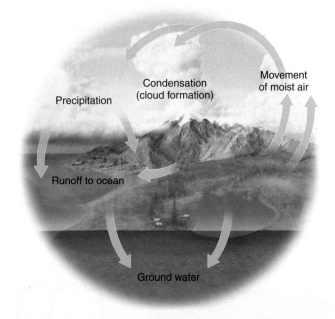

Condensation (cloud formation)

Movement of moist air

Precipitation

Runoff to ocean

Ground water

• Precipitation is the fall of liquid or solid water from the atmosphere to the Earth's land or ocean surface.

2 Humidity 102

• Humidity describes the amount of water vapor present in air.

• Warm air can hold much more water vapor than cold air.

• **Specific humidity** measures the mass of water vapor in a mass of air, in grams of water vapor per kilogram of air. **Relative humidity** measures water vapor in the air as the percentage of the maximum amount of water vapor that can be held at the given air temperature.

• Specific humidity is greatest near the equator and lower near the poles, as shown in the diagram.

Global specific humidity • Figure 4.6

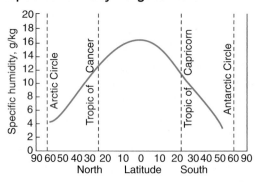

• Condensation occurs at the **dew-point temperature**.

3 Adiabatic Processes 105

• When an air parcel moves upward in the atmosphere, it encounters lower pressure and so expands and cools. As shown in the diagram, when it descends, it encounters higher pressure and so is compressed and warms. These are **adiabatic processes** because the temperature change occurs as a result of a change in pressure rather than from the application or withdrawal of energy from the surroundings.

Adiabatic cooling and heating • Figure 4.7

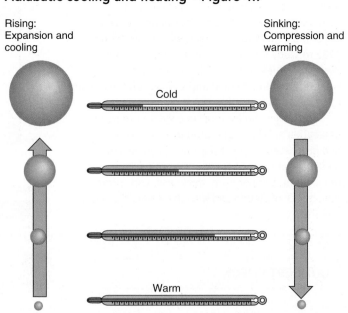

Rising: Expansion and cooling

Sinking: Compression and warming

Cold

Warm

- The **dry adiabatic lapse rate** describes the rate at which an unsaturated parcel of air cools as it rises and warms as it descends. When condensation or deposition is occurring, the cooling rate is described with the **moist adiabatic lapse rate**.

4 Clouds and Fog 108

- Clouds are composed of droplets of water or crystals of ice that form on a **condensation nucleus**.

- Clouds, as we see in this photo, typically occur in layers, as stratiform clouds, or in globular masses, as cumuliform clouds.

**Cloud families and types: Stratiform clouds •
Figure 4.10**

- **Fog** occurs when a cloud forms at ground level.

5 Precipitation 111

- **Precipitation** forms in warm clouds by the coalescence of water droplets and in cool clouds by ice crystallization.

- Precipitation from clouds occurs as rain, hail, snow, and sleet.

- In **orographic precipitation**, air moves up and over a mountain barrier. As it moves up, it is cooled adiabatically, and rain forms. As it descends the far side of the mountain, it is warmed, producing a rain shadow effect.

- **Convective precipitation** occurs when a surface is heated unequally and an air parcel becomes warmer and less dense than the surrounding air, as shown in the diagram. Because it is less dense, it rises. As it moves upward, it cools, and condensation with precipitation may occur.

**Four types of atmospheric lifting: Convective lifting •
Figure 4.17**

6 Human Impacts on Clouds and Precipitation 119

- Acid deposition of sulfate and nitrate particles can acidify soils and lakes, causing the death of fish and trees. The photo shows the impact of **acid rain** on a marble statue.

Effects of acid rain • Figure 4.20

(continued on next page)

- Because global warming, produced by increasing CO_2 levels in the atmosphere, increases the evaporation of surface water, atmospheric moisture levels increase. This tends to enhance the greenhouse effect.

- As the climate warms, more clouds are likely to form. Although clouds enhance the greenhouse effect, they also increase our planet's albedo. Because the albedo effect is stronger, clouds are presently cooling the planet. But this cooling effect will become less effective as cloud cover increases.

- Increased moisture could also reduce temperatures by increasing the amount and duration of snow cover.

Key Terms

- acid rain 119
- adiabatic process 105
- advection fog 110
- condensation nucleus 108
- condense 98
- convective precipitation 116
- deposition 98
- dew-point temperature 104
- dry adiabatic lapse rate 106
- evaporate 98
- evapotranspiration 100
- fog 110
- humidity 102
- hydrologic cycle 100
- hydrosphere 98
- lifting condensation level 106
- moist adiabatic lapse rate 106
- orographic precipitation 116
- precipitation 100
- radiation fog 110
- relative humidity 102
- specific humidity 104
- sublimation 98
- transpiration 100

Critical and Creative Thinking Questions

1. What effects would colder and warmer climatic periods have on the hydrologic cycle?

2. Many cities in the southwestern United States get their drinking water from the Colorado River. Using the percentages in the diagram, determine what percentage of the *total* fresh water on Earth (not just of surface and atmospheric water) streams and rivers make up. Because this reservoir is such a small percentage compared to the total hydrologic cycle, what does this tell us about its long-term sustainability as a water source when subjected to changes in the hydrologic cycle?

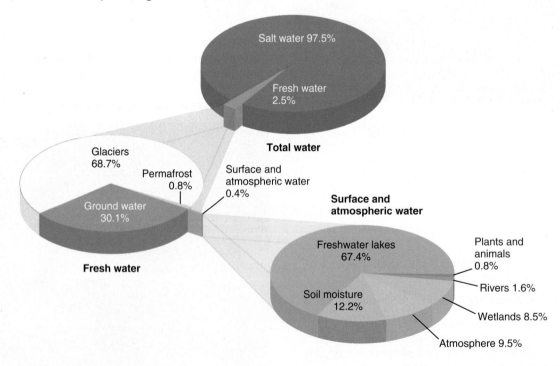

3. Describe how adiabatic processes control the rise and descent of a hot-air balloon.

4. Describe one way precipitation in your region typically forms, including both the process of atmospheric lifting and the formation of precipitation within clouds.

5. Look at the map. What do you notice about the precipitation pattern in the United States? How does the precipitation pattern affect water resource issues in the western versus eastern parts of the country?

Mean annual precipitation, cm

More than 300	200 – 300	150 – 200	100 – 150	60 – 100	40 – 60	20 – 40	10 – 20	Less than 10	Non-land area	No data available

More than 118	79 – 118	59 – 79	39 – 59	24 – 39	16 – 24	8 – 16	4 – 8	Less than 4	Non-land area	No data available

Mean annual precipitation, in.

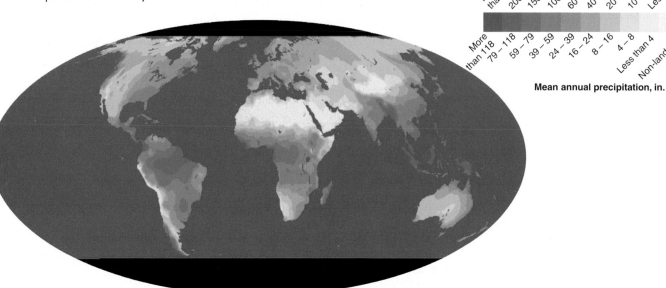

6. Describe the most common forms of air pollution and the damage that air pollution can cause. What measures are being taken, or could be taken, in your area to reduce air pollution?

7. What role do scientists predict that clouds will have in the future in regard to climate warming?

What is happening in this picture?

Landsat 7 satellite users may be frustrated when clouds obscure the Earth's surface from view. However, geographers have found that some images from the satellite can help illuminate processes that form and influence clouds. This image was captured September 15, 1999, over Alexander Selkirk Island, Chile, where Peak Inocentes rises 1319 m (4327 ft), inducing a spectacular series of swirls as the cloud formation moves north. The island's namesake was marooned for four years on the island and was the inspiration for *Robinson Crusoe*.

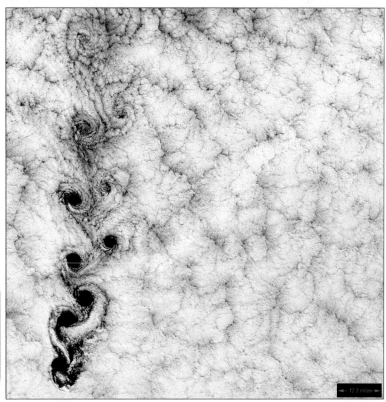

Think Critically

1. If the clouds are moving north, on which side of the island would you expect precipitation to fall?
2. How would you explain the dark holes in the swirls in terms of humidity and lapse rates?

Self-Test

(Check your answers in Appendix xx.)

1. Of the freshwater reservoirs listed here, the one with the smallest size is _____.

   a. atmospheric water

   b. ground water

   c. freshwater lakes

   d. soil water

2. The movement of water among the great global reservoirs constitutes the _____.

   a. water-factor cycle

   b. hydraulic cycle

   c. hydrologic cycle

   d. evaporation–precipitation cycle

3. The diagram shows the three states of water. Identify and label the six processes through which water changes state. In which of these processes is latent heat absorbed, and in which is it released?

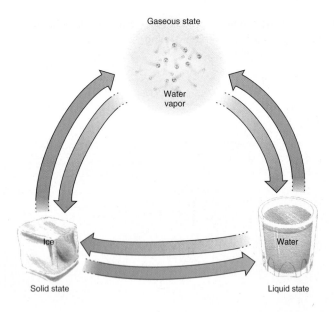

Gaseous state

Water vapor

Solid state

Ice

Liquid state

Water

4. Relative humidity _____.

   a. is the total amount of water vapor present in the air

   b. is responsible for life on Earth

   c. depends on the volume of water present in the air unrelated to temperature

   d. is the amount of water vapor in the air compared to the amount it could hold

5. The _____ of the air represents the actual quantity of water vapor held by the air.

   a. relative humidity

   b. saturation level

   c. specific humidity

   d. absolute saturation level

6. Using information from the diagram, calculate the relative humidity of a parcel of air at 20°C that contains 7.5 g of water per kilogram of air.

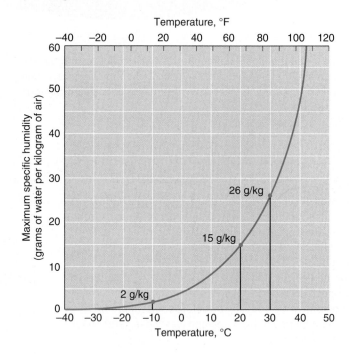

7. Because rising air cools less rapidly when condensation is occurring as a result of the release of latent heat, the _____ has a lesser absolute value than the _____.
   a. dry adiabatic lapse rate; moist adiabatic lapse rate
   b. dry adiabatic lapse rate; environmental adiabatic lapse rate
   c. environmental adiabatic lapse rate; moist adiabatic lapse rate
   d. moist adiabatic lapse rate; dry adiabatic lapse rate

8. What type of cloud is shown in the photograph? Do such clouds form at high, middle, or low levels in the sky?

9. The flat surface of cloud bottoms results when water vapor reaches the _____.
   a. environmental lapse level
   b. troposphere level
   c. lifting condensation level
   d. sublimation level

10. _____ is the type of precipitation that forms when layers of ice build up on ice pellets that are suspended in the strong updrafts of thunderstorms.
   a. Freezing rain
   b. Snow
   c. Hail
   d. Sleet

11. _____ precipitation is a result of the lifting of air over a highland area.
   a. Convective
   b. Orographic
   c. Convergent
   d. Frontal

12. _____ lifting of air is due to surface heating.
   a. Convective
   b. Orographic
   c. Frontal
   d. Adiabatic

13. Acid deposition is produced by the release of sulfur dioxide and _____ into the air.
   a. carbon monoxide
   b. ozone
   c. nitric oxide
   d. sulfuric acid

14. Increasing clouds due to climate change _____ global surface temperatures.
   a. cools
   b. warms
   c. could both warm or cool
   d. has no effect on

15. Water vapor in the atmosphere _____.
   a. leads to cooler days and warmer evenings, on average
   b. releases large amounts of heat to the stratosphere
   c. transports internal energy toward the equator
   d. acts as a greenhouse gas and absorbs longwave radiation

THE PLANNER ✓

Review your Chapter Planner on the chapter opener and check off your completed work.

Global Atmospheric and Oceanic Circulation

Energy from the Sun combined with the effects of the Earth's rotation generates the circulation patterns of our oceans and atmosphere. These powerful forces of nature have influenced all aspects of our lives, from where we live to what foods we can grow. We can track the progress of our civilization based to a large degree on harnessing the winds and currents through early exploratory sailing and commercial trading. For centuries, Dutch farmers captured wind to power their milling of grains, and their nation led the world's oceanic trading enterprise with large sailing fleets. Before the advent of steam ships, knowledge of the global circulation patterns was instrumental in the development of commercial transport.

Today, society is returning to wind energy as a practical alternative to fossil fuels, which are diminishing in supply and have been linked to climate change. By 2050, as much as one-third of the world's electricity could come from wind energy, such as this wind farm along the San Gorgonio Pass near Palm Springs, California. In the coming decades, meteorologists, engineers, and policymakers will all be actively engaged in deciding how best to harness this force of nature.

CHAPTER OUTLINE

Wind farms capture strong winds that funnel through the San Gorgonio Pass in California.

CHAPTER PLANNER ✓

- ❏ Study the picture and read the opening story.
- ❏ Scan the Learning Objectives in each section:
  p. 130 ❏ p. 132 ❏ p. 137 ❏ p. 147 ❏ p. 150 ❏
- ❏ Read the text and study all figures and visuals. Answer any questions.

Analyze key features

- ❏ Process Diagram p. 133 ❏ p. 138 ❏ p. 143 ❏ p. 144 ❏ p. 147 ❏ p. 149 ❏
- ❏ What a Geographer Sees, p. 142
- ❏ Where Geographers Click, p. 149
- ❏ Geography InSight, p. 153
- ❏ Video Explorations, p. 154
- ❏ Stop: Answer the Concept Checks before you go on.
  p. 131 ❏ p. 136 ❏ p. 146 ❏ p. 148 ❏ p. 154 ❏

End of chapter

- ❏ Review the Summary and Key Terms.
- ❏ Answer the Critical and Creative Thinking Questions.
- ❏ Answer What is happening in this picture?
- ❏ Complete the Self-Test and check your answers.

Atmospheric Pressure

LEARNING OBJECTIVES

1. **Explain** the origin of atmospheric pressure.

2. **Describe** how atmospheric pressure varies with altitude.

We live at the bottom of a vast ocean of air—the Earth's **atmosphere**. Like the water in the ocean, the air in the atmosphere is constantly pressing on the Earth's surface beneath it and on everything that it surrounds.

The atmosphere exerts pressure because gravity pulls the gas molecules of the air toward the Earth. Gravity is an attraction among all masses—in this case, between gas molecules and the Earth's vast bulk.

> **atmospheric pressure** The pressure exerted by the atmosphere because of the force of gravity acting on the overlying column of air.

Atmospheric pressure is produced by the weight of a column of air above a unit area of the Earth's surface. At sea level, about 1 kilogram of air presses down on each square centimeter of surface (1 kg/cm²)—about 15 pounds on each square inch of surface (15 lb/in.²).

Measuring Atmospheric Pressure

The basic metric unit of pressure is the **pascal (Pa)**. At sea level, the average pressure of air is 101,320 Pa. Many atmospheric pressure measurements are reported in bars and **millibars (mb)** (1 bar = 1000 mb = 10,000 Pa). In this book, we use the millibar as the metric unit of atmospheric pressure. Standard sea-level atmospheric pressure is 1013.2 mb.

> **barometer** An instrument that measures atmospheric pressure.

A **barometer** measures atmospheric pressure based on the same principle as drinking soda through a straw. When using a straw, you create a partial vacuum in your mouth by lowering your jaw and moving your tongue. The pressure of the atmosphere then forces soda up through the straw.

The oldest, simplest, and most accurate instrument for measuring atmospheric pressure, the mercury barometer, works very much like a straw (**Figure 5.1**).

Because the mercury barometer is so accurate and is used so widely, atmospheric pressure is commonly expressed using the height of the column in centimeters or inches. The chemical symbol for mercury is Hg, and standard sea-level pressure is expressed as 76.0 cm Hg (30 in. Hg).

This value corresponds to 1013.2 mb or 101,320 Pa. In this book, we use *in. Hg* as the English unit for atmospheric pressure.

Atmospheric pressure at a single location on the Earth's surface varies only slightly from day to day. On a cold, clear winter day, the sea-level barometric pressure may be as high as 1030 mb (30.4 in. Hg), while in the center of a storm system of rising warm air, pressure may be about 5% lower, perhaps 980 mb (28.9 in. Hg). These differences arise largely because cold air is denser than warm air, so a column of cold air has a higher pressure at the bottom than a column of warm air.

Changes in atmospheric pressure are associated with traveling weather systems. Scientists can measure atmospheric conditions, including pressure, by using weather

Mercury barometer • Figure 5.1

Atmospheric pressure pushes the mercury upward into the tube, balancing the pressure exerted by the weight of the mercury column. As atmospheric pressure changes, the level of mercury in the tube rises and falls.

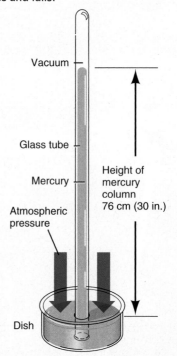

Vacuum

Glass tube

Mercury

Atmospheric pressure

Height of mercury column 76 cm (30 in.)

Dish

Radiosonde • Figure 5.2

A meteorologist in Nova Scotia launches a radiosonde to obtain a series of measurements for key atmospheric parameters, including pressure, altitude, GPS location, temperature, relative humidity, and wind speed and direction. Twice each day, the U.S. National Weather Service launches 100 of the 800 radiosondes worldwide under agreement with the World Meteorological Organization. Satellite sensors are calibrated with the radiosonde data to provide a snapshot of the atmosphere above us.

balloons that carry sensors called radiosondes up into the atmosphere (**Figure 5.2**).

Atmospheric Pressure and Altitude

If you have felt your ears pop during an elevator ride in a tall building, or when you're on an airplane that is climbing or descending, you've experienced a change in atmospheric pressure related to altitude. The pressure at any elevation in the atmosphere depends on the mass of air molecules above that elevation—the more mass, the more weight, and so the higher the pressure. Pressure increases as you go down and decreases as you go up. Because the density of air depends on pressure (as well as temperature), we also find that the density of air decreases upward, along with atmospheric pressure (**Figure 5.3**).

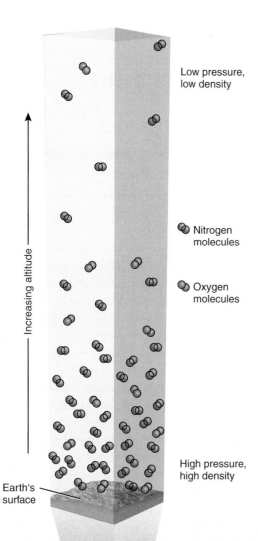

Low pressure, low density

Nitrogen molecules

Oxygen molecules

High pressure, high density

Earth's surface

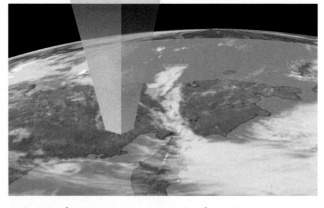

Atmospheric pressure, air density, and altitude • Figure 5.3 _____

Close to the surface of the Earth, air molecules are packed together more tightly as a result of the weight of the air column above, resulting in higher atmospheric pressure and a greater density. At higher altitudes, with fewer gas molecules above, pressure and density are lower.

CONCEPT CHECK **STOP**

1. **How** do mercury barometers measure atmospheric pressure?

2. **What** is the relationship between altitude and atmospheric pressure?

Wind Speed and Direction

LEARNING OBJECTIVES

1. **Describe** how pressure gradients drive wind.
2. **Explain** the Coriolis effect.
3. **Describe** the effect of friction on wind speed and direction.

Wind is defined as air moving horizontally over the Earth's surface. Air motions can also be vertical, but these are known by other terms, such as *updrafts* and *downdrafts*. Wind direction is identified by the direction from which the wind comes; a west wind blows from west to east, for example. Wind maps are used to depict the prevailing winds and are used in designing airport runways and locations for wind farms. The increasing importance of renewable energy to replace fossil fuels has greatly stimulated interest in identifying windy locations. It is estimated that 50% of the U.S. landscape and near offshore waters are appropriate for placing wind turbine generators for creating electricity.

Like all other motion, the movement of wind is defined by its direction and velocity, which can be tracked by a **wind vane** and an **anemometer**, respectively (**Figure 5.4**). The direction and velocity of wind are determined by three key factors: pressure gradients, the Coriolis effect, and friction.

Combination cup anemometer and wind vane • Figure 5.4

The anemometer and wind vane observe wind speed and direction, which are displayed on the meter below. The wind vane and anemometer are mounted outside, with a cable from the instrument leading to the meter, which is located indoors.

Pressure Gradients

Wind is caused by differences in atmospheric pressure between one region and another. A difference in pressure with distance or elevation, or **pressure gradient**, creates a force that tends to push air in the direction of the gradient (**Figure 5.5**). On a weather map,

> **pressure gradient**
> Change of atmospheric pressure measured along a line at right angles to the isobars.

Isobars and a pressure gradient • Figure 5.5

Because atmospheric pressure is higher at Oklahoma City than at Nashville, the pressure gradient creates a force that tends to push air toward Nashville. Although the force pushes directly toward Nashville, the motion of the air will actually spiral outward from Oklahoma City and inward toward Nashville, in response to friction and the Coriolis effect, shown in Figures 5.7–5.9.

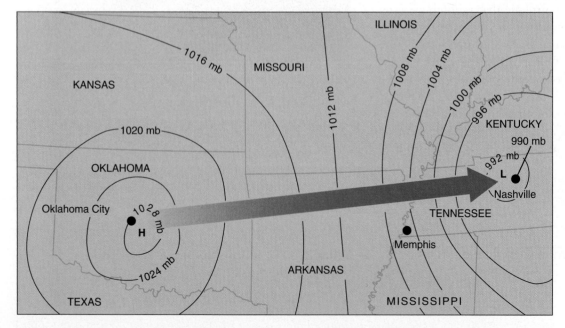

Ask Yourself

Where would you find the greatest pressure gradient on this map?
a. Oklahoma City
b. Southwestern Missouri
c. Memphis
d. Nashville

PROCESS DIAGRAM

Development of a pressure gradient • Figure 5.6

THE PLANNER ✓

Unequal heating of the Earth's surface leads to a pressure gradient and causes wind.

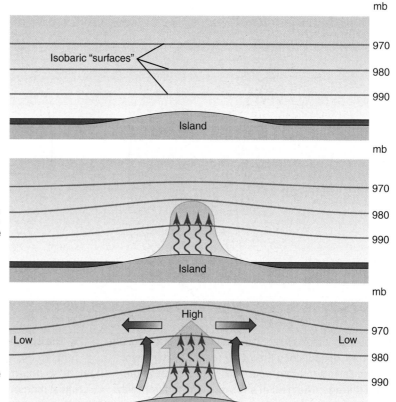

1 Uniform atmosphere (heated equally)
Imagine a uniform atmosphere above a ground surface. The isobaric levels (actually isobaric surfaces in three dimensions) are parallel with the ground surface.

2 Uneven heating
Imagine now that the underlying ground surface, an island, is warmed by the Sun, with cool ocean water surrounding it. The warm surface air rises and mixes with the air above, warming the column of air above the island.

3 Pressure gradient
The warmer air over the island expands, pushing the isobaric surfaces upward. The result is that a pressure gradient is created and air at higher pressure (above the island) flows toward lower pressure (above the ocean).

Think Critically
If the island were in the Arctic and covered by glacial ice, would the pressure gradient be the same or different?

---

isobar A line on a map drawn through all points having the same atmospheric pressure.

a pressure gradient is shown by a series of **isobars**. Widely spaced isobars indicate a more gradual change in pressure and closely spaced isobars indicate a steep slope for the change in pressure. Steeper gradients create a stronger pressure force.

Pressure gradients can arise as a result of uneven heating of the atmosphere (**Figure 5.6**). Low latitudes receive more insolation and have higher temperatures than high latitudes, creating a general pressure gradient that drives global circulation, as you will see in greater detail later in the chapter. Terrain differences and land cover also cause uneven heating and generate pressure gradients.

The Coriolis Effect

We've seen that a pressure gradient force will tend to push air from high to low pressure. However, as soon as the air starts to move, it is affected by the **Coriolis effect**, named for the French scientist who first identified it. The Coriolis effect describes how the Earth's rotation affects the motion of objects, including fluids such as air and water, as they move over the Earth's surface.

To understand the Coriolis effect, imagine yourself riding a rocket launched from the North

Coriolis effect An effect of the Earth's rotation that acts like a force to deflect a moving object on the Earth's surface to the right in the northern hemisphere and to the left in the southern hemisphere.

The Coriolis effect acts like a force to deflect winds and currents.

a. Imagine you are riding a rocket launched from the North Pole toward New York, aimed along the 74° W longitude meridian. Because of the Earth's rotation, your path would curve to the right, toward Chicago. The path of a rocket launched from New York toward the North Pole would also curve to the right, toward Greenland, for the same reason.

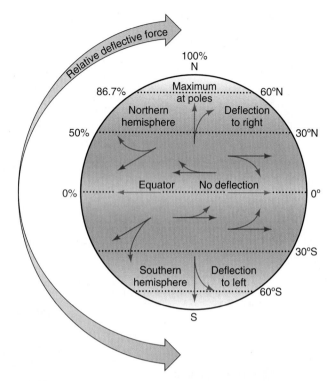

b. The Coriolis effect appears to deflect winds and ocean currents to the right in the northern hemisphere and to the left in the southern hemisphere. Blue arrows show the direction of initial motion, and red arrows show the direction of motion apparent to the Earth observer.

Pole, heading in the direction of New York (**Figure 5.7**). While on your flight, however, the Earth will be rotating, moving to your left, and soon you will be off your course to New York. Before long, you will be over Chicago, not New York. Your rocket has behaved as if a force has pushed it to the right. If you repeat this thought experiment by riding a rocket northward from the South Pole, you will find that the Earth below you moves to your right, so your apparent direction will be leftward.

In other words, the Coriolis effect causes an object in the northern hemisphere to move as if a force were pulling it to the right, and in the southern hemisphere, objects move as if they were pulled to the left. This apparent deflection doesn't depend on direction of motion—it occurs whether the object is moving toward the north, south, east, or west. The strength of the Coriolis effect increases with the speed of the motion and with latitude.

The force of the pressure gradient and the Coriolis effect combine to determine the direction of the wind.

The pressure gradient pushes the air from high pressure to low, at a right angle to the isobars, while the Coriolis effect pushes the air rightward, changing the direction of motion. As shown in **Figure 5.8**, the pressure gradient and the Coriolis effect exactly balance each other when the direction of motion is parallel to the isobars. We call this type of airflow—parallel to the isobars—the **geostrophic wind** and it describes airflow at upper levels in the atmosphere.

> **geostrophic wind** Wind at high levels above the Earth's surface moving parallel to the isobars, at a right angle to the pressure gradient.

The Frictional Force

Closer to the Earth's surface, a third force affects the speed and direction of wind. As wind in the lower troposphere moves over the Earth's surface, friction creates a force that opposes the motion of the wind. In general terms, a rougher surface, with mountains, trees, or

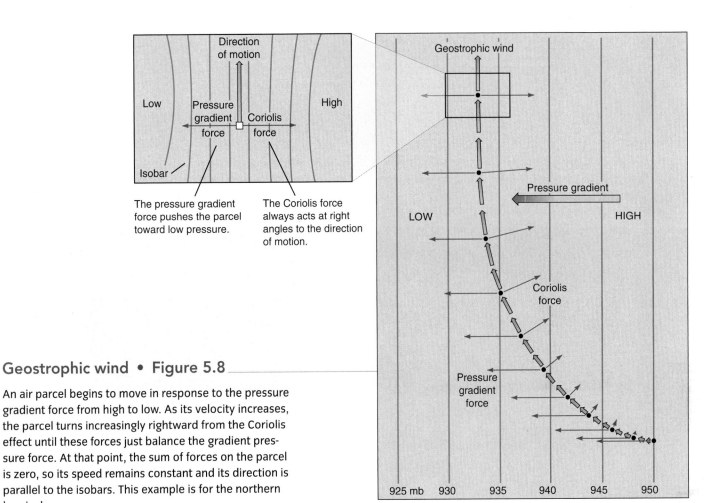

Geostrophic wind • Figure 5.8

An air parcel begins to move in response to the pressure gradient force from high to low. As its velocity increases, the parcel turns increasingly rightward from the Coriolis effect until these forces just balance the gradient pressure force. At that point, the sum of forces on the parcel is zero, so its speed remains constant and its direction is parallel to the isobars. This example is for the northern hemisphere.

Figure 5.8 labels:
Direction of motion
Low — High
Pressure gradient force
Coriolis force
Isobar
The pressure gradient force pushes the parcel toward low pressure.
The Coriolis force always acts at right angles to the direction of motion.

Geostrophic wind
Pressure gradient
LOW — HIGH
Coriolis force
Pressure gradient force
925 mb   930   935   940   945   950

buildings, has a greater frictional drag on wind than a smooth surface. The frictional force is greatest close to the surface, and it decreases with altitude.

Changes in land cover, including new buildings, deforestation, or new forest growth, can increase the effect of friction on surface winds. For example, in the northern hemisphere, surface wind velocity has been reduced by as much as 15% since the 1980s due to changes in land cover, especially forest regrowth in abandoned agricultural regions of Eastern Europe and Western Asia. Planners and developers of wind energy resources are carefully monitoring these changes to determine optimal locations for renewable energy development.

When the frictional force is added to the pressure gradient force and the apparent force of the Coriolis effect, it causes winds near the surface to move across the isobars at an angle (**Figure 5.9**). As a result, the air moves toward low

Balance of forces on a parcel of surface air • Figure 5.9

A parcel of air in motion near the surface is subjected to three forces. The sum of these three forces produces motion toward low pressure but at an angle to the pressure gradient. This example is for the northern hemisphere.

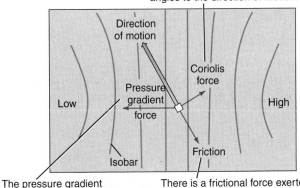

The Coriolis force always acts at right angles to the direction of motion.

Direction of motion
Pressure gradient force
Low — High
Coriolis force
Friction
Isobar

The pressure gradient force pushes the parcel toward low pressure.

There is a frictional force exerted by the ground surface that is proportional to the wind speed and always acts in the opposite direction to the direction of motion.

Cyclones and anticyclones • Figure 5.10

Close to the surface, the interaction between the pressure gradient, the Coriolis effect, and friction results in air spiraling into low-pressure centers and out of high-pressure centers.

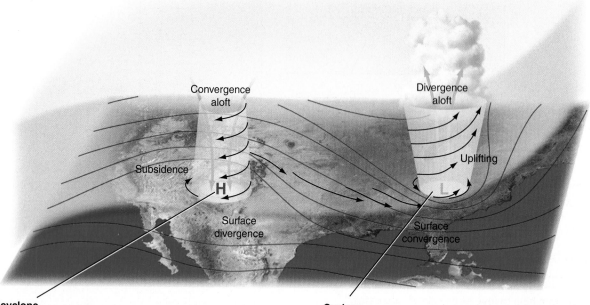

Anticyclone
As air spirals around and out of surface high-pressure centers, it produces divergence of air from the region. Air from above descends to replace the air that is removed. This air warms as it descends, inhibiting condensation and precipitation.

Cyclone
As air spirals around and into a surface low-pressure center, it produces convergence of air into the region. This convergence forces air at the surface to rise and cool adiabatically, leading to condensation and precipitation.

---

cyclone A center of low atmospheric pressure where surface air converges into a spiral and is uplifted to the upper troposphere.

anticyclone A center of high atmospheric pressure where upper troposphere air spirals down to the surface and diverges outward.

pressure in a spiraling, converging motion. This inward spiral toward a central low pressure forms a **cyclone**, also called a **low-pressure system** (**Figure 5.10**). This inward spiraling motion is called *convergence*, and it also causes the air at the center to rise. In a high-pressure area, air spirals downward and outward to form an **anticyclone**, also called a **high-pressure system**. This outward spiraling motion is called *divergence*. In the northern hemisphere, cyclones rotate counterclockwise and anticyclones rotate clockwise. In the southern hemisphere, these directions are reversed.

Low-pressure centers (cyclones) are often associated with cloudy or rainy weather, whereas high-pressure centers (anticyclones) are often associated with fair weather. Why is this? When air is forced upward, it is cooled according to the adiabatic principle, allowing condensation and precipitation to occur. So, cloudy and rainy weather often accompanies the inward and upward air motion of cyclones. In contrast, in anticyclones, the air sinks and spirals outward. When air descends, it is warmed by the adiabatic process, so condensation can't occur. That is why anticyclones are often associated with fair weather.

Cyclones and anticyclones can be 1000 kilometers (about 600 miles) across, or more. A fair weather system—an anticyclone—may stretch from the Rockies to the Appalachians. Cyclones and anticyclones can remain more or less stationary, or they can move, sometimes rapidly, to create regional weather disturbances.

CONCEPT CHECK

1. **Why** doesn't air always move in the direction of the pressure gradient?

2. **How** does geostrophic wind arise?

3. **Why** are anticyclones and cyclones associated with different weather patterns?

Global Wind and Pressure Patterns

LEARNING OBJECTIVES

1. **Explain** how Hadley cells drive tropical circulation.
2. **Describe** surface wind patterns in the midlatitudes.
3. **Explain** pressure patterns in the higher latitudes.
4. **Describe** patterns of global circulation in the upper air.

In an idealized model of the Earth, warm air over the equator would expand, rise and form a belt of low pressure (**Figure 5.11**). Cooler, denser surface air over the poles would move along the pressure gradient into the low-pressure area at the equator. At upper levels, rising air at the equator would move toward the poles, replacing the surface air moving equatorward and forming a looping circulation. However, the rotation of the Earth, the change in insolation patterns with the seasons, and the unique arrangement of continents and oceans all disrupt this simple model.

Tropical Circulation

The Earth's surface is heated more strongly by the Sun near the equator than are surfaces at higher latitudes. This creates a zone of rising warm air and low pressure at the equator known as the **equatorial trough**. At low levels, air in both hemispheres moves toward the equatorial trough, where it converges, rises to upper levels, and then turns poleward as a return flow. However, the Coriolis effect deflects this upper airflow eastward, and it eventually descends at about a 30° latitude. This completes

Global surface winds on an ideal Earth • Figure 5.11

We can see what global surface winds and pressures would be like on an ideal Earth, without the disrupting effects of oceans and continents and the variation of the seasons. Surface winds are shown on the disk of the Earth, and the cross section at the right shows winds aloft.

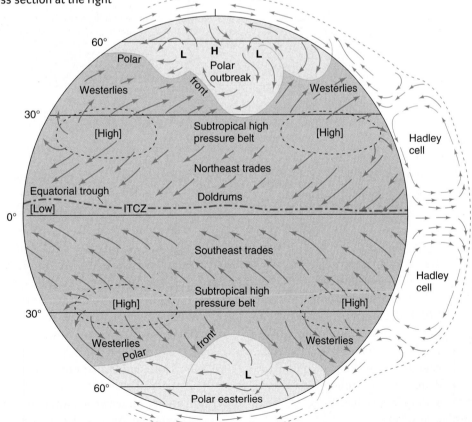

Hadley cells • Figure 5.12

Heat energy and moisture drive the Hadley cell dynamics along the equator, moving air masses in a westerly polar direction.

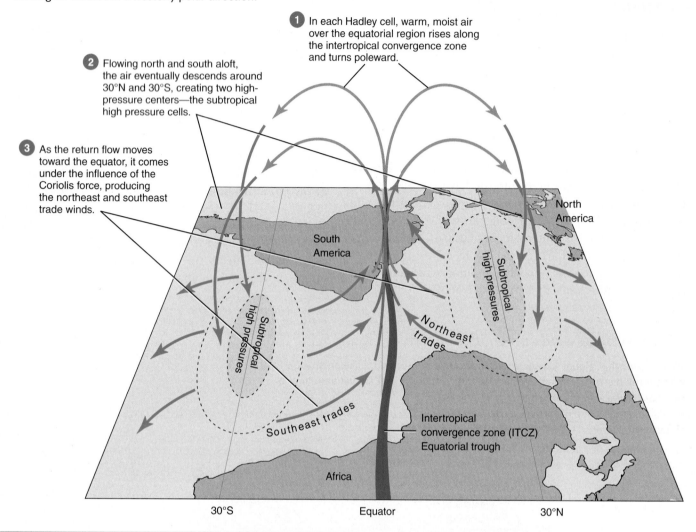

1 In each Hadley cell, warm, moist air over the equatorial region rises along the intertropical convergence zone and turns poleward.

2 Flowing north and south aloft, the air eventually descends around 30°N and 30°S, creating two high-pressure centers—the subtropical high pressure cells.

3 As the return flow moves toward the equator, it comes under the influence of the Coriolis force, producing the northeast and southeast trade winds.

North America

South America

Subtropical high pressures

Subtropical high pressures

Northeast trades

Southeast trades

Intertropical convergence zone (ITCZ) Equatorial trough

Africa

30°S   Equator   30°N

a circulation loop called a **Hadley cell**, named for the English meteorologist who first conceived it (**Figure 5.12**).

Intertropical convergence zone The narrow zone circling the atmosphere near the equatorial trough is called the **intertropical convergence zone (ITCZ)**. Because the air is converging and moving upward, surface winds are light and variable. This region is also known as the doldrums for its tendency to leave sailing ships becalmed for days and weeks without sufficient wind.

Hadley cell A low-latitude atmospheric circulation cell with rising air over the equatorial trough and sinking air over the subtropical high-pressure belts.

intertropical convergence zone (ITCZ) A zone of convergence of air masses along the equatorial trough.

The ITCZ shifts along with the seasons, following the zone of highest insolation toward each of the tropics in turn. The shift is moderate in the western hemisphere. Over the oceans, the ITCZ moves only a few degrees north from January to July. In South America, the ITCZ lies across the Amazon in January and swings northward by about 20°.

In Asia, however, there is a shift of about 40° latitude between winter and summer. This large shift occurs because of the contrast between the Indian Ocean and the vast Eurasian continent to the north. In the winter, central Asia is intensely cold, creating a great dome of high pressure that causes an airflow from the northeast across southern Asia and the Indian Ocean. This flow enhances the normal migration of the ITCZ to the south of the equator.

In the summer, intense heating of the interior high deserts causes the air to warm and rise, creating a strong low-pressure center over the interior. This low creates a southwesterly flow across south Asia, which draws the ITCZ well to the north. This movement of the ITCZ and the change in the pressure pattern with the seasons create a reversing wind pattern in Asia known as the **monsoon** (Figure 5.13).

Asian monsoon • Figure 5.13

The Asiatic monsoon winds alternate in direction from January to July, responding to reversals of barometric pressure over the large Asian continent.

a. January

In the winter, there is a strong outflow of dry, continental air from the north across China, Southeast Asia, India, and the Middle East. During this winter monsoon, dry conditions prevail.

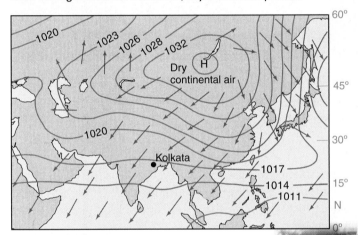

b. July

In the summer, warm, humid air from the Indian Ocean and the southwestern Pacific moves northward and eastward into Asia, bringing rainstorms, some lasting weeks, like this one in Kolkata.

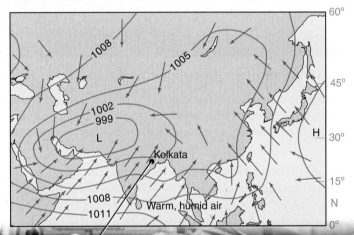

Put It Together

Review Figure 5.9 and answer this question. Considering the direction of the winds compared to the isobars, which statement is most correct?

a. Because this area is near the equator, the Coriolis effect has no influence on these winds.

b. Because the pressure gradients are great, friction has no influence on these winds.

c. Because some of the winds are over the ocean, neither the Coriolis effect nor friction has an influence on these winds.

d. Because the wind direction arrows cross the isobars at oblique angles, both the Coriolis effect and friction are important for these winds.

Summer thunderstorm on the Arizona desert • Figure 5.14

Summer in the deserts of the American Southwest occasionally sees incursions of warm, moist maritime tropical (mT) air, leading to clouds and thundershowers. Intense rainfall from these storms can create hazardous flash floods.

A monsoon effect is also felt in North America, although to a lesser extent. In North America during summer, warm, moist air from the Gulf of Mexico tends to move northward and westward into the southern United States. Southwestern states, such as Nevada, Arizona, and New Mexico, receive the major portion of their rainfall in dramatic summer deluges from this monsoon effect (**Figure 5.14**). In winter, the airflow across North America generally reverses, and dry continental air from Canada moves south and eastward.

Subtropical high-pressure cells
As air descends on the poleward side of the Hadley cell circulation, surface pressures are high. This produces two **subtropical high-pressure belts**, each centered at about 30° latitude. Two, three, or four very large and stable anticyclones form within these belts (**Figure 5.15**). In the northern hemisphere, they include two large high-pressure cells that flank North America—the Hawaiian high and the Azores high. In the southern hemisphere, similar high-pressure cells flank South America and Southern Africa.

You have seen that insolation is most intense when the Sun is directly overhead, and you also know that the latitude at which the Sun is directly overhead changes from solstice to solstice. This cycle causes the Hadley cells to intensify and move poleward during their high-Sun season, which also moves the subtropical high-pressure belts poleward.

When the Hawaiian and Azores highs intensify and move northward in the summer, the east and west coasts of North America feel their effects. On the west coast, dry, descending air from the Hawaiian high dominates, so fair weather and rainless conditions prevail. On the east coast, air from the Azores high flows across the continent from the southeast. These winds travel long distances across warm, tropical ocean surfaces before reaching land, picking up moisture and getting warmer. So they generally produce hot, humid weather for the central and eastern United States. In winter, these two anticyclones weaken and move to the south—leaving North America's weather to the mercy of colder winds and air masses from the north and west.

Trade winds
Winds around the subtropical high-pressure centers spiral outward and move toward equatorial as well as middle latitudes. The winds moving equatorward are the dependable **trade winds** that drove the sailing ships of merchant traders. North of the equator, they blow from the northeast and are called the northeast trade winds. South of the equator, they blow from the southeast and are the southeast trade winds.

> **subtropical high-pressure belt** An area of persistent high atmospheric pressure centered at about 30° N and 30° S latitude.

Midlatitude Circulation

Poleward of the subtropical highs, air spirals outward, producing the **westerly winds**. These are southwesterly winds in the northern hemisphere and northwesterly winds in the southern hemisphere.

Between about 30° and 60° latitude, the pressure and wind patterns become more complex. This is a zone of conflict between air bodies with different characteristics: Masses of cool, dry air move into the region, heading eastward and equatorward. Meanwhile, circulation around the subtropical high-pressure cells moves warm, moist air poleward. The boundary between these air masses of different characteristics is known as the **polar front**. As their motions ebb and flow, the polar front moves, and midlatitude weather changes.

> **polar front** The boundary between cold polar air masses and warm subtropical air masses.

> **jet stream** A high-speed airflow in a narrow band within the upper-air westerlies and along certain other global latitude zones at high altitudes.

Jet streams Within the upper-air westerlies, **jet streams** reach great speeds in narrow zones at high altitudes, beginning around the tropopause. They occur where atmospheric pressure gradients are strong. Along a jet stream, the air moves in pulses along broadly curving tracks. The greatest wind speeds occur in the center of the jet stream, with velocities decreasing away from it.

The **polar-front jet stream** (or simply the polar jet stream) follows in an irregular pattern along the boundary between cold polar air and warm subtropical air. The polar jet stream is generally located between 35° and 65° latitude in both hemispheres. It is typically found at altitudes of 10 to 12 km (about 30,000 to 40,000 feet), and wind speeds in the jet stream range from 75 to as much as 125 meters/second (about 170 to 280 miles/hour).

The polar jet stream tends to shift closer to the equator during the winter. For example, over the United States, the polar jet stream moves south as far as the southernmost states during the winter months, where it influences the movement of cold air masses. Because it affects the path of surface storms moving easterly, weather forecasters closely monitor its location. During the summer months, this jet stream tends to remain in its northernmost pathway and has minimal impact on weather in the continental United States.

Subtropical high-pressure cells • Figure 5.15

Over the oceans, surface winds spiral outward from the subtropical high-pressure cells, feeding the trade winds and westerlies. These subtropical cells also exist in the Pacific and Indian Oceans.

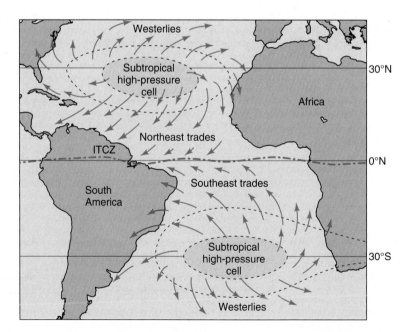

Ask Yourself

In the days of sailing ships, which pattern of navigation made the most sense, considering prevailing wind directions?

a. United States to Africa to England back to the United States
b. United States to England to Africa back to the United States
c. United States to England back to the United States
d. United States to Africa back to the United States

WHAT A GEOGRAPHER SEES

Jet Streams and Air Travel

A New York resident flying home after a visit to Dubai might be unpleasantly surprised to find that the westward flight home is noticeably longer than the original eastbound flight. The traveler might blame the pilot, but a geographer would know that the real blame goes to the jet stream, which pushes against the plane as it travels west.

Pilots have been aware of strong eastward headwinds since the early days of aviation, but today, with the help of satellite and astronaut photos, such as this one (Figure a), we can see evidence of the path of the subtropical jet stream in the form of a band of cirrus clouds, at about 25° N latitude. These clouds have formed on the equatorward edge of the jet stream, which is moving from west to east at an altitude of about 12 km (40,000 ft) at a rate of 300 km/hr (186 mi/hr) (Figure b).

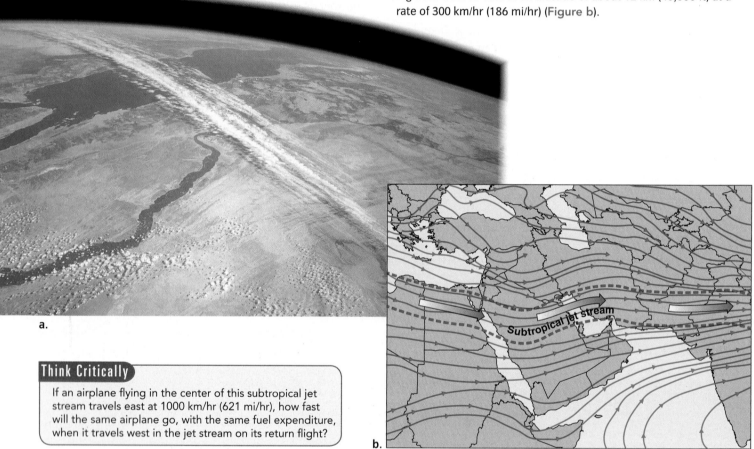

a.

b.

Think Critically

If an airplane flying in the center of this subtropical jet stream travels east at 1000 km/hr (621 mi/hr), how fast will the same airplane go, with the same fuel expenditure, when it travels west in the jet stream on its return flight?

The **subtropical jet stream** occurs at the tropopause, just above subtropical high-pressure cells in the northern and southern hemispheres. There, westerly wind speeds can reach 100 to 110 m/s (about 215 to 240 mph), in association with the increase in velocity that occurs as an air parcel moves poleward from the equator. Air transportation travel times can vary significantly, depending on whether airplanes fly with or against the jet streams (see *What a Geographer Sees*).

Jet stream disturbances While the jet stream is typically a region of confined, high-velocity westerly winds found in the midlatitudes, the jet stream can actually contain broad wave-like undulations called **jet stream disturbances**. These are also sometimes called **Rossby waves**, after Carl-Gustaf Rossby, the Swedish-American meteorologist who first observed and studied them.

Many processes can produce disturbances in the eastward flow of the jet stream. One of the most important is related to *baroclinic instability*, which makes the atmosphere unstable to small disturbances in the jet stream and allows these disturbances to grow over time. (The word *baroclinic* comes from the prefix *baro-*, meaning "pressure," and the suffix *-cline*, meaning "gradient.")

The development of these disturbances is shown in **Figure 5.16**. For a period of several days or weeks, the jet stream flow may be fairly smooth. Then, an undulation develops. As the undulation grows, warm air pushes poleward, and a tongue of cold air is brought to the south. Eventually, the cold tongue is pinched off, leaving a pool of cold air at a latitude far south of its original location. This cold pool may persist for some days or weeks,

slowly warming with time. Because of its cold center, it will contain low pressures aloft. In addition, the cold air in the core will descend and diverge at the surface, creating surface high pressure. Similarly, a warm air pool will be pinched off far to the north of its original location. Within the core of the warm pool will be rising air, with convergence and low pressure at the surface, as well as high pressure aloft.

Jet stream disturbances • Figure 5.16

THE PLANNER

Disturbances form in the upper-air westerlies of the northern hemisphere, marking the boundary between cold polar air and warm tropical air.

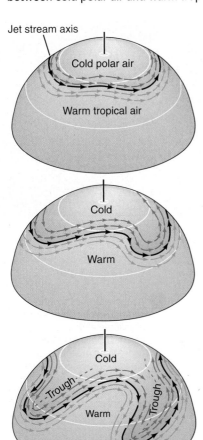

Jet stream axis

Cold polar air

Warm tropical air

1 The flow of air along the front will be fairly smooth for several days or weeks, but then it will begin to undulate.

Cold

Warm

2 As the undulation becomes stronger, disturbances in the jet stream begin to form. Warm air pushes poleward, and cold air is brought to the south.

Cold

Trough

Trough

Warm

3 The disturbance becomes stronger and more developed. A tongue of cold air is brought to the south and forms a low-pressure trough at upper elevations.

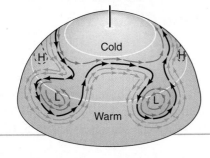

Cold

H

H

L

L

Warm

4 Eventually, the tongue is pinched off, leaving a pool of cold air at a latitude far south of its original location. The pool of cold air forms an upper-air cyclone of cold air that can persist for some days or a week.

Growth of disturbances in the jet stream • Figure 5.17

Jet stream disturbances grow because the curving circulation around regions of cool air and warm air tends to make the warm regions even warmer and the cool regions even cooler, which in turn makes the circulation stronger.

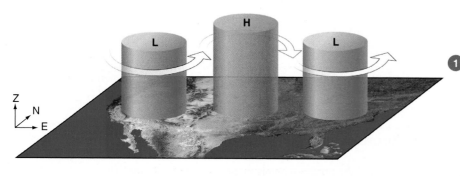

1 Pressure disturbances in the upper atmosphere are associated with gradients in temperature. Here, the warmer air column (red) produces high pressures aloft while the cooler air columns (blue) produce lower pressures. The pressure patterns produce disturbances in the jet stream, which circulates geostrophically around the pressure centers.

2 Looking down from above, the change in temperature from one location to another results in a disturbance in the global-scale north–south temperature gradient between low and high latitudes, as shown by the isotherms. The counterclockwise flow around the low-pressure center aloft brings warm, tropical air from the south into the vicinity of the warm air column, which makes the region even warmer. Conversely, the high-pressure center aloft brings cold air from the north into the vicinity of the cold air column, making that region even cooler.

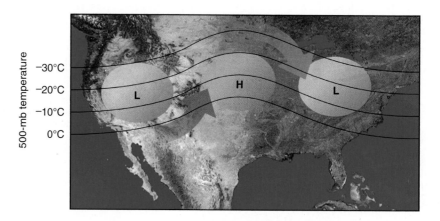

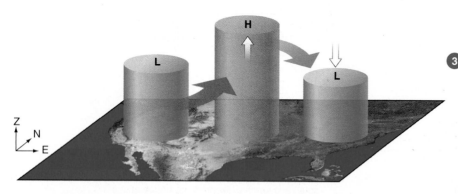

3 As the warm air column continues to warm, it expands, producing even higher pressures aloft. In addition, the cold air column continues to cool, thereby decreasing pressures further. The pressure gradient between the cold-air region and the warm-air region intensifies, and so the wind speeds also intensify.

What causes a jet-stream disturbance to grow over time? This growth is actually a complicated process and involves the interaction of the upper-air circulations—formed by the high- and low-pressure centers—with the global temperature gradient between the warmer air of low latitudes and colder air of higher latitudes. This growth process is described in **Figure 5.17**.

High-Latitude Circulation

The northern hemisphere has two large continental masses separated by oceans and an ocean at the pole. The southern hemisphere has a large ocean with a cold, ice-covered continent at the center. These differing land–water patterns strongly influence the development of high- and low-pressure centers with the seasons (**Figure 5.18**).

Polar circulation • Figure 5.18

Continents are colder in winter and warmer in summer than oceans at the same latitude. We know that cold air is associated with surface high pressure and warm air with surface low pressure. So, continents have high pressure in winter and low pressure in summer. Values greater than average barometric pressure are shown in red, and lower values are in green.

a. January In northern hemisphere winter, air spirals outward from the strong Siberian high and its weaker cousin, the Canadian high. The Icelandic low and Aleutian low spawn winter storm systems that move southward and eastward on to the continents. In the southern hemisphere summer, air converges toward a narrow low pressure belt around the south polar high.

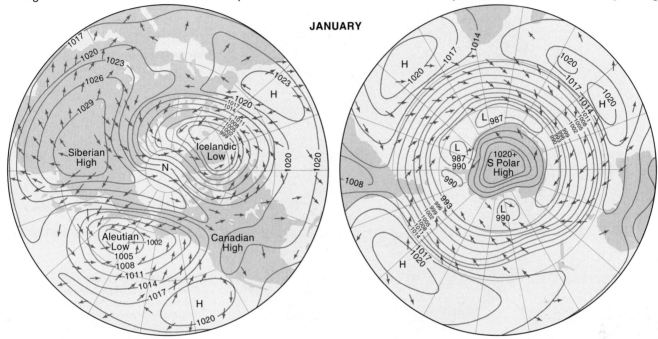

JANUARY

b. July In northern hemisphere summer, the continents show generally low-surface pressure, and high pressure builds over the oceans. The strong Asiatic low brings warm, moist Indian Ocean air over India and Southeast Asia. A weaker low forms over the deserts of the southwestern United States and northwestern Mexico. Outspiraling winds from the Hawaiian and Azores highs keep the west coasts of North America and Europe warm and dry, and the east coasts of North America and Asia warm and moist. In the southern hemisphere summer, the pattern is little different from winter, although pressure gradients are stronger.

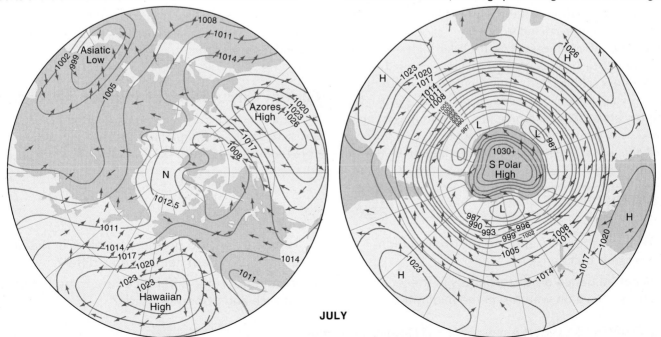

JULY

Circulation in the upper air has four major features: weak equatorial easterlies, tropical high-pressure belts, upper-air westerlies, and polar lows.

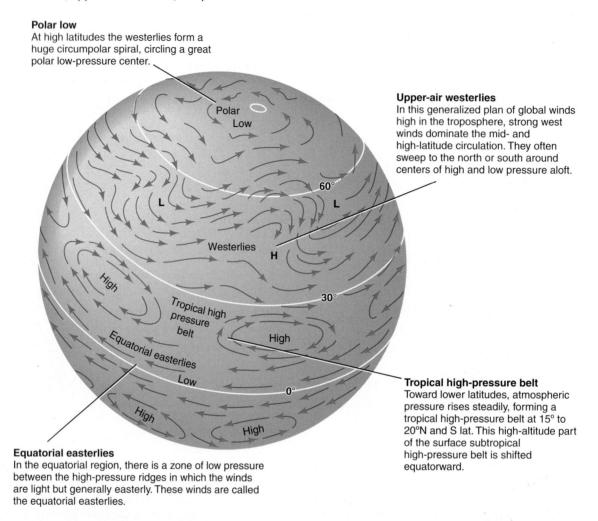

Polar low
At high latitudes the westerlies form a huge circumpolar spiral, circling a great polar low-pressure center.

Upper-air westerlies
In this generalized plan of global winds high in the troposphere, strong west winds dominate the mid- and high-latitude circulation. They often sweep to the north or south around centers of high and low pressure aloft.

Tropical high-pressure belt
Toward lower latitudes, atmospheric pressure rises steadily, forming a tropical high-pressure belt at 15° to 20°N and S lat. This high-altitude part of the surface subtropical high-pressure belt is shifted equatorward.

Equatorial easterlies
In the equatorial region, there is a zone of low pressure between the high-pressure ridges in which the winds are light but generally easterly. These winds are called the equatorial easterlies.

The higher latitudes of the southern hemisphere have a polar continent surrounded by a large ocean. There is a permanent anticyclone—the South Polar high—centered over Antarctica because the continent is covered by ice and is always cold. Easterly winds spiral outward from the high-pressure center. Surrounding the high is a band of deep low pressure, with strong, inward-spiraling westerly winds. As early mariners sailed southward, they encountered this band, where wind strength intensifies toward the pole. Because of the strong prevailing westerlies, they named these southern latitudes the *roaring forties*, *flying fifties*, and *screaming sixties*.

Global Circulation at Higher Altitudes

At higher altitudes in the troposphere, circulation patterns are somewhat simpler and smoother than those closer to the surface (**Figure 5.19**).

CONCEPT CHECK

1. **How** do the ITCZ and the subtropical high-pressure belts arise?

2. **What** influence do disturbances in the jet stream have on weather in the midlatitudes?

3. **How** and why do pressure patterns in the northern and southern hemispheres differ?

4. **What** important differences are there between surface and upper-air winds?

Local Winds

LEARNING OBJECTIVES

1. **Describe** the topographic situations that result in daily cycles of airflow.

2. **Explain** the pressure conditions that lead to Chinook and Santa Ana winds.

So far we have focused on global air circulation, but local areas also have distinctive wind patterns. Pressure differences in the atmosphere can result from differences in surface terrain, such as mountains, valleys, and water.

Daily Cycles of Winds

Coastal areas experience a daily cycle of sea and land breezes as a result of uneven heating of land and water surfaces. The contrast between the land surface, which heats and cools rapidly, and the ocean surface, which has a more uniform temperature, induces pressure gradients to create sea and land breezes (**Figure 5.20**).

During the day, the Sun heats the land more than it heats the water. Warm air over the land expands and rises, creating a lower-pressure area. Surface air over the sea then moves into the lower-pressure area over the land, creating a daytime **sea breeze**. Where coastal bluffs occur,

Sea and land breezes • Figure 5.20

THE PLANNER

Uneven warming over sea and land surfaces leads to pressure gradients and wind.

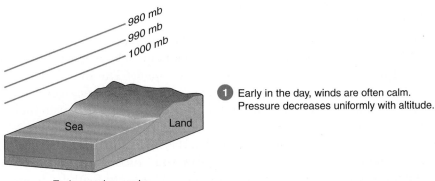

980 mb
990 mb
1000 mb

1 Early in the day, winds are often calm. Pressure decreases uniformly with altitude.

Sea   Land

Early morning—calm

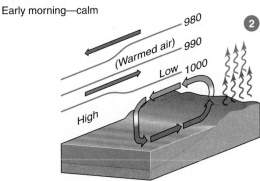

980
(Warmed air) 990
Low 1000
High

2 Later in the day, the Sun has warmed the land and the air above it, causing the air to rise. The warmed air moves from land back toward the ocean at higher altitudes. Closer to the surface, air over the ocean moves to replace the air that has risen over the land, creating a sea breeze.

Afternoon—sea breeze

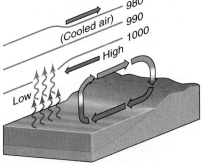

980
(Cooled air) 990
1000
High
Low

3 At night, radiation cooling over land creates a reversed situation, with higher pressure at the surface and lower pressure aloft, creating a land breeze.

Night—land breeze

Valley and mountain breezes • Figure 5.21

Uneven warming over valleys and mountains generates valley and mountain breezes.

a. During the day, mountain hill slopes are heated intensely by the Sun, making the air expand and rise, creating a valley breeze. An air current moves up valleys from the plains below—upward over rising mountain slopes, toward the summits.

b. At night, the hill slopes are chilled by radiation, setting up a reversed pattern of circulation. The cooler, denser hill slope air moves valleyward, down the hill slopes to the plain below.

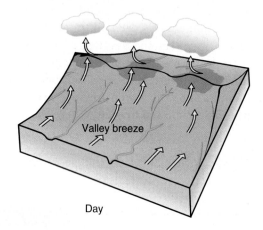

Valley breeze

Day

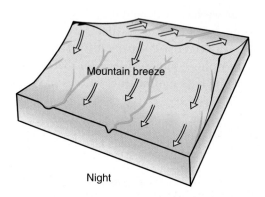

Mountain breeze

Night

hang gliding enthusiasts congregate to take advantage of these consistent swift winds.

During the night, the process is reversed. The land cools more quickly than the water, so the cooling air over the land creates a higher-pressure area. The air in this higher-pressure area moves toward the lower-pressure area over the water, causing a **land breeze**.

A similar cycle of daily airflow exists in hilly or mountainous areas. During the day, the air over the mountain slopes, especially those facing south, is heated more intensely than the air over the valley floor. The air over the mountain rises as it is warmed, creating a low-pressure area. The cooler air over the valley floor flows up the slope into the low-pressure area as a **valley breeze**. Hang gliding enthusiasts also congregate in areas where launching from mountain slopes allows for extended rides on the valley breezes. At night, the cycle is reversed, and air flows down the mountain into the valley as a **mountain breeze** (Figure 5.21).

Other Topographic Winds

Local wind systems result when a steep pressure gradient develops on either side of a mountain, with a high-pressure area on the windward side of the mountain and a low-pressure area on the leeward side of the mountain. Such a situation produces a warm, dry wind called a **Chinook**

wind (Figure 5.22). When channeled into valleys on the leeward side of a mountain, Chinook winds can raise local temperatures very rapidly. The dry air has great evaporating ability, and it can make a snow cover rapidly sublimate in winter or turn a dry brush cover to tinder in summer.

If you're from Southern California, you're probably familiar with the **Santa Ana winds**, fierce, searing winds that often drive raging wildfire into foothills communities. These winds result from the air draining from high-pressure masses created over the Great Basin between the Rockies and the Sierra Nevada mountains. As the dry air flows downhill toward the lower coastal areas, it warms adiabatically and dries to rapidly desiccate the landscape, leading to wildfire hazards (see *Where Geographers Click*).

CONCEPT CHECK

1. **Why** does a sea breeze flow toward land during the day?

2. **Why** are the Santa Ana and Chinook winds associated with fires?

Chinook winds • Figure 5.22

Local topography, such as that found in parts of the Rocky Mountains, can give rise to a warm, dry Chinook wind on the eastern slopes.

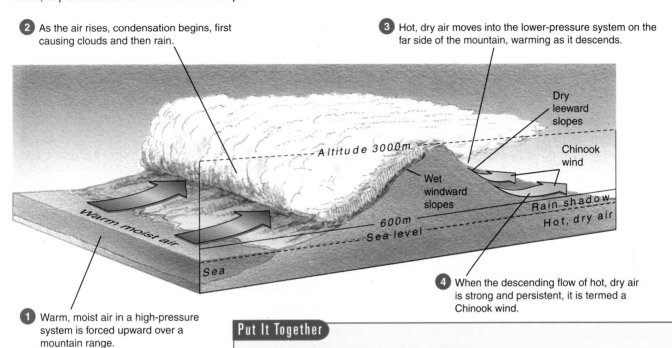

2 As the air rises, condensation begins, first causing clouds and then rain.

3 Hot, dry air moves into the lower-pressure system on the far side of the mountain, warming as it descends.

Dry leeward slopes

Chinook wind

Altitude 3000m

Wet windward slopes

Rain shadow

Warm moist air

600m

Sea level

Hot, dry air

Sea

1 Warm, moist air in a high-pressure system is forced upward over a mountain range.

4 When the descending flow of hot, dry air is strong and persistent, it is termed a Chinook wind.

Put It Together

Review Figure 5.11 and answer this question.
For north–south mountain ranges in midlatitude regions (30° to 45° latitude), dry regions will be found on the _____ side in the northern hemisphere and on the _____ in the southern hemisphere.

a. east; east
b. west; west
c. east; west
d. west; east

Where Geographers CLICK

Fire Weather and Santa Ana Winds

www.noaawatch.gov/themes/fire.php

NOAA's National Weather Service (NWS) tracks fire weather conditions such as those produced by the Santa Ana winds. The NWS uses data from a variety of satellites and sensors, which can all be accessed from this Web site. This image of southern California and northern Mexico from NASA's Aqua satellite thermal sensor shows Santa Ana conditions, in which continental dry air descends and moves westward across the coastline. Adiabatic heating makes the air very hot and dry, creating a fire weather hazard for the region.

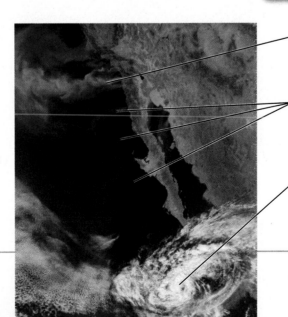

Wildfires have already begun in some areas, as is apparent from the smoke drifting off the Southern California coast.

The desiccation of the local landscape is evident from the dust blowing off the Baja California peninsula.

A low-pressure air mass is pushing up from the south tropical seas with moisture, providing a visible contrast to the clear dry Santa Ana air.

Oceanic Circulation

LEARNING OBJECTIVES

1. **Describe** the general pattern of global ocean surface currents.

2. **Explain** how ocean circulation is driven by energy differences at different latitudes.

3. **Describe** El Niño and La Niña ocean current changes in the Pacific.

T he circulation pattern of the atmosphere is closely related to the circulation of the world's oceans. Together, atmospheric and oceanic circulations transfer energy and moisture around the globe. Cyclical changes in these circulation patterns can affect our weather and climate in dramatic ways.

Ocean Currents

An **ocean current** is any persistent, dominantly horizontal flow of ocean water. Surface currents are driven by prevailing winds. Energy of motion is transferred from

> **ocean current** A persistent, dominantly horizontal flow of water.

Ocean current gyres • Figure 5.23

Great surface gyres occur in each hemisphere, driven by prevailing winds. In the northern hemisphere, two clockwise gyres occur that dominate the northern Pacific (not displayed) and Atlantic systems. In the southern hemisphere, an extra gyre (not displayed) in the Indian Ocean provides three counterclockwise gyres.

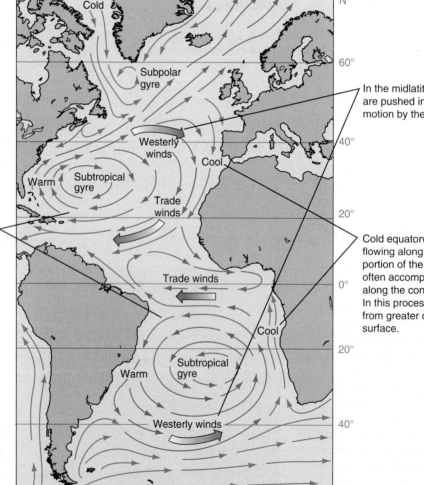

In the tropical regions, ocean currents flow westward, pushed by the northeast and southeast trade winds.

In the midlatitudes, currents are pushed in a slow eastward motion by the westerly winds.

Cold equatorward currents flowing along the eastern portion of the ocean basins are often accompanied by upwelling along the continental margins. In this process, colder water from greater depths rises to the surface.

Warm surface waters in the tropics move poleward, cooling on the way. Now cold and dense, these waters sink at higher latitudes, flow equatorward, and eventually upwell to the surface at far distant locations to cool the surrounding regions and complete the circuit.

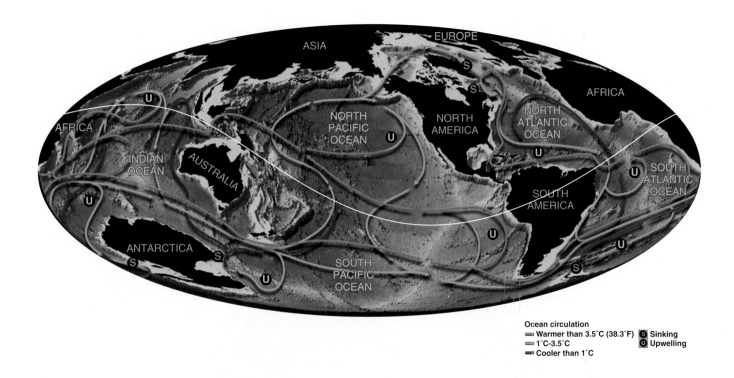

Ocean circulation
— Warmer than 3.5°C (38.3°F) **S** Sinking
— 1°C-3.5°C **U** Upwelling
— Cooler than 1°C

the wind to the water by the friction of the air blowing over the water surface. Because of the Coriolis effect, however, the actual direction of water drift is deflected about 45° from the direction of the driving wind.

As **Figure 5.23** shows, between continents, where the flow of water is uninterrupted, the ocean circulates in two large circular movements, called **gyres**, which are related to the movements of air around the subtropical high-pressure cells. The frictional drag of prevailing winds is the primary driver of these gyres. Near the equator, the trade winds generate currents that move surface water westward. In the midlatitudes, the westerlies generate currents that move the surface water eastward. As a result, gyres move clockwise in the northern hemisphere and counterclockwise in the southern hemisphere.

Deep currents move ocean waters in a slow circuit across the floors of the world's oceans. Coupled with these deep currents are very broad and slow surface currents on which the more rapid surface currents, described above, are superimposed. **Figure 5.24** diagrams

this slow flow pattern, which links all of the world's oceans. It is referred to as *thermohaline circulation*.

In this circulation, Atlantic surface water moves northward, through the tropics, and becomes saltier, and therefore denser, by evaporation. The dense, salty water cools as it moves northward, becoming even denser, and eventually it sinks along the northern edge of the Atlantic, creating a slow but steady bottom flow. Eventually the cold water upwells at far distant locations, as described in the figure.

Circulation and Energy Transfer

Global circulation of the atmosphere and oceans moves energy from low to high latitudes in a flow that is essential to sustaining the global energy balance. Recall from Chapter 2 that the Earth has a global energy budget of incoming and outgoing radiation. In places where radiant energy flows in faster than it flows out, net radiation is positive, providing an energy surplus. In other places, net radiation can be negative. For the entire Earth and atmosphere, the net radiation is zero over the year.

Warm ocean water and warm, moist air move poleward, while cooler water and cooler, drier air move toward the equator.

a. Between about 40° N and 40° S, incoming solar radiation exceeds outgoing longwave radiation throughout the year, leading to an energy surplus. Closer to the poles, outgoing longwave radiation exceeds incoming solar radiation, leading to an energy deficit.

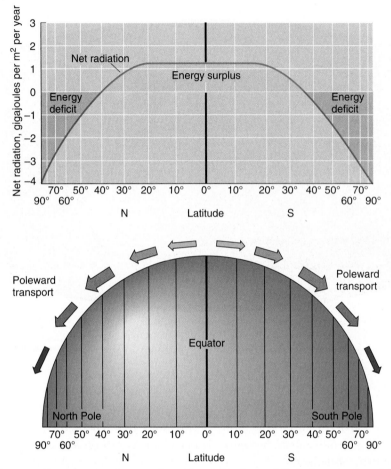

b. Because there is an energy surplus at low latitudes and there is an energy deficit at high latitudes, energy flows from low latitudes to high latitudes by means of oceanic and atmospheric circulation.

We saw earlier that solar energy input varies strongly with latitude. In order to maintain the Earth's energy balance, absorbed solar energy is moved from regions of excess to regions of deficit, carried by ocean currents and atmospheric circulation (**Figure 5.25**).

This poleward energy transfer, driven by the imbalance in net radiation between low and high latitudes, is the power source for broad-scale atmospheric circulation patterns and ocean currents. Without this circulation, low latitudes would heat up and high latitudes would cool down until a radiative balance was achieved, leaving the Earth with much more extreme temperature contrasts—very different from the planet we are familiar with now.

Cycles in Atmospheric and Oceanic Circulation

Sometimes current patterns change, producing changes in the weather that can last many years (**Figure 5.26**).

The most notable of these cyclic changes in ocean currents occurs every few years, when there is a shift in average barometric pressure between the western and east-central Pacific, leading to an **El Niño–Southern Oscillation (ENSO)** event. Peruvian fishers whose anchovy harvests plummeted around Christmas every few years originally gave the Spanish name El Niño to this disastrous weather phenomenon, in seasonal tribute to the *Christ Child*, or *little boy*. During an El Niño event, Pacific

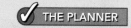
Because the heat capacity of water is so high, sea-surface temperature has an important effect on both weather and climate. Periodic changes in sea-surface temperature patterns have predictable effects on weather patterns around the globe.

a. Sea-surface temperature

Ocean temperatures are generally warm near the equator and cold at high latitudes, but winds and currents distort this simple picture.

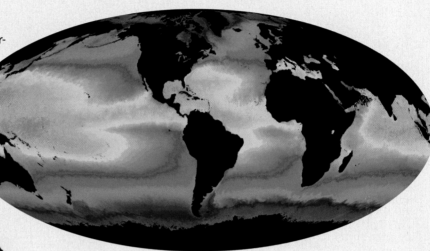

Sea-surface temperature, °C

35.5 34 32 30 28 26 24 22 20 18 16 14 12 10 8 6 4 2 0 -1.55

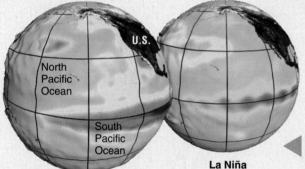

El Niño   La Niña

b. El Niño–La Niña

El Niño is the appearance of unusually warm water in the eastern equatorial Pacific (left). El Niño is often followed by La Niña, characterized by colder-than-normal ocean temperatures across the equatorial Pacific (right).

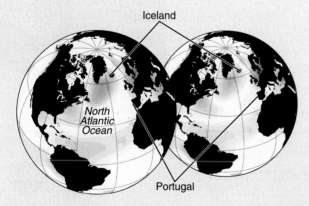

c. North Atlantic Oscillation

When the air pressure over Portugal becomes greater than that over Iceland, the westerly wind in the Atlantic becomes stronger and brings warm and wet marine air to northern Europe for a mild winter (left). The opposite brings a colder and drier winter (right).

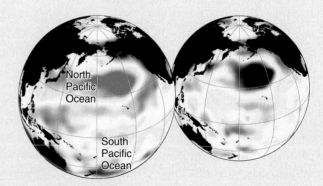

d. Pacific Decadal Oscillation

The Pacific Decadal Oscillation (PDO) is a long-lived climate pattern of Pacific Ocean temperatures. The extreme warm (left) or cool (right) phases can last up to two decades.

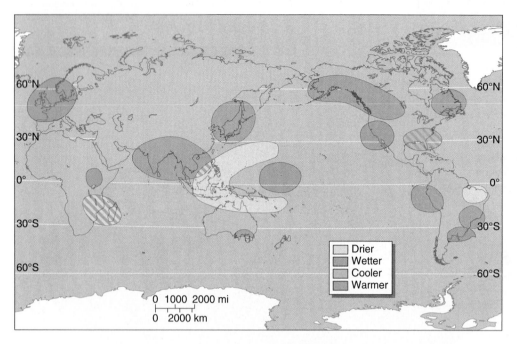

	Drier
	Wetter
	Cooler
	Warmer

0 1000 2000 mi
0   2000 km

El Niño events drastically alter the climate, even in many areas far from the Pacific Ocean. As a result, some areas are drier, some wetter, some cooler, and some warmer than usual. Typically, northern areas of the contiguous United States are warmer during the winter, whereas southern areas are cooler and wetter.

surface currents shift into a characteristic pattern. Upwelling of cold, nutrient-rich ocean currents along the Peruvian coast ceases, trade winds weaken, and a weak equatorial eastward current develops. Temperature and precipitation patterns change around the globe, bringing floods, droughts, heat waves, and cold outbreaks (**Figure 5.27**). El Niño cycles have historically occurred every three to eight years; however, since 2002, they have increased in frequency to every one to two years. Scientists are monitoring changes in the ENSO patterns carefully to determine what effects global climate change may have on this important weather phenomenon. Currently, U.S. climate change scientists hypothesize that the increased frequency is caused by the rise in sea-surface temperatures.

In contrast to El Niño is **La Niña**, in which Peruvian coastal upwelling is enhanced, trade winds strengthen, and cooler-than-average water is carried far westward in an equatorial plume. Together El Niño and La Niña have a dramatic impact on global weather patterns, affecting all aspects of life, including industry, tourism, and agriculture (see *Video Explorations*).

We have covered the basics of the global atmospheric and oceanic circulation patterns that frame the Earth's weather systems. Natural cycles that have existed for centuries are currently behaving in fascinating ways and indicate that we can expect increases in the frequency and intensity of weather events. Climate scientists are daily amassing information about our planet to better understand how the future will unfold. In Chapter 6, we will explain these weather systems in more detail, and in Chapter 7, we will address global climates and climate change.

Video Explorations

El Niño and La Niña

WILEY PLUS Video

Climatic disturbances caused by El Niño and La Niña affect the entire planet. View images of the phenomenon from a satellite measuring the height of the ocean. Observe changes in sea level and water temperatures that can create droughts, fires, monsoons, hurricanes, mudslides, and other weather disturbances.

CONCEPT CHECK STOP

1. **What** are the driving forces of deep currents as compared to surface currents?

2. **What** area of the globe has an energy surplus?

3. **What** effects does El Niño have on weather patterns in the United States?

Summary

1 Atmospheric Pressure 130

- The term **atmospheric pressure** describes the weight of air pressing on a unit of surface area. Atmospheric pressure is measured using a **barometer**, shown in the diagram.

Mercury barometer •
Figure 5.1

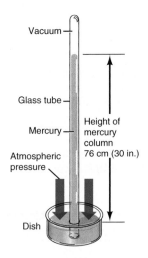

- Atmospheric pressure decreases rapidly as altitude increases.

2 Wind Speed and Direction 132

- **Isobars** indicate areas of equal pressure on a map.
- Air motion is produced by **pressure gradients**. Pressure gradients form when air is unevenly heated.
- The **Coriolis effect** deflects wind motion, as this diagram shows.

The Coriolis effect • Figure 5.7

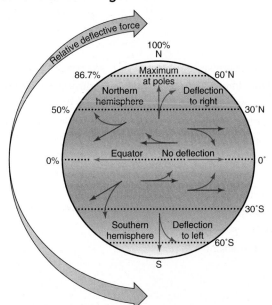

- At upper levels of the atmosphere, where the pressure gradient force and the Coriolis effect are in balance, **geostrophic winds** flow parallel to the isobars.
- Near the Earth's surface, friction slows down winds and decreases the Coriolis effect.
- **Cyclones** are centers of low pressure and convergence, and **anticyclones** are centers of high pressure and divergence.

3 Global Wind and Pressure Patterns 137

- **Hadley cells** develop because the equatorial and tropical regions receive more isolation than the higher latitudes.
- Hadley cell circulation drives the northeast and southeast trade winds, the convergence and lifting of air at the **intertropical convergence zone (ITCZ)**, and the sinking and divergence of air in the **subtropical high-pressure belts**, as shown in this diagram.

Subtropical high-pressure cells • Figure 5.15

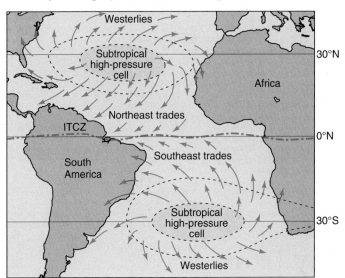

- The **polar front** is a zone of conflict between air masses with different pressure and temperature characteristics.
- **Jet streams** are high-speed airflows that occur in narrow zones at higher altitudes. The **polar-front jet stream** and **subtropical jet stream** are concentrated westerly wind streams with high wind speeds.
- **Jet stream disturbances** develop in upper-air westerlies, bringing cold, polar air equatorward and warmer tropical air poleward.

4 Local Winds 147

- Local winds are generated by local pressure gradients.

- **Sea breezes** and **land breezes** are caused by unequal heating and cooling of land and water surfaces.

- **Mountain breezes** and **valley breezes**, shown in this diagram, **Santa Ana winds**, and **Chinook winds** are other examples of local winds.

Valley and mountain breezes • Figure 5.21

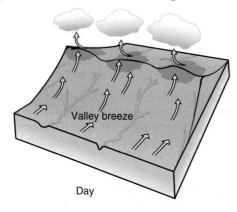

Valley breeze

Day

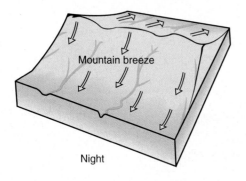

Mountain breeze

Night

5 Oceanic Circulation 150

- **Ocean currents** are driven at the surface by the frictional drag of winds. The trade winds push ocean water westward near the equator, and the westerlies push ocean waters eastward in the midlatitudes, forming huge **gyres** of rotating water.

- Deeper currents are driven by differences in temperature and salinity, which cause currents to rise or sink.

- Global circulation is driven by the transfer of absorbed solar energy from the equator, where there is an energy surplus, to the poles, where there is an energy deficit.

- Patterns of oceanic circulation undergo periodic changes that bring changes in weather patterns around the world. Equatorial oceans are warmer during **El Niño–Southern Oscillation (ENSO)** and cooler during **La Niña**, as shown in the diagram.

Cyclic patterns of oceanic circulation: El Niño–La Niña • Figure 5.26

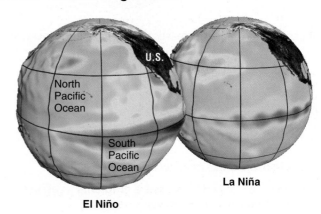

La Niña

El Niño

Key Terms

Critical and Creative Thinking Questions

1. How does air pressure change as you walk up a mountain, and what is the implication for the amount of oxygen you absorb with each breath?

2. What type of weather systems dominate the area where you live? Do you notice a consistent pattern of cyclones and anticyclones? Is your weather influenced by prevailing westerly winds? Are you subject to any local winds (such as mountain breezes or Santa Ana winds)?

3. An island directly on the equator might experience a particular annual pattern of wet and dry weather. There would be two rainy seasons (spring and fall) and two dry seasons (summer and winter). How could the migration of the ITCZ explain this pattern?

4. An airline pilot is planning a nonstop flight from Los Angeles to Sydney, Australia. What general wind conditions can the pilot expect to find in the upper atmosphere as the airplane travels? What jet streams will be encountered? Will they slow or speed the aircraft on its way?

5. Predict some changes in weather that might occur on land in South America caused by changes in ocean temperatures associated with El Niño conditions.

6. On the map, notice the latitude of Europe compared to that of North America. Why does England have a relatively warm climate compared to cities at similar latitudes in Canada (for example, Winnipeg)?

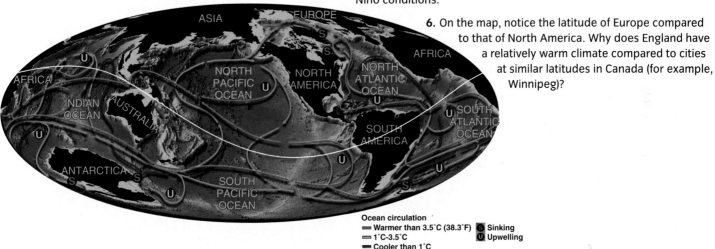

Ocean circulation
- Warmer than 3.5°C (38.3°F) **S** Sinking
- 1°C-3.5°C **U** Upwelling
- Cooler than 1°C

What is happening in this picture?

This striking image, acquired by a NASA satellite, shows land- and sea-surface temperature for a week in April. The eastern coast of North America is on the left, and the Atlantic Ocean is on the right. The color scale ranges from deep red (hot) to violet (cold). The Gulf Stream, bringing warm waters northward, stands out as a tongue of red and orange, extending from the Caribbean along the coasts of Florida and the Carolinas.

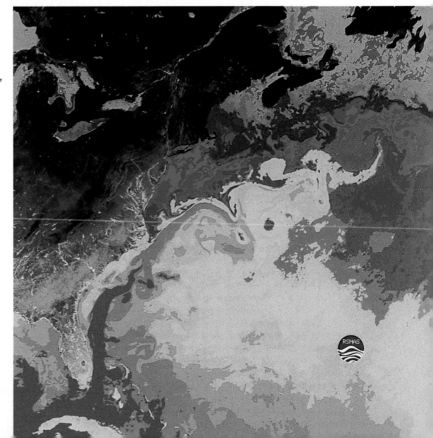

Think Critically

1. The Labrador Current can be seen in the purple to blue colors bringing cold water southward from Canada to Maine. What explanation can be given for the direction of this current, moving clockwise to the left, compared to the Gulf Stream Current, moving to the right off the North Carolina coast?

2. Looking at the land, note the violet colors of the northern Great Lakes. What might this indicate?

Self-Test

(Check your answers in the Appendix.)

1. A barometer is an instrument used to measure _____.

a. air pressure

b. hydraulic pressure

c. tectonic pressure at earthquake fault zones

d. glacial compression under ice sheets

2. The boiling point of water lowers at higher elevations because _____.

a. water is less dense at higher elevations

b. air is denser at higher elevations

c. air pressure is lower at higher elevations

d. upward water pressure is much greater

3. The map shows isobars over Oklahoma City and Nashville. Label the high- and low-pressure centers and draw an arrow showing the direction of the pressure gradient.

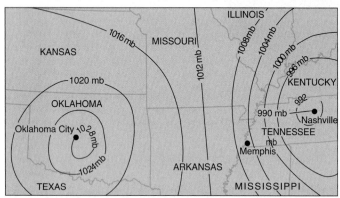

4. The Coriolis effect is _____.

a. a result of the Earth's rotation from east to west

b. a result of the Earth's rotation from the west to the east and causes objects to curve to the right in the northern hemisphere

c. a result of the Earth's rotation from the west to the east and causes objects to curve to the left in the northern hemisphere

d. unrelated to other physical phenomena on the Earth

5. At upper levels in the atmosphere, as a parcel of air moves in response to a pressure gradient, it is turned progressively to the right in the northern hemisphere until the pressure gradient and Coriolis forces balance to produce the _____.

a. geostrophic wind

b. tropospheric wind

c. upper-air westerlies

d. equatorial easterlies

6. Which force becomes relatively less important as altitude in the atmosphere increases?

a. Coriolis effect

b. frictional force

c. centrifugal force

d. pressure gradient

7. Add the following labels to the force arrows in the diagram: "Pressure gradient force," "Coriolis force," and "Friction."

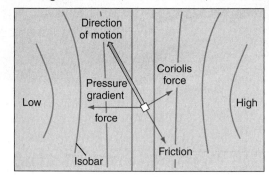

8. Cloudy and rainy weather is often associated with the inward and upward convergence of air within _____.

a. anticyclones

b. cold fronts

c. warm fronts

d. cyclones

9. In the Hadley cell convection loop, air rises at the ITCZ and descends in the _____.

a. polar high-pressure cells

b. subpolar low-pressure cells

c. subtropical high-pressure cells

d. polar low-pressure cells

10. The movement of the ITCZ and the change in the pressure pattern with the seasons create a reversing wind pattern in Asia known as the _____, where cool, dry airflow from the northeast dominates during the low-Sun season and warm, moist airflow from the southwest dominates during the high-Sun season.

a. northeast trades

b. southeast trades

c. monsoon

d. westerlies

11. Jet streams are _____.

a. narrow zones at a high altitude in which wind streams reach great speeds—faster than the speed of sound

b. narrow zones at a high altitude in which wind streams sometimes reach speeds over 150 miles per hour

c. rivers of wind that exist only along the equator and travel at fairly high velocities

d. rivers of wind that flow from east to west at high latitudes.

12. The diagram shows the formation of jet stream disturbances in the upper atmosphere. Label (a) the jet stream axis, (b) the regions of high pressure, (c) the regions of low pressure, and (d) the newly formed upper-air cyclones.

13. A land breeze generally occurs _____.

a. at night, when the land cools below the surface temperature of the sea

b. when strong winds blow in from the sea over the land

c. only during certain restricted seasons

d. during the day, when the land heats above the surface temperature of the sea

14. For water to sink as part of the thermohaline circulation, it must become _____ and _____.

a. more fresh, cooler

b. more fresh, warmer

c. more salty, cooler

d. more salty, warmer

15. El Niño is the name given to the warm phase of an effect associated with the reversal of _____.

a. ocean currents in the northern hemisphere

b. ocean currents in the southern ocean

c. wind flow patterns along the ITCZ in the Indian Ocean

d. ocean currents in the equatorial Pacific Ocean

THE PLANNER ✓

Review your Chapter Planner on the chapter opener and check off your completed work.

6 Weather Systems

In 2005, Hurricane Katrina laid waste to the city of New Orleans and much of the Louisiana and Mississippi Gulf coasts. The result was devastation unparalleled in U.S. disaster history. Total losses from the winds and flooding were estimated at more than $125 billion. The official death toll of more than 1800 only hints at the social devastation caused by the disaster.

New Orleans is a city that is particularly vulnerable to hurricane flooding. The city was built largely on the floodplain delta of the Mississippi River, and most of its land area has slowly sunk below sea level, as underlying river sediments have compacted through time. Levees protect the city from flooding from the Mississippi River and from nearby lakes and canals. But these were overtopped under the assault of Katrina's sustained winds, which struck early on August 29 and caused the Gulf of Mexico waters to surge upriver.

Over the next two days, the water level rose until 80% of the city was covered with water at depths of up to 6 meters (20 feet). Such events remind us that we remain at the mercy of the weather.

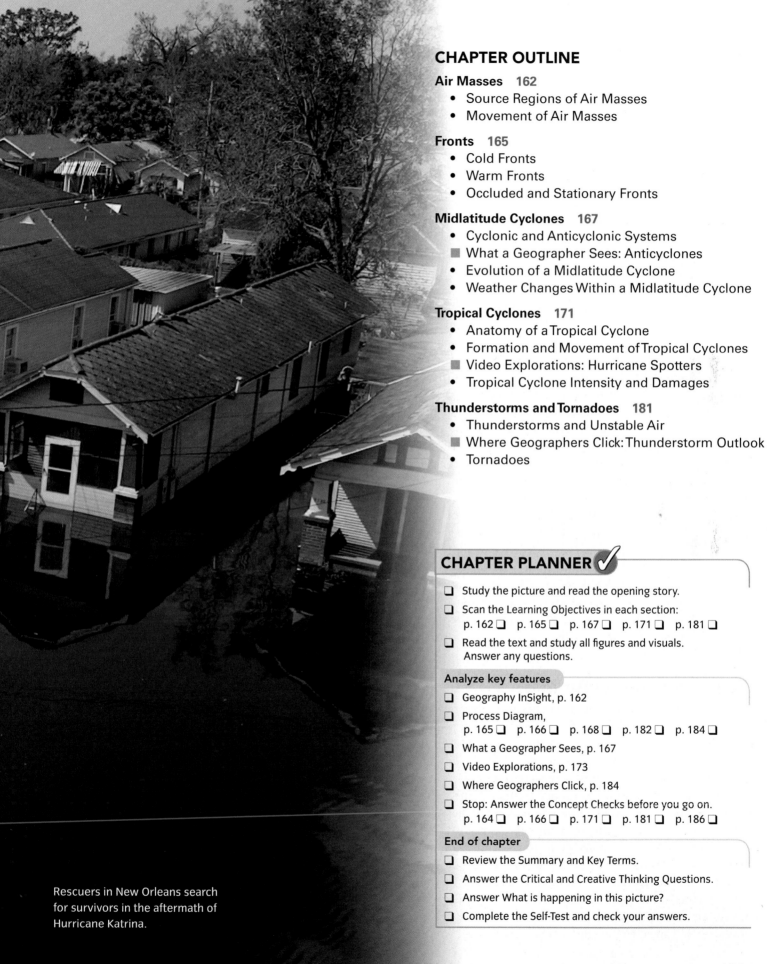

CHAPTER OUTLINE

CHAPTER PLANNER ✓

Rescuers in New Orleans search for survivors in the aftermath of Hurricane Katrina.

Air Masses

LEARNING OBJECTIVES

1. **Explain** how air masses are classified.
2. **Describe** the kinds of changes moving air masses bring to a region.

So far you have learned how energy is transferred around the globe through atmospheric and oceanic circulation. Now let's take a closer look at how that circulation leads to distinctive weather patterns.

Weather refers to the state of the atmosphere at one location and at one time. We typically determine weather in terms of factors such as temperature, moisture, precipitation, and winds. Weather conditions change from day to day or even from hour to hour, and they can be different from one neighborhood to the next. In contrast, climate considers the average temperature, moisture, precipitation, and winds for a broader region over longer periods of time.

Weather systems are patterns of atmospheric circulation that lead to distinctive weather events, such as cyclones or thunderstorms. Weather systems are often associated with the motion of **air masses**—large bodies of air with fairly uniform temperature and moisture characteristics. An air mass can be several thousand kilometers or miles across and can extend upward to the top of the troposphere. We characterize each air mass by its temperature and moisture characteristics.

> **air mass** An extensive body of air in which temperature and moisture characteristics are fairly uniform over a large area.

Source Regions of Air Masses

Air masses pick up their characteristics in source regions, where the air moves slowly or stagnates. The nature of the underlying surface—continent or ocean—usually determines the moisture content. We classify air masses by the latitude and surface type of their source regions (**Figure 6.1**).

Air mass temperature can range from extremely frigid temperatures for cA air masses to balmy and hot temperatures for mE. Humidity of an air mass can range from near zero for the cA air mass to as much as 99% for the mE air mass. Maritime equatorial air can hold about 200 times as much moisture as continental arctic air. We will be referring to the source region nomenclature of Figure 6.1 for air masses throughout upcoming chapters as we address the full spectrum of climate characteristics.

Geography InSight

Air masses form when large bodies of air acquire the temperature and moisture characteristics of the underlying surface conditions. In the center of this idealized globe is a continent, which produces continental (c) air masses, surrounded by oceans, which produce maritime (m) air masses.

a. The maritime polar (mP) air mass originates over midlatitude oceans. It holds less water vapor than the maritime tropical air mass, so the mP air mass yields only moderate precipitation. Much of this precipitation occurs over mountain ranges on the western coasts of continents.

b. The maritime tropical (mT) air mass and maritime equatorial (mE) air mass originate over warm oceans in the tropical and equatorial zones. They are quite similar in temperature and water vapor content. With their high values of specific humidity, both can produce heavy precipitation.

c. The continental arctic (cA) air mass and continental antarctic (cAA) air mass originate near or over the poles. They are extremely cold and hold almost no water vapor.

d. The continental polar (cP) air mass originates over North America and Eurasia in the subarctic zone. It has low specific humidity and is very cold in winter. This GOES (geostationary operational environmental satellite) image shows a cP air mass across the eastern United States, displacing moist air into Gulf and Atlantic water bodies.

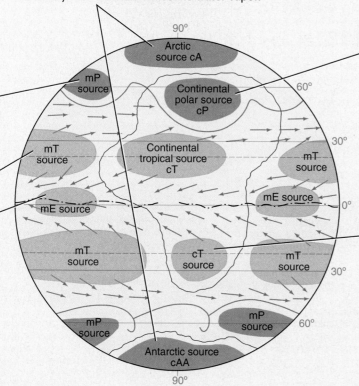

e. The continental tropical (cT) air mass has its source region over subtropical deserts of the continents. Although this air mass may have a substantial water vapor content, it tends to be stable and has low relative humidity when heated strongly during the daytime.

Air Mass	Symbol	Source region
Arctic	A	Arctic Ocean and fringing lands
Antarctic	AA	Antarctica
Polar	P	Continents and oceans, lat. 50–60° N and S
Tropical	T	Continents and oceans, lat. 20–35° N and S
Equatorial	E	Oceans close to equator

We also use two subdivisions to specify the type of underlying surface:

Air Mass	Symbol	Source region
Maritime	m	Oceans
Continental	c	Continents

Air masses are categorized with a two- or three-letter code, in which the first letter indicates the surface type and the second letter or letters indicate the general latitude of the source region. Combining the two types of labels produces a list of six important types of air masses: maritime equatorial (mE), maritime tropical (mT), continental tropical (cT), maritime polar (mP), continental arctic (cA), and continental antarctic (cAA).

Air masses acquire temperature and moisture characteristics in their source regions and then move across North America, where they have a strong influence on weather.

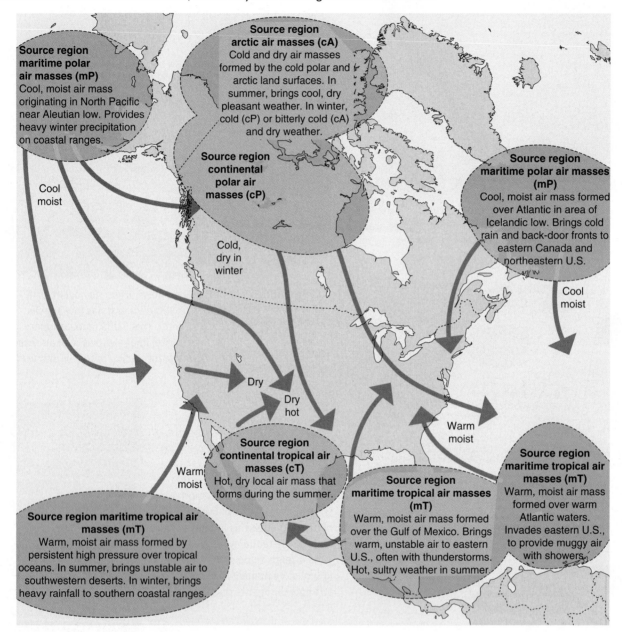

Movement of Air Masses

Sometimes air masses stay over their source region for a long time, in which case meteorologists characterize them by their pressure systems. But often they are driven out of their source region by pressure differences. Remember from Chapter 5 that air tends to move from a high-pressure area toward a low-pressure area.

When an air mass moves to a new area, its properties change due to the influence of the new surface environment. For example, the air mass may lose energy to a cold surface or take up water vapor from a warm ocean.

When an air mass arrives, it brings weather changes to its destination region, including precipitation, storms, and changes in temperature and humidity (**Figure 6.2**).

CONCEPT CHECK   STOP

---

1. **What** are the characteristics of a cP air mass?

2. **What** causes air masses to move away from their source region?

Fronts

LEARNING OBJECTIVES

1. **Explain** how frontal lifting occurs at a cold front.
2. **Describe** what happens at a warm front, in comparison to a cold front.
3. **Define** occluded and stationary fronts.

There is often a sharply defined boundary, or **front**, between a given air mass and neighboring air masses, differentiated by temperature and moisture content. A front serves as the leading edge of an air mass, like a bumper on a car. Meteorologists often explain weather dynamics by identifying the major fronts as they move across the landscape.

Cold Fronts

A **cold front** forms when a cold air mass invades a zone occupied by a warm air mass. Because the colder air mass is denser than the warmer air mass, it remains in contact with the ground.

> **front** The surface or boundary of contact between two different air masses.
>
> **cold front** A moving weather front along which a cold air mass moves underneath a warm air mass, causing the warm air mass to lift rapidly.

It typically moves more rapidly than the warm air mass. The forward speed of the front is based on a combination of its density and the steepness of the pressure gradient. As it moves forward, the cold air mass forces the warmer air mass to rise above it (**Figure 6.3**). Recall from Chapter 4 that this type of frontal lifting is one of the ways that clouds and precipitation form. As the warmer air mass rises, it cools, and when it cools to the condensation level, clouds form, and then rain.

As the cold front advances, characteristic weather patterns occur ahead of, behind, and at the front. Before the front arrives, temperatures are still warm, but high cirrus clouds form in the sky, winds shift, and barometric pressure drops, indicating that a front is arriving. As the front arrives, there is a sharp rise in pressure, and temperatures drop rapidly. Winds are gusty and variable. A dense wall of cumulonimbus clouds produces heavy precipitation, sometimes including hail and lightning. Once the front has passed through the area, the colder air mass brings cooler temperatures and higher pressures. The rain lets up, and clouds begin to clear.

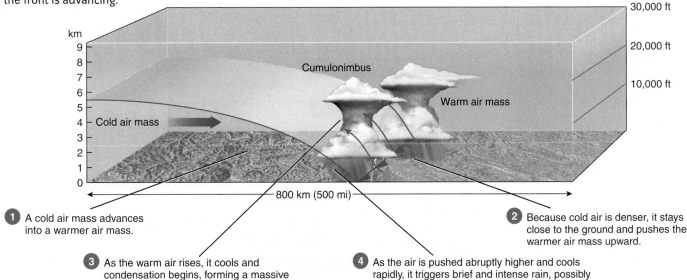

Frontal lifting at a cold front • Figure 6.3

THE PLANNER

A cold front forms where a cold air mass advances into an area occupied by a warm air mass. The cold front is indicated with a blue line, with triangles projecting in the direction the front is advancing.

1. A cold air mass advances into a warmer air mass.

2. Because cold air is denser, it stays close to the ground and pushes the warmer air mass upward.

3. As the warm air rises, it cools and condensation begins, forming a massive line of cumulus, or globular clouds (shown here greatly enlarged).

4. As the air is pushed abruptly higher and cools rapidly, it triggers brief and intense rain, possibly accompanied by thunderstorms, lightning, or hail.

Frontal lifting at a warm front • Figure 6.4

Warm fronts result when warm air advances toward a cold air mass. On a map, the warm front is indicated with a red line, with half circles projecting in the direction the front is advancing.

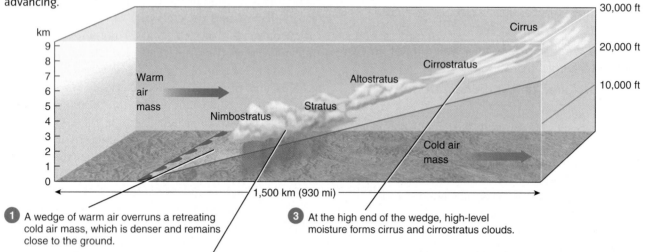

1 A wedge of warm air overruns a retreating cold air mass, which is denser and remains close to the ground.

3 At the high end of the wedge, high-level moisture forms cirrus and cirrostratus clouds.

2 Warm, moist air at the low end of the wedge is uplifted and forms nimbostratus clouds—large, dense, blanket-like clouds that often produce precipitation. If the warm air is stable, the precipitation is steady. If the warm air is unstable, heavy showers or thunderstorms can occur.

Ask Yourself

Which type of clouds first indicates an approaching warm front?

a. nimbostratus
b. stratus
c. altostratus
d. cirrus

Warm Fronts

> **warm front** A moving weather front along which a warm air mass slides over a cold air mass, producing stratiform clouds and precipitation.

A **warm front** develops when warm air moves into a region of colder air (**Figure 6.4**). Warm fronts involve the same process of frontal lifting, but because the lifting is more gradual, the weather patterns associated with warm fronts are more moderate.

Warm fronts also bring characteristic weather patterns ahead of, behind, and right at the front. As a warm front approaches, temperatures are still cool, but pressures begin to drop, and cirrus, cirrostratus, and altostratus clouds form in the sky. As the front arrives, temperatures rise, stratus clouds form, and light precipitation begins. Pressures are steady or falling, and winds are variable.

Occluded and Stationary Fronts

> **occluded front** A weather front along which a fast-moving cold front overtakes a warm front, forcing the warm air mass aloft.

Cold fronts normally move along the ground faster than warm fronts. So, when both types are in the same neighborhood, a cold front can overtake a warm front. The result is an **occluded front**. (*Occluded* means "closed" or "shut off.") The colder air of the fast-moving cold front remains next to the ground, forcing both the warm air and the less cold air ahead to rise over it. The warm air mass is lifted completely free from contact with the ground.

As an occluded front approaches, temperatures are cool, with light to moderate precipitation and falling pressures. As the front arrives, nimbostratus or cumulonimbus clouds form, temperatures fall, and precipitation increases. Pressure remains low but steady. Once the front has passed, temperatures are cold and pressures rise. Cumulus clouds begin to clear and light precipitation clears.

There is a fourth type of front, known as a **stationary front**, in which two air masses are in contact, but there is little or no relative motion between them. Stationary fronts often arise when a cold or warm front stalls and stops moving forward. This type of front is indicated on a map by a series of blue triangles and red semicircles pointing in opposite directions.

CONCEPT CHECK	STOP

1. **What** weather conditions are associated with a cold front?

2. **What** weather conditions are associated with a warm front?

3. **Why** do cold air masses tend to overtake warm air masses?

Midlatitude Cyclones

LEARNING OBJECTIVES

1. **Differentiate** cyclones and anticyclones.
2. **Describe** the origins and evolution of a midlatitude cyclone.
3. **Explain** the weather changes associated with midlatitude cyclones.

Air masses are set in motion by wind systems—typically, cyclones and anticyclones that involve masses of air moving in a spiral.

Cyclonic and Anticyclonic Systems

As we saw in Chapter 5, when the Coriolis force, the pressure gradient force, and friction interact, air spirals inward and converges in a **cyclone**, while air spirals outward and diverges in an **anticyclone**. Most types of cyclones and anticyclones are large features that move slowly across the Earth's surface, bringing changes in the weather as they move.

In a cyclone, the air converges and rises. As the air rises, the pressure of the surrounding air decreases, causing the air to expand and cool adiabatically. If the air is moist, this cooling can cause condensation or deposition, leading to precipitation.

In an anticyclone, the air diverges and descends. As the air descends, it encounters higher surrounding pressures, causing the air to warm adiabatically. Warming air does not lead to condensation, so skies are fair, except for occasional puffy cumulus clouds that sometimes develop in a moist surface air layer. For this reason, we often call anticyclones *fair-weather systems*. Toward the center of an anticyclone, winds are light and variable. We find anticyclones in the midlatitudes, typically associated with ridges or domes of clear, dry air that move eastward and equatorward. Geostationary satellites provide images for monitoring anticyclones around the globe (see *What a Geographer Sees*).

WHAT A GEOGRAPHER SEES

Anticyclones

You might appreciate these fair skies (Figure a) as a fortunate period of good weather, perfect for outdoor recreation. A geographer would attribute this fair weather to an anticyclone. This GOES geostationary satellite image shows a large anticyclone centered over eastern North America (Figure b). Anticyclones typically bring fair weather, and you can see that skies in this area are cloudless.

There is a boundary between clear sky and clouds running across the Gulf of Mexico and the Florida peninsula. A geographer would identify this line as the leading edge of a cool, dry air mass. Farther inland, you can also see some puffy clouds lying over the Appalachians at **A**. These are orographic cumulus clouds. At **B**, low clouds and fog, generated when moist air crosses the cool California ocean current, line up along the continental coastline. At **C**, the two lines of clouds intersecting in an inverted V-shape most likely mark a cold front on the left and a warm front on the right.

a.

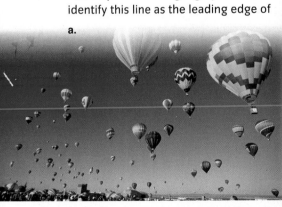

b.

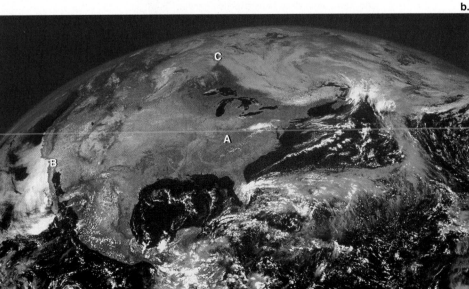

Life history of a midlatitude cyclone • Figure 6.5

In the northern hemisphere, a cyclone normally forms at a polar front and moves eastward as it develops, propelled by prevailing westerly winds aloft.

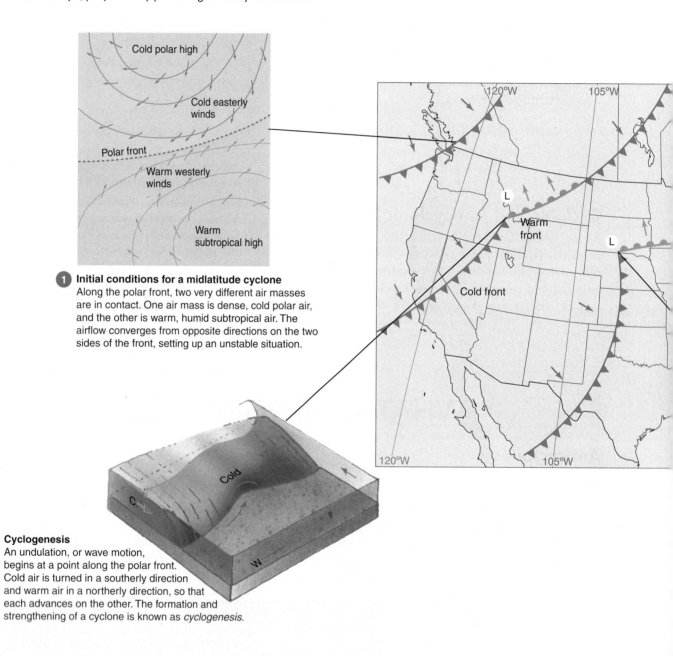

1 Initial conditions for a midlatitude cyclone
Along the polar front, two very different air masses are in contact. One air mass is dense, cold polar air, and the other is warm, humid subtropical air. The airflow converges from opposite directions on the two sides of the front, setting up an unstable situation.

2 Cyclogenesis
An undulation, or wave motion, begins at a point along the polar front. Cold air is turned in a southerly direction and warm air in a northerly direction, so that each advances on the other. The formation and strengthening of a cyclone is known as *cyclogenesis*.

Evolution of a Midlatitude Cyclone

A **midlatitude cyclone**, sometimes called a wave cyclone, is a dominant weather system in middle and high latitudes. It is a large inspiral of air that repeatedly forms, intensifies, and

> **midlatitude cyclone** A traveling cyclone of the midlatitudes that involves repeated interaction of cold and warm air masses along sharply defined fronts.

dissolves along the polar front (**Figure 6.5**). Midlatitude cyclones are large features, spanning 1000 kilometers (about 600 miles) or more. These are the *lows* that meteorologists show on weather maps that typically last 3 to 6 days.

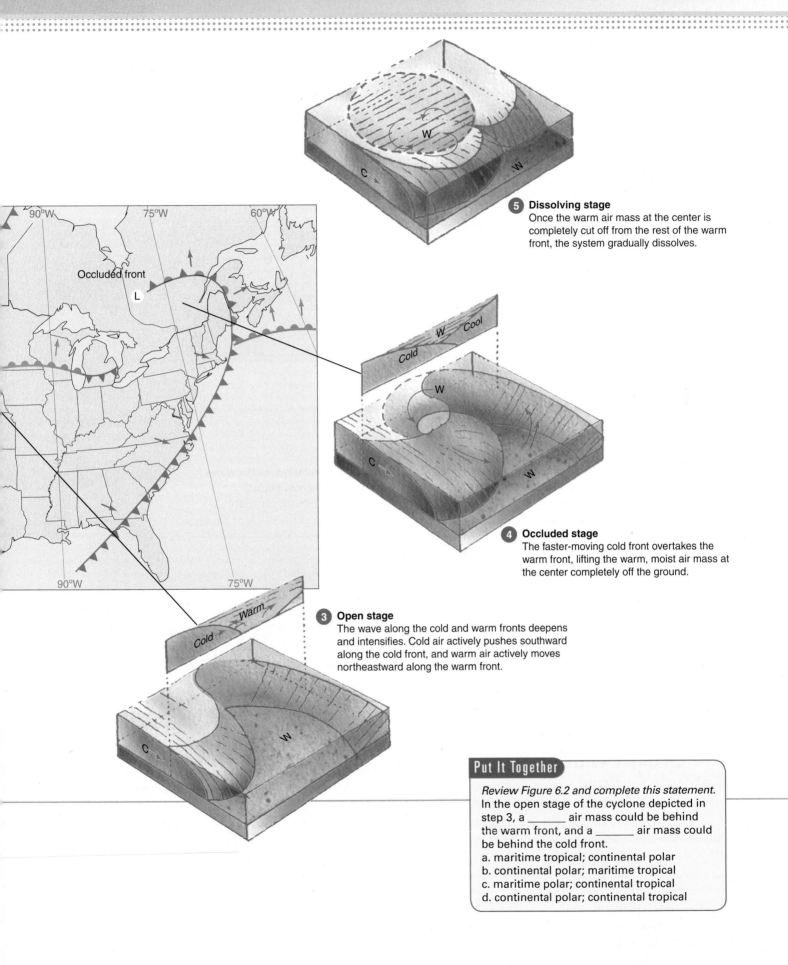

⑤ Dissolving stage
Once the warm air mass at the center is completely cut off from the rest of the warm front, the system gradually dissolves.

④ Occluded stage
The faster-moving cold front overtakes the warm front, lifting the warm, moist air mass at the center completely off the ground.

③ Open stage
The wave along the cold and warm fronts deepens and intensifies. Cold air actively pushes southward along the cold front, and warm air actively moves northeastward along the warm front.

Put It Together

Review Figure 6.2 and complete this statement. In the open stage of the cyclone depicted in step 3, a _____ air mass could be behind the warm front, and a _____ air mass could be behind the cold front.
a. maritime tropical; continental polar
b. continental polar; maritime tropical
c. maritime polar; continental tropical
d. continental polar; continental tropical

Characteristic weather patterns develop during the open and occluded stages of a midlatitude cyclone.

a. Open stage

There is a broad layer of clouds along the warm front, bringing steady rain. Along the cold front there are cumulonimbus clouds associated with thunderstorms. These yield heavy rains but only along a narrow belt.

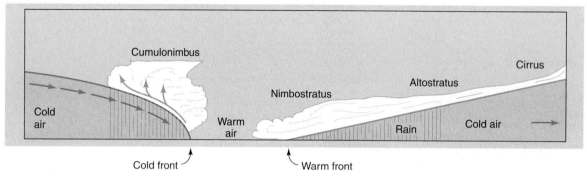

b. Occluded stage

The warm air mass is lifted well off the ground and yields heavy precipitation. As the warm air loses its moisture, the cold air continues to completely displace the now occluded warm air from the surface. Clouds in the upper occluded warm air eventually dissipate with the loss of moisture.

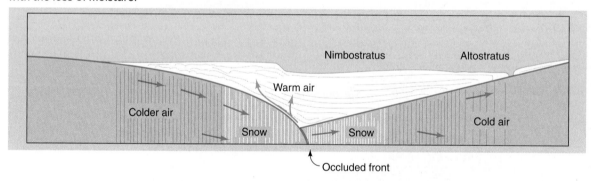

Weather Changes Within a Midlatitude Cyclone

Each stage in the evolution of a midlatitude cyclone is associated with distinctive weather patterns (**Figure 6.6**). As a midlatitude cyclone forms and gains strength, a cold front and a warm front begin to move, and precipitation begins. In the open stage, precipitation zones along the two fronts are strongly developed. The zone along the warm front is wider than the zone along the cold front. Finally, in the occluded stage, precipitation intensifies as the cold front overtakes the warm front.

Many cyclones are weak and pass overhead with little more than a period of cloud cover and light precipitation. But when the temperature and pressure difference between cold and warm fronts is significant, the pressure gradient is steep. As a result, the inspiraling motion is strong, and intense winds and heavy rain or snow can accompany the cyclone. In this case, we call the disturbance a **cyclonic storm** (**Figure 6.7**).

> **cyclonic storm**
> An intense weather disturbance within a moving cyclone that generates strong winds, cloudiness, and precipitation.

Nor'easter • Figure 6.7

Intense midlatitude cyclones can be powerful storms. Shown here is wave damage from a type of midlatitude cyclone called a *northeaster*, or *nor'easter*, which struck Saco, Maine, in April, 2007. A nor'easter forms off the Atlantic coast and heads north along the eastern seaboard, bringing strong winds from the northeast, along with high tides and high surf.

CONCEPT CHECK STOP

1. **Why** is precipitation associated with cyclones but not with anticyclones?

2. **Why** do midlatitude cyclones typically originate along the polar front?

3. **How** do weather patterns change as a cyclone moves from the open stage to the occluded stage?

Tropical Cyclones

LEARNING OBJECTIVES

1. **Describe** the characteristics of tropical cyclones.

2. **Explain** the conditions that contribute to the formation of tropical cyclones.

3. **Describe** the damages that occur when tropical cyclones make landfall.

So far, we have discussed weather systems of the midlatitudes and poleward. Weather systems of the tropical and equatorial zones show some basic differences from those of the midlatitudes. Upper-air winds are often weak, so air mass movement is slow and gradual. Air masses are warm and moist, and different air masses tend to have similar characteristics, so there aren't such clearly defined fronts. And without these fronts, there are no large, intense wave cyclones like those in the midlatitudes. On the other hand, the high moisture content leads to intense convectional activity in low-latitude maritime air masses, capable of fueling powerful and destructive **tropical cyclones**. Because these air masses are very moist, only slight convergence and uplifting are needed to trigger precipitation.

> **tropical cyclone**
> An intense traveling cyclone of tropical and subtropical latitudes, accompanied by high winds and heavy rainfall.

Anatomy of a Tropical Cyclone

Tropical cyclones are known as **hurricanes** in the western hemisphere, **typhoons** in the western Pacific off the coast of Asia, and cyclones in the Indian Ocean. This type of storm develops over oceans in 8° to 15° N and S latitudes but not closer to the equator.

An intense tropical cyclone is an almost circular storm center of extremely low pressure. Because of the very strong pressure gradient, winds spiral inward at high speed. Convergence, uplift, and condensation are intense, producing very heavy rainfall. The latent heat released by condensation warms the converging air, causing even more

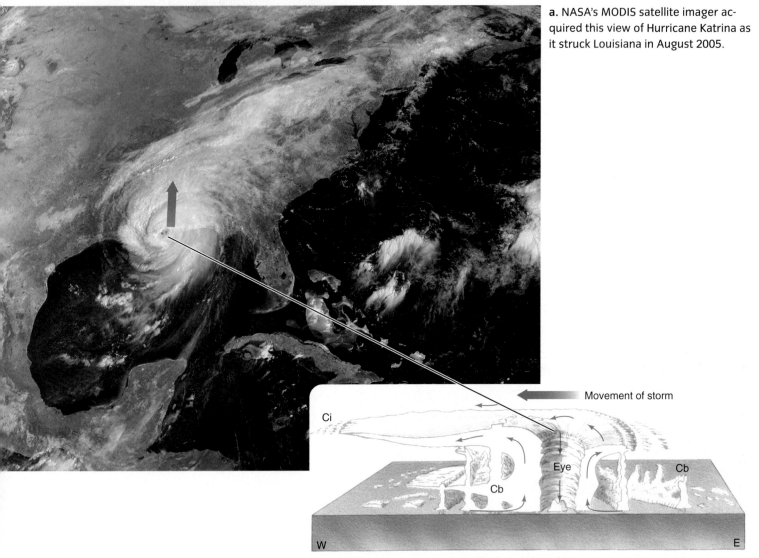

a. NASA's MODIS satellite imager acquired this view of Hurricane Katrina as it struck Louisiana in August 2005.

b. In this schematic diagram, cumulonimbus (Cb) clouds in concentric rings rise through dense stratiform clouds. Cirrus (Ci) clouds fringe out ahead of the storm. The width of the diagram represents about 1000 km (about 600 mi).

uplift and condensation, fueling the growth of the storm. The storm's diameter may be 150 to 500 km (about 100 to 300 mi). Wind speeds can range from 30 to 50 m/s (about 65 to 135 mi/hr) and are sometimes much higher. Barometric pressure in the storm center commonly falls to 950 mb (28.1 in. Hg) or lower.

A characteristic feature of a well-developed tropical cyclone is its central **eye**, in which clear skies and calm winds prevail (**Figure 6.8**). The eye is a cloud-free vortex produced by the intense spiraling of the storm. In the eye, air descends from high altitudes and is adiabatically warmed. As the eye passes over a site, calm prevails, and the sky clears. It may take about half an hour for the eye to pass, after which the storm strikes with renewed ferocity,

but with winds blowing from the opposite direction. Wind speeds are highest along the cloud wall of the eye.

Formation and Movement of Tropical Cyclones

Scientists do not know the exact formation mechanism, but they do know that in order for a tropical cyclone to form, four conditions need to be in place: low pressure, a weak Coriolis force, high humidity, and warm temperatures. Understanding how each of these factors contributes to the formation of tropical storms helps meteorologists predict where and when such storms are likely to occur and what paths they will take (see *Video Explorations*).

Video Explorations

Hurricane Spotters

Scientists are using new satellite sensors, including the microwave radar on NASA's SeaWinds satellite, to locate tropical storms with 4 days of advance warning. This satellite's ability to detect winds on the sea surface has added to the National Hurricane Center's arsenal of science tools for predicting storm activity. This video shows how early results from the satellite have caused meteorologists to reconsider their previous understanding of the process of cyclone formation.

Low pressure Typically, a tropical cyclone originates as a slow-moving band of low pressure, which then intensifies and grows into a deep, circular low. Tropical cyclones often begin as an **easterly wave**—a slowly moving trough of low pressure within the belt of tropical easterlies (trades). These waves occur in latitudes 5° to 30° N and S over oceans, but not over the equator itself (**Figure 6.9**).

A weaker Coriolis force At the latitude of tropical cyclone development, the Coriolis force is weaker than in the midlatitudes. Recall from Chapter 5 that at upper levels in the atmosphere, the Coriolis force balances the pressure gradient around a cyclone, and the magnitude of this balancing Coriolis force is proportional to the wind speed. With a weaker Coriolis force, the winds must move faster to balance the pressure gradient. So given a midlatitude cyclone and a tropical cyclone with equal central low pressures, winds around the tropical cyclone will be faster.

At the equator, the Coriolis force disappears. This explains why tropical cyclones do not form at latitudes between about 5° N and 5° S. Without a Coriolis force to cause winds to turn around a low pressure center, they do not converge to form cyclones.

High humidity Tropical cyclones always form over oceans because the air over oceans has high water vapor content. In the western hemisphere, hurricanes originate in the Atlantic, off the west coast of Africa; in the Caribbean Sea; or off the west coast of Mexico. Curiously, tropical cyclones almost never form in the South Atlantic or southeast Pacific regions. As a result, South America is not threatened by these severe storms. In the Indian Ocean, cyclones originate both north and south of the equator, moving north and east to strike India, Pakistan, and Bangladesh, as well as south and west to strike the eastern coasts of Africa and Madagascar. Typhoons of the western Pacific also form both north and south of the equator, moving into northern Australia, Southeast Asia, China, and Japan.

An easterly wave passing over the West Indies • Figure 6.9

There is a zone of weak low pressure at the surface, under the axis of the wave. The wave travels westward at a rate of about 100 km (62 mi) per day. Surface air flow converges on the eastern, or rear, side of the wave axis. This convergence causes the moist air to be lifted, producing scattered showers and thunderstorms.

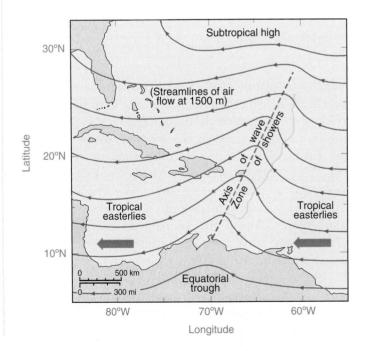

Ocean temperatures and cyclone formation • Figure 6.10

Tropical cyclones are most likely to occur in areas of greatest heating. Dotted lines show where the sea-surface temperature can be greater than 26.5°C (81°F). Cyclones last until they move over cooler waters or hit land. When a cyclone encounters warmer waters, as Hurricane Katrina did in 2005 in the Gulf of Mexico, it picks up energy and intensifies.

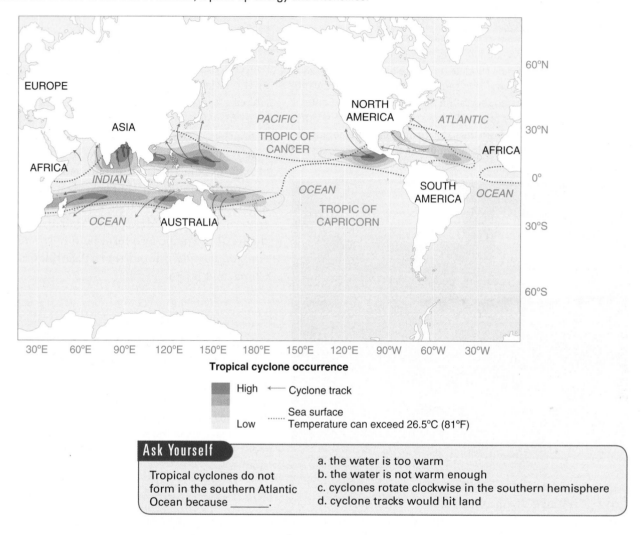

Tropical cyclone occurrence

High ← Cyclone track

Low ⋯⋯ Sea surface Temperature can exceed 26.5°C (81°F)

Ask Yourself

Tropical cyclones do not form in the southern Atlantic Ocean because _____.

a. the water is too warm
b. the water is not warm enough
c. cyclones rotate clockwise in the southern hemisphere
d. cyclone tracks would hit land

Warm temperatures High sea-surface temperatures, over 26.5°C (81°F), are required for tropical cyclones to form (**Figure 6.10**). As a result, tropical cyclones occur only during certain seasons, when ocean temperatures are warmest. For hurricanes of the North Atlantic, the season runs from May through November, with maximum frequency in late summer or early autumn. In the southern hemisphere, the season is roughly the opposite. These periods follow the annual migrations of the intertropical convergence zone (ITCZ) to the north and south with the seasons.

Even slight changes in sea-surface temperatures can lead to an increased number of cyclones. Since 1995, sea-surface temperatures in the northern Atlantic from August to October have averaged about 0.5°C (about 1°F) warmer than during 1970–1994, providing more latent heat to fuel cyclones. The result has been the present period of increased hurricane activity (**Figure 6.11**). By 2005, the average number of named storms had increased to 13 per year, compared to 8.6 during 1970–1994. The number of hurricanes had increased from 5 to 7.7 per year, with 3.6 major hurricanes, compared to 1.5 in prior years. In 2010, 19 storms were named, with a total of 12 hurricanes. This trend of increasing storms and hurricanes follows the trend of increasing sea-surface temperatures. From these trends, climate scientists hypothesize that increased global warming can be expected to fuel increased cyclone activity and intensity.

Movement of tropical cyclones Once cyclogenesis begins, the storm moves westward through the trade-wind belt, often intensifying as it travels. It can then curve north-west, north, and northeast, steered by winds aloft. Tropical cyclones can penetrate well into the midlatitudes. Along the southeastern coast of the United States, many tropical cyclones approach from the east and then turn and head back out to sea.

South Florida is particularly at the mercy of Atlantic and Gulf hurricanes. They can attack the peninsula from both the east and west sides or pass across the peninsula. The location of warm waters on both the Gulf and Atlantic sides of the peninsula provides the energy to maintain hurricane intensity. In 1992, Hurricane Andrew struck the east coast of Florida near Miami. It was the second-most-damaging storm to occur in the United States, claiming dozens of lives and more than $35 billion in property damage, measured in today's dollars. In 2004, four major hurricanes affected Florida—Charley, Frances, Ivan, and Jeanne. Taken together, the storms destroyed more than 25,000 homes in Florida, with another 40,000 homes sustaining major damage.

Meteorologists now use satellite images to track cyclones, along with a host of aircraft and unstaffed hurricane sounders and sea buoys (**Figure 6.12** on the next page). Cyclones are often easy to identify by their distinctive pattern of inspiraling bands of clouds and a clear central eye. The cyclone eye provides clues to meteorologists regarding the intensity of the storm and whether it will reach hurricane status.

For convenience, tropical cyclones are named as they are tracked by weather forecasters. Male and female

Tropical cyclones and rising sea-surface temperatures • Figure 6.11

These figures show hurricane tracks for 1985–1994 (blue) and for 1995–2004 (red).
The thicker the line of the track, the more intense the hurricane.

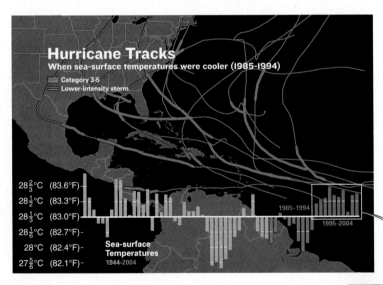

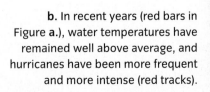

b. In recent years (red bars in Figure **a.**), water temperatures have remained well above average, and hurricanes have been more frequent and more intense (red tracks).

a. The bar chart shows sea-surface temperatures in the western Atlantic for both periods, as well as sea-surface temperatures going back to 1944. Bars above the line are warmer than average and bars below the line are cooler. In 1985–1994 (blue bars), cooler water temperature prevailed, and there were fewer hurricanes (blue tracks).

Conditions inside a hurricane • Figure 6.12

To understand how hurricanes work and to improve forecasts, researchers need detailed information from the heart of the storms. During the 2005 hurricane season, the most active on record, scientists investigated hurricanes from top to bottom (this one shown in cross section) with satellites, airplanes, and new kinds of instrumented probes.

DATA
GATHERERS
FROM TOP
TO BOTTOM

▲
Satellites
500 to
22,000 mi

Ⓐ In space
Satellites track a storm's shape and position and use heat-sensing infrared instruments to map its eye and most powerful updrafts.

Ⓑ In the storm
The Hurricane Rainband and Intensity Change Experiment (RAINEX) was the first to send NOAA and National Science Foundation aircraft on simultaneous flights through hurricanes, deploying three P-3 aircraft with Doppler radar through hurricane rain bands. The data showed how these rings of thunderstorms interact with the eyewall, where a hurricane's winds are strongest, to intensify or weaken a storm.

Dropsondes
Dropped from planes, these probes relay measurements of pressure, wind speed and direction, humidity, and temperature as they fall to the sea.

▲
G-IV jet aircraft
42,000 ft

P-3 propeller aircraft
8,000, 12,000, and 14,000 ft
▼

Ⓒ Close to the water
In September 2005, an unmanned aircraft called Aerosonde flew into the core of tropical storm Ophelia just 1,200 feet above the waves, monitoring how heat from the ocean was transferred to the storm.

Rain band   *Rain band*   *Eyewall*

Aerosonde
1,200 ft
▼

Hurricane's path

Storm surge

Ⓓ In the ocean
Hurricanes Katrina and Rita strengthened dramatically when they crossed the Loop Current in the Gulf of Mexico. Ocean probes showed that the Loop Current's warmth extended to a depth of 300 feet, increasing the supply of heat to the storms. As a hurricane nears land, its winds pile up a destructive hill of water called storm surge.

Deeper layers of warm water

Ocean probes
Dropped from planes, these probes then sink, measuring conditions to a depth of over 3,000 feet.

Cool water

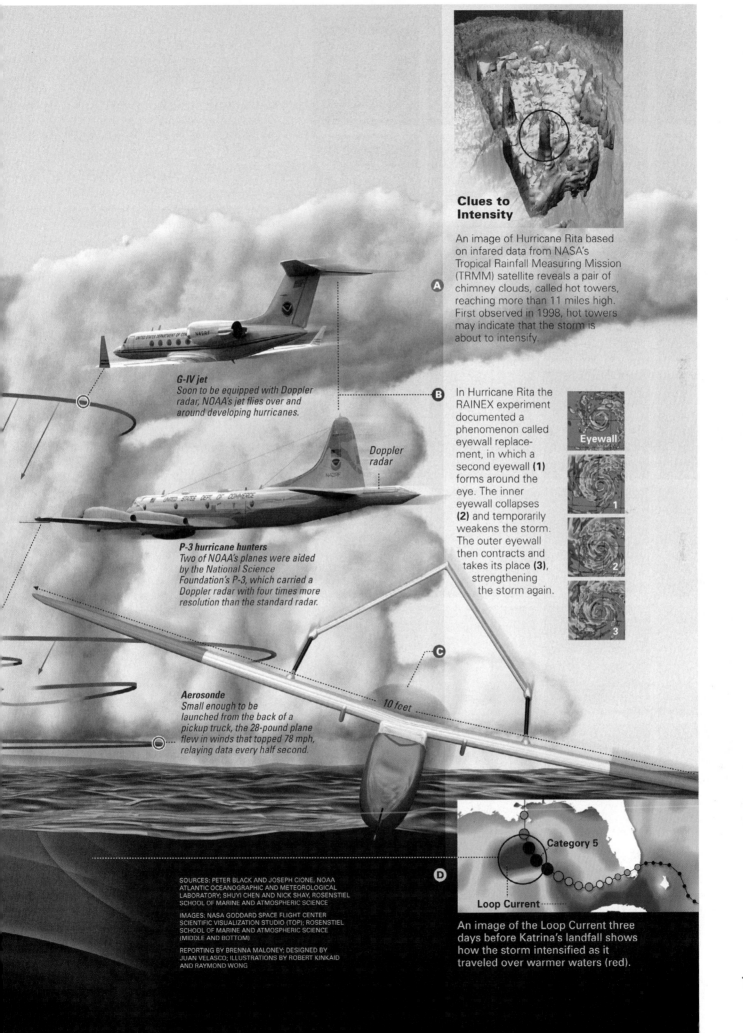

Clues to Intensity

An image of Hurricane Rita based on infared data from NASA's Tropical Rainfall Measuring Mission (TRMM) satellite reveals a pair of chimney clouds, called hot towers, reaching more than 11 miles high. First observed in 1998, hot towers may indicate that the storm is about to intensify.

A

B

In Hurricane Rita the RAINEX experiment documented a phenomenon called eyewall replacement, in which a second eyewall **(1)** forms around the eye. The inner eyewall collapses **(2)** and temporarily weakens the storm. The outer eyewall then contracts and takes its place **(3)**, strengthening the storm again.

Eyewall

1

2

3

G-IV jet
Soon to be equipped with Doppler radar, NOAA's jet flies over and around developing hurricanes.

Doppler radar

P-3 hurricane hunters
Two of NOAA's planes were aided by the National Science Foundation's P-3, which carried a Doppler radar with four times more resolution than the standard radar.

C

Aerosonde
Small enough to be launched from the back of a pickup truck, the 28-pound plane flew in winds that topped 78 mph, relaying data every half second.

10 feet

D

Category 5

Loop Current

An image of the Loop Current three days before Katrina's landfall shows how the storm intensified as it traveled over warmer waters (red).

SOURCES: PETER BLACK AND JOSEPH CIONE, NOAA ATLANTIC OCEANOGRAPHIC AND METEOROLOGICAL LABORATORY; SHUYI CHEN AND NICK SHAY, ROSENSTIEL SCHOOL OF MARINE AND ATMOSPHERIC SCIENCE

IMAGES: NASA GODDARD SPACE FLIGHT CENTER SCIENTIFIC VISUALIZATION STUDIO (TOP); ROSENSTIEL SCHOOL OF MARINE AND ATMOSPHERIC SCIENCE (MIDDLE AND BOTTOM)

REPORTING BY BRENNA MALONEY; DESIGNED BY JUAN VELASCO; ILLUSTRATIONS BY ROBERT KINKAID AND RAYMOND WONG

Hurricanes of 2005 • Figure 6.13

The year 2005 was the most active year on record for Atlantic hurricanes. Shown here are the 27 named storms that occurred during the official season, June 1 to November 30. A twenty-eighth storm, Zeta, was observed during December and into January 2006. Katrina was the costliest Atlantic storm on record. Wilma was the most intense, at one point observed with a central pressure of 882 mb (26.05 in. Hg) and winds of 83 m/s (185 mi/hr).

names are alternated in an alphabetical sequence renewed each season. Different sets of names are used within distinct regions, such as the western Atlantic, western Pacific, and Australian regions. Names are reused, but the names of infamous storms that cause significant damage or destruction are retired from further use. **Figure 6.13** provides a gallery of Atlantic hurricanes of 2005.

Tropical Cyclone Intensity and Damages

Tropical cyclones can be tremendously destructive. Islands and coasts feel the full force of the high winds and flooding as tropical cyclones move onshore. The devastation caused by cyclones depends on conditions in the area where they hit land and also on their intensity.

Cyclone intensity Cyclone intensity is evaluated on the **Saffir-Simpson scale** (**Figure 6.14**). Cyclone intensity can change as the cyclone moves along its path, losing intensity over cooler water or land or gaining intensity as it passes over warmer water.

For example, Hurricane Katrina, which originated southeast of The Bahamas, crossed Florida as a moderate category 1 storm. As it moved into the Gulf of Mexico, it gained energy from the warm waters of the Gulf and intensified to a category 5 storm. Rapid changes in the intensity of a storm can make it difficult for community leaders to prepare for disaster and to convince local populations to evacuate.

The degree of damage and loss of life caused by cyclones cannot be predicted based solely on the category of the storm. The topography of the area, local building codes, and the effectiveness of warnings

1. Minimal damage
Winds 33–42 m/s (74–95 mi/hr)
Storm surge 1.2–1.6 m (4–5 ft)

2. Moderate damage
Winds 43–49 m/s (96–110 mi/hr)
Storm surge 1.7–2.6 m (6–8 ft)

Small trees down,
roof damage

3. Extensive damage
Winds 50–58 m/s (111–130 mi/hr)
Storm surge 2.7–3.8 m (9–12 ft)

Moderate to heavy
damage to homes,
many trees down

4. Extreme damage
Winds 59–69 m/s (131–155 mi/hr)
Storm surge 3.9–5.4 m (13–18 ft)

Major damage
to all structures

5. Catastrophic damage
Winds > 70 m/s (155 mi/hr)
Storm surge > 5.5 m (18 ft)

Severe damage
to all structures

Saffir-Simpson scale of tropical cyclone intensity • Figure 6.14

The Saffir-Simpson scale ranks cyclone intensity on the basis of wind speed, central
pressure, and storm surge, from category 1 (weak) to 5 (devastating).

and evacuations all factor into the amount of devastation caused by a storm. Storm damages are typically caused by high winds and flooding related both to intense precipitation and to storm surges. Removal of the coastal vegetation and mangroves that serve as natural buffers against storm energy exposes coastlines to even greater storm damage.

Precipitation Tropical cyclones also produce a large amount of rainfall, which can produce freshwater flooding, raising rivers and streams out of their banks. On steep slopes, soil saturation and high winds can topple trees and produce disastrous earthflows and landslides (**Figure 6.15**). Central America was severely affected when Hurricane Mitch stalled offshore for a few days in 1998, producing 190 cm (75 inches) of rainfall and killing more than 19,000 people in mudslides and floods.

Earthflow • Figure 6.15

This entire neighborhood in Tegucigalpa, Honduras, slumped downhill (moving from left to right in this photo) during Hurricane Mitch (1998). Torrential rains saturated the soil, allowing gravity to take effect.

Storm surges Because the atmospheric pressure at the center of a cyclone is so low, sea level rises toward the center of the storm. High winds create a damaging surf and push water toward the coast, raising sea level even higher. Saltwater waves attack the shore at points far inland of the normal tidal range.

Low pressure, winds, and the underwater shape of a bay floor can combine to produce a sudden rise of water level, known as a **storm surge**, that carries ocean water and surf far inland (**Figure 6.16**). At Galveston, Texas, in 1900, a sudden storm surge generated by a severe hurricane flooded the low coastal city and drowned about 6000 people—the largest death toll in a natural disaster yet experienced within the United States.

> **storm surge** A rapid rise of coastal water level accompanying the onshore arrival of a tropical cyclone.

The storm surge was responsible for much of the flooding and destruction along the Gulf Coast during and after Hurricane Katrina in 2005, mentioned in the chapter opener. The city of New Orleans was particularly hard hit by flooding because of its vulnerable low-lying position at or below sea level and the reliance on a decaying levee system to protect the city from flood waters from both the ocean and surrounding lakes. Following the storm, as levees were overtopped and canal walls failed, 80% of the city was covered with water at depths of up to 6 m (20 ft) (**Figure 6.17**). The Gulf coasts of Mississippi and Louisiana were also hard hit, with a coastal storm surge bringing salt water as high as 8 m (25 ft), penetrating from 10 to 20 km (6–12 mi) inland.

Storm surge • Figure 6.16

By raising the sea level through a combination of high winds and low surface pressures, storm surge can inundate low-lying areas and subject them to heavy surf.

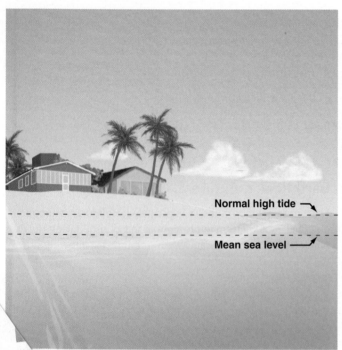

Normal conditions
Structures above high tide are not subject damage from high tide are not ing waves. break-

Storm surge
A storm surge combined with a high tide lifts the sea level so that structures are subjected to continuous pounding of heavy surf.

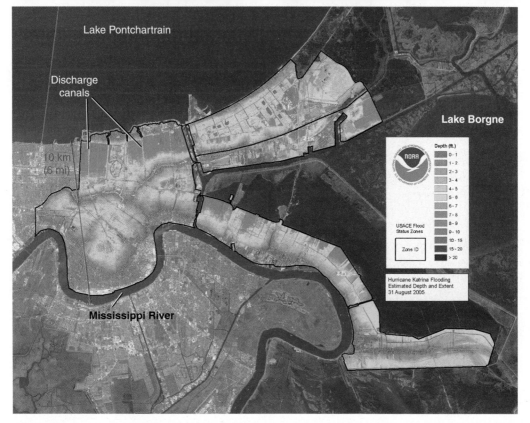

Flooding in New Orleans • Figure 6.17

This image shows the extent of flooding during and following the passage of Hurricane Katrina. The depth of the flooding can be gauged by the scale on the right.

CONCEPT CHECK   STOP

1. **What** are weather conditions inside the eye of a tropical cyclone?

2. **What** effect is global warming likely to have on the formation of tropical cyclones?

3. **What** causes tropical cyclones to gain or lose intensity as they travel?

Thunderstorms and Tornadoes

LEARNING OBJECTIVES

1. **Explain** the conditions that cause thunderstorms.

2. **Describe** the formation of a tornado.

he same principles that help us understand the formation of hurricanes can explain the occurrence of more local weather events, such as thunderstorms and tornadoes. Although thunderstorms and tornadoes are smaller in scale and don't usually affect such broad regions as tropical cyclones, they occur more frequently and can cause devastating damage.

Thunderstorms and Unstable Air

With intense heating of the ground during sunny summer days, a warm air layer can form in the lower atmosphere. The air temperature in this layer decreases more strongly than usual with elevation, which makes the air unstable. Under these conditions, a heated parcel of rising air stays warmer (and less dense) than the surrounding air, even though the parcel is cooled at the dry adiabatic rate. This

PROCESS DIAGRAM

THE PLANNER

Each stage in the development of a thunderstorm has
characteristic vertical winds and precipitation.

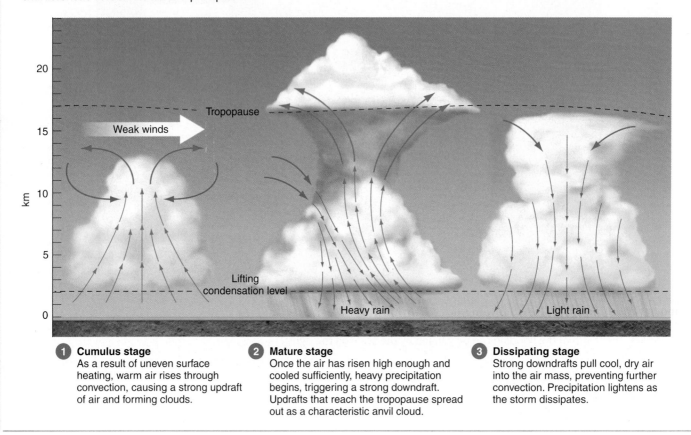

1 Cumulus stage
As a result of uneven surface
heating, warm air rises through
convection, causing a strong updraft
of air and forming clouds.

2 Mature stage
Once the air has risen high enough and
cooled sufficiently, heavy precipitation
begins, triggering a strong downdraft.
Updrafts that reach the tropopause spread
out as a characteristic anvil cloud.

3 Dissipating stage
Strong downdrafts pull cool, dry air
into the air mass, preventing further
convection. Precipitation lightens as
the storm dissipates.

means that it continues to rise until condensation occurs.
And because it cools more slowly as it rises with condensation occurring (moist adiabatic rate), it will stay even
warmer than the surrounding air.

The result is that once condensation starts to occur,
the rising motion continues even
more strongly, producing towering
columns of cumulus clouds with
strong updrafts. As the process continues, dense cumulonimbus clouds
or **thunderstorms**, including heavy
rain, lightning, and even hail or tornadoes, can form (**Figure 6.18**).

thunderstorm An
intense local storm
associated with a tall,
dense cumulonim-
bus cloud in which
there are very strong
updrafts of air.

Severe thunderstorms In a severe thunderstorm, large
amounts of cooler, drier surrounding air enter the cloud from
the upwind side, creating a strong downdraft. The advancing
air pushes large volumes of moist surface air upward, feeding
the convection. As the updraft reaches higher levels, the rate
of rise slows. At such high altitudes, the winds are typically strong, dragging the cloud top downwind and giving

the thunderstorm cloud its distinctive shape—resembling
a blacksmith's anvil (**Figure 6.19**).

An essential component of a severe thunderstorm
is **wind shear**—a change in wind velocity with height—
that keeps the cool, dry air entering from the upwind side
while the warm, moist, rising air stays on the downwind
side of the cell.

The heavy rains and powerful wind gusts that accompany thunderstorms are sometimes violent enough to
topple trees and raise the roofs of weak buildings.

Lightning Thunderstorms can produce lightning and
hail. As thunderstorms create strong updrafts and downdrafts, clouds build up tremendous static electric charges
that can trigger lightning. The exact mechanism is not
completely understood, but generally, negative charges
are carried downward, and positive charges are carried
upward. When the separation of charges reaches a thresh-
old, the molecules and atoms of the atmosphere become
conductive, and electric current flows along a narrow
path. The atmosphere is heated explosively to produce

Anvil cloud • Figure 6.19

This photo, taken by space shuttle astronauts, shows an anvil cloud approximately 50 km (30 mi) wide over Africa.

Think Critically

In which direction is this thunderstorm moving, and how do you know?

light and create a shock wave in the air that we hear as thunder (**Figure 6.20**). Most lightning occurs within the storm cloud; only about 20% of lightning strikes connect to the Earth's surface.

Forecasting thunderstorms

Hot summer air masses in the central and southeastern United States are often unstable, making thunderstorms very common in these regions during the summer. Severe thunderstorms can bring lightning, hail, and other hazardous conditions, so forecasters continually work to improve their models for predicting storms. Forecasters at the National Oceanic and Atmospheric Administration (NOAA) closely monitor changes in the atmosphere by studying current and past data from satellite imagery, radars, surface weather stations, weather balloons, wind profilers, and lightning-detection networks.

The severe weather forecast process begins with an assessment of where both severe and non-severe thunderstorms are expected to occur around the United

Lightning • Figure 6.20

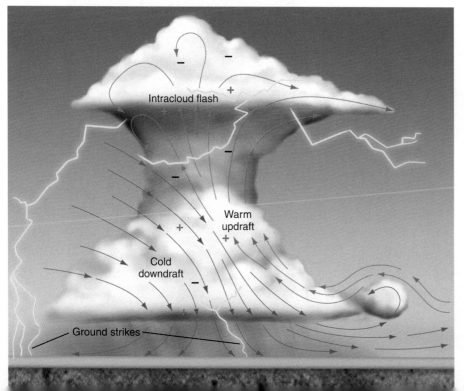

Turbulent upward and downward motions (blue and red arrows) of air, water, and ice pellets in a thunderstorm create areas of positive and negative static charge within the cloud that are discharged by lightning. Most lightning occurs within clouds. In a ground strike, a negative charge in the cloud above generates a positive charge in the ground below by attraction. When the electrical potential between the cloud and the ground is large enough, a lightning strike occurs.

Thunderstorms and Tornadoes **183**

Where Geographers CLICK

Thunderstorm Outlook

www.spc.noaa.gov/products/exper/enhtstm/

Visit this Web site to explore NOAA's experimental enhanced-resolution thunderstorm probability maps, which show the areas at risk of thunderstorms and the probability that thunderstorms will occur, given as a percentage. In this example, an approaching cold front brought three tornadoes to Missouri, in the region with a 70% risk.

States, and areas are ranked according to the risk. Many local National Weather Service (NWS) offices make emergency staffing decisions based on such assessments. As time progresses, a severe weather threat often becomes better defined over an area, and if needed, NOAA issues a severe thunderstorm warning to alert the public, aviators, and local NWS offices (see *Where Geographers Click*).

Tornadoes

Severe thunderstorms can sometimes result in tornadoes. A **tornado** is a small but intense cyclonic vortex in which air spirals at tremendous speed (**Figure 6.21**). Tornadoes occur as parts of cumulonimbus clouds traveling in advance of a cold front. They are often associated with thunderstorms spawned by fronts in midlatitudes, but they can also occur inside tropical cyclones and across ocean surfaces.

tornado A small, very intense wind vortex with extremely low air pressure in the center, formed below a dense cumulonimbus cloud in proximity to a cold front.

PROCESS DIAGRAM

Formation of a tornado • Figure 6.21

 THE PLANNER

1 Wind shear
Higher winds moving faster than lower winds (wind shear) initiates a horizontal rolling motion, or vortex.

High speed winds aloft (1,000–3,000 m, 3300–9800 ft)

Low speed winds near surface (100–500 m, 330–1640 ft)

2 Mesocyclone formation
Localized updrafts lift the rotating column of air. This forms a spinning funnel of air that develops into a thunderstorm with a strong rotation, called a mesocyclone.

3 Tornado formation
The upward and inward spiraling motion of the air intensifies, and the convergence of air causes the funnel of air to spin more rapidly, just as ice skaters speed their spins by drawing their arms inward. A tornado is born.

Tornado

184

The tornado appears as a dark funnel cloud hanging from the base of a dense storm cloud. At its lower end, the funnel may be 100 to 450 m (about 300 to 1500 ft) in diameter. The funnel appears dark because of condensing moisture, dust, and debris swept up by the wind. As the tornado moves across the countryside, the funnel writhes and twists.

Wind speeds in a tornado run as high as 100 m/s (about 225 mi/hr), exceeding known speeds of any other storm. Only the strongest buildings constructed of concrete and steel can withstand these violent winds. The **Enhanced Fujita Scale** is commonly used to describe tornado strength and expected damages (**Figure 6.22**).

The Enhanced Fujita Scale for tornadoes • Figure 6.22

The Enhanced Fujita Scale, implemented since 2007, is used to rate the severity of tornadoes. It is based on the type of damage left in the wake of the tornado.

EF0. Light damage
29–38 m/s (65–85 mi/hr)

Chimneys are damaged, tree branches broken, shallow-rooted trees toppled.

EF1. Moderate damage
39–49 m/s (86–110 mi/hr)

Roof surfaces peeled off, some tree trunks broken, garages and trailer homes destroyed.

EF2. Considerable damage
50–60 m/s (111–135 mi/hr)

Roofs damaged, trailer homes and outbuildings destroyed, damage from airborne debris, large trees uprooted or snapped.

EF3. Severe damage
61–73 m/s (136–165 mi/hr)

Roofs and walls torn from structures, small buildings destroyed, non-reinforced masonry buildings destroyed, forest uprooted.

EF4. Devastating damage
74–89 m/s (166–200 mi/hr)

Homes destroyed, some lifted distance from foundation, cars blown some distance, debris includes large objects.

EF5. Incredible damage
>90 m/s (>200 mi/hr)

Strong homes lifted from foundations, concrete structures damaged, damage from debris the size of automobiles, trees debarked.

Frequency of occurrence of strong tornadoes in the United States • Figure 6.23

The data shown in this map span a 48-year period. Only strong tornadoes are counted (EF3 or above on the Enhanced Fujita Scale). Most tornadoes occur in the central and southeastern states; they are rare over mountainous and forested regions. They are almost unknown west of the Rocky Mountains and are relatively less frequent on the eastern seaboard.

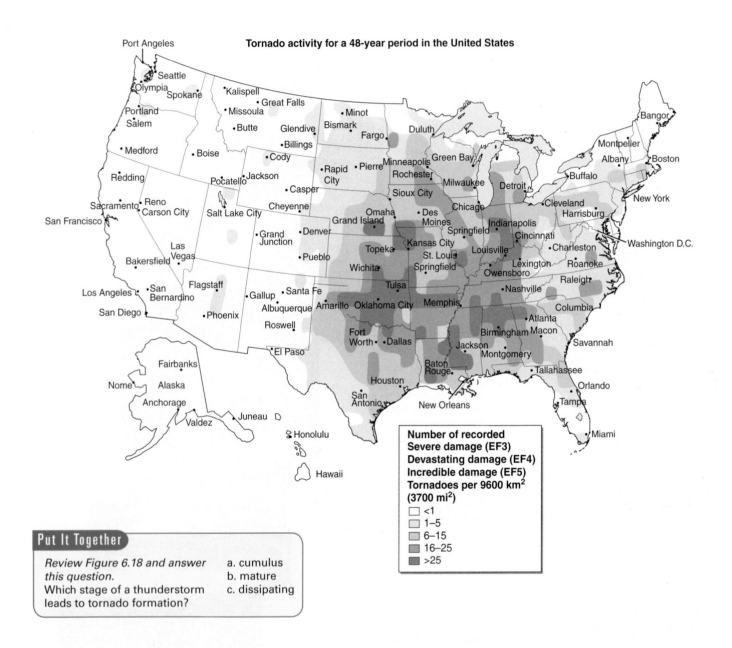

Tornado activity for a 48-year period in the United States

**Number of recorded
Severe damage (EF3)
Devastating damage (EF4)
Incredible damage (EF5)
Tornadoes per 9600 km²
(3700 mi²)**

☐ <1
☐ 1–5
☐ 6–15
☐ 16–25
☐ >25

Put It Together

Review Figure 6.18 and answer this question.
Which stage of a thunderstorm leads to tornado formation?

a. cumulus
b. mature
c. dissipating

The conditions are most favorable for tornadoes when a cold front of maritime polar air lifts warm, moist maritime tropical air—most commonly in spring and summer. Tornadoes are most frequent and violent in the United States (**Figure 6.23**). They also occur in Australia in substantial numbers and are occasionally reported from other midlatitude locations. The NWS maintains a tornado-forecasting and warning system. Whenever weather conditions favor tornado development, NWS alerts the danger area and activates systems for observing and reporting any tornado.

CONCEPT CHECK   **STOP**

1. **What** causes the formation of an anvil cloud?
2. **How** is the strength of a tornado described?

Summary

1 Air Masses 162

- **Weather** refers to the state of the atmosphere at a single location and time, whereas climate refers to regional weather conditions over longer periods of time.

- **Weather systems** are patterns of atmospheric circulation that lead to distinctive weather events. Weather systems are often associated with the movement of air masses.

- **Air masses** are large bodies of air with relatively uniform temperature and moisture characteristics. Air masses are distinguished by the latitude and surface type of their source regions, as shown in the diagram.

Source regions of air masses • Figure 6.1

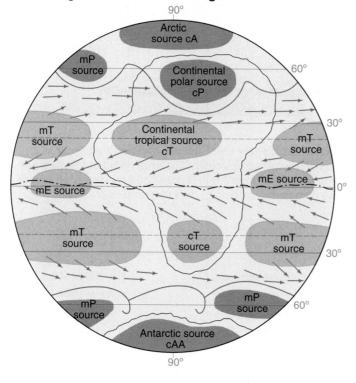

- A **cold front**, as shown in the diagram, forms when advancing cold air abruptly forces warm air to rise above it. This frontal lifting typically causes clouds to form and can bring intense rain and storms.

Frontal lifting at a cold front • Figure 6.3

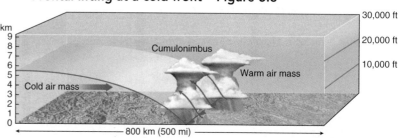

- A **warm front** develops when an advancing warm air mass is gradually forced up over a colder air mass, typically causing the formation of clouds and steady precipitation.

- In an **occluded front**, a cold front overtakes a warm front, pushing a pool of warm, moist air above the surface. **Stationary fronts** occur when two air masses are in contact, with little or no relative motion between them.

3 Midlatitude Cyclones 167

- An **anticyclone** is typically a fair-weather system of high pressure.

- **Midlatitude cyclones** form in the midlatitudes, at the boundary between cool, dry air masses and warm, moist air masses, as shown in the diagram.

Life history of a midlatitude cyclone • Figure 6.5

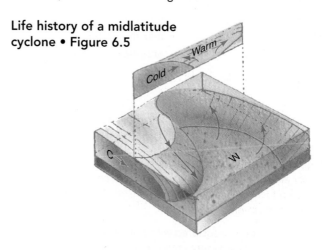

2 Fronts 165

- **Fronts** are the boundaries between air masses. Distinctive weather patterns develop ahead of, right along, and behind a front.

- When temperature and pressure differences are significant, **cyclonic storms**, including nor'easters, can bring intense winds and heavy rain or snow.

4 Tropical Cyclones 171

- **Tropical cyclones** develop over very warm tropical oceans and can intensify to become vast inspiraling systems of very high winds with very low central pressures in the **eye** of the storm, as shown in the diagram.

Anatomy of a tropical cyclone • Figure 6.8

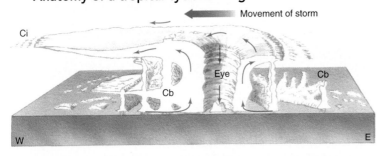

- Tropical cyclones are known as **hurricanes** in the western hemisphere, **typhoons** in the western Pacific, and **cyclones** in the Indian Ocean.

- Tropical cyclones can be very destructive, bringing high winds, **storm surges**, and heavy rainfall to coastal areas.

5 Thunderstorms and Tornadoes 181

- If air is warm, moist, and unstable, as shown in the diagram, **thunderstorms** can form, generating hail and lightning.

Development of a thunderstorm • Figure 6.18

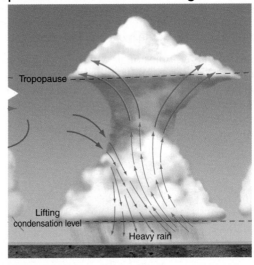

- **Tornadoes** are very small, intense cyclones that occur as a part of thunderstorm activity. Their high winds can be very destructive. Tornado strength is classified based on wind speeds and damages according to the **Enhanced Fujita Scale**.

Key Terms

- air mass 162
- anticyclone 167
- cold front 165
- cyclone 167
- cyclonic storm 170
- easterly wave 173

- Enhanced Fujita Scale 185
- eye 172
- front 165
- hurricane 171
- midlatitude cyclone 168
- occluded front 166

- Saffir-Simpson scale 178
- stationary front 166
- storm surge 180
- thunderstorm 182
- tornado 184
- tropical cyclone 171

- typhoon 171
- warm front 166
- weather 162
- weather system 162
- wind shear 182

Critical and Creative Thinking Questions

1. Often, one air mass source region will dominate weather in an area for an extended period. What air mass commonly dominates your local area, and how does the weather change when another air mass displaces it?

2. Why are cold fronts associated with severe storms and warm fronts associated with more steady rains?

3. Why do you think tornadoes are uncommon in mountainous areas?

4. What happens to a tropical cyclone that follows a track leading toward the equator?

5. What impact of a tropical cyclone usually leads to the most loss of life? Why are developing countries such as Bangladesh especially vulnerable to tropical cyclones?

6. How can damage patterns on the ground show that a tornado occurred rather than just intense winds?

7. Consider the diagram. How is the direction of travel of tropical weather systems fundamentally different from that of the low pressure systems that are common in the majority of the United States? What implication does this have for hurricane tracks?

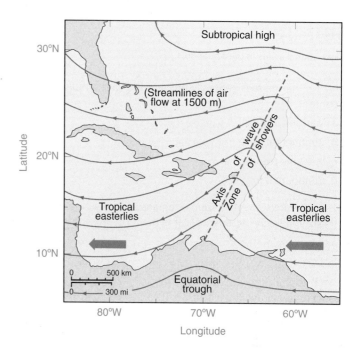

What is happening in this picture?

The photograph shows a line of cumulus clouds advancing from left to right. The clouds were formed when warm, moist air was pushed aloft. Sailors refer to these lines as squall lines due to the extreme air turbulence accompanied with flash downpours of rain. These occur most frequently in near-shore areas.

Think Critically

1. Why was the warm air forced upward?

2. The clouds mark an advancing front. What type of front might this be?

Self-Test

(Check your answers in the Appendix.)

1. _____ air masses generally possess the lowest moisture content.

   a. Maritime tropical

   b. Continental polar

   c. Maritime tropical

   d. Continental tropical

2. The diagram shows the source regions that give global air masses their characteristics. Label the following sources.

   a. cA

   b. cAA

   c. cP

   d. cT

   e. mE

   f. mP

   g. mT

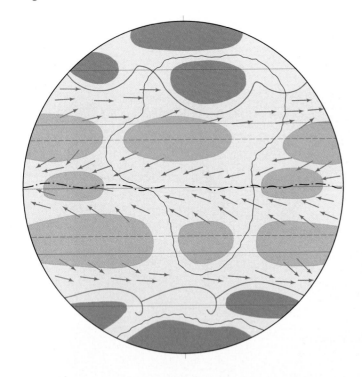

3. A(n) _____ forms when a cold air mass overtakes a warm air mass.

   a. warm front

   b. occluded front

   c. cold front

   d. stationary front

4. In _____, convergence and uplift typically cause condensation and precipitation, while subsidence in _____ causes the air to warm, producing clear conditions.

   a. cyclones; warm fronts

   b. anticyclones; cyclones

   c. anticyclones; warm fronts

   d. cyclones; anticyclones

5. The diagram shows a front that is formed when a cold front has overtaken a warm front. What is the type of front called?

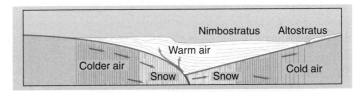

6. A(n) _____ is a center of high pressure and is generally responsible for fair weather.

   a. anticyclone

   b. cyclone

   c. trade wind

   d. midlatitude storm front

7. Heavy rains along a narrow belt are typically found along _____ fronts associated with _____.

    a. cold; cyclones

    b. warm; cyclones

    c. cold; anticyclones

    d. warm; anticyclones

8. Tropical weather tends to be _____.

    a. based on westerly waves

    b. based on divergence of air masses

    c. convectional in nature

    d. dominated by continental polar air masses

9. The diagram shows two anticyclones in contact on the polar front. Mark the region where a low-pressure trough forms. What type of weather system forms under these conditions?

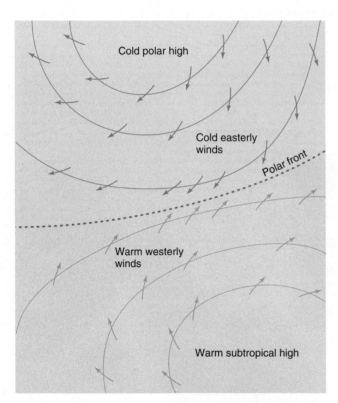

10. Hurricanes and typhoons generally develop within the _____ latitudinal zones.

    a. 20° to 30° N and S

    b. 15° to 25° N and S

    c. 8° to 15° N and S

    d. 30° to 45° N and S

11. The _____ continent is rarely threatened by hurricanes or typhoons.

    a. North American

    b. South American

    c. African

    d. Eurasian

12. A _____ is a sudden rise of water level caused by a hurricane.

    a. storm surge

    b. flood

    c. tsunami

    d. tidal flood

13. Thunderstorms are caused by _____ air that rises.

    a. cool, dry

    b. warm, dry

    c. cool, moist

    d. warm, moist

14. A _____ is a small but intense cyclonic vortex with very high wind speeds.

    a. hurricane

    b. tornado

    c. typhoon

    d. cyclone

15. Which area of the United States experiences tornadoes most frequently?

    a. Midwest

    c. New England

    b. Southwest

    d. Pacific Northwest

THE PLANNER

Review your Chapter Planner on the chapter opener and check off your completed work.

7 Global Climates and Climate Change

Polar bears evolved 200,000 years ago from brown bears to be uniquely suited to the climate and environment of the high arctic. With their white fur coats and thick layers of fat close to the skin, they are well insulated against the cold of the arctic winter. Roaming the ice sheets of the Arctic Ocean, they hunt and devour the ringed and bearded seals that also inhabit arctic ice.

Today, polar bear populations are under stress from the environmental effects of climate change. In the past 25 years or so, the extent and thickness of arctic sea ice have both diminished substantially. Sea ice breakup is now happening earlier in the season, reducing the time during which the bears can hunt for seals. Furthermore, if snow cover is reduced or melts earlier, the dens where cubs are birthed can be breached, threatening their survival.

As human-induced climate change progresses, climates and environments may shift too quickly for some species to adapt. The overall population size of the polar bear is expected to decline by more than 30% in the next 35 to 50 years. Whether the polar bear will survive into the 22nd century is in considerable doubt.

CHAPTER OUTLINE

CHAPTER PLANNER ✓

- ❏ Study the picture and read the opening story.
- ❏ Scan the Learning Objectives in each section:
  p. 194 ❏   p. 199 ❏   p. 206 ❏   p. 215 ❏   p. 223 ❏   p. 229 ❏
- ❏ Read the text and study all figures and visuals. Answer any questions.

Analyze key features

- ❏ Geography InSight p. 196 ❏   p. 202 ❏
- ❏ What a Geographer Sees, p. 214
- ❏ Video Explorations, p. 227
- ❏ Where Geographers Click, p. 231
- ❏ Process Diagram, p. 232
- ❏ Stop: Answer the Concept Checks before you go on.
  p. 199 ❏   p. 205 ❏   p. 214 ❏   p. 223 ❏   p. 228 ❏   p. 233 ❏

End of chapter

- ❏ Review the Summary and Key Terms.
- ❏ Answer the Critical and Creative Thinking Questions.
- ❏ Answer What is happening in this picture?
- ❏ Complete the Self-Test and check your answers.

A polar bear and her cub walk across the ice near Churchill, Manitoba, Canada.

Keys to Climate

LEARNING OBJECTIVES

1. **Explain** the factors that contribute to global temperature patterns.
2. **Discuss** the reasons for regional variation in precipitation patterns.

In Chapter 6 we talked about weather and weather systems. Whereas weather describes precipitation, winds, and temperatures at a particular time and place, **climate** is the average weather of a region. International conventions now use 30 years as the minimum period for defining a climate. Temperature and precipitation are two of the key measures of climate. To establish the

> **climate** The annual cycle of prevailing weather conditions at a given place, based on statistics taken over an extended period.

climate of a region, climatologists begin by answering three basic questions: (1) What is the mean annual temperature? (2) What is the mean annual precipitation? and (3) What is the seasonal variation in temperature and precipitation?

Global Temperature Patterns

As we saw in Chapter 3, there are three primary factors that control the magnitude and range of temperatures experienced at a location through the year: latitude, location, and elevation (**Figure 7.1**). Latitude controls insolation—the flow of solar energy received by the Earth and atmosphere—and determines its variation through the year. A coastal location moderates temperature variation compared

Factors controlling temperature • Figure 7.1

A variety of factors influence global temperature patterns, including latitude, location, and elevation.

a. Latitude

Insolation varies with latitude. The annual cycle of temperature at any place depends on its latitude. Near the equator, temperatures are warmer and the annual range is low. Toward the poles, such as here in Nunavut, Canada, temperatures are colder and the annual range is greater.

c. Elevation

It is cooler at higher elevations, such as here at Kluane National Park, Canada, than at sea level because the atmosphere cools with elevation at the average environmental temperature lapse rate.

b. Location

Ocean-surface temperatures vary less with the seasons than temperatures of land surfaces. This means that regions near the coast, such as Cornwall, England, show a smaller annual variation in temperature, while inland regions show a larger variation.

to a continental interior location. And temperatures are generally lower at higher elevations. **Figure 7.2** shows how latitude and location influence overall temperatures and the annual temperature cycle. At coastal locations, the annual temperature cycle varies much less than at continental locations, and continental locations have much colder overall temperatures at higher latitudes.

Global temperature patterns • Figure 7.2

In general, average annual temperatures are highest at the equator and progressively cooler toward the poles. The location of a region—coastal or continental—also plays a role in determining temperature.

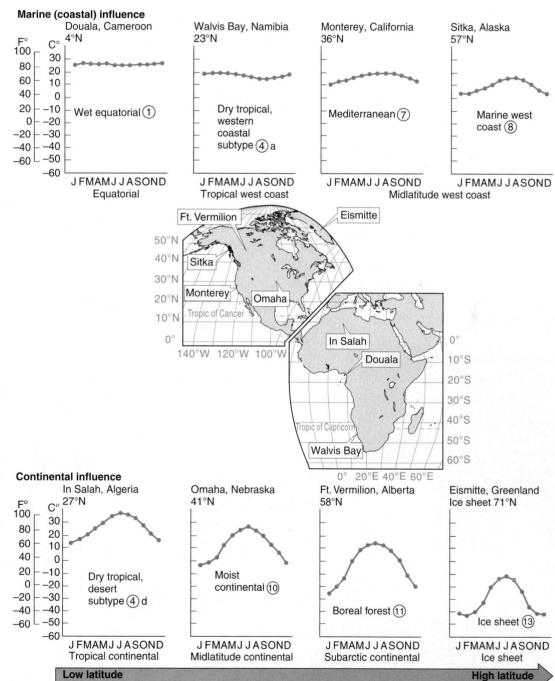

Marine (coastal) influence

a. In coastal locations, temperatures are more moderate and seasonal variations are minimal, with the least seasonal variation at the equator and the greatest toward the poles.

Douala, Cameroon 4°N — Wet equatorial ① — Equatorial

Walvis Bay, Namibia 23°N — Dry tropical, western coastal subtype ④ a — Tropical west coast

Monterey, California 36°N — Mediterranean ⑦ — Midlatitude west coast

Sitka, Alaska 57°N — Marine west coast ⑧

Continental influence

b. In continental locations, temperatures are more extreme and seasonal variations are more pronounced, with the least seasonal variation at the equator and the greatest toward the poles.

In Salah, Algeria 27°N — Dry tropical, desert subtype ④ d — Tropical continental

Omaha, Nebraska 41°N — Moist continental ⑩ — Midlatitude continental

Ft. Vermilion, Alberta 58°N — Boreal forest ⑪ — Subarctic continental

Eismitte, Greenland Ice sheet 71°N — Ice sheet ⑬ — Ice sheet

Low latitude ➝ **High latitude**

Ask Yourself

A location between Sitka, Alaska, and Ft. Vermilion, Alberta, if located at a similar elevation to both sites, would most likely have a _____.

a. greater mean annual temperature than Sitka
b. greater mean annual temperature than Ft. Vermilion
c. greater annual temperature range than Ft. Vermilion
d. greater annual temperature range than Sitka

Global Precipitation Patterns

Along with temperature, the other key measure of climate is precipitation. Mean annual precipitation varies widely across the globe, ranging from nearly zero to more than 500 cm (200 in.) (**Figure 7.3**). Global precipitation patterns are in large part determined by air mass characteristics and movements, which in turn are affected by latitude, topography, air masses, persistent pressure patterns, and atmospheric and oceanic circulation.

Geography InSight

Factors controlling global precipitation

• **Figure 7.3**

Annual precipitation patterns are determined by a variety of factors. This global map of mean annual precipitation uses *isohyets*—lines drawn through all points having the same annual precipitation—labeled in centimeters and inches.

a. Air masses
Air masses that form over continental regions are drier, while air masses from marine locations such as Baffin Island, in Canada, shown here, are moister and support more precipitation. In regions where dry and moist air masses come into contact, fronts can form, leading to precipitation.

b. Prevailing wind and ocean currents
On the western portion of midlatitude continents, such as the California coast shown here, the prevailing westerly winds bring warm, moist, but stable air off the ocean and onto the continent, resulting in drier summers. On the eastern portion, the winds from the summer subtropical high (Bermuda) bring moist air from the waters off the east coast, resulting in significant precipitation.

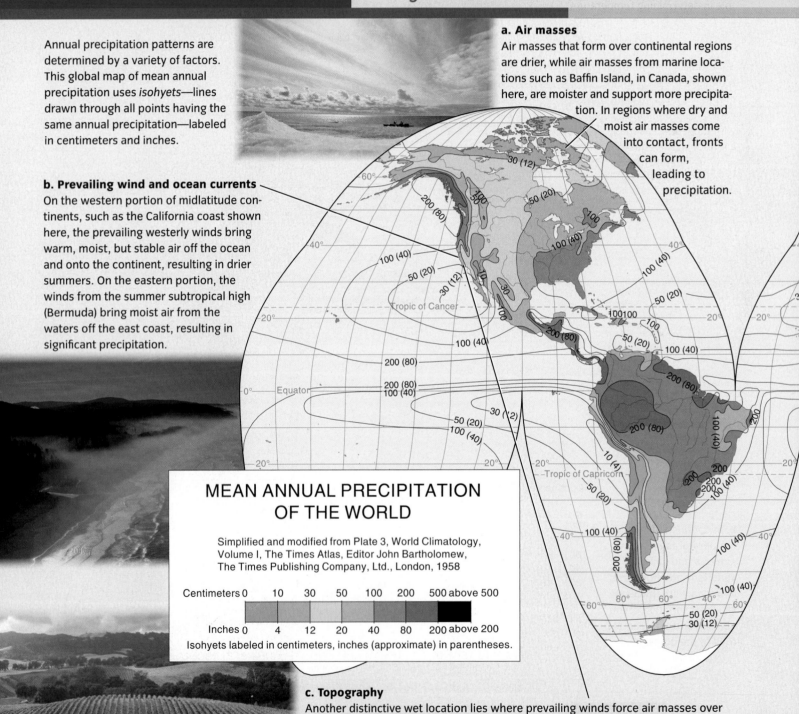

MEAN ANNUAL PRECIPITATION OF THE WORLD

Simplified and modified from Plate 3, World Climatology, Volume I, The Times Atlas, Editor John Bartholomew, The Times Publishing Company, Ltd., London, 1958

Centimeters 0	10	30	50	100	200	500 above 500

Inches 0	4	12	20	40	80	200 above 200

Isohyets labeled in centimeters, inches (approximate) in parentheses.

c. Topography
Another distinctive wet location lies where prevailing winds force air masses over hills and mountain ranges, such as at this winery in California, producing abundant orographic precipitation. On the leeward side, adiabatic warming of air produces warm and dry conditions.

Dry and moist climates Regions can be broadly classified as moist or dry, according to the amount of precipitation they receive compared with the total evaporation. In dry climates, vegetation is sparse or absent. The total annual evaporation of moisture from the soil and from plant foliage greatly exceeds the annual precipitation. Generally speaking, streams in dry climates dry up in the summer months.

Dry climates range from very dry deserts nearly devoid of plant life to moister regions that support some grasses or shrubs. Dry climates can be either semiarid (or

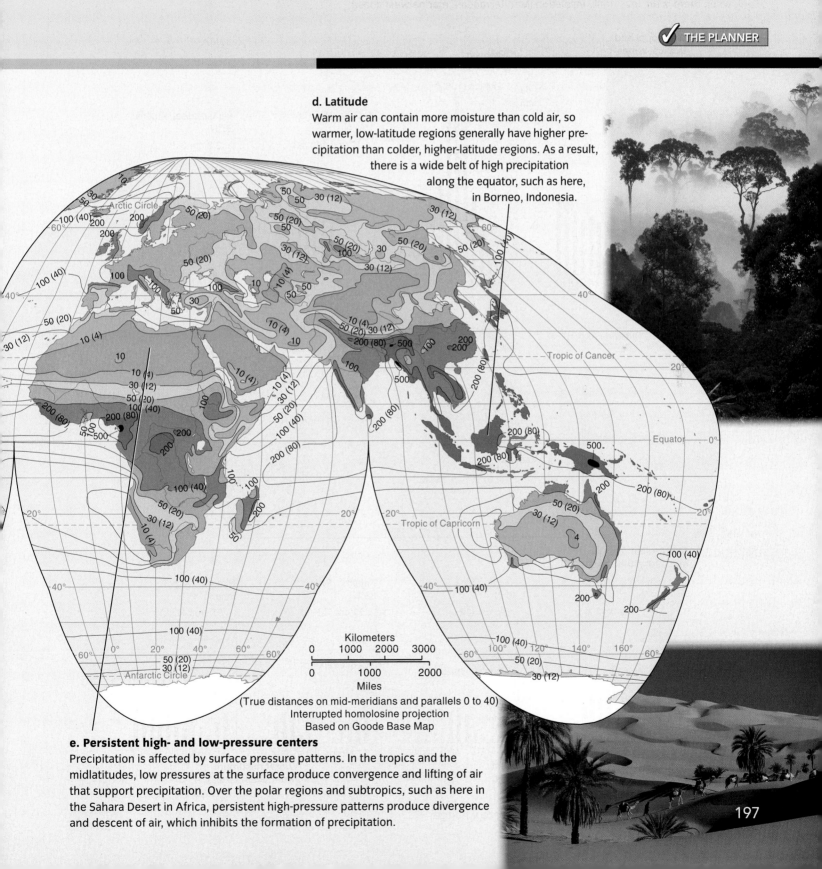

d. Latitude
Warm air can contain more moisture than cold air, so warmer, low-latitude regions generally have higher precipitation than colder, higher-latitude regions. As a result, there is a wide belt of high precipitation along the equator, such as here, in Borneo, Indonesia.

Kilometers
0 1000 2000 3000

Miles
0 1000 2000

(True distances on mid-meridians and parallels 0 to 40)
Interrupted homolosine projection
Based on Goode Base Map

e. Persistent high- and low-pressure centers
Precipitation is affected by surface pressure patterns. In the tropics and the midlatitudes, low pressures at the surface produce convergence and lifting of air that support precipitation. Over the polar regions and subtropics, such as here in the Sahara Desert in Africa, persistent high-pressure patterns produce divergence and descent of air, which inhibits the formation of precipitation.

197

There are three broad seasonal patterns of precipitation. Precipitation can be relatively uniform over the course of the year, ranging from very wet to very dry (wet equatorial to tropical desert). Precipitation can be at its maximum during the summer (or season of high Sun), in which daily insolation is at its peak (Asiatic monsoon, wet-dry tropical, moist subtropical, moist continental). Precipitation can reach its maximum during the winter or cooler season (season of low Sun), when there is the least daily insolation (Mediterranean, marine west coast).

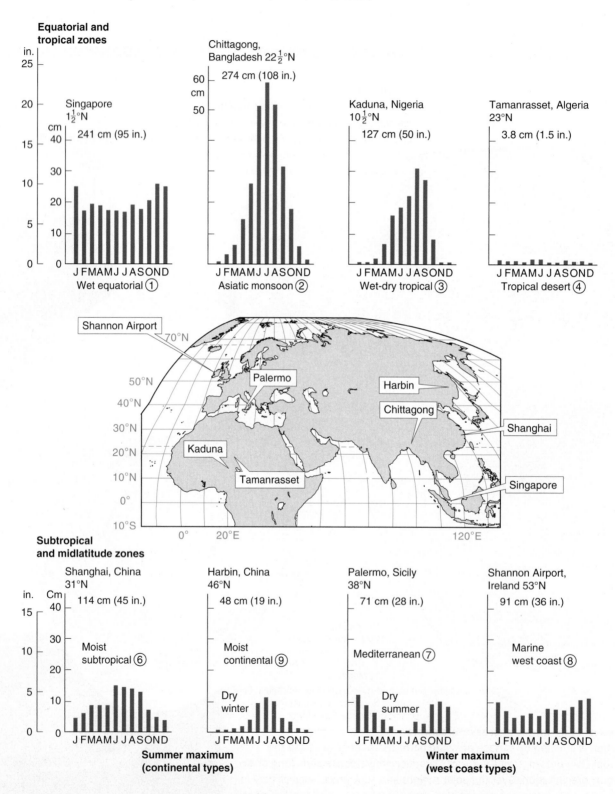

steppe, pronounced "step") or arid. **Semiarid climates** have enough precipitation to support sparse grasses and shrubs. **Arid climates** are extremely dry and can support little or no vegetation. Subtle shifts in weather patterns can have a profound effect on the expansion of deserts in these climates.

Moist climates have enough rainfall to keep the soil moist through much of the year and sustain the year-round flow of the larger streams. Moist climates support forests or prairies of dense tall grasses.

Seasonal patterns of precipitation Some climates cannot be accurately described as either dry or moist climates. They alternate between a very wet season and a very dry season, so we will refer to them here as *wet-dry climates.* Monthly precipitation values can tell us if there is a pattern of alternating dry and wet seasons (**Figure 7.4**).

The natural vegetation, soils, crops, and human use of the land are all different in a region of alternating dry and wet seasons compared with a place where precipitation is more uniform throughout the year. It also makes a great deal of difference whether the wet season coincides with a season of higher temperatures or with a season of lower temperatures. A warm season that is also wet enhances the growth of both native plants and crops. On the other hand, a warm season that is dry places great stress on growing plants.

CONCEPT CHECK

1. **How** and why do the annual cycles of air temperature recorded by weather stations at Douala, Cameroon, and Ft. Vermilion, Alberta, differ?

2. **What** climate factors determine how hospitable regions are for the growth of natural vegetation?

Climate Classification

LEARNING OBJECTIVES

1. **Explain** how climatologists classify climate around the world.

2. **Describe** how air mass movement and frontal zones define three broad regional categories of climates.

W e can describe the climate of a region pretty accurately by using mean monthly values of air temperature and precipitation. A **climograph** combines graphs of monthly temperature and precipitation for an observing weather station (**Figure 7.5**).

> **climograph** A graph on which two or more climate variables are plotted for each month of the year.

Reading a climograph • Figure 7.5

A climograph is a figure that combines graphs of monthly temperature and precipitation for an observing station. Shown here is a climograph for Timbo, Guinea, not far from the equator in Africa. We will see this climograph again as an example of the tropical wet-dry climate.

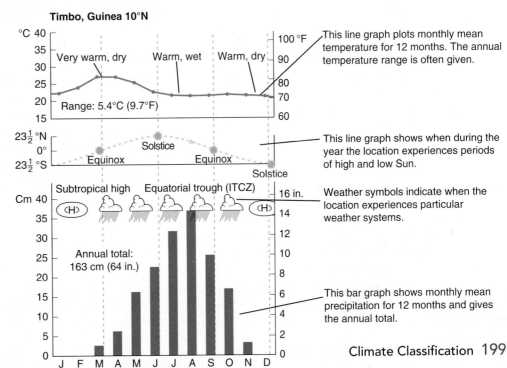

This line graph plots monthly mean temperature for 12 months. The annual temperature range is often given.

This line graph shows when during the year the location experiences periods of high and low Sun.

Weather symbols indicate when the location experiences particular weather systems.

This bar graph shows monthly mean precipitation for 12 months and gives the annual total.

Climate Classification **199**

Collecting climate data • Figure 7.6

A network of meteorological sensors collect data for the key weather variables air temperature, wind speed and direction, solar radiation, relative humidity, barometric pressure, and precipitation. These meteorological field data create a comprehensive picture of the Earth's daily weather. Climatologists analyze these data, in combination with data collected by weather satellites, to define the climate parameters for a given region over a period of 30 years or more.

a. The National Weather Service operates a network of automated stations along with the voluntary Cooperative Observer Stations, as pictured here, for approximately 4000 data collection stations across the country.

Climographs reflect the monthly data collected at field stations, as averaged over many years—often several decades. Most frequently collected are data describing temperature, wind, humidity, pressure, and precipitation, but other weather variables are also measured at many sites (**Figure 7.6**).

Climate Classification Systems

Climatology is the science of analyzing climate—weather over the long term—as it varies over time and around the globe. Climatologists are often computer modelers with strong backgrounds in physics and math, and they use computer programs to simulate the Earth's weather at various time scales. They study past and present climate systems to predict future trends. Based on our current understanding of global warming trends, climate change will be a major challenge for society as a whole over the next century. Careers in climatology represent a strategic position from which to address this huge challenge.

Geographers and applied climatologists have devised a number of classification systems to group the climates of individual locations into distinctive climate types. Like all other scientific models, scientific classification systems do not provide a perfect mirror of the real world, but they do provide a useful method to summarize, communicate, and exchange information. For example, Linnaeus created a classification system for describing plants and animals by genus and species (for example, *Homo sapiens*) that gives biologists a universal framework for their discussion of species. In chemistry, the periodic table is another example of a classification system; in this system, elements are arranged by their atomic

b. Ocean buoys, both moored and drifting, collect meteorological data for coastal and open-ocean areas at more than 1500 locations. This National Oceanic and Atmospheric Administration (NOAA) buoy is being serviced by technicians, ferried by the buoy tender in the background.

c. Government meteorologists prepare to launch a weather balloon into the atmosphere, one of hundreds launched twice daily across the United States. Each balloon carries instrument packages called radiosondes (white boxes) and a parachute (red, on the ground) for return of the equipment.

properties. However, not all such systems enjoy universal acceptance. Because classification systems reflect the methods and values of the scientists who create them, different systems can be more or less appropriate for different purposes.

One climate classification system still in wide use was developed by the Austrian climatologist Vladimir Köppen in 1918 and modified by Rudolf Geiger and Wolfgang Pohl in 1953. Primarily designed to capture variability of vegetation around the globe, it uses a system of letters to label climates and defines the climates by mean annual precipitation and temperature, as well as precipitation in the driest month. Köppen's climate system is descriptive and empirical, relying on annual and monthly values of temperature and precipitation at stations throughout the world. *Empirical* systems are based on experimental evidence. By using descriptions based on the reality of where different types of plants grow, we can

gain an understanding of the climate features that support such vegetation growth patterns. The Köppen climate system is easy and convenient to use, given monthly weather data, but it is not directly related to the underlying processes that differentiate the Earth's climates.

In comparison, the classification used in this chapter is designed to be understood and explained by the characteristics of air masses, their movements, and their interactions along frontal boundaries—that is, by the weather patterns various regions experience throughout the year. This system is actually based on monthly variation in soil moisture, but it captures these causes of climate variation quite well. In this book we will discuss 13 distinctive climate types (**Figure 7.7** on the next page). The types follow quite naturally from the principles governing temperature and precipitation that we have discussed in prior chapters.

The classification system used in this book recognizes 13 distinctive climates, based on the weather experienced by a given region throughout the year. Each climate is characterized by unique weather patterns and often by identifiable vegetation types. For convenience, the climates are numbered in the following graphs and maps according to the legend on this map.

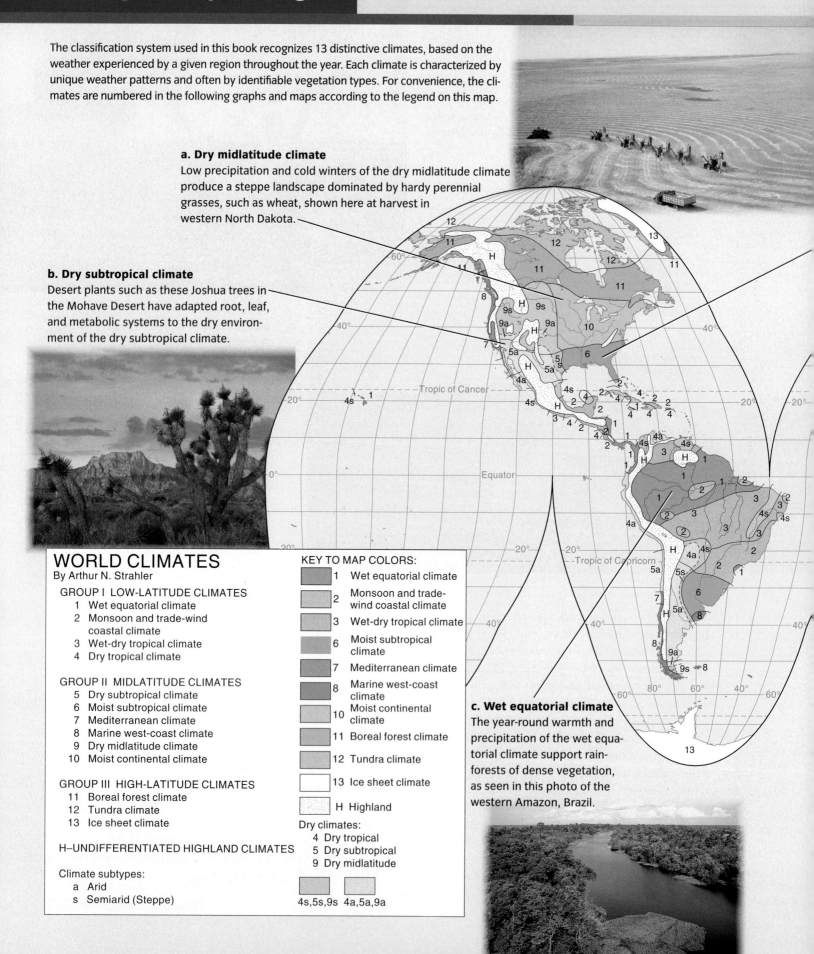

a. Dry midlatitude climate
Low precipitation and cold winters of the dry midlatitude climate produce a steppe landscape dominated by hardy perennial grasses, such as wheat, shown here at harvest in western North Dakota.

b. Dry subtropical climate
Desert plants such as these Joshua trees in the Mohave Desert have adapted root, leaf, and metabolic systems to the dry environment of the dry subtropical climate.

c. Wet equatorial climate
The year-round warmth and precipitation of the wet equatorial climate support rainforests of dense vegetation, as seen in this photo of the western Amazon, Brazil.

WORLD CLIMATES
By Arthur N. Strahler

GROUP I LOW-LATITUDE CLIMATES
 1 Wet equatorial climate
 2 Monsoon and trade-wind coastal climate
 3 Wet-dry tropical climate
 4 Dry tropical climate

GROUP II MIDLATITUDE CLIMATES
 5 Dry subtropical climate
 6 Moist subtropical climate
 7 Mediterranean climate
 8 Marine west-coast climate
 9 Dry midlatitude climate
 10 Moist continental climate

GROUP III HIGH-LATITUDE CLIMATES
 11 Boreal forest climate
 12 Tundra climate
 13 Ice sheet climate

H–UNDIFFERENTIATED HIGHLAND CLIMATES

Climate subtypes:
 a Arid
 s Semiarid (Steppe)

KEY TO MAP COLORS:
 1 Wet equatorial climate
 2 Monsoon and trade-wind coastal climate
 3 Wet-dry tropical climate
 6 Moist subtropical climate
 7 Mediterranean climate
 8 Marine west-coast climate
 10 Moist continental climate
 11 Boreal forest climate
 12 Tundra climate
 13 Ice sheet climate
 H Highland

Dry climates:
 4 Dry tropical
 5 Dry subtropical
 9 Dry midlatitude

4s,5s,9s   4a,5a,9a

d. Moist subtropical climate
The native vegetation of the moist subtropical climate is forests of broadleaf deciduous and evergreen trees and shrubs, shown here in the southeastern United States.

e. Dry tropical climate
Although dry tropical deserts are largely extremely arid, there are broad zones at the margins that are semiarid. These steppes have a short wet season when grasses grow and support nomadic grazing cultures, such as those of Iran.

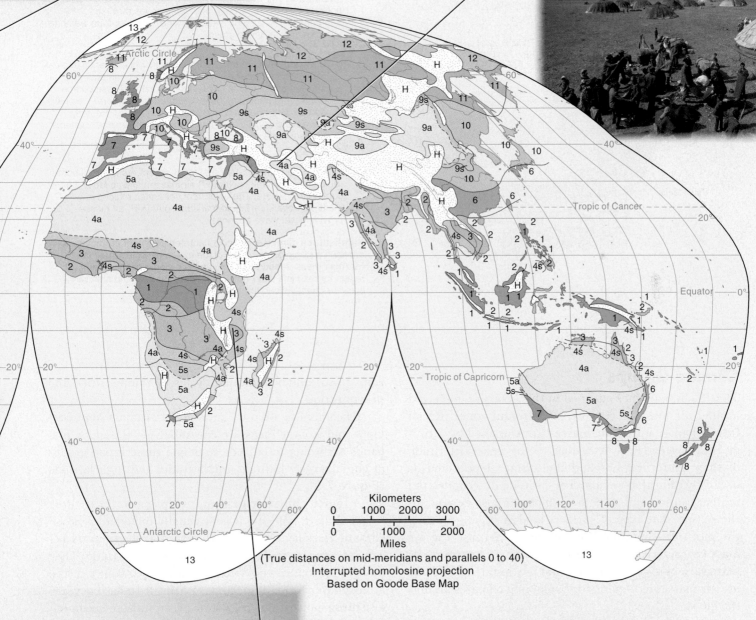

Kilometers
0   1000 2000 3000
0   1000   2000
Miles
(True distances on mid-meridians and parallels 0 to 40)
Interrupted homolosine projection
Based on Goode Base Map

f. Wet-dry tropical climate
The native vegetation of the wet-dry tropical climate must survive alternating seasons of very dry and very wet weather. This situation produces a savanna environment of sparse vegetation, as seen here in southern Kenya.

Climate Classification **203**

Climate groups and air mass source regions • Figure 7.8

Using a map of air mass source regions for a simplified hypothetical continent, we can identify five global bands associated with three major climate groups: low latitude, midlatitude, and high latitude. Each group has a set of distinctive climates with unique characteristics that are explained by the movements of air masses and frontal zones.

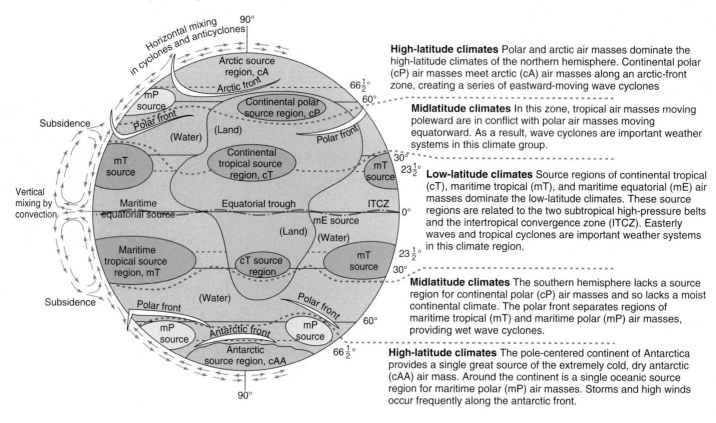

High-latitude climates Polar and arctic air masses dominate the high-latitude climates of the northern hemisphere. Continental polar (cP) air masses meet arctic (cA) air masses along an arctic-front zone, creating a series of eastward-moving wave cyclones

Midlatitude climates In this zone, tropical air masses moving poleward are in conflict with polar air masses moving equatorward. As a result, wave cyclones are important weather systems in this climate group.

Low-latitude climates Source regions of continental tropical (cT), maritime tropical (mT), and maritime equatorial (mE) air masses dominate the low-latitude climates. These source regions are related to the two subtropical high-pressure belts and the intertropical convergence zone (ITCZ). Easterly waves and tropical cyclones are important weather systems in this climate region.

Midlatitude climates The southern hemisphere lacks a source region for continental polar (cP) air masses and so lacks a moist continental climate. The polar front separates regions of maritime tropical (mT) and maritime polar (mP) air masses, providing wet wave cyclones.

High-latitude climates The pole-centered continent of Antarctica provides a single great source of the extremely cold, dry antarctic (cAA) air mass. Around the continent is a single oceanic source region for maritime polar (mP) air masses. Storms and high winds occur frequently along the antarctic front.

Climate Groups

The 13 climate types outlined in this book are organized around air mass movements and frontal zones. Recall from Chapter 6 that air masses are classified by a two- or three-letter code, according to the general latitude of their source regions and the surface type—land or ocean—within that region. The latitude determines the temperature of the air mass, which can also depend on the season. The type of surface, land or ocean, controls the moisture content of the air mass. Because the air mass characteristics control the two most important climate variables—temperature and precipitation—we can use air masses as a guide for explaining climates around the globe.

We also know that frontal zones are regions in which air masses are in contact. When unlike air masses are in contact with one another, cyclonic precipitation is likely to develop. Because the position of frontal zones changes with the seasons, seasonal movements of frontal

zones also influence annual cycles of temperature and precipitation.

By combining what we know about air mass source regions and frontal zones, we can subdivide the globe into bands according to latitude to define three broad groups of climates: low latitude, midlatitude, and high latitude (**Figure 7.8**).

A fourth climate type is an exception to these three broad climate types, so it isn't usually included in broad climate classification schemes. **Highland climates** occupy mountains and high plateaus at any latitude. They tend to be cool to cold because air temperatures in the atmosphere normally decrease with altitude. They are also often moist from orographic precipitation, becoming wetter with elevation. Rain shadows can occur on sheltered sides of mountains and high plateaus. Highland areas usually derive their annual temperature cycle and the times of their wet and dry seasons from the climate of the surrounding lowland (**Figure 7.9**).

Highland climates are generally cooler and wetter than lowland climates in the same region.

a. New Delhi, the capital city of India, is situated in the Ganges lowland. Here, summer is oppressively hot.

b. Shimla, a mountain refuge from the hot weather, is located at about 2200 m (about 7200 ft), in the foothills of the Himalayas.

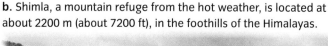

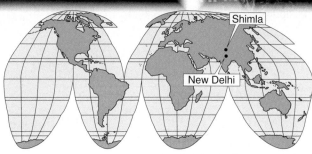

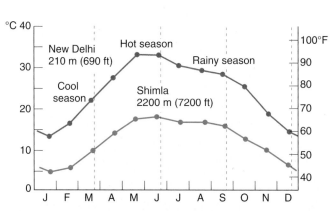

c. The annual temperature cycles for the two cities are quite similar in shape, with January as the minimum month for both. However, the hot-season temperature averages over 32°C (90°F) in New Delhi, while Shimla enjoys a pleasant 18°C (64°F).

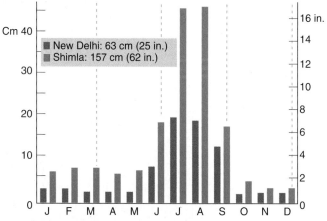

d. The annual rainfall cycles are also similar in shape, but Shimla receives more than double the rainfall of New Delhi.

CONCEPT CHECK 🛑 STOP

1. **What** does it mean to say that a classification system is empirical?

2. **Why** are highland climates an exception to the three broad groups of climates?

Low-Latitude Climates

LEARNING OBJECTIVES

1. **Describe** the global distribution and features of low-latitude climates.

2. **Explain** how air mass movements cause the weather patterns of low-latitude climates.

The low-latitude climates lie for the most part between the tropics of Cancer and Capricorn, occupying all of the equatorial zone (10° N to 10° S), most of the tropical zone (10–15° N and S), and part of the subtropical zone. The low-latitude climate regions include the equatorial

Global distribution of wet equatorial and monsoon and trade-wind coastal climates • Figure 7.10

The wet equatorial climate region ① lies between 10° N and S and includes the Amazon lowland of South America, the Congo Basin of equatorial Africa, and the East Indies, from Sumatra to New Guinea. The monsoon and trade-wind coastal climate ② occurs between 5° and 25° N and S.

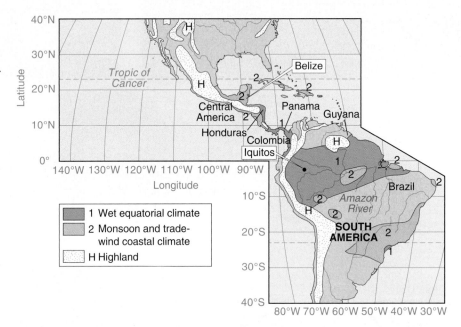

1 Wet equatorial climate
2 Monsoon and trade-wind coastal climate
H Highland

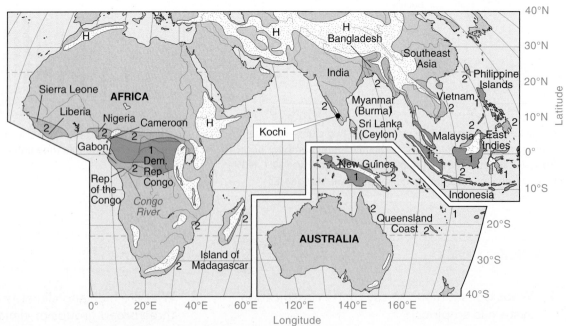

trough of the intertropical convergence zone (ITCZ), the belt of tropical easterlies (northeast and southeast trades), and large portions of the oceanic subtropical high-pressure belt. There are four low-latitude climates: wet equatorial, monsoon and trade-wind coastal, wet-dry tropical, and dry tropical. We will now look at each one in detail, beginning with the wet equatorial and monsoon and trade-wind coastal climates (**Figure 7.10**).

The Wet Equatorial Climate

The **wet equatorial climate** is controlled by the intertropical convergence zone (ITCZ)

> **wet equatorial climate** The moist climate of the equatorial zone, with abundant precipitation and uniformly warm temperatures throughout the year.

and is dominated by warm, moist maritime equatorial (mE) and maritime tropical (mT) air masses that yield heavy convectional rainfall (**Figure 7.11**). It is designated with the number ① in the maps and diagrams throughout this chapter. This climate has uniformly warm temperatures and abundant precipitation in all months, although there may be a period with less rainfall when the ITCZ shifts toward the opposite tropic. Rainforest is the natural vegetation cover. Streams flow abundantly throughout most of the year, and the riverbanks are lined with dense forest vegetation.

Profile of the wet equatorial climate • Figure 7.11

Iquitos, Peru (lat. 3° S), located in the upper Amazon lowland, shows mean monthly temperature and precipitation patterns that are characteristic of the warm, rainy, wet equatorial climate.

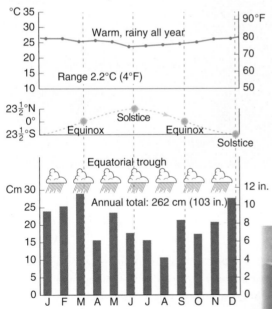

There are uniformly warm temperatures throughout the year, with mean monthly and mean annual temperatures close to 27°C (81°F).

There's a large amount of precipitation every month, but there is a seasonal rainfall pattern, with heavier rain when the ITCZ migrates into the region.

The climograph for Belize City, a Central American east-coast city (lat. 17° N), demonstrates the warm temperatures and seasonally heavy rainfall that are characteristic of a trade-wind coastal climate.

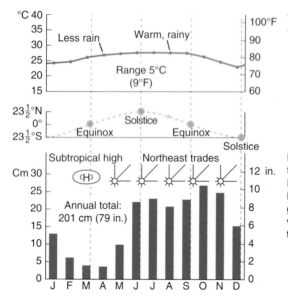

Temperatures are warm throughout the year.

Rainfall is abundant from June through November, when the ITCZ is nearby. Following the December solstice, rainfall is greatly reduced, with minimum values in March and April, when the ITCZ is farthest away.

The Monsoon and Trade-Wind Coastal Climate

monsoon and trade-wind coastal climate The moist climate of low latitudes that shows a strong rainfall peak in the high-Sun season and a short period of reduced rainfall in the low-Sun season.

Like the wet equatorial climate, the **monsoon and trade-wind coastal climate** ② experiences warm temperatures throughout the year and high annual rainfall. However, rainfall has a distinctly seasonal rhythm.

The warmest temperatures occur during clear skies in the high-Sun season (when the Sun is nearest to overhead at noon), just before the clouds and rain of the wet season appear. Minimum temperatures are at the time of low Sun. Rainfall here shows a strong seasonal pattern. In the high-Sun season, the ITCZ is nearby, so monthly rainfall is greater. In the low-Sun season, when the ITCZ has migrated to the other hemisphere, the region is dominated by subtropical high pressure, so there is less monthly rainfall.

Trade-wind coasts As its name suggests, the monsoon and trade-wind coastal climate type is produced by two different situations. On trade-wind coasts, heavier rainfall is produced during the high-Sun season as a result of the seasonal migration of the ITCZ (**Figure 7.12**). During the high-Sun season, moisture-laden maritime tropical (mT) and maritime equatorial (mE) air masses are moved onshore onto narrow coastal zones by trade winds or by monsoon circulation patterns. As the warm, moist air passes over coastal hills and mountains, the orographic effect touches off convectional shower activity. The east coasts of land masses experience this trade-wind effect because the trade winds blow from east to west. Trade-wind coasts are found along the east sides of Central and South America, the Caribbean Islands, Madagascar (Malagasy), Southeast Asia, the Philippines, and northeast Australia.

Monsoon climate A similar seasonal pattern of heavy rainfall occurs in monsoon coastal climates (**Figure 7.13**). In the Asiatic summer monsoon, the monsoon circulation brings mE air onshore. Because the onshore monsoon winds blow from southwest to northeast, the western coasts of land masses are exposed to this moist air flow. Western India and Myanmar (formerly Burma) are examples of regions with this type of climate. Moist air also penetrates well inland in Bangladesh, providing very heavy monsoon rains.

Because of abundant rainfall, rainforest is the natural vegetation of much of the monsoon and trade-wind coastal climate. In the drier areas, a monsoon forest develops that is adapted to the seasonal dry period.

Profile of the monsoon climate • Figure 7.13

Kochi, on the southwestern tip of India, has the weak temperature cycle but strong precipitation cycle of the monsoon climate. The photo shows downtown Kochi in the rainy season. Warm mE air masses moving in from the ocean provide frequent showers at this time.

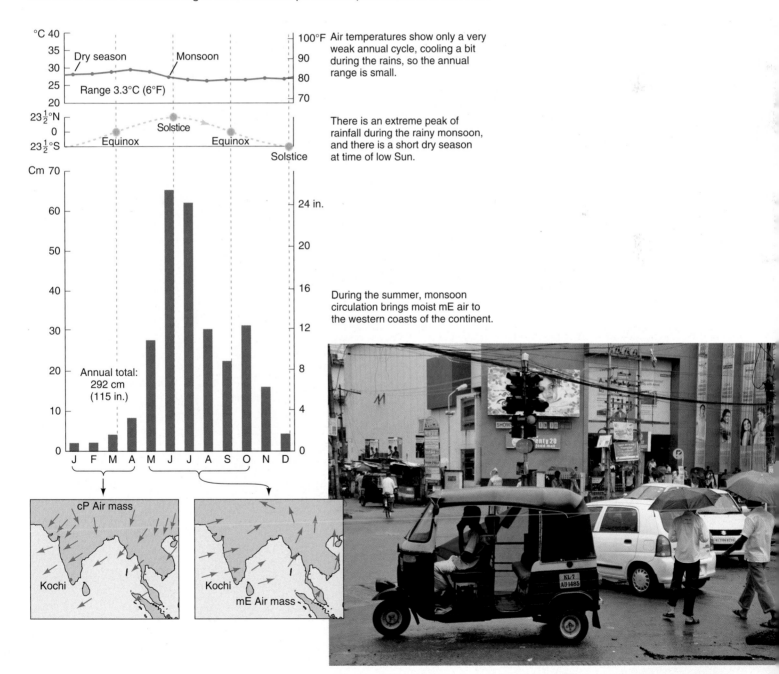

Air temperatures show only a very weak annual cycle, cooling a bit during the rains, so the annual range is small.

There is an extreme peak of rainfall during the rainy monsoon, and there is a short dry season at time of low Sun.

During the summer, monsoon circulation brings moist mE air to the western coasts of the continent.

Global distribution of the wet-dry tropical climate • Figure 7.14

The wet-dry tropical climate ③ lies between 5° and 20° N and S in Africa and the Americas and between 10° and 30° N in Asia. In Africa and South America, the climate occupies broad bands poleward of the wet equatorial and monsoon and trade-wind coastal climates.

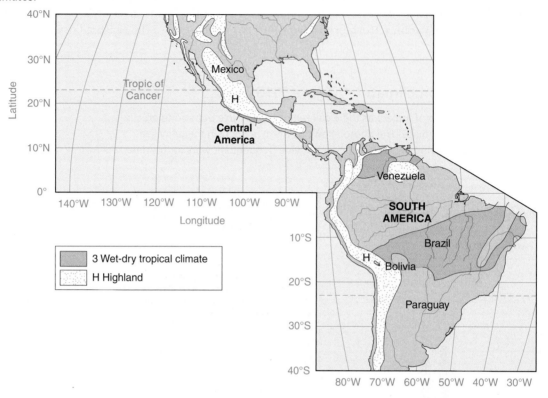

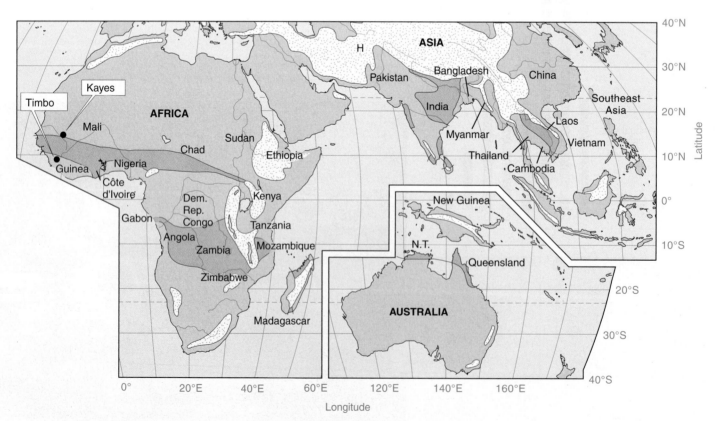

The Wet-Dry Tropical Climate

wet-dry tropical climate The climate of the tropical zone characterized by a very wet season alternating with a very dry season.

As we move farther poleward, the seasonal cycles of rainfall and temperature become stronger in the **wet-dry tropical climate** ③ (**Figure 7.14**).

The wet-dry tropical climate is produced by an alternating dominance of the wet ITCZ and dry subtropical high pressure (**Figure 7.15**). During the high-Sun season, the ITCZ is nearby and moist maritime tropical (mT) and maritime equatorial (mE) air masses dominate, creating the wet season. As the ITCZ migrates toward the opposite tropic, subtropical high pressure moves into the region and dry continental tropical (cT) air masses prevail, creating the dry season. Cooler temperatures in the dry season give way to a very hot period before the rains begin.

In central India and Vietnam, Laos, and Cambodia, mountain barriers tend to block some flows of warm, moist mE and mT air provided by trade and monsoon winds. This reduces overall rainfall and enhances the dry season, yielding a wet-dry tropical climate.

Profile of the wet-dry tropical climate • Figure 7.15

Timbo, Guinea (lat. 10° N), is typical of the wet-dry tropical climate region of western Africa.

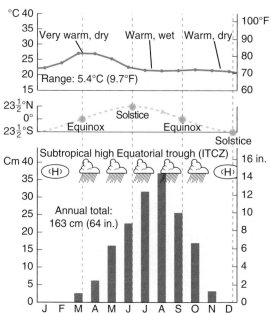

In February and March, insolation increases and air temperature rises sharply. When the rains set in, the cloud cover and evaporation of rain cause temperatures to drop.

The rainy season begins just after the March equinox and peaks when the ITCZ has migrated to its most northerly position.

Monthly rainfall peaks in August, but as the low-Sun season arrives, the ITCZ moves south and precipitation decreases. December through February are practically rainless, when subtropical high pressure dominates the climate, and stable, subsiding continental tropical (cT) air pervades the region.

Global distribution of the dry tropical, dry subtropical, and dry midlatitude climates • Figure 7.16

Nearly all of the dry tropical climate ④ lies between latitudes 15° and 25° N and S. The dry subtropical climate ⑤ is a poleward and eastward extension of the dry tropical climate, with cooler temperatures. The dry midlatitude climate ⑨ occurs in continental interiors, far from the sources of moist air masses.

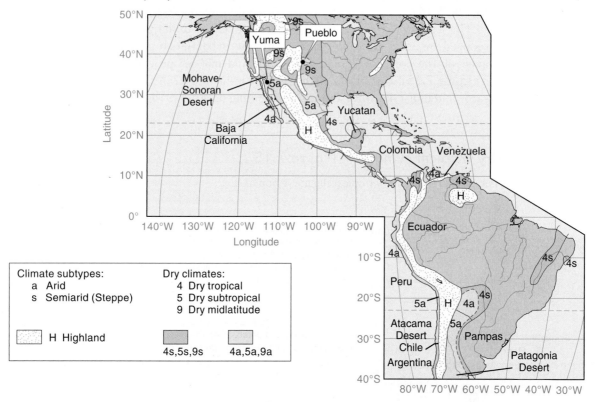

Climate subtypes:
a Arid
s Semiarid (Steppe)

Dry climates:
4 Dry tropical
5 Dry subtropical
9 Dry midlatitude

H Highland

4s,5s,9s   4a,5a,9a

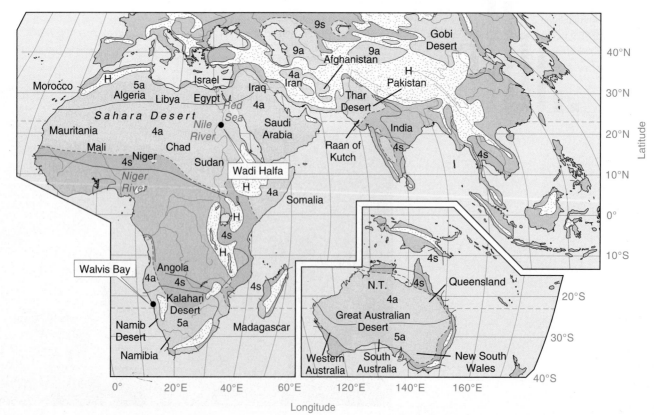

The Dry Tropical Climate

The **dry tropical climate** ④ is found in the center and east sides of subtropical high-pressure cells (**Figure 7.16**). As we saw in Chapter 5, air descends and warms adiabatically in these regions, inhibiting condensation, so rainfall is very rare and occurs only when unusual weather conditions move moist air into the region. Skies are clear most of the time, so the Sun heats the surface intensely, keeping air temperatures high. During the high-Sun period, heat is extreme. During the low-Sun period, temperatures are cooler. Given the dry air and lack of cloud cover, the daily temperature range is very large (**Figure 7.17**).

The driest areas of the dry tropical climate are near the tropics of Cancer and Capricorn. Rainfall increases

dry tropical climate The climate of the tropical zone, with high temperatures and low rainfall.

toward the equator. Traveling equatorward, we encounter regions that have short rainy seasons when the ITCZ is near, until finally the climate grades into the wet-dry tropical type. The boundary line between climate types can shift as a result of even small changes in climate or land use.

In recent years, large areas of wet-dry tropical climate in Asia, Africa, and Australia are becoming deserts, through a process called **desertification** (see *What a Geographer Sees* on the next page).

The largest region of dry tropical climate is the Sahara–Saudi Arabia–Iran–Thar desert belt of North Africa and southern Asia, which includes some of the driest regions on Earth. Another large region is the desert of central Australia. The west coast of South

Profile of the dry tropical climate • Figure 7.17

Wadi Halfa, on the Nile River in Sudan (lat. 22° N), almost on the Tropic of Cancer, has a climograph that is typical of the high temperatures and low rainfall of the dry tropical climate.

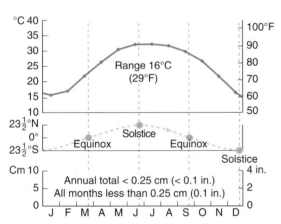

There is a strong annual temperature cycle with a very hot period at the time of high Sun. Daytime maximum air temperatures are frequently between 43° and 48°C (about 110° to 120°F) in the warmer months. There is a comparatively cool season at the time of low Sun.

There is too little rainfall to show on the climograph. Over a 39-year period, the maximum rainfall recorded in a 24-hour period at Wadi Halfa was only 0.75 cm (0.3 in.)

Put It Together

Review Figures 7.3 and 7.16 and answer this question.
Starting from Wadi Halfa, in which direction would you travel to encounter increased rainfall most quickly?

a. North
b. East
c. South
d. West

WHAT A GEOGRAPHER SEES

Desertification

A casual reader might recognize this portrait of Africa as a map of the amount of vegetation cover of the continent. But a geographer would be drawn to the transitional Sahel region, outlined in blue. South of the Sahel, rainfall and vegetation increase toward the equatorial rainforests of central Africa. North of the Sahel is the Sahara Desert, with virtually no vegetation. In the past, settlement, grazing, and agriculture were possible in the Sahel, but today are increasingly difficult. As a result of periods of drought and poor land use practices, such as overgrazing and overpopulation of marginal lands, the region is now threatened with desertification, as sand blows in from the Sahara and covers former croplands. Desertification is a particular threat in areas such as this one, where poverty and civil strife make populations especially vulnerable to climate change.

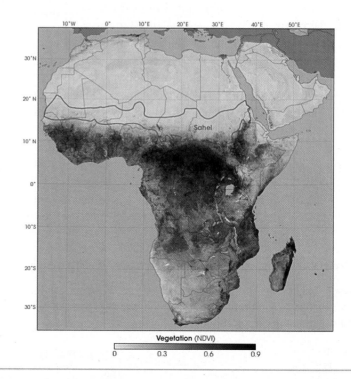

Vegetation (NDVI)
0   0.3   0.6   0.9

America, including portions of Ecuador, Peru, and Chile, also exhibits the dry tropical climate, but these western coastal desert regions, along with those on the west coast of Africa, are strongly influenced by cold ocean currents and the upwelling of deep, cold water, just offshore. The cool water moderates coastal zone temperatures, reducing the seasonality of the temperature cycle (**Figure 7.18**).

CONCEPT CHECK STOP

1. **What** causes the seasonal rainfall patterns in the monsoon and trade-wind coastal climates?

2. **Why** do western coastal deserts experience moderated temperatures compared to other dry tropical climates?

Profile of the western coastal desert climate • Figure 7.18

Walvis Bay, Namibia (lat. 23° S), situated along the desert coast of western Africa near the Tropic of Capricorn, has a climograph typical of the west-coast deserts that constitute a subcategory of the dry tropical climate.

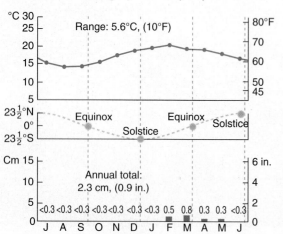

The monthly temperatures are remarkably cool for a location that is nearly on the Tropic of Capricorn.

Because of the coastal location, the annual range of temperatures is also small–only 5°C (9°F). Coastal fog is a persistent feature of this climate, providing most of the environmental moisture.

Midlatitude Climates

LEARNING OBJECTIVES

1. **Describe** the characteristic weather patterns of the six midlatitude climates.
2. **Explain** the causes of the weather patterns in each climate.

The midlatitude climates almost fully occupy the land areas of the midlatitude zone and a large proportion of the subtropical latitude zone. They also extend into the subarctic latitude zone along the western edge of Europe, reaching to latitude 60°. Unlike the low-latitude climates, which are about equally distributed between northern and southern hemispheres, nearly all of the midlatitude climate area is in the northern hemisphere. In the southern hemisphere, the land area poleward of latitude 40° is so small that the climates are dominated by a great southern ocean.

In the northern hemisphere, the midlatitude climates lie between two groups of very dissimilar air masses that interact intensely. Tongues of maritime tropical (mT) air masses enter the midlatitude zone from the subtropical zone, where they meet and conflict with tongues of maritime polar (mP) and continental polar (cP) air masses along the polar-front zone.

The midlatitude climates include the poleward halves of the great subtropical high-pressure systems and much of the belt of prevailing westerly winds. As a result, weather systems, such as traveling cyclones and their fronts, characteristically move from west to east. This global air flow influences the distribution of climates from west to east across the North American and Eurasian continents.

There are six midlatitude climate types, ranging from those with strong wet and dry seasons to those with uniform precipitation. Temperature cycles for these climate types are also quite varied. We will now examine each of these climate types in more detail.

The Dry Subtropical Climate

The **dry subtropical climate** ⑤ is simply a poleward extension of the dry tropical climate, characterized by the warm, dry weather of the poleward side of the subtropical high-pressure cells. As latitude increases, insolation decreases somewhat and becomes more seasonal. This tends to reduce overall temperatures and increase the contrast between summer and winter temperatures (**Figure 7.19**). As a result, the

> **dry subtropical climate** The dry climate of the subtropical zone, transitional between the dry tropical climate and the dry midlatitude climate.

Profile of the dry subtropical climate • Figure 7.19

Yuma, Arizona (lat. 33° N), has a climograph that is typical of an arid region in the dry subtropical climate.

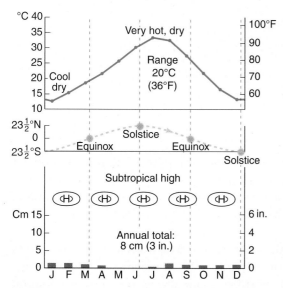

There is a strong seasonal temperature cycle, with a dry hot summer and freezing temperatures in December and January.

Precipitation is small in all months but has peaks in late winter and late summer. The August maximum is caused by the invasion of maritime tropical (mT) air masses, which bring thunderstorms to the region. Higher rainfalls from December through March are produced by midlatitude wave cyclones following a southerly path. Two months, May and June, are nearly rainless.

Global distribution of the moist subtropical climate • Figure 7.20

The moist subtropical climate ⑥ includes most of the southeastern United States, from the Carolinas to eastern Texas. Southern China, Taiwan, and southernmost Japan are also included. In South America, this climate includes parts of Uruguay, Brazil, and Argentina. In Australia, it consists of a narrow band between the eastern coastline and the eastern interior ranges.

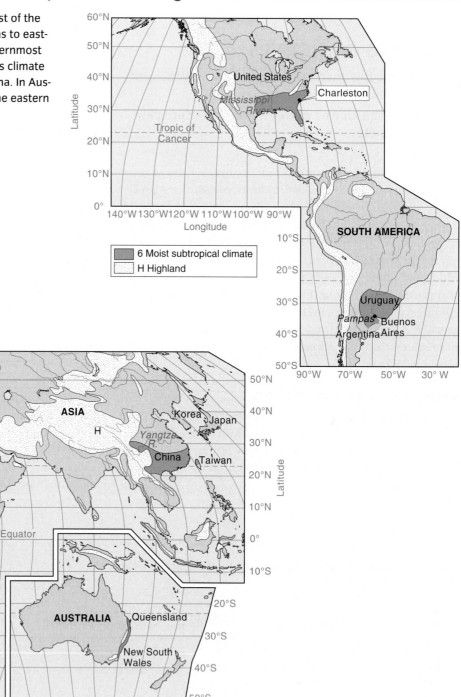

lower-latitude portions of this climate have a distinct cool season, and the higher-latitude portions have a cold season. The cold season occurs at a time of low Sun and is caused in part by the invasion of cold continental polar (cP) air masses from higher latitudes. Wave cyclones occasionally move into the subtropical zone in the low-Sun season, producing precipitation. There are both arid and semiarid subtypes in this climate.

The dry subtropical climate is found in a broad band of North Africa, connecting with the Near East. Southern Africa and southern Australia also contain this climate. A band of dry subtropical climate occupies Patagonia, in South America.

In North America, the Mojave and Sonoran deserts of the American Southwest and northwest Mexico are regions of dry subtropical climate. At the northern margin, in the interior Mojave Desert of southeastern California at about 34° N, the great summer heat is comparable to that in the Sahara Desert, but low Sun brings a winter season unseen in the tropical deserts. Here, cyclonic precipitation can occur in most months, including the cool low-Sun months.

The Moist Subtropical Climate

On the eastern side of continents that lie in latitudes of about 20° to 35° N and S, the subtropical high-pressure cells provide an eastward flow of maritime tropical (mT) air to the **moist subtropical climate** ⑥ (**Figure 7.20**). The flow

> **moist subtropical climate** The moist climate of the subtropical zone, characterized by abundant rainfall and a strong seasonal temperature cycle.

of warm, moist air dominates this climate, providing warm temperatures, high humidity, and high precipitation (**Figure 7.21**). Nearby warm ocean currents, like the Gulf Stream, also add moisture and warmth.

In summer, this mT air often becomes unstable, providing abundant convective precipitation. Occasional tropical cyclones add to the summer rainfall total. There is also plenty of winter precipitation, produced by midlatitude cyclones. Continental polar (cP) air masses frequently invade this climate region in winter, bringing spells of subfreezing weather. In Southeast Asia, this climate is characterized by a strong monsoon effect, with much more rainfall in the summer than in the winter.

Profile of the moist subtropical climate • Figure 7.21

Charleston, South Carolina (lat. 33° N), located on the eastern seaboard, has a climograph that is typical of the moist subtropical climate, with a mild winter and a warm summer and year-round precipitation.

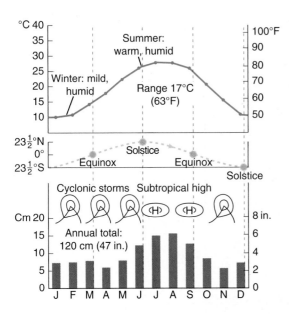

There is a strongly developed annual temperature cycle, with a large annual range. Winters are mild, with the January mean temperature well above the freezing mark.

Ample precipitation falls in every month, with a definite summer maximum.

Global distribution of the Mediterranean and marine west-coast climates • Figure 7.22

In North America, the Mediterranean climate ⑦ is found in central and Southern California. In the southern hemisphere, it occurs along the coast of Chile, in the Cape Town region of South Africa, and along the southern and western coasts of Australia. In Europe, this climate type surrounds the Mediterranean Sea, giving the climate its name. The general latitude range of marine west-coast climate ⑧ is 35° to 60° N and S.

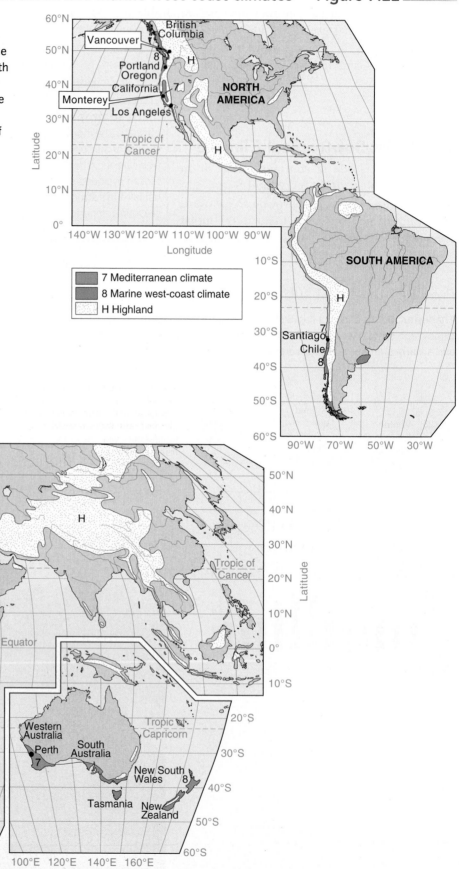

The Mediterranean Climate

Mediterranean climate The climate type of the midlatitudes, characterized by the alternation of a very dry summer and a mild, rainy winter.

The **Mediterranean climate** ⑦, found between latitudes 30° and 45° N and S (**Figure 7.22**), is unique because it has a wet winter and a very dry summer. This is because the climate is located along the west coasts of continents, just poleward of the dry eastern side of the subtropical high-pressure cells. When the subtropical high-pressure cells move poleward in summer, they enter the Mediterranean climate region. Dry continental tropical (cT) air dominates, producing the dry summer season. In winter, the subtropical high-pressure cells weaken and move toward the equator. Moist mP air mass invades, with cyclonic storms that generate ample rainfall.

The Mediterranean climate spans arid to humid climates, depending on location. Generally, the closer an area is to the tropics, the stronger the influence of subtropical high pressure will be, and thus the drier the climate. The temperature range is moderate, with warm to hot summers and mild winters. Coastal zones between latitudes 30° and 35° N and S, such as Southern California, show a smaller annual range, with very mild winters (**Figure 7.23**).

The native vegetation of the Mediterranean climate environment is adapted to survive through the long summer drought. Shrubs and trees are typically equipped with small, hard, or thick leaves that resist water loss through transpiration.

Profile of the Mediterranean climate • Figure 7.23

Monterey, California, has a typical Mediterranean precipitation pattern, with an almost rainless summer. The temperature range here is moderated by the coastal location, with westerly winds moving across cool currents just offshore. Coastal upwelling of bottom waters brings nutrients to surface waters, supporting extensive fisheries.

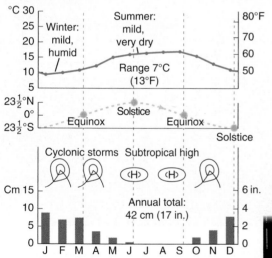

At Monterey, the annual temperature cycle is weak because of the coastal location.

The summer is very dry. Rainfall drops to nearly zero for four consecutive summer months but rises to substantial amounts in the rainy winter season.

Profile of the marine west-coast climate • Figure 7.24

Vancouver, British Columbia (lat. 49° N), has a climograph typical of the marine west-coast climate. The conifer forests seen lining the nearby mountains thrive on the abundant orographic precipitation provided by the moist marine air.

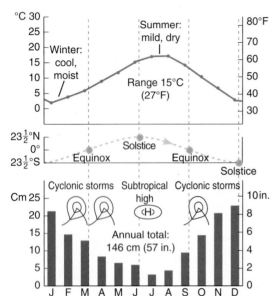

The annual temperature range is small, and winters are very mild for this latitude.

The annual precipitation total is large, with most precipitation falling in winter.

The Marine West-Coast Climate

marine west-coast climate The cool, moist climate of west coasts in the mid-latitude zone, usually with abundant precipitation and a distinct winter precipitation maximum.

Like the Mediterranean climate, the **marine west-coast climate** ⑧ has a winter precipitation maximum. Located on midlatitude west coasts poleward of the Mediterranean climate, it is also influenced by subtropical high pressure in the summer, but not as strongly. This makes the summers drier than the winters. These locations receive the prevailing westerlies from a large ocean, and there are frequent cyclonic storms involving cool, moist mP air masses that provide ample precipitation in all months. Where the coast is mountainous, the orographic effect causes large amounts of precipitation annually. The annual temperature range is comparatively small for midlatitudes. The marine influence keeps winter temperatures milder than at inland locations at equivalent latitudes (**Figure 7.24**).

In North America, the marine west-coast climate occupies the western coast from Oregon to northern British Columbia and southwesternmost Alaska. In western Europe, the British Isles, northern Portugal, and much of France fall under this climate category. In the southern hemisphere, this climate includes New Zealand and the southern tip of Australia, as well as the island of Tasmania and the Chilean coast south of 35° S.

The Dry Midlatitude Climate

dry midlatitude climate The dry climate of the midlatitude zone, with a strong annual temperature cycle and cold winters.

The **dry midlatitude climate** ⑨ is almost exclusively limited to the interior regions of North America and Eurasia, where sources of moist maritime air are far away. Much of the area lies within the rain shadow of mountain ranges on the west or south. The mountain ranges effectively block the eastward flow of maritime air masses. In winter, continental polar (cP) air masses dominate. In summer, solar heating creates a local dry continental air mass, but occasionally maritime air masses invade, causing convective rainfall. The annual temperature cycle is strongly developed, with a large annual range. Summers are warm to hot, but winters are cold to very cold (**Figure 7.25**). The latitude range of this climate is 35° to 55° N.

The largest expanse of the dry midlatitude climate is in Eurasia, stretching from the Black Sea region to the Gobi Desert and northern China. True arid deserts and extensive areas of highlands occur in the central portions of this region. In North America, the dry western interior regions, including the Great Basin, Columbia Plateau, and the Great Plains, are of the semiarid, or steppe, subtype. A small area of dry midlatitude climate is found in southern Patagonia, near the tip of South America.

Profile of the dry midlatitude climate • Figure 7.25

Pueblo, Colorado (lat. 38° N), just east of the Rocky Mountains, has a climograph typical of the dry midlatitude climate, with a marked summer maximum of rainfall in the summer months.

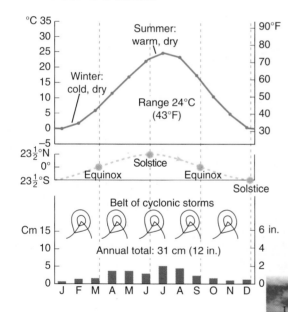

The temperature cycle has a large annual range, with warm summers and cold winters. January, the coldest winter month, has a mean temperature just below freezing.

Most of the annual precipitation is convective summer rainfall, which occurs when moist maritime tropical (mT) air masses invade from the south and produce thunderstorms. In the winter, snowfall is light.

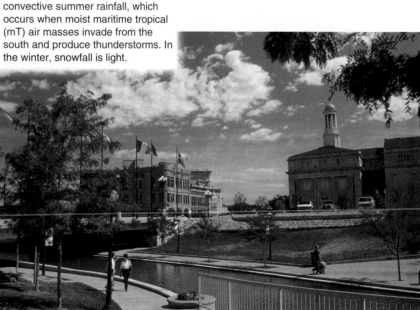

Global distribution of the moist continental climate • Figure 7.26

The moist continental climate ⑩ is restricted to the northern hemisphere, between latitudes 30° and 55° N in North America and Asia. Most of the north and eastern half of the United States, from Tennessee to the Canadian border and somewhat beyond, lies in this climate region. In Asia, it occurs in northern China, Korea, and Japan. Most of central and eastern Europe in latitudes 45° to 60° N has a moist continental climate.

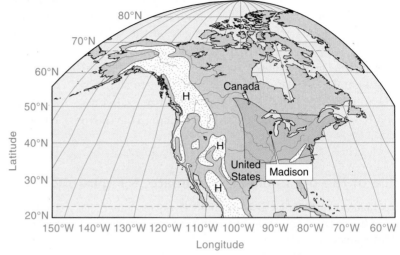

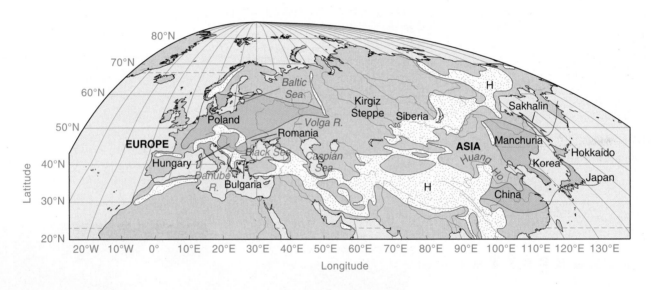

The Moist Continental Climate

moist continental climate The moist climate of the midlatitudes with strongly defined winter and summer seasons and adequate precipitation throughout the year.

The **moist continental climate** ⑩ lies in the midlatitudes at the polar-front zone, where polar and tropical air masses meet. As a result, day-to-day weather is highly variable. Located in the midlatitudes in the central and eastern parts of North America and Eurasia (**Figure 7.26**), this climate can experience wave cyclones at any time of year. Precipitation is ample throughout the year, increasing in summer when maritime tropical (mT) air masses invade. Seasonal

temperature contrasts are strong. Cold winters are dominated by continental polar (cP) and continental arctic (cA) air masses from subarctic source regions (**Figure 7.27**).

Most of the eastern half of the United States from Tennessee to the north, as well as the southernmost strip of eastern Canada, lies in the moist continental climate zone. China, Korea, and Japan have a moist continental climate region that has more summer rainfall and a drier winter than North America. This is an effect of the monsoon circulation, which moves moist maritime tropical (mT) air across the eastern side of Asia in summer and dry continental polar southward through the region in winter. In Europe, the moist continental climate lies in a higher latitude belt

Profile of the moist continental climate • Figure 7.27

Madison, Wisconsin (lat. 43° N), has a moist continental climate. Much of the winter precipitation is snow, as shown in this photo, which remains on the ground for long periods.

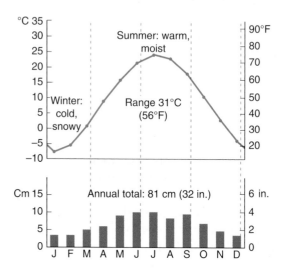

Winters are cold—with three consecutive monthly means well below freezing—and summers are warm, making the annual temperature range very large.

There is ample precipitation in all months, and the annual total is large. There is a summer maximum of precipitation when the maritime tropical (mT) air mass invades, and convective precipitation forms along moving cold fronts and squall lines.

(45° to 60° N) and receives precipitation from mP air masses coming from the North Atlantic. This climate does not occur in the southern hemisphere, since there are no large land masses or source regions for cP air masses.

Forests are the dominant natural vegetation cover throughout most of the climate, although tall, dense grasses are the natural cover where the climate grades into drier climates, such as the dry midlatitude climate.

CONCEPT CHECK STOP

1. **What** are the similarities and differences between the six climate types of the midlatitude climate group?

2. **What** makes the Mediterranean climate unique?

High-Latitude Climates

LEARNING OBJECTIVES

1. **Describe** the characteristic weather patterns of the three high-latitude climates.

2. **Explain** the causes of the weather patterns in each of the three high-latitude climates.

By and large, the high-latitude climates occupy the northern subarctic and arctic latitude zones of the northern hemisphere. They also extend southward into the midlatitude

zone as far south as about latitude 47° in eastern North America and eastern Asia. The boreal forest climate and tundra climate are absent in the southern hemisphere, which lacks a high-latitude land mass between 50° and 70° S. However, the ice sheet climate is present in both hemispheres, in Greenland and Antarctica.

The high-latitude climates coincide closely with the belt of prevailing westerly winds that circles each pole. In the northern hemisphere, this circulation sweeps maritime polar (mP) air masses, formed over the northern oceans, into conflict with continental polar (cP) and continental arctic (cA) air masses on the continents. Jet-stream disturbances form in the westerly flow, bringing lobes of warmer, moister air poleward into the region in exchange for colder, drier air that is pushed equatorward. As a result of these processes, wave cyclones are frequently produced along the arctic-front zone.

The Boreal Forest Climate

boreal forest climate The cold climate of the subarctic zone in the northern hemisphere, with long, severe winters and generally low annual precipitation.

The **boreal forest climate** ⑪ occupies the source region for cP air masses, which are cool or cold, dry, and stable. As a result, temperatures are cool or cold, particularly in winter, and precipitation is low. In North America, the boreal forest climate stretches from central and western Alaska across

the Yukon and Northwest Territories to Labrador on the Atlantic coast. In Europe and Asia, it reaches from the Scandinavian Peninsula eastward across all of Siberia to the Pacific (**Figure 7.28**).

The boreal forest climate is a continental climate with long, bitterly cold winters and short, cool summers (**Figure 7.29**). Colder cA air masses very commonly invade the region. The annual range of temperature is greater here than in any of the other climates, and the range is greatest in Siberia, in Russia.

Precipitation increases substantially in summer, when maritime air masses penetrate the continent with wave cyclones, but the total annual precipitation is small. Although much of the boreal forest climate is moist, large areas in western Canada and Siberia have low annual precipitation and are therefore cold and dry.

The land surface features of much of the region of boreal forest climate were shaped beneath the great ice

Global distribution of the boreal forest climate • Figure 7.28

The boreal forest climate ⑪ ranges from 50° to 70° N latitude.

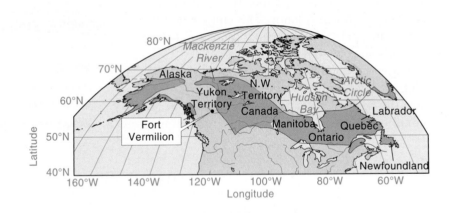

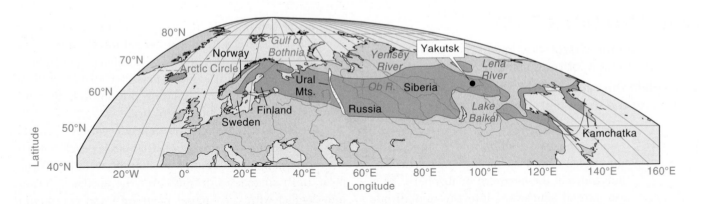

11 Boreal forest climate

Profile of the boreal forest climate • Figure 7.29

Fort Vermilion, Alberta (lat. 58° N), and Yakutsk, Siberia (lat. 62° N), both have climographs characteristic of the boreal forest climate, with extreme winter cold and a very great annual range in temperature. A snow cover remains over solidly frozen ground, shown in this photo from Yakutsk, through the entire winter.

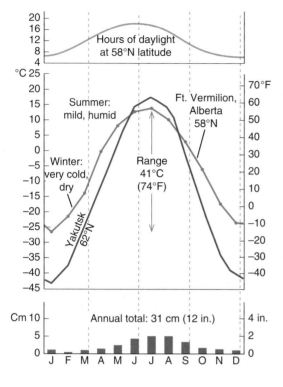

Temperatures are below freezing for more than half the year. The summers are short and cool.

Precipitation has a marked annual cycle with a summer maximum, but the total annual precipitation is small.

sheets of the last ice age. Erosion by the moving ice exposed hard bedrock over vast areas and created numerous shallow rock basins that are now lakes (**Figure 7.30**).

The dominant upland vegetation of the boreal forest climate region is boreal forest, consisting largely of needleleaf trees. Although the growing season in the boreal forest climate is short, cultivating crops there is possible. Farming is largely limited to lands surrounding the Baltic Sea, bordering Finland and Sweden. Crops grown in this area include barley, oats, rye, and wheat. The needleleaf forests provide paper, pulp, cellulose, and construction lumber.

Lakes in a boreal forest • Figure 7.30

Much of the boreal forest consists of low but irregular topography, formed by continental ice sheets during the last ice age. Low depressions scraped out by the moving ice are now occupied by lakes. Alaska's Mulchatna River is in the foreground of this aerial photo.

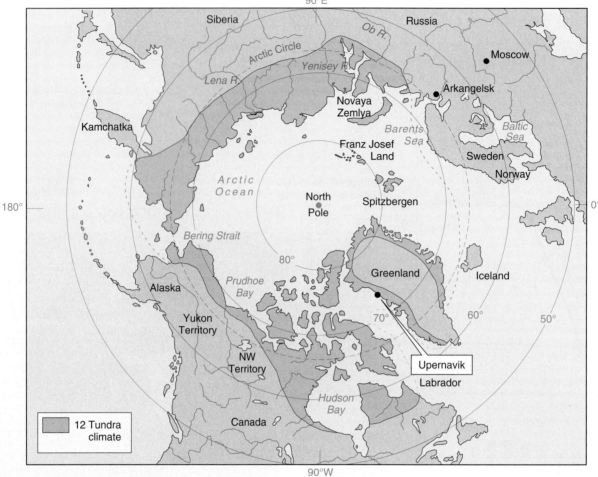

Global distribution of the tundra climate • Figure 7.31

The latitude range for the tundra climate ⑫ is 60° to 75° N and S, except for the northern coast of Greenland, where tundra occurs at latitudes above 80° N.

The Tundra Climate

The **tundra climate** ⑫ is a cold maritime climate of the polar and arctic regions. It occupies arctic coastal fringes, ringing the Arctic Ocean and extending across the island region of northern Canada (**Figure 7.31**). This climate includes the Alaska North Slope, the Hudson Bay region, and the Greenland coast in North America. In Eurasia, this climate type occupies the northernmost fringe of the Scandinavian Peninsula and Siberian coast. The Antarctic Peninsula (not shown in Figure 7.31) also belongs to this climate.

This climate is dominated by polar (cP and mP) and arctic (cA) air masses. Winters are long and severe. The nearby ocean water moderates winter temperatures so they don't fall to the extreme lows found in the continental interior. There is a very short mild season, but many climatologists do not recognize this as a true summer (**Figure 7.32**).

> **tundra climate** The cold climate of the arctic zone, which has eight or more months of frozen ground.

The term *tundra* describes both an environmental region and a major class of vegetation. Hardy perennial grasses and sedges, accompanied by low, frost-resistant shrubs, are arranged sparsely on rocky soils. Peat bogs are numerous. In some places, a distinct tree line—roughly along the 10°C (50°F) isotherm of the warmest month—separates the boreal forest and tundra.

Because of the cold temperatures experienced in the tundra and northern boreal forest climate zones, the ground is typically frozen to great depth. This perennially frozen ground, or *permafrost*, prevails over the tundra region. Temperature increases in the Arctic associated with global warming have thawed the upper permafrost layer at some locations, leaving unstable soils that are unable to support roads and buildings. Oil exploration pressures have added to the challenges facing those charged with protecting the tundra ecosystems (see *Video Explorations*).

Upernavik, located on the west coast of Greenland (lat. 73° N), has a climograph typical of the tundra climate, with a short mild period of above-freezing temperatures and little annual precipitation.

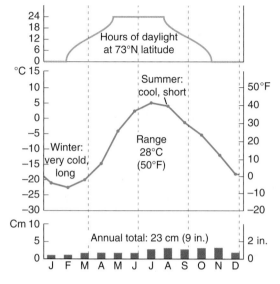

The long winter is very cold, but the annual temperature range is not large, given the high latitude.

Total annual precipitation is low. In July, the sea-ice cover melts and the ocean water warms, raising the moisture content of the local air mass, increasing precipitation.

Video Explorations

Alaska: The Price of Progress

The Arctic National Wildlife Refuge is a large wilderness area located on the northern edge of our continent. It has been the focus of controversy in recent years because underneath this area lies a major oil reservoir, and it has been proposed that the United States begin drilling. Some argue that the United States must reduce its reliance on oil from the Middle East and note that drilling in Alaska could generate $80 billion. However, others believe that the threat to wildlife in this unique tundra reserve is too great. This video explores both sides of this controversial issue.

Profile of the ice sheet climate • Figure 7.33

The ice sheet climate is always below freezing and bitterly cold in winter.

a. Antarctica

Snow and ice accumulate at higher elevations here in the Dry Valleys region of Victoria Land, Antarctica. Most of Antarctica is completely covered by ice sheets of the polar ice cap.

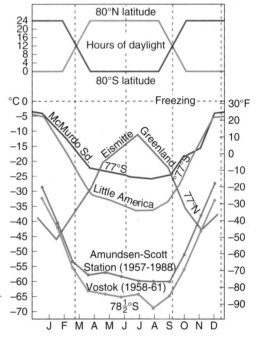

b. Temperature graphs for five ice sheet stations

Eismitte is on the Greenland ice cap; the other stations are in Antarctica. Temperatures in the interior of Antarctica are far lower than at any other place on Earth. A low of −89.2°C (−128.6°F) was observed in 1983, at Vostok, about 1300 km (about 800 mi) from the South Pole at an altitude of about 3500 m (11,500 ft). At the pole (Amundsen-Scott Station), July, August, and September have averages of about −60°C (−76°F). Temperatures are considerably higher, month for month, at the Little America research base (now abandoned) in Antarctica because it is located close to the Ross Sea and is at a low altitude.

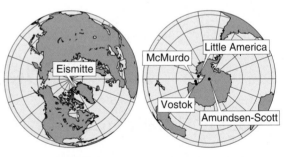

The Ice Sheet Climate

ice sheet climate
The severely cold climate found on the Greenland and Antarctic ice sheets.

The **ice sheet climate** ⑬ coincides with the source regions of arctic (A) and antarctic (AA) air masses, situated on the vast, high ice sheets of Greenland and Antarctica and over polar sea ice of the Arctic Ocean (**Figure 7.33**). Mean annual temperature in the ice sheet climate is much lower than that of any other climate, with no monthly mean above freezing. Strong temperature inversions, caused by radiative loss from the surface, develop over the ice sheets. In Antarctica and Greenland, the high surface altitude of the ice sheets intensifies the cold. Strong cyclones with blizzard winds are frequent. There is very little precipitation, almost all occurring as snow, but the snow accumulates because of the continuous cold. The latitude range for this climate is 65° to 90° N and S.

CONCEPT CHECK

1. **What** are the similarities and differences between the three climate types of the high-latitude climate group?

2. **What** explains the variations in weather patterns between climates of each type?

Climate Change

LEARNING OBJECTIVES

1. **Explain** the effects of climate change on temperature and precipitation.

2. **Describe** the effects of climate change on extreme weather events.

3. **Identify** the major challenges posed by climate change.

So far in this chapter, we have focused on the characteristics of the Earth's major climates as we experience them today. However, these climate characteristics have changed many times over the course of the Earth's history. Based on a variety of field data, scientists now agree that our planet has experienced a series of climate shifts over the course of its recent geologic history, alternating between colder periods of glaciation and milder interglacial periods, such as the present (**Figure 7.34**).

Plant and animal species are in dynamic equilibrium with the environment, which means that as the environment

Temperature and CO₂ records • Figure 7.34

Scientific data from ice cores and other field measurements portray a cycle of climate change that includes ice ages and interglacial periods over the past few hundred thousand years and the past century (inset graph). These graphs demonstrate the correlation between historical CO_2 levels and temperature cycles and show that today's levels of CO_2 are much higher than in the past.

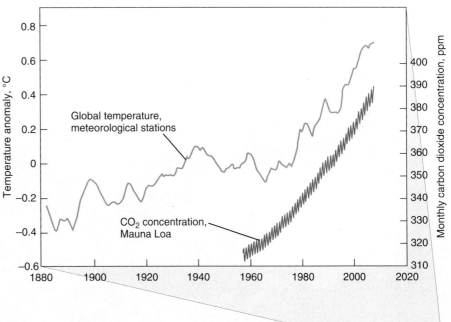

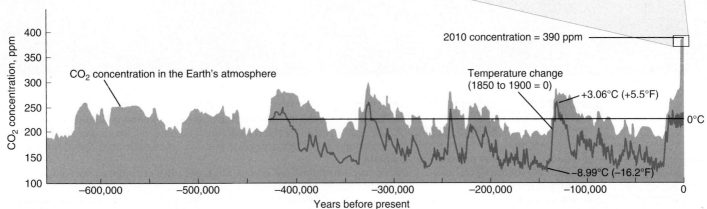

changes, species must adjust by adapting to the new environments or moving to keep pace with the changing areas of their older environments. For example, the polar bear (*Ursus maritimus*) evolved 200,000 years ago from its brown bear cousin by adapting to the icy arctic landscape. During the last glaciation, many common plant species moved equatorward and then migrated back during the first few centuries of interglacial climate.

Since the last glaciation ended approximately 10,000 years ago, our human species has thrived in the relatively calm and mild current distribution of climates. However, the Earth is entering a new era of climate change that has been triggered to a significant degree by the increases in carbon dioxide (CO_2) and other greenhouse gases in our atmosphere, discussed in detail in Chapter 3.

Shifting Climate Characteristics

As we established at the beginning of this chapter, climates are characterized by their annual patterns of temperature and precipitation. Both temperature and precipitation are expected to change as a result of global warming. However, climate change will not bring a simple or uniform poleward shift of boundaries between the different climate zones discussed in this chapter. Instead, there will be great variability in the effects experienced in different regions. All land regions will get warmer, but with more variability in temperature and precipitation.

Temperature Temperature increases are expected to be greatest at the poles due to the circulation patterns that move energy away from the equator. Based on current trends, scientists expect late-summer arctic sea ice to nearly disappear by the latter part of the century (**Figure 7.35**). Warming at the poles influences the pressure zones that push air masses into the midlatitudes and control weather patterns. These shifting weather patterns will affect every aspect of our modern lives, from agricultural output to human health and natural disasters.

Polar climate • Figure 7.35 _____

The ice sheet climate is experiencing the greatest rate of warming compared to all climates. Measurable impacts in the Arctic and in Antarctica include permafrost melting, ice loss, and seasonal shifts. These images show the loss of perennial, or year-round, sea ice in the Arctic between 1979 and 2003. The ice is declining at a rate of 9% per decade.

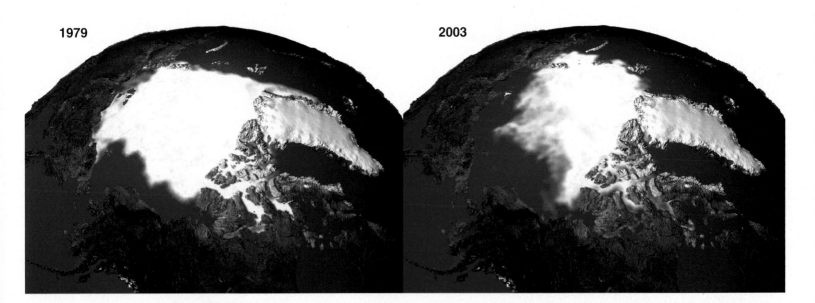

1979 **2003**

> **Put It Together**
>
> *Review Figures 7.28 and 7.29 and answer this question.*
> If the Arctic continues to warm, which of the following transformations is most likely?
>
> a. The area covered by the tundra climate will expand in all directions.
> b. The area covered by the boreal forest climate will expand in all directions.
> c. The boreal forest climate will move into areas once considered tundra climate.
> d. The tundra climate will move into areas once considered boreal forest climate.

Where Geographers CLICK

Global Monitoring of Climate Change

http://iridl.ldeo.columbia.edu/maproom/.Global/

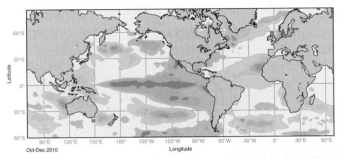

Oct-Dec 2010

This Web site from the International Research Institute for Climate and Society provides select mapping options to display the status of global Earth climate monitoring that can be generated based on the user's selection of key parameters. Based on scientific data from satellites and ground stations, this Web site can create maps to define current conditions, anomalies from decadal averages, and detailed regional maps of our changing climates.

Precipitation Global warming will trigger changes in average annual precipitation around the world. Scientists generally expect increased precipitation and evaporation in some areas and drier conditions in others. The prevailing factors that influence global precipitation, such as ocean currents and pressure cells, are shifting and will shift further. For example, the subtropical high-pressure cells will strengthen and move poleward, reducing precipitation in subtropical zones. In turn, rainfall will increase at higher latitudes and in some areas of the tropics. However, a lot of uncertainty still remains about just where and how precipitation patterns will change.

Climate scientists are now actively measuring shifts in precipitation, atmospheric temperatures, atmospheric circulation, and ocean temperatures that create the Earth's climates (see *Where Geographers Click*). Their data will help to calibrate and confirm predictions of climate change.

Weather Variability

Climate change will affect more than the temperature and precipitation patterns of different climates around the world. The frequency of extreme weather events is also predicted to change. For example, increases in extreme temperature events are predicted for California. In the western mountain ranges, overall precipitation will be down, but winter snowfall extremes will be more likely.

There is also sound evidence of an increase in extreme weather events. Variations in storm paths and intensities reflect variations in major features of the atmospheric circulation patterns. Tropical storm and hurricane frequencies vary from year to year, but data show substantial increases in intensity and duration of these severe storms since the 1970s. The recent increased frequency of the El Niño-Southern Oscillation (ENSO) cycles, which is thought to be linked to global warming, has brought record floods and droughts across all continents. The frequency of weather-related natural disasters has also increased, as cyclone activity and above-average rainfall has been triggered by warmer sea and land temperatures (**Figure 7.36**). Insurance companies reported property damage in 2010 due to natural disasters to be $130 billion. The increasing frequency and intensity of weather events is a trend that will require international attention to mitigate disasters.

Frequency of natural disasters • Figure 7.36

This graph shows that natural disasters have been occurring more frequently in recent years. Some of this increase is probably due to population growth and improved information and reporting. However, the occurrence of weather-related natural disasters, such as floods and cyclones, has increased significantly while the incidence of earthquakes has remained about the same, suggesting that some of the increase is related to climate changes.

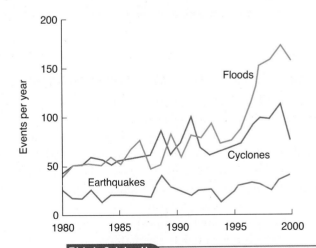

Think Critically

What role do population growth and infrastructure development play in the damages caused by natural disasters?

Global climate modeling • Figure 7.37

A global climate model allows scientists to project different scenarios for the Earth's future climate, based on selected targets for future emissions of greenhouse gases. The model is calibrated for reliability using historical data to ensure that it correctly predicts the temperature and moisture fluctuations, as recorded by the global network of weather stations. Current data are then used to set the initial meteorological conditions for the model runs that simulate climate scenarios for 50 and 100 years into the future.

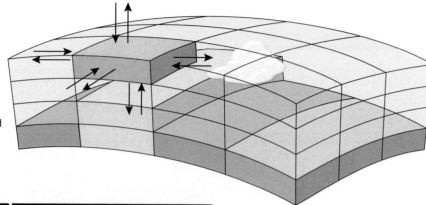

1 Scientist divide the Earth into a grid of modeling cells. Each cell represents a certain space (typically hundreds of kilometers) and time (hours or days) of the Earth's oceans, land, and atmosphere.

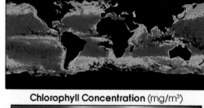

Chlorophyll Concentration (mg/m³)

.01 .03 .1 .3 1 3 10 30 60

Sea Surface Temperature (°C)

-2   45

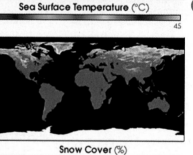

Cloud Fraction

0.0 0.2 0.4 0.6 0.8 1.0

Snow Cover (%)

0 20 40 60 80 100

2 Global monitoring satellites and field data provide the real-life conditions in each cell. Major data fields include atmosphere, oceans, land and ice, but many global climate models include only atmosphere and ocean interactions.

Future Challenges and Adaptations

As our climate changes and global temperatures increase, humans, plants, and animals will all have to face new challenges and develop new strategies to adapt. Over the past few years, climate research scientists have used **global climate models** to investigate the Earth's past and future climates. These are complex computer programs that re-create and model the physical, chemical, and biological conditions in the Earth's atmosphere, oceans, land, and ice. Global climate models are calibrated to determine how well they re-create the Earth's existing climate by using historic records from past decades. The ability of these models to re-create our known weather patterns establishes their reliability for future projections.

Once a global climate model has been proven to be reliable, scientists can project what the Earth's future climate might be. Global climate models are run under different sets of assumptions for greenhouse gas emissions—for example, whether emissions stay the same, decrease as called for by international goals, or increase according to current trends (**Figure 7.37**). About a dozen major climate change models have been included in the studies by the Intergovernmental Panel on Climate Change (IPCC). No single model gives a definitive picture of the Earth's future climate, but together these climate projections show a range of possibilities and suggest how our policy decisions now may affect future conditions.

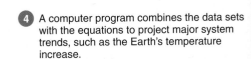

Biological	Chemical	Physical
Vegetation	Aerosols	Precipitation
		Wind
Chlorophyll	Greenhouse gases	Clouds
		Temperature
Moisture	Ozone	Pressure

3 Mathematical equations predict the real-world mixing interactions and processes that go on in each modeling cell. These include calculations for physical, chemical, and biological processes.

4 A computer program combines the data sets with the equations to project major system trends, such as the Earth's temperature increase.

5 Different scenarios for the Earth's temperature increase are generated based on various assumptions for future greenhouse gas emissions, such as no change in emissions, low emissions growth, or increased emissions growth. The global climate model is able to project global trends, but it cannot predict conditions within a given modeling cell.

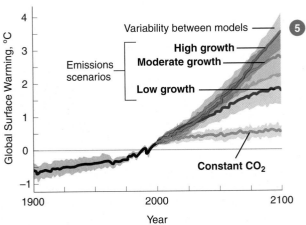

Climate scientists are careful to point out that the Earth is much more complicated than the models and mathematical equations that simulate the Earth–ocean–atmosphere system. It is beyond even today's state-of-the-science to predict our future climate with great accuracy. However, the convergence of global climate models makes it possible to make predictions that are more likely to be correct in some areas than in others.

The big question facing communities and governments around the world is how human societies will be affected by coming climate change and how to adapt to that change. For physical geographers and other scientists, the coming decades will provide great challenges and rewards for their professional careers, as they work to understand and mitigate the negative effects of environmental change.

CONCEPT CHECK

1. **Which** region is expected to experience the most significant warming?
2. **What** is the effect of more frequent ENSO cycles on global climate?
3. **How** are global climate models used to predict climate change?

Summary

1 Keys to Climate 194

- **Climate** is the average weather of a region, based on annual patterns of monthly averages of temperature and precipitation.

- Temperature depends on latitude, which determines the annual pattern of insolation. It also depends on location—continental or maritime—which enhances or moderates the annual insolation cycle. Finally, it depends on elevation, which decreases with altitude in the atmosphere.

- Global precipitation patterns are largely determined by air mass characteristics and movements.

- In dry climates, precipitation is largely evaporated from soil surfaces and transpired by vegetation. There are two subtypes: **arid** (driest) and **semiarid**, or steppe (a little wetter). In moist climates, precipitation exceeds evaporation and transpiration. In wet-dry climates, strong wet and dry seasons alternate.

- Annual precipitation falls into three patterns: uniform, low-Sun (winter) maximum, and high-Sun (summer) maximum, such as the pattern shown in the graph.

Seasonal precipitation patterns • Figure 7.4

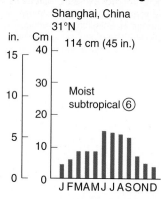

2 Climate Classification 199

- Climate classification systems use mean monthly measurements of temperature and precipitation to assign locations to climate types.

- Climate types can be classified into three groups, arranged by latitude: low-latitude climates, midlatitude climates, and high-latitude climates.

- **Highland climates** are cool and generally wet, and they derive their characteristics from surrounding lowlands.

3 Low-Latitude Climates 206

- The **wet equatorial climate** is warm to hot, with abundant rainfall from the ITCZ. This is the steamy climate of the Amazon and Congo basins.

- The **monsoon and trade-wind coastal climate** is warm to hot, with a very wet rainy season. It occurs in coastal regions that are influenced by trade winds or a monsoon circulation. The climates of western India, Myanmar, and Bangladesh are good examples.

- The **wet-dry tropical climate** is warm to hot, with very distinct wet seasons when the ITCZ is nearby and dry seasons when subtropical high pressure dominates, as shown in the climograph. This type includes interior regions of Vietnam, Laos, and Cambodia and occurs in Africa and the Americas as well.

Profile of the wet-dry tropical climate • Figure 7.15

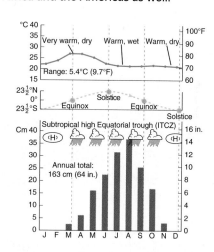

- The **dry tropical climate** describes the world's hottest deserts—extremely hot in the high-Sun season and a little cooler in the low-Sun season, with little or no rainfall. The Sahara Desert, Saudi Arabia, and the central Australian desert are examples of this climate. Drought, overgrazing, and poor agricultural practices cause the expansion of deserts, called **desertification**.

4 Midlatitude Climates 215

- The **dry subtropical climate** includes desert regions. But they are found farther poleward than the dry tropical climate and so aren't as hot. This type includes the hottest part of the American Southwest desert.

- The **moist subtropical climate** is dominated by eastward flows of maritime tropical air masses, providing hot and humid summers, mild winters, and ample rainfall year-round, as shown in the climograph. This type includes the southeastern regions of the United States and China.

Profile of the moist subtropical climate • Figure 7.21

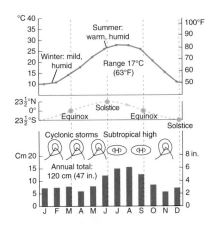

- The **Mediterranean climate** is marked by a hot, dry summer when subtropical high pressure dominates and a rainy winter from mP air in cyclonic storms. Southern and central California, Spain, southern Italy, Greece, and the coastal regions of Lebanon and Israel are prime examples.

- The **marine west-coast climate** has a warm, largely dry summer, when subtropical high pressure is nearby, and a cool, wet winter, when moist mP air dominates. Climates of this type include the Pacific Northwest—coastal Oregon, Washington, and British Columbia.

- The **dry midlatitude climate** is a dry climate of midlatitude continental interiors where sources of moisture are far away. The steppes of central Asia and the Great Plains of North America are familiar locales with this climate—warm to hot in summer, cold in winter, and with low annual precipitation.

- The **moist continental climate** lies in the polar-front zone, where polar and tropical air masses meet. This is the climate of the eastern United States and lower Canada—cold in winter, warm in summer, with ample precipitation through the year.

5 High-Latitude Climates 223

- The **boreal forest climate** occurs in the source region for continental polar air masses. This is a snowy climate with short, cool summers and long, bitterly cold winters as

shown in the climograph. Northern Canada, Siberia, and central Alaska are regions of boreal forest climate.

Profile of the boreal forest climate • Figure 7.29

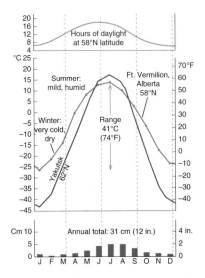

- **Tundra climate** is a cold maritime climate of polar and arctic regions. Winters are long and severe, but temperatures are somewhat moderated by the nearby Arctic Ocean. This is the climate of the coastal arctic regions of Canada, Alaska, Siberia, and Scandinavia.

- The **ice sheet climate** occurs in the source region of arctic and antarctic air masses in Greenland and Antarctica. It is bitterly cold. Temperatures can drop below −50°C (−58°F) during the sunless winter months. Even during summer days of 24-hour sunlight, temperatures remain well below freezing.

6 Climate Change 229

- Humans have enjoyed a 10,000-year interglacial period of mild climate. However, the Earth is entering a new climate era, in part due to increased global warming as a result of the combustion of fossil fuels.

- As global warming proceeds, all land areas are projected to be warmer. Some areas will become wetter and some drier.

- Weather will become more variable, with an increased frequency of rare, extreme events, such as floods, droughts, and severe storms.

- Climate scientists use **global climate models** to help predict how our climate will change.

Key Terms

Critical and Creative Thinking Questions

1. Why do scientists classify climates according to precipitation instead of relative humidity?

2. Why is the annual temperature cycle of the wet equatorial climate so uniform? The wet-dry tropical climate has two distinct seasons. What factors produce the dry season and the wet season?

3. What is the climate classification for your current location? Discuss whether your recent weather is typical for that classification and explain why your classification is still correct, even if recent weather does not match what is expected.

4. Adapting to future climate change is difficult. Even if adapting to changes in mean temperature and precipitation are feasible, give three reasons that adapting to increases in extreme weather events is difficult.

What is happening in this picture?

Heat from this house, near Fairbanks, Alaska, combined with the effect of blocking the access of cold winter air to the land surface, has caused the permafrost underneath the house to thaw. The released water turned the soil to soft mud, no longer capable of supporting the weight of the structure.

Think Critically

1. Relate the trends in climate change to the impacts felt by polar societies.
2. What kinds of adaptations would you recommend for communities perched on permafrost?

Self-Test

(Check your answers in the Appendix.)

1. Which of the following is *not* a primary factor that influences global temperature patterns?
   a. longitude
   b. latitude
   c. location
   d. elevation

2. Which of the following factors would most likely lead to a relatively wet climate?
   a. location at high latitude
   b. location on the western edge of a midlatitude continent
   c. location in a persistent low-pressure center
   d. location on the leeward side of a mountain range

3. There are three types of monthly patterns of precipitation: (1) uniformly distributed precipitation, (2) a precipitation maximum during the high-Sun season, and (3) _____.
   a. a precipitation maximum during the late low-Sun season
   b. a precipitation minimum during the late high-Sun season
   c. a precipitation maximum during the low-Sun season
   d. a precipitation minimum during the early high-Sun season

4. What information is necessary for producing a climograph?
   a. one year of observations of temperature and relative humidity
   b. many years of observations of temperature and relative humidity
   c. one year of observations of temperature and air pressure
   d. many years of observations of temperature and air pressure
   e. many years of observations of temperature and precipitation

5. In contrast to the wet equatorial climate, the dry tropical climate is characterized by _____.

a. uniform temperatures all year

b. a low annual variation in solar insolation

c. no annual variation in solar insolation

d. a very strong temperature cycle

6. The dry tropical climates are caused by _____.

a. a lack of moisture

b. their placement in relationship to the Prime Meridian

c. their distance from the equator and the Arctic Circle

d. stationary subtropical cells of high pressure

7. The _____ air masses have no influence on low-latitude climates.

a. cT   b. mE   c. cP   d. mT

8. A geographic feature common to dry climates is _____.

a. no permanently flowing streams

b. a surface covered with dry lakes

c. a complete lack of precipitation

d. high surface winds

9. In which climate are you most likely to see the tree shown in this photo?

a. moist subtropical   c. Mediterranean climate

b. dry subtropical   d. wet-dry tropical

10. Which midlatitude climate has the greatest range of annual precipitation?

a. dry subtropical   c. Mediterranean climate

b. moist subtropical   d. dry midlatitude

11. The tundra climate is _____.

a. colder than the boreal forest climate

b. colder than the ice sheet climate

c. drier than the boreal forest climate

d. moderated by nearby ocean

12. Which climate is found in the regions colored green on these maps?

13. Identify which climographs represent (1) the dry midlatitude climate at Pueblo, Colorado; (2) the tundra climate of Upernavik, Greenland; and (3) the wet-dry tropical climate of Timbo, Guinea.

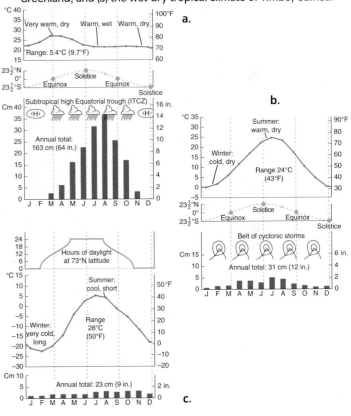

14. Where are temperature increases expected to be the greatest for scenarios of future climate change?

a. low-latitude climates   c. high-latitude climates

b. midlatitude climates   d. maritime climates

15. Global climate models predict that all _____ will receive _____ precipitation.

a. areas; more   c. dry areas; more

b. wet areas; less   d. areas; more variable

THE PLANNER ✓

Review your Chapter Planner on the chapter opener and check off your completed work.

The Earth from the Inside Out

Hidden in the cliffs of Lalibela, in Ethiopia, is a monastic complex with splendid—if unconventional—architecture. Many centuries ago, a volcanic eruption deposited layers of volcanic ash on this mountainside. The ash solidified into a porous red rock, which medieval Ethiopian stonemasons used to make 11 churches. But they didn't build these churches stone by stone from the ground up. Instead, they carved the churches out of the rock.

The most famous church is Bet Giorgis, or "House of Saint George," which stands 13 meters high in a well of rock. Stonemasons first dug out a trench, isolating a column of rock, and then hollowed the cross-shaped church out of this monolith. As they chiseled out the chambers, doors, and windows, they threw out more than enough rock to make a whole new church. Legend says that the Ethiopian king, Lalibela, after whom the site is named, was told to build the complex in a dream. Archeologists think that the construction took around 100 years and was completed in the 13th century. The churches at Lalibela are now designated as a United Nations Educational, Scientific, and Cultural Organization (UNESCO) World Heritage Site.

The processes that craft the Earth's natural landscape are slow and complicated. But they have given us the materials we build with and have sculpted the beautiful natural landscapes that surround us.

Bet Giorgis, or "House of Saint George," is one of several churches and buildings carved from uniform, fine-grained rock in Lalibela, Ethiopia.

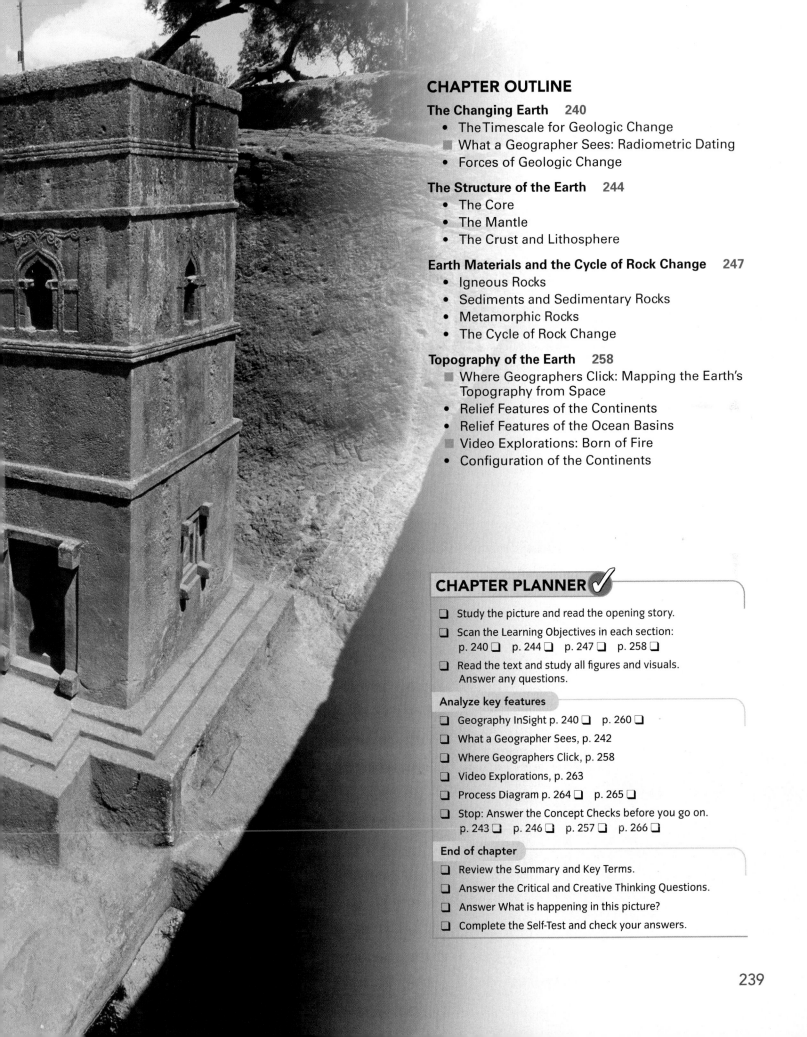

CHAPTER OUTLINE

CHAPTER PLANNER ✓

- ❑ Study the picture and read the opening story.
- ❑ Scan the Learning Objectives in each section:
  p. 240 ❑   p. 244 ❑   p. 247 ❑   p. 258 ❑
- ❑ Read the text and study all figures and visuals. Answer any questions.

Analyze key features

- ❑ Geography InSight p. 240 ❑   p. 260 ❑
- ❑ What a Geographer Sees, p. 242
- ❑ Where Geographers Click, p. 258
- ❑ Video Explorations, p. 263
- ❑ Process Diagram p. 264 ❑   p. 265 ❑
- ❑ Stop: Answer the Concept Checks before you go on.
  p. 243 ❑   p. 246 ❑   p. 257 ❑   p. 266 ❑

End of chapter

- ❑ Review the Summary and Key Terms.
- ❑ Answer the Critical and Creative Thinking Questions.
- ❑ Answer What is happening in this picture?
- ❑ Complete the Self-Test and check your answers.

The Changing Earth

LEARNING OBJECTIVES

1. **Describe** the geologic timescale.

2. **Distinguish** between endogenic and exogenic processes.

This chapter begins the portion of the book dedicated to **geomorphology**, that is, the study of the shape of the Earth's features and how they change over time. Geographers build upon the foundations of geomorphology to understand the differences in the distribution of humans, plant and animal species, and natural resources.

Uniformitarianism is the idea that the same geologic processes we can observe today have operated since the beginning of the Earth's history. This means that the

Geography InSight   Geologic time   •   Figure 8.1

The geologic timescale encompasses a conceptual framework essential for the study and comprehension of the incremental changes that have shaped today's Earth. Nearly all the landscape features visible to us today were formed within the Paleozoic to Cenozoic Eras. Geographers study the evolution of the landscape and biodiversity using the geologic timescale as a universal tool for common reference and understanding.

AGES OF THE COSMOS
Billions of years ago

13.7 13   12   11   10   9   8   7   6   5   4   3   2   1   0

First galaxies form — Spiral galaxies emerge — Solar system forms — Present

First stars form
Big Bang

AGES OF EARTH
Millions of years ago

4500   4000   3500   3000   2500   2000   1500   1000   600

PRISCOAN   ARCHAEAN   PROTEROZOIC

←— P R E C A M B R I A N —→

Priscoan
The earliest-known rocks on the Earth date from this period. Using radiometric dating, scientists have dated these 4.3-billion-year-old rocks in Canada's Northwest Territories.

Proterozoic
Around 2.8 to 2.4 billion years ago, photosynthetic bacteria, growing as layers and mats in sunlit shallow water, began to add oxygen to the atmosphere. The rock shown here is formed from layers of these fossilized microbes built up over time.

same cycles and forces that shape the Earth today can help us understand how the Earth has changed since its earliest history. In other words, the present is the key to the past.

The Timescale for Geologic Change

In order to talk about the time frame of geologic events, we use the **geologic timescale** (Figure 8.1). Geologists divide the 4.5 billion years since the Earth formed into eons, eras, and periods. Eons are vast chunks of time divided into eras, which in turn are divided into periods. The divisions are broad in the early parts of the Earth's history and much narrower in the past few hundred million years,

which we know more about. Many of the names of the divisions describe the world or ecosystems at the time. For example, Paleozoic means "old life" and includes early land plants, insects, reptiles, and amphibians; Mesozoic means the era of "middle life" dominated by the dinosaurs. Period names often come from rock formations, such as the coals of the Carboniferous period.

Geologists use **radiometric dating** to determine the ages of different rocks. This technique uses what we know about physics and the rate of radioactive decay of rock elements to determine the ages of rocks. Many geologic timescale divisions were originally drawn based on the dating of rock layers. As scientists use advances in

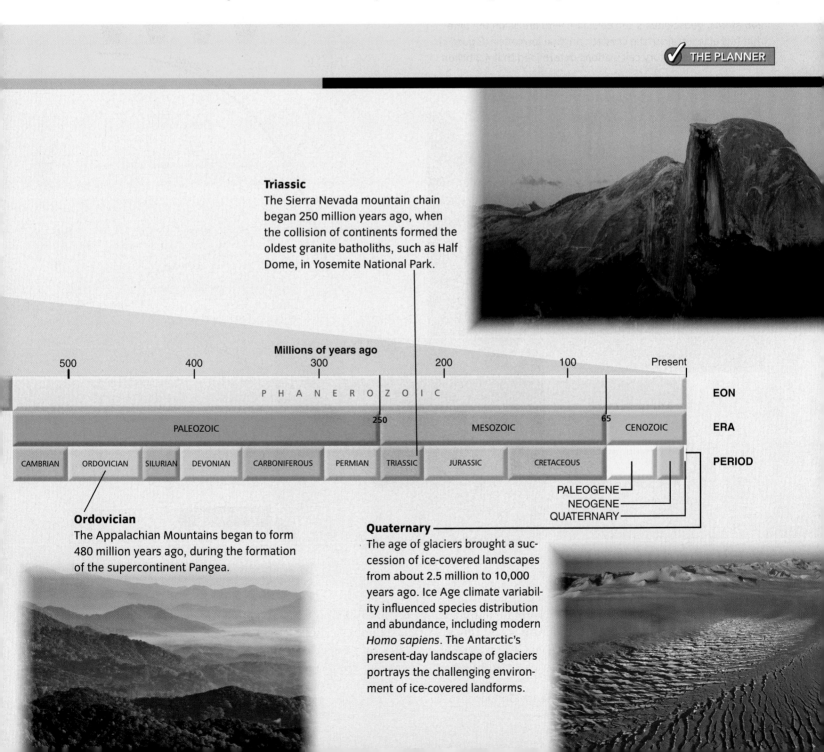

THE PLANNER

Triassic
The Sierra Nevada mountain chain began 250 million years ago, when the collision of continents formed the oldest granite batholiths, such as Half Dome, in Yosemite National Park.

Millions of years ago

| 500 | 400 | 300 | 200 | 100 | Present |

PHANEROZOIC — EON

PALEOZOIC — 250 — MESOZOIC — 65 — CENOZOIC — ERA

CAMBRIAN | ORDOVICIAN | SILURIAN | DEVONIAN | CARBONIFEROUS | PERMIAN | TRIASSIC | JURASSIC | CRETACEOUS — PERIOD

PALEOGENE
NEOGENE
QUATERNARY

Ordovician
The Appalachian Mountains began to form 480 million years ago, during the formation of the supercontinent Pangea.

Quaternary
The age of glaciers brought a succession of ice-covered landscapes from about 2.5 million to 10,000 years ago. Ice Age climate variability influenced species distribution and abundance, including modern *Homo sapiens*. The Antarctic's present-day landscape of glaciers portrays the challenging environment of ice-covered landforms.

WHAT A GEOGRAPHER SEES

Radiometric Dating

This outcrop of metamorphosed sandstone and gneiss in western Australia (**Figure a**) provides a stunning backdrop for nature photography and hikers. But what a geographer sees is that these exposed rock formations represent some of the oldest rocks on the planet. Tiny crystals of the mineral zircon that are found in these rocks (**Figure b**) act as atomic time capsules by incorporating radioactive atoms of uranium-238 into their crystal structure as they are formed. These atoms are slowly transformed by radioactive decay into lead-206 at a known rate. By measuring the current ratio of uranium-238 atoms to lead-206 atoms, geoscientists can calculate with precision the time that has elapsed since the crystal's original formation (**Figure c**). Results from laboratory calculations determined that 4.3 billion years have passed since this rock's formation. Accurate dating of ancient rock formations like this one helps scientists refine their understanding of the Earth's age and subsequent Earth-changing processes.

a.

b.

c.

Uranium-238 decays to Lead-206 (half-life = 4.5 billion years)

No. of atoms

0 / 100

50 / 50

75 / 25

87.5 / 12.5

Daughter atoms

Parent atoms

— Growth of daughter atoms
— Decline of parent atoms

Time (half-lives)
0   1   2   3   4   5

Ask Yourself

A rock that contained 50% U-238 and 50% lead-206 would be _____ billion years old.

radiometric techniques to date the layers or individual rock formations, a more accurate picture is emerging of the Earth's geologic history (see *What a Geographer Sees*).

The vast scale of the Earth's history is hard to grasp. If you think of the history of the Earth since its formation as spanning a single 24-hour day, you can see the long time frame of geologic history and the relatively short period of evolving life-forms (**Figure 8.2**).

The Earth's history on a 24-hour clock
• Figure 8.2

If the Earth's history were compressed into a single 24-hour day, the last 2 hours and 50 minutes would delineate the proliferation of life.

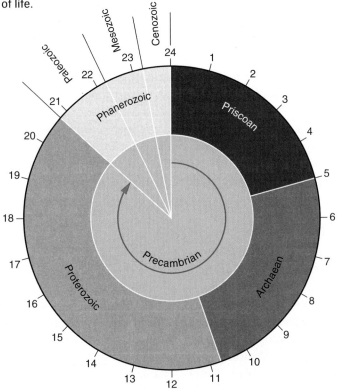

Forces of Geologic Change

The Earth's surface is constantly changing as old crust is broken down and new crust is formed. Volcanic and tectonic activity brings fresh rock to the planet's surface. We call these internal processes, or **endogenic processes**, because they work from within the Earth. External processes, or **exogenic processes**, such as weathering by wind and water, work at the Earth's surface. They lower continental surfaces by removing and transporting mineral matter through the action of running water, waves and currents, glacial ice, and wind.

The specific shapes of the Earth's surface are called **landforms**. We distinguish between landforms that have been freshly created and those that have been shaped by weathering, erosion, and mass wasting. **Initial landforms** are newly created by volcanic and tectonic activity. Exogenic processes wear down initial landforms to create **sequential landforms** (Figure 8.3). By examining landforms, we can begin to understand what kinds of Earth processes formed and shaped them. Geomorphology's major processes and landforms are the focus of this chapter through Chapter 15.

Initial and sequential landforms • Figure 8.3

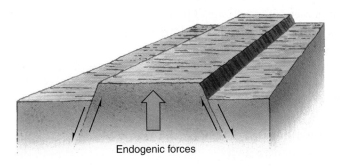

a. Crustal activity, an endogenic process, produces an initial landform in the shape of this mountain block.

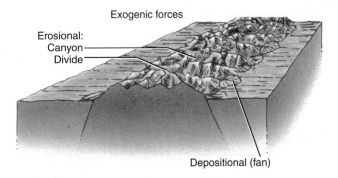

b. Erosion and deposition of sediments, which are exogenic processes, carve the mountain into a sequential landform.

CONCEPT CHECK STOP

1. **How** do geologists know the age of rocks?

2. **How** does an initial landform become a sequential landform?

The Structure of the Earth

LEARNING OBJECTIVES

1. **Contrast** the inner and outer core.
2. **Differentiate** the upper and lower mantle.
3. **Describe** the relationship of the lithosphere and the asthenosphere.

Before we begin to discuss how the Earth's surface is changing, we need to establish what we know about the structure of the Earth and how this affects what we see at the surface.

Our planet is almost spherical, bulging slightly at the equator, with a radius of approximately 6400 kilometers (about 4000 miles) and an equatorial circumference of approximately 40,000 km (24,900 mi). How can we know what lies deep within the Earth?

Scientists gain knowledge about the Earth's interior from a variety of observations. Active lava flows from volcanoes show us what lies below the Earth's crust. We can study exposed rock layers in road cuts and canyons to learn how sediments and other rock formations are laid down over tens of thousands and millions of years to create a record of past climates and Earth-forming events. We can study the composition of the Earth near the surface with deep drilling. To explore the Earth at even deeper levels, we can observe the paths of earthquake waves traveling through the center of the Earth. A combination of field sampling and laboratory testing, in conjunction with scientific publications and vibrant discussions among Earth scientists, has given us increasing clarity about the hidden world of our planet's structure.

Our planet contains a central core with several layers, or shells, surrounding it (**Figure 8.4**). The densest matter is at the center, and each layer above it is increasingly less dense and cooler up to the surface.

The Core

core The spherical central mass of the Earth, composed largely of iron; consists of an outer liquid zone and an inner solid zone.

The Earth's central **core** is about 3500 km (about 2200 mi) in radius and is very hot—somewhere between 3000°C and 5000°C (about 5400°F to 9000°F).

We know from measurements of earthquake waves passing through the Earth that the core consists of two distinct layers. The outer core is liquid, as demonstrated by the fact that energy waves suddenly change behavior when they reach this boundary. In contrast, the inner core is solid and made mostly of iron, with some nickel. The inner core remains solid despite the high temperatures because of the extreme pressure of all the Earth materials surrounding it.

The liquid iron core creates a magnetic field as the fluid flows around the solid core and interacts with the Earth's existing magnetic field. This process in turn creates a dynamic energy condition that maintains the Earth's perpetual magnetic field.

The Mantle

The core is surrounded by the **mantle**—a shell about 2900 km (about 1800 mi) thick, made of mafic (a word formed from "ma," for *ma*gnesium-bearing, and "fic," from *fer*ric, or iron-bearing) silicate minerals. Mantle temperatures range from about 2800°C (about 5100°F) near the core to about 1800°C (about 3300°F) near the crust. The mantle is the largest of the Earth's layers, making up more than 80% of the Earth's total volume.

mantle The rock layer of the Earth beneath the crust and surrounding the core, composed of mafic igneous rock containing silicate minerals.

Like the core, the mantle can be subdivided further into different zones, characterized by different temperatures and compositions. The lower mantle is hotter than the upper mantle, but it is largely rigid, as a result of intense pressure at this depth.

In the upper mantle, although the temperature is lower, the pressure is also lower, so rocks are less rigid. This softer, plastic part of the mantle is the **asthenosphere**. This is the most fluid layer of the mantle, where molten material forms in hotspots as a result of heat generated by endogenic radioactive decay. Processes occurring in this region provide the energy to drive earthquake and volcanic activity at the Earth's surface, along with continuous motions that deform the Earth's crust.

The Crust and Lithosphere

The thin, outermost layer of our planet is the Earth's **crust**. The Earth's crust is separated from the mantle by the boundary called the **Moho** (short for Mohorovičić

crust The outermost solid layer of the Earth, composed largely of silicate materials.

discontinuity). At the Moho, seismic waves indicate that a sudden change in the density of materials occurs. The crust, composed of varied rocks and minerals, ranges from about 7 to 40 km (about 4 to 25 mi) thick and contains the continents and ocean basins. It is the source of soil on the lands, of salts of the sea, of gases of the atmosphere, and of all the water of the oceans, atmosphere, and lands.

The crust that lies below ocean floors—oceanic crust—consists almost entirely of mafic rocks. The continental crust consists of two continuous zones—a lower zone of dense mafic rock and an upper zone of lighter felsic rock. Felsic rock (a word formed from *fel*dspar and *si*lica) is composed of silicates of aluminum, sodium, potassium, and calcium. Another key distinction between continental and oceanic crust is that the crust is much thicker beneath the continents—35 km (22 mi) on average—than it is beneath the ocean floors, where it is typically 7 km (4 mi).

Geologists use the term **lithosphere** to describe an outer Earth shell of rigid, brittle rock, including the crust and also the cooler, upper part of the mantle. The lithosphere ranges in thickness from 60 to 150 km (40 to 95 mi). It is thickest under the continents and thinnest under the ocean basins.

You can think of the lithosphere on top of the asthenosphere as a hard, brittle shell resting on a soft, plastic underlayer. Because the asthenosphere is soft and plastic, the rigid lithosphere can easily move over it.

The lithosphere floats on the plastic asthenosphere, much like an iceberg floats in water, in an equilibrium state

The structure of the Earth • Figure 8.4

Seismic studies and other direct and indirect observations have given scientists a good idea of the internal structure of the Earth.

a. The Earth can be divided into three major compositional layers: the **core**; the **mantle**; and the **crust**. Temperature and pressure increase with depth.

c. The composition and thickness of the crust varies depending on whether the crust is below an ocean or on a continent. Oceanic crust is thinner and denser than continental crust.

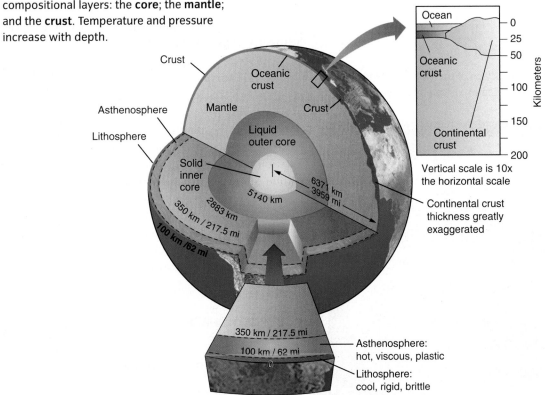

b. The compositional boundaries between layers do not necessarily coincide with the boundaries of zones that differ in strength. For example, the upper mantle includes both a soft, plastic asthenosphere and also the rigid lithosphere. The lithosphere includes the rigid portion of the upper mantle as well as the crust above it.

Isostasy floats the world • Figure 8.5

Mountains float on the viscous asthenosphere much like an iceberg floats in water by displacing its weight. As the mountain erodes and weighs less, it rises, gaining back elevation until the erosion process again reduces its mass.

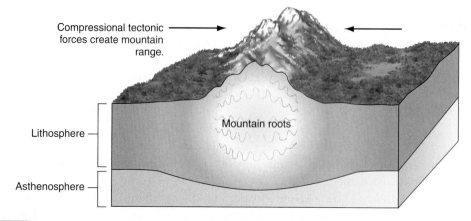

Compressional tectonic forces create mountain range.

Lithosphere

Mountain roots

Asthenosphere

Ask Yourself

If a large series of glaciers formed in this mountain range, the elevation of the surface underneath the glacier would _____.

a. increase
b. decrease
c. stay the same
d. stay the same but shift laterally

known as **isostasy** (**Figure 8.5**). As erosion reduces the height and weight of a mountain block over time, the block becomes lighter and floats higher on the asthenosphere by the process of isostasy. This causes the block to rise, making the mountains higher again and subject to more erosion. The process continues until the mountains are a range of low hills with shallow mountain roots.

The lithospheric shell is divided into large pieces called **lithospheric plates** (**Figure 8.6**). A single plate can be as large as a continent and can move independently of the plates that surround it—like a great slab of floating ice on the polar sea. Lithospheric plates can separate from one another at one location, while elsewhere they may collide in crushing impacts that raise great mountains. The major relief features of the Earth—its continents and ocean basins—were created by the continuous movements of these plates in geologic timescales on the surface of the Earth, which we will discuss in greater detail throughout the upcoming chapters.

> **lithospheric plate**
> A segment of lithosphere moving as a unit, sliding on top of the asthenosphere in contact with adjacent lithospheric plates along plate boundaries.

Lithospheric shell • Figure 8.6

Lithospheric plates cover the surface of the globe, moving incrementally through geologic timescales, driven by convectional currents in the asthenosphere.

CONCEPT CHECK **STOP**

1. **Why** is the inner core solid despite its high temperatures?
2. **Which** portion of the mantle is least rigid and why?
3. **What** allows lithospheric plates to move over the asthenosphere?

Earth Materials and the Cycle of Rock Change

LEARNING OBJECTIVES

1. **Discuss** the formation and properties of igneous rocks.

2. **Compare** the formation and properties of the three classes of sedimentary rocks.

3. **Explain** the formation of metamorphic rocks.

4. **Outline** the steps of the cycle of rock change.

The most abundant elements in the Earth's crust are oxygen, silicon, aluminum, iron, calcium, sodium, potassium, and magnesium (**Figure 8.7**). They exist in a variety of rock combinations and are formed through physical and chemical processes, both endogenic and exogenic. Oxygen, the most abundant element, readily combines with many of these elements in the form of oxides. Crustal elements can form chemical compounds that we recognize as **minerals** (**Figure 8.8**). A mineral is a naturally formed solid, inorganic substance with a characteristic crystal structure and chemical composition.

Rocks are usually composed of two or more minerals. Often many different minerals are present, but a few rock varieties are made almost entirely of one mineral. Most rock in the Earth's crust is extremely old, dating back many millions of years, but rock is also being formed at this very hour, as active volcanoes emit lava that solidifies on contact with the atmosphere or ocean.

Crust elements • Figure 8.7

The Earth's crust is composed of the principal elements listed. Oxygen, which is bound with other elements as oxides, represents approximately half of the crustal volume. If we include water (H_2O) located in the crust, ground water, and aquifers, the oxygen total is much higher.

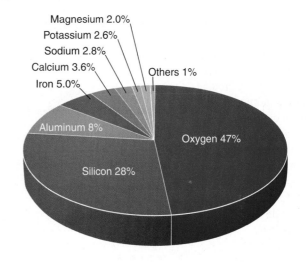

Magnesium 2.0%
Potassium 2.6%
Sodium 2.8%
Calcium 3.6%
Iron 5.0%
Others 1%
Aluminum 8%
Oxygen 47%
Silicon 28%

Quartz crystals • Figure 8.8

Quartz, or silicon dioxide, is a very common mineral. Under unusual circumstances, quartz is found as regular six-sided crystals, shown here in this sample from Venezuela. Usually it appears as a clear or light-colored mineral, and it is often present in sediments such as beach sand or river gravel.

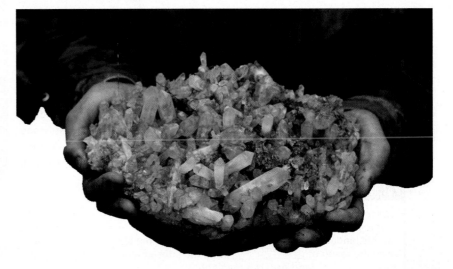

Rocks fall into three major classes: *igneous*, *sedimentary*, and *metamorphic* rocks. Each class of rock has unique properties and structures that affect how the rocks are shaped into landforms. Different types of rocks are worn down at different rates. Some are easily eroded, whereas others are much more resistant. For example, we usually find weak rocks under valleys and strong rocks under hills, ridges, and uplands (**Figure 8.9**).

Landforms and rock resistance • Figure 8.9

Weaker rock erodes more rapidly than resistant rock, leaving the resistant rock standing as mountains, ridges, or belts of hills.

b. Granite is an igneous rock that is rich in quartz, which is very resistant to erosion and decay. Here a climber ascends a steep granite slope in the White Mountains of New Hampshire.

a. Igneous rock, consisting of crystals formed by cooling of molten rock, is often resistant to erosion, forming mountains and high plateaus. Most metamorphic rocks are also resistant to erosion, typically forming hills and hill belts. Sedimentary rocks can be strong or weak, depending on their composition. Strong sedimentary rocks form ridges or hills, while weak ones form valleys.

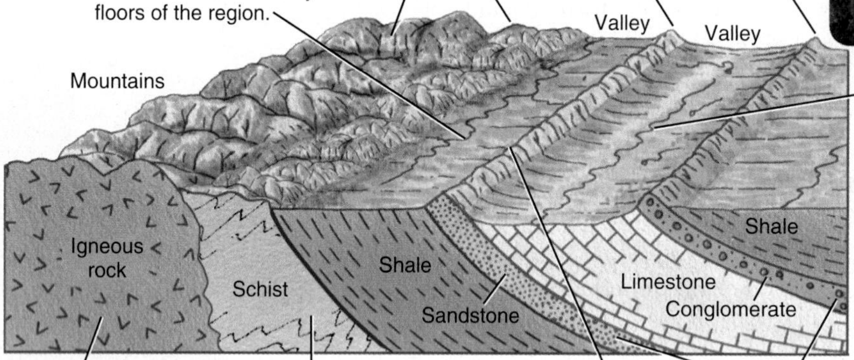

Shale is a weak sedimentary rock that is easily eroded and forms the low valley floors of the region.

Hills Ridge Ridge

Valley Valley

Mountains

Igneous rock

Schist

Shale

Sandstone

Limestone
Conglomerate

Shale

Limestone is dissolved by carbonic acid in rain and surface water, also forming valleys in humid climates. In arid climates limestone is a resistant rock and usually forms ridges and cliffs.

The igneous rocks are resistant–typically forming uplands or mountains rising above adjacent areas of shale and limestone.

Metamorphic rocks vary in resistance. Metamorphic rocks are sometimes found in contact with igneous rock that provides the heat for the metamorphic process.

Sandstone and conglomerate are typically resistant and form ridges or uplands.

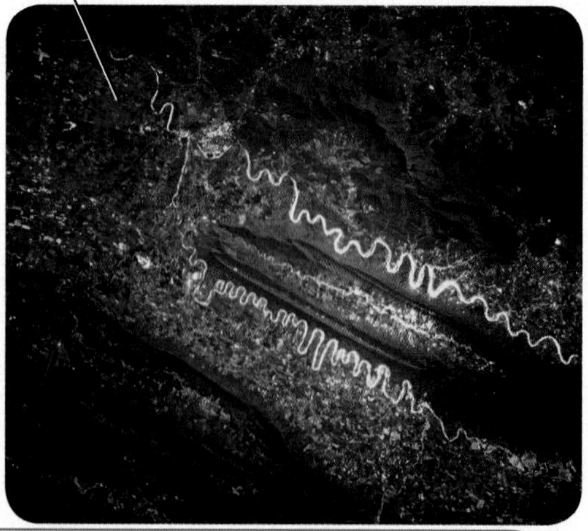

c. Sandstone is a resistant rock, and sandstone layers often stand out as ridges along the landscape. In this image of Virginia's Shenandoah Valley, taken from the Space Shuttle, sandstone ridges alternate with softer rocks of limestone and shale that are extensively farmed.

Put It Together

Review the section Forces of Geologic Change and complete this statement.
In this landscape, the igneous rock landform most likely was the result of _____, and the subsequent _____ landform is a result of _____.

a. exogenic processes; sequential; endogenic processes
b. endogenic processes; sequential; exogenic processes
c. exogenic processes; initial; endogenic processes
d. endogenic processes; initial; exogenic processes

Lava is molten rock that reaches the Earth's surface. It can flow out from the volcanic vent as a thick liquid or be ejected violently as fine volcanic ash and coarser particles, sometimes as large as boulders. These photos are from recent eruptions of Kilauea volcano in Hawaii Volcanoes National Park.

a. During a volcanic eruption, red-hot lava is sometimes ejected high into the air, forming a fire fountain. In the foreground is the congealed surface of fresh lava. Underneath, the lava is still very hot and may be moving downhill under its black crust. ▼

▲ b. During a large eruption, lava can run down the flanks of the volcano as a river of fire. At Kilauea, some of these lava flows reach the sea.

Igneous Rocks

Igneous rocks are formed when molten material, or **magma**, solidifies. The magma moves upward from pockets a few kilometers below the Earth's surface, through fractures in older solid rock. There the magma cools, forming rocks of mineral crystals.

Magma that solidifies below the Earth's surface and remains surrounded by older, preexisting rock is called **intrusive igneous rock**. Because intrusive rocks cool slowly, they develop mineral crystals that are visible to the eye. If the magma reaches the surface and emerges as lava, it forms **extrusive igneous rock** (**Figure 8.10**). Extrusive igneous rocks cool very rapidly on the land

> **igneous rock** Rock formed from the cooling of magma.
>
> **magma** Rock in a mobile, high-temperature molten state.

surface or ocean bottom and thus show crystals of only microscopic size. You can see formation of extrusive igneous rock today where volcanic processes are active.

Most igneous rock consists of silicate minerals—chemical compounds that contain silicon and oxygen atoms. These rocks also contain mostly metallic elements. The mineral grains in igneous rocks are very tightly interlocked, so the rock is normally very hard. Quartz (see Figure 8.8), which is made of silicon dioxide (SiO_2), is the most common mineral of all rock classes. It is quite hard and resists chemical breakdown, as can be seen when viewed in its greatly reduced state as beach sand.

Silicate minerals and igneous rocks • Figure 8.11

Only the most common silicate mineral groups are listed, along with six common igneous rocks, both volcanic and plutonic.

Grain size

Extrusive rocks

Intrusive rocks

Felsic

Silica content

Mafic

Rhyolite lies at the felsic, high-silica end of the scale. It is usually pale, ranging from nearly white to shades of gray, yellow, red, or lavender.

Granite, the intrusive equivalent of rhyolite, is common because felsic magmas usually crystallize before they reach the surface. It is found most often in continental crust, especially in the cores of mountain ranges.

Andesite is an intermediate-silica rock. It is usually light to dark gray, purple, or green.

Diorite is the intrusive equivalent of andesite, an intermediate-silica rock.

Basalt, a mafic rock, is dominant in oceanic crust and the most common igneous rock on Earth. It typically has a dark gray, dark green, or black color.

Gabbro is the intrusive equivalent of basalt, a low-silica rock.

Figure 8.11 shows some other common minerals and the intrusive and extrusive rocks that are made from them. Silicate minerals in igneous rocks are classed as felsic, which are light colored and less dense, and mafic, which are dark colored and denser. Felsic rock contains mostly felsic minerals, and mafic rock contains mostly mafic minerals. Ultramafic rock is dominated by mafic minerals rich in magnesium and iron and is the densest of the three rock types.

Plutons • Figure 8.12

Igneous intrusions of hot magma cool to form plutons. The largest of these intrusions are batholiths and stocks. When magma pushes its way vertically between fractures in the rock, it cools to form a dike, and when it is horizontal, it forms a sill. Sometimes the overlaying rock bulges upward to form a laccolith. Other plutons are the channels of extinct volcanoes.

Volcanic neck

Volcanic pipe

Laccolith

Dike

Dike

Sill

Xenoliths

Batholith

Put It Together

Review the beginning of this section, Igneous Rocks, and complete this statement.
The grain sizes in the rock formed by the surface lava flow would be _____ compared to the grain sizes found in the rock formed in a sill.

Intrusive igneous rock, formed deep below the Earth's surface, cools to form **plutons** (**Figure 8.12**). The largest plutons are **batholiths**. Smaller plutons include sills, dikes, and laccoliths. Sometimes batholiths contain xenoliths—fragments of surrounding rock that are incorporated into the batholith without melting.

Batholiths are eventually uncovered by erosion and appear at the surface, usually as hilly or mountainous uplands (**Figure 8.13**). A good example is the Idaho batholith, a granite mass exposed over an area of about 40,000 sq km (about 16,000 sq mi)—a region almost as large as New Hampshire and Vermont combined. Another example is the Sierra Nevada batholith of California, which makes up most of the high central part of that great mountain block. The Canadian shield includes many batholiths.

Batholiths • Figure 8.13

Underlying much of the Earth's continents are large expanses of igneous intrusive cooled magma, called *batholiths*.

Exposed batholith

Batholith

b. Sugar Loaf Mountain, which rises above the city of Rio de Janeiro, Brazil, is a small granite dome-like projection of a batholith that lies below the ground and harbor.

a. Batholiths appear at the land surface after long-continued erosion has removed thousands of meters of overlying rocks. Small projections of the intrusion appear first and are surrounded by older rock.

Sediments and Sedimentary Rocks

Now let's turn to the second rock class, the **sedimentary rocks**. Sedimentary rocks are made from layers of mineral particles found in other rocks (igneous, sedimentary, and metamorphic) that have been released by weathering. They also include rocks made from newly formed organic matter, both plant biomass and invertebrates. Most inorganic minerals in sedimentary rocks are from igneous rocks.

> **sedimentary rock**
> Rock formed from the accumulation of sediment.

Rocks exposed at the Earth's surface are broken down into fragments of many sizes in a process called weathering. Physical weathering divides rock into ever-smaller pieces, and chemical weathering alters the chemical composition of mineral grains through exposure to oxygen and water. When weathered rock or mineral particles are transported by air, water, or glacial ice, we call them sediment. Streams and rivers carry sediment to lower levels, where it builds up. Sediments usually accumulate on shallow seafloors bordering continents, but they also collect in inland valleys, lakes, and marshes. Wind can also transport sediments.

Over geologic timescales, the sediments become compacted and harden to form sedimentary rock, with distinctive visible characteristics. Sediments are generally deposited by wind and water in layers, called **strata**. The study of thickness and patterns of these strata can tell us much about the history of an area, including the past presence of oceans and rivers as well as the history of faulting and deformation. In the Grand Canyon, for example, light-colored layers indicate long drought periods, while dark-colored strata indicate periods of more moist conditions.

There are three major classes of sediment: clastic sediment, chemically precipitated sediment, and organic sediment (**Figure 8.14**).

Some common sedimentary rock types • Figure 8.14

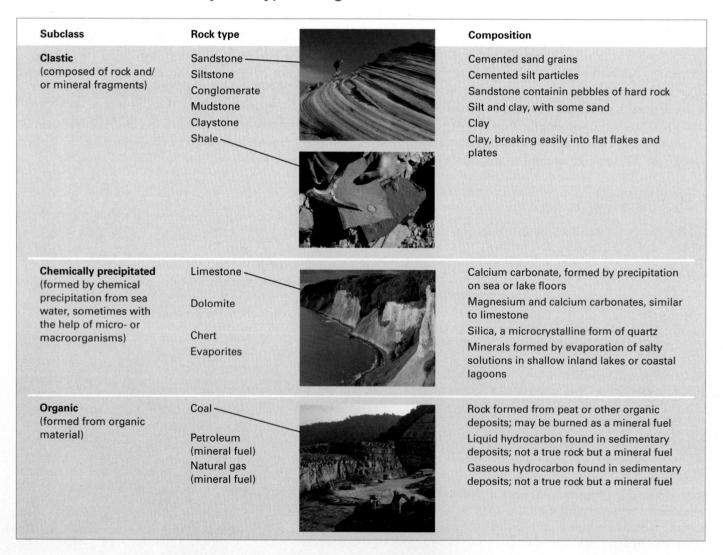

Subclass	Rock type	Composition
Clastic (composed of rock and/or mineral fragments)	Sandstone	Cemented sand grains
	Siltstone	Cemented silt particles
	Conglomerate	Sandstone containin pebbles of hard rock
	Mudstone	Silt and clay, with some sand
	Claystone	Clay
	Shale	Clay, breaking easily into flat flakes and plates
Chemically precipitated (formed by chemical precipitation from sea water, sometimes with the help of micro- or macroorganisms)	Limestone	Calcium carbonate, formed by precipitation on sea or lake floors
	Dolomite	Magnesium and calcium carbonates, similar to limestone
	Chert	Silica, a microcrystalline form of quartz
	Evaporites	Minerals formed by evaporation of salty solutions in shallow inland lakes or coastal lagoons
Organic (formed from organic material)	Coal	Rock formed from peat or other organic deposits; may be burned as a mineral fuel
	Petroleum (mineral fuel)	Liquid hydrocarbon found in sedimentary deposits; not a true rock but a mineral fuel
	Natural gas (mineral fuel)	Gaseous hydrocarbon found in sedimentary deposits; not a true rock but a mineral fuel

a. Sandstone

Sandstone is composed of sand-sized particles, normally grains of eroded quartz, that are cemented together in the process of rock formation. This example is the Navajo sandstone, which is found on the Colorado Plateau in Utah and Arizona. It was originally deposited in layers by moving sand dunes.

b. Conglomerate

Conglomerate is a sedimentary rock of coarse particles of many different sizes. These climbers are working their way through some beds of conglomerate on their way to the top of Shipton's Arch, Xinjiang, China. On the left, the softer sediment between hard cobbles has eroded away, leaving the rounded rocks sticking out.

c. Shale

Shale is a rock formed mostly from very fine-textured particles—silt and clay—deposited in calm water. It is typically gray or black in color and breaks into flat plates, as shown here. Some shale deposits contain fossils, like these ancient trilobites, marine animals of Cambrian age, found near Antelope Springs, Utah.

Global Locator

Xinjiang, China

NATIONAL GEOGRAPHIC

Clastic sediment Sediment that is made up of inorganic rock and mineral fragments (clasts) is called clastic sediment. Clastic sediments can come from igneous, sedimentary, or metamorphic rocks, and so they can include a very wide range of minerals. Quartz and feldspar usually dominate clastic sediment.

The size of clastic sediment particles determines how easily and how far they are transported by wind and water currents. Fine particles are easily suspended in slowly moving fluids, while coarser particles are moved only by stronger currents of water or air. When the fluid velocity decreases, coarser particles settle out first. In this way, particles of different sizes are sorted by the fluid motion.

When layers of clastic sediment build up, the lower strata are pushed down by the weight of the sediments above them. This pressure compacts the sediments, squeezing out excess water. Dissolved minerals recrystallize in the spaces between mineral particles in a process called cementation. **Figure 8.15** shows some important varieties of clastic sedimentary rocks.

Chemically precipitated sediment Chemically precipitated sediments are made of solid inorganic mineral compounds that separate out from saltwater solutions, or from the silica and calcium carbonate shells of microorganisms. Rock salt is an example of chemically precipitated sediment. One of the most common sedimentary rocks formed by chemical precipitation is limestone (**Figure 8.16**). Sediments of the tiny shells of microscopic organisms (diatoms and foraminifera, for example) can form limestone layers tens of meters thick. Limestone can also be formed from coarse fragments of shells or corals.

Organic sediment Organic sediment is made up of the tissues, or biomass, of plants and animals. Peat is an example of an organic sediment. This soft, fibrous, brown or black substance accumulates in bogs and marshes where the acidic water conditions prevent the decay of plant or animal remains.

Peat is a form of hydrocarbon—a compound of hydrogen, carbon, and oxygen. Hydrocarbon compounds are the most important type of organic sediment—one that we increasingly depend on for fossil fuel. They formed from the remains of plants or tiny zooplankton and algae that built up over millions of years and were compacted under thick layers of inorganic clastic sediment. Hydrocarbons can be solid (peat and coal), liquid (petroleum), or gas (natural gas). Coal is the only hydrocarbon that is a rock (**Figure 8.17**). We often find natural gas and petroleum in open interconnected pores in a thick sedimentary rock layer, such as in a porous sandstone (**Figure 8.17b**).

Chalk cliffs provide a record of climate change • Figure 8.16

The white chalk cliffs of Rugen Island in Germany (**a**) are composed of layers of calcium carbonate from the shells of ancient microorganisms. These microorganisms form their tiny calcium carbonate shells in different shapes, depending on the temperature and acidity of the ocean water (**b**). By examining the fossil shells of each layer under a microscope, scientists can observe the shifts in ocean water temperatures over tens of thousands of years to provide a record of the Earth's climate changes in past millennia.

b.

a.

Knowledge of sedimentary rock formations helps geologists search for fossil fuels.

a. Strip mine

In strip mining, including the controversial practice of mountain-top removal, layers of coal are mined by removing overlying rock, allowing direct access to the coal deposit. This strip mine is located in Man, West Virginia.

Think Critically

What are the environmental consequences of the following?
1. strip mining for coal
2. extracting oil and natural gas with wells

b. Trapping of oil and gas

This idealized cross-section shows an oil pool on a dome structure in sedimentary strata. Well A will draw gas, well B will draw oil, and well C will draw water. Here the impervious rock is shale and the reservoir rock is sandstone.

The Earth's fossil fuels have accumulated over many millions of years. But our industrial society is consuming them ever more rapidly. These fuels are nonrenewable resources; as they become scarcer, they will become more costly. Once they are gone, there will be no more. Even if we wait another thousand years for more fossil fuels to be created, the amount we gain will scarcely be measurable in comparison to the stores produced in the geologic past.

The amount of carbon currently being released into the atmosphere from the burning of fossil fuels is causing alarm among climate change scientists. Rapid release of geologically stored carbon is pushing the atmospheric levels of CO_2, a greenhouse gas, to the highest recorded levels in over 600,000 years. The majority of scientists believe that these increases are responsible for increasing global temperatures, which in turn are creating weather systems of greater variability and magnitude.

Metamorphic Rocks

The mountain-building, or orogenic, processes of the Earth's crust involve tremendously high pressures and temperatures. These extreme conditions alter igneous and sedimentary rock, transforming the rock so completely in texture and structure that we have to reclassify it as **metamorphic rock** (**Figure 8.18**). In many cases, the mineral components of the parent

> **metamorphic rock** Rock altered in physical or chemical composition by heat, pressure, or other endogenic processes taking place at a substantial depth below the surface.

rock are changed into different mineral varieties. In some cases, the original minerals may recrystallize.

Metamorphic rocks tend to be more resistant to weathering than their parent rocks because the heat and pressure welds their mineral grains together and may transform their minerals to stronger forms. For example, extreme heat and pressure transform shale into

Some common metamorphic rock types • Figure 8.18

Rock Type	→	Description
Slate	→	Shale exposed to heat and pressure that splits into hard flat plates
Schist	→	Shale exposed to intense heat and pressure that shows evidence of shearing
Quartzite	→	Sandstone that is "welded" by a silica cement into a very hard rock of solid quartz
Marble	→	Limestone exposed to heat and pressure, resulting in larger, more uniform crystals
Gneiss	→	Rock resulting from the exposure of clastic sedimentary or intrusive igneous rocks to heat and pressure

 a. Schist

When shale is exposed to heat and pressure for long periods, the minerals in the shale recrystallize and grow together to form a stronger rock called schist. Shown is Granite Gorge, Grand Canyon National Park.

b. Quartzite

Under heat and pressure, sandstone recrystallizes to form a very strong rock of connected quartz grains called quartzite. Seneca Rocks, West Virginia, provides a striking example.

c. Marble

Marble can provide a useful and beautiful building material. The marble in this quarry in Vermont was formed from limestone by heat and pressure during the creation of the Appalachian mountain chain.

The cycle of rock change • Figure 8.19

The cycle of rock change has been active since our planet became solid and internally stable. It continuously forms and re-forms rocks of all three major classes. Not even the oldest igneous and metamorphic rocks that we have found are the "original" rocks of the Earth's crust, for they were recycled eons ago.

slate or schist, sandstone into quartzite, and limestone into marble.

Metamorphism can occur in several different situations. When magma intrudes into deep crustal rocks, the heat can transform the adjacent rock in a process called contact metamorphism. For example, marble forms when heat and pressure in the Earth's crust causes limestone to recrystallize and form larger, more uniform crystals of calcite.

Regional metamorphism is associated with large areas of tectonic activity from the collision of lithospheric plates. This form of metamorphism is responsible for the more common rocks, such as slate, schist, and gneiss, that are associated with major mountain ranges.

The Cycle of Rock Change

Rocks are constantly being transformed from one class to another in the cycle of rock change, which recycles crustal minerals over many millions of years (**Figure 8.19**). This cycle, operating over the geologic timescale, involves various physical, chemical, and biological processes that create, transform, and recycle the Earth's crust.

In the surface environment, rocks weather into fragments that form sediments through erosion and deposition. As sediments accumulate, they are buried and compressed by the weight of sediments above and are cemented by mineral-rich solutions to form sedimentary rock. In the deep-Earth environment, igneous or sedimentary rock is transformed by heat and pressure into metamorphic rock, which may ultimately melt to form magma. When the magma cools, either on or near the surface, it forms igneous rock that can be weathered into sediment, completing the cycle.

CONCEPT CHECK

1. **Why** are many batholiths eventually exposed?

2. **What** can strata tell us about the history of a region?

3. **How** and why are the properties of metamorphic rocks different from the rocks from which they were formed?

4. **Why** can't we find rocks that date from the formation of the Earth?

Topography of the Earth

LEARNING OBJECTIVES

1. **Describe** the major relief features of the continents.
2. **Explain** the ocean basin's major relief features.
3. **Outline** the development of the theory of plate motion.

Having outlined the structure of the Earth and the composition of the Earth's crust, we will now explore the way the crust is distributed into major surface features, called its topography. Modern technology allows scientists to map much of the Earth's topography through remote sensing (see *Where Geographers Click*). The most obvious topographic patterns on the Earth's surface are its relief features—mountain chains, midoceanic ridges, high plateaus, and ocean trenches, for example. In upcoming chapters we will discuss in greater detail the processes responsible for these features.

Relief Features of the Continents

We divide continents into two types of regions: active mountain-making belts and inactive regions of old, stable rock.

Alpine chains Active mountain-making belts are narrow zones that are usually found along the margins of lithospheric plates. We call these belts alpine chains because they are characterized by high, rugged mountains, such as the Alps of Central Europe. Even today, alpine mountain-building continues in many places.

The mountain ranges in the active belts grow through one of two very different geologic processes. First is volcanism, in which massive accumulations of volcanic rock are formed by extrusion of magma. Many lofty mountain ranges consist of chains of volcanoes built of extrusive igneous rocks. For example, the Cascade Range in the Pacific Northwest region of the United States contains a chain of volcanic mountains, including Mt. St. Helens, Mt. Baker, and Mt. Hood.

Where Geographers CLICK

Mapping the Earth's Topography from Space

www2.jpl.nasa.gov/srtm/

The Shuttle Radar Topography Mission (SRTM) collected elevation data for most of the planet to create a high-resolution digital topographic database of the Earth. Mounting a specially modified radar system onboard the space shuttle *Endeavour* during an 11-day mission in February 2000, NASA scientists acquired images measuring the local shape of the Earth over the Earth's land surface between 60° N and 60° S latitude. They then assembled these measurements into a global database of elevations on a grid scale of 30 m for the United States and 90 m for the rest of the world. Visit this site for spectacular flyovers and images of places far and near, including the relief features mentioned in this chapter.

This simulated image of Mt. St. Helens shows how detailed topographic information can be used to display a life-like landscape. The synthesized view also shows Mt. Adams and Mt. Hood. Color is keyed to elevation, with green to brown to white showing increasing elevation.

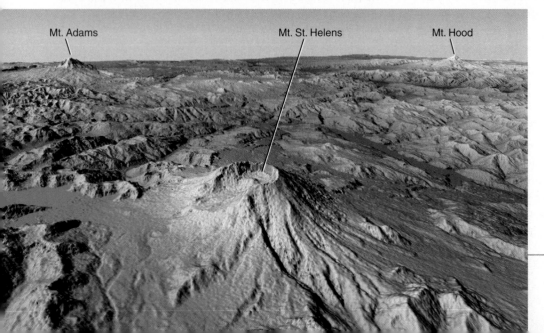

Mt. Adams   Mt. St. Helens   Mt. Hood

Continental shields • Figure 8.20

a. Map of continental shields

Shields are areas of ancient rocks that have been eroded to levels of low relief. Mountain roots are shown with a brown line. The areas of oldest rock are circled in red.

b. Canadian shield

Continental glaciers stripped the Canadian shield of its sediments during the Ice Age, leaving a landscape of low hills, rock outcrops, and many lakes, shown here in Lac La Ronge Provincial Park, Saskatchewan.

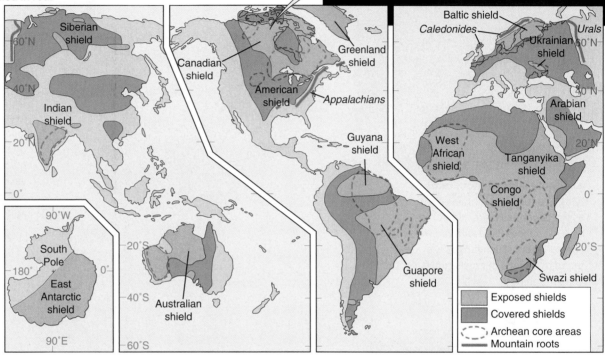

The second mountain-building process is tectonic activity—the breaking and bending of the Earth's crust under internal Earth forces. Tectonic activity usually occurs when lithospheric plates come together in collisions, as we will see in more detail in the next chapter. Crustal masses that are raised by tectonic activity create mountains and plateaus. A well-known example is the Himalaya mountain range, which resulted from the collision of the subcontinent of India with the continent of Asia. In some instances, volcanism and tectonic activity combine to produce a mountain range.

Continental shields Belts of recent and active mountain-making account for only a small portion of the continental crust. Most of the rest is much older, comparatively inactive rock that we call continental shields and mountain roots.

Continental shields are regions of low-lying igneous and metamorphic rocks that are resistant to weathering and that have formed the basis for continent building since ancient times. The core areas of some shields are made of rock dating back to the Archean eon, 2.5 to 3.9 billion years ago. Shields may be either exposed or covered by layers of sedimentary rock (**Figure 8.20**).

Remains of older mountain belts lie within the shields in many places. These mountain roots are mostly formed of ancient sedimentary rocks that have been intensely bent and folded and, in some locations, changed into metamorphic rocks. Thousands of meters of overlying rocks have been removed from these old tectonic belts so that only the lowermost structures remain. Roots appear as chains of long, narrow ridges, rarely rising over 1000 m (about 3280 ft) above sea level. The Appalachian mountain chain and its extensions to the Taconic, Green, and White Mountains provide a familiar example.

Continental landforms Along with the alpine ranges, the continents feature six different major types of landforms, as shown in **Figure 8.21** on the next page. The global pattern of landforms is largely related to plate tectonic activity, as we will see in greater detail in Chapter 9.

At the global scale, there are six major types of continental landforms. Tectonic activity creates mountain belts, crustal uplift creates high plateaus, and spreading creates widely spaced mountains. Crustal subsidence creates depressions. In stable regions, long-continued erosion yields low plateaus, hills, and plains. Some global regions are covered by ice sheets or ice caps.

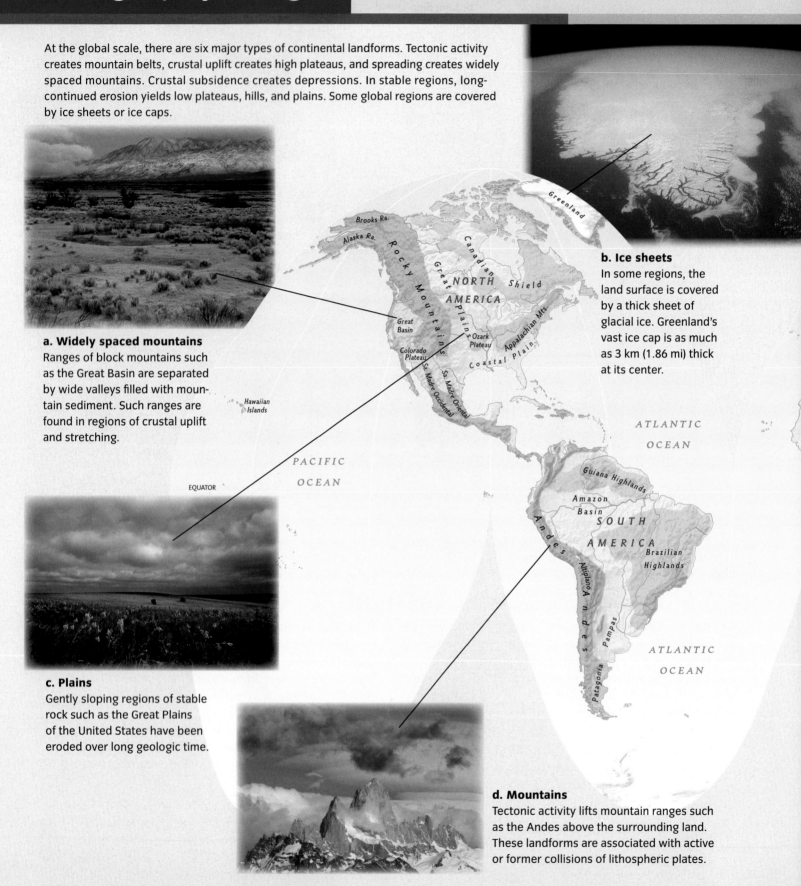

b. Ice sheets
In some regions, the land surface is covered by a thick sheet of glacial ice. Greenland's vast ice cap is as much as 3 km (1.86 mi) thick at its center.

a. Widely spaced mountains
Ranges of block mountains such as the Great Basin are separated by wide valleys filled with mountain sediment. Such ranges are found in regions of crustal uplift and stretching.

c. Plains
Gently sloping regions of stable rock such as the Great Plains of the United States have been eroded over long geologic time.

d. Mountains
Tectonic activity lifts mountain ranges such as the Andes above the surrounding land. These landforms are associated with active or former collisions of lithospheric plates.

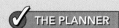
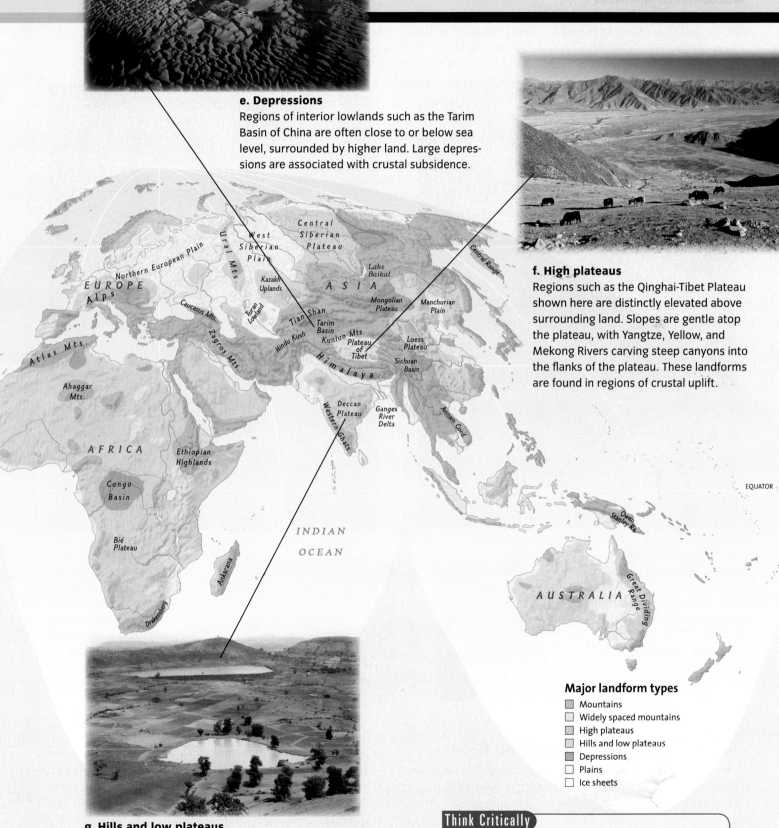

e. Depressions
Regions of interior lowlands such as the Tarim Basin of China are often close to or below sea level, surrounded by higher land. Large depressions are associated with crustal subsidence.

f. High plateaus
Regions such as the Qinghai-Tibet Plateau shown here are distinctly elevated above surrounding land. Slopes are gentle atop the plateau, with Yangtze, Yellow, and Mekong Rivers carving steep canyons into the flanks of the plateau. These landforms are found in regions of crustal uplift.

EUROPE
Alps
Northern European Plain
Ural Mts.
West Siberian Plain
Central Siberian Plateau
ASIA
Kazakh Uplands
Lake Baikal
Central Range
Caucasus Mts.
Turan Lowland
Zagros Mts
Tian Shan
Mongolian Plateau
Manchurian Plain
Atlas Mts.
Hindu Kush
Tarim Basin
Kunlun Mts.
Plateau of Tibet
Loess Plateau
Ahaggar Mts.
Himalaya
Sichuan Basin
Annam Cord.
AFRICA
Ethiopian Highlands
Western Ghats
Deccan Plateau
Ganges River Delta
Congo Basin
Bié Plateau
Ankarana
INDIAN OCEAN
EQUATOR
Owen Stanley Ra.
Drakensberg
AUSTRALIA
Great Dividing Range

Major landform types
- ▢ Mountains
- ▢ Widely spaced mountains
- ▢ High plateaus
- ▢ Hills and low plateaus
- ▢ Depressions
- ▢ Plains
- ▢ Ice sheets

g. Hills and low plateaus
The Deccan Plateau is an example of a low plateau formed by long-continued weathering and erosion of an ancient outpouring of volcanic lava. Low hills and plateaus also occur where extensive layers of thick, wind-borne or glacial sediments are undergoing erosion.

Think Critically
The last great continental glaciers retreated around 10,000 years ago. Which landforms would likely be the first to be covered by ice sheets in the next ice age?

Topography of the Earth **261**

Using precise radar measurements of the height of the ocean surface, it is possible to infer the depth of the water and therefore construct a map of undersea topography. Deeper regions are shown in tones of purple, blue, and green, while shallower regions are in tones of yellow and reddish brown. Data were acquired by the U.S. Navy Geosat satellite altimeter.

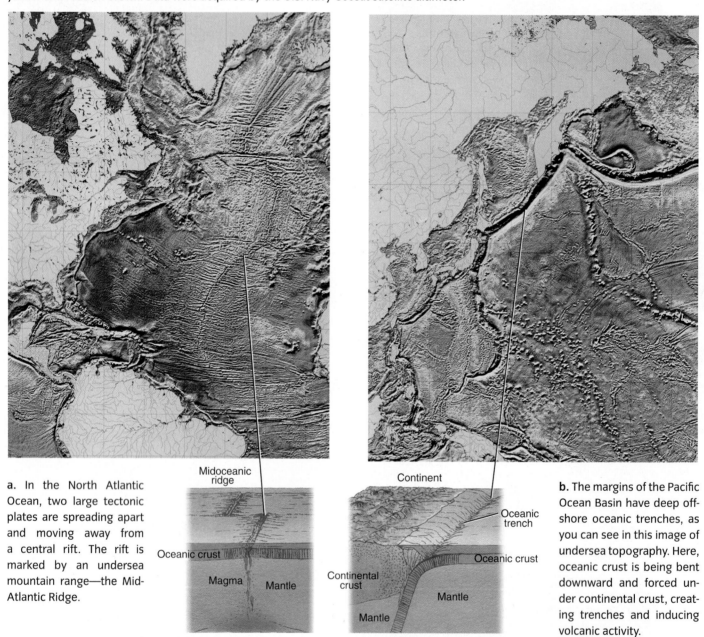

a. In the North Atlantic Ocean, two large tectonic plates are spreading apart and moving away from a central rift. The rift is marked by an undersea mountain range—the Mid-Atlantic Ridge.

Midoceanic ridge

Oceanic crust

Magma Mantle

Continent

Oceanic trench

Oceanic crust

Continental crust

Mantle

Mantle

b. The margins of the Pacific Ocean Basin have deep offshore oceanic trenches, as you can see in this image of undersea topography. Here, oceanic crust is being bent downward and forced under continental crust, creating trenches and inducing volcanic activity.

Relief Features of the Ocean Basins

Oceans make up 71% of the Earth's surface. Relief features of oceans are quite different from those of the continents. Much of the oceanic crust is less than 60 million years old, while the great bulk of the continental crust is over 1 billion years old. The young age of the oceanic crust is quite remarkable. We will see later that plate tectonic theory explains this age difference.

Just like continental topography, undersea topography is largely determined by tectonic activity (**Figure 8.22**). Where two plates are spreading apart, a midoceanic ridge occurs. In areas where two lithospheric plates collide, the ocean floor shows deep trenches where one plate is being pushed under the other. These areas, such as the Pacific Basin, are characterized by earthquakes and volcanic activity, as we will discuss further in Chapter 9.

Video Explorations

Born of Fire

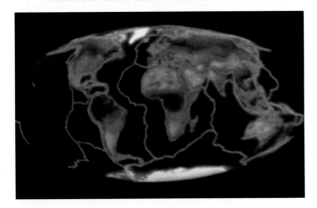

Though undetectable to us, the continents move an inch or more each year. Over the course of the planet's history, this motion has led to a cycle of collisions and splitting. The ocean floors and continents that form the Earth's crust, or lithosphere, are in continuous motion. The contact that occurs along the plate margins is often marked by earthquakes and volcanic eruption. This video provides a dynamic close-up of the Earth's internal dynamics and the energy of our lava- and magma-driven world and demonstrates the underlying tectonic processes that form and re-form our world.

In areas where plates spread, or move apart, a mid-oceanic ridge of submarine hills divides the basin in about half. Precisely in the center of the ridge, at its highest point, is a narrow trench-like rift. The location and form of this rift suggest that the crust is being pulled apart along the line of the rift.

Where the crust pulls apart at the rift, magma wells up and hardens to fill the void, creating new oceanic crust. The discovery that the youngest area of the Earth's crust was created along midoceanic ridges was an important clue to our modern understanding that the lithospheric plates are in motion.

Configuration of the Continents

The long formation history of the Earth's surface features is driven by the movement of lithospheric plates sliding over the hot viscous asthenosphere (see *Video Explorations*). If the plates are in motion, then the configuration of our continents must have changed many times over the history of the Earth. The way the continents look today is but a brief snapshot of the Earth's dynamic form in geologic timescales.

The theory of continental drift As good navigational charts became available, geographers could see the close correspondence between the outlines of the eastern coast of South America and the western coastline of Africa. In the early 20th century, German meteorologist and geophysicist Alfred Wegener proposed the first full-scale scientific theory describing the breakup of a single supercontinent, which he named **Pangea**, into multiple continents that drifted apart in a process he called continental drift. Wegener suggested that Pangea existed intact as early as about 300 million years ago, in the Carboniferous period. Wegener supported his theory by demonstrating that similar fossils and present-day plant species were found on separate continents, as if they had once been adjacent (**Figure 8.23**). Unfortunately, Wegener's explanation of the physical process that caused the continents to separate was incomplete, and most scientists of the time rejected his theory.

Wegener's Pangea • Figure 8.23

Alfred Wegener's 1915 map fits together the continents that today border the Atlantic Ocean Basin. The sets of dashed lines show the fit of Paleozoic tectonic structures between Europe and North America and between southernmost Africa and South America.

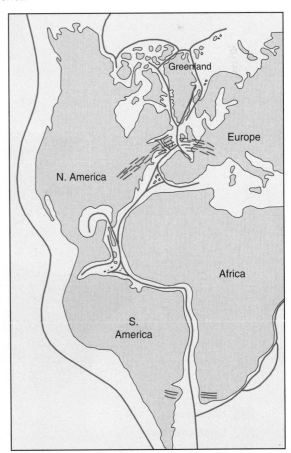

Seafloor spreading In the 1960s, seismologists discovered the mechanism responsible for the motion of the plates—seafloor spreading (**Figure 8.24**). Scientists studied patterns of magnetism on the seafloor and later obtained sediment samples to determine the relative age of ocean crust. They found that the crust was youngest along the seafloor spreading zones and got progressively older as it spread away from the midoceanic ridges. This evidence showed that new crust was formed by magma upwelling along the ocean ridge, while older crust cooled and sank back down into the mantle at ocean trenches. As the tectonic plates are spread apart at spreading centers, the continents of the plates move apart. The theory that describes our current understanding of how the plates move is known as plate tectonics. Chapter 9 discusses different types of plate movements and what happens on the Earth's surface where plates meet.

Seafloor spreading • Figure 8.24

Lava extrudes along the midoceanic ridge, forming new oceanic crust. When lava cools, it becomes magnetized in polarity with the Earth's magnetic field. Periodically, in the range of tens of thousands of years, the polarity of the Earth's magnetic field is reversed, and the polarity of crust that forms during that period is also reversed. Teams of geoscientists have mapped the successive ages—magnetic bands—of the newly formed ocean crust and discovered symmetry in the age of crust on both sides of the midoceanic ridge.

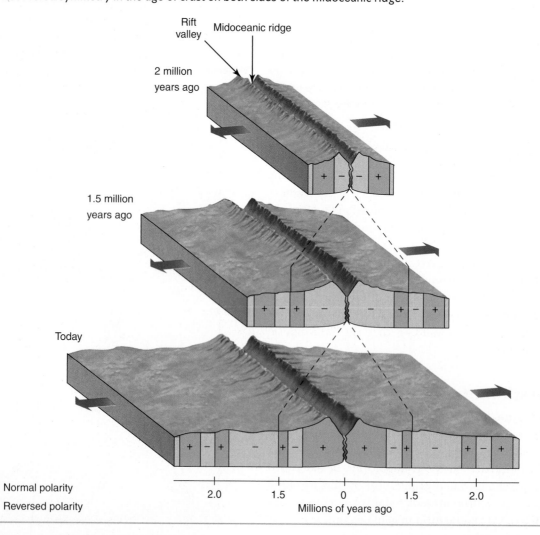

History of the continents • Figure 8.25

THE PLANNER

The Earth's lithospheric plates have twisted and turned, collided, and then broken apart over the past 600 million years to create a geography of the Earth as we know it today.

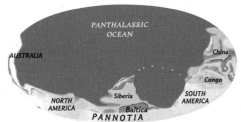

1 **600 million years ago**
A supercontinent, known as Rodinia, split apart, and oceans filled the basins. Fragments collided, thrusting up mountain ranges. Glaciers spread, twice covering the equator. A new polar supercontinent, Pannotia, formed.

2 **500 million years ago**
A breakaway chunk of Pannotia moved north, splitting into three masses—Laurentia (North America), Baltica (northern Europe), and Siberia. In shallow waters, the first multicellular animals with exoskeletons appeared, and the Cambrian explosion of life took off.

3 **300 million years ago**
Laurentia collided with Baltica and later with Avalonia (Britain and New England). The Appalachian Mountains arose along the edge of the supercontinent Pangea.

4 **200 million years ago**
Dinosaurs roamed Pangea, which stretched nearly from pole to pole and almost encircled Tethys, the oceanic ancestor of the Mediterranean Sea. The immense Panthalassic Ocean surrounded the supercontinent.

5 **100 million years ago**
Pangea broke apart. The Atlantic poured in between Africa and the Americas. India split away from Africa, and Antarctica and Australia were stranded near the South Pole.

6 **50 million years ago**
Moving continental fragments collided—Africa into Eurasia, pushing up the Alps, and India into Asia, raising the Plateau of Tibet. With dinosaurs now extinct, birds and once-tiny mammals began to evolve rapidly.

7 **Present day**
Formation of the Isthmus of Panama and the split of Australia from Antarctica changed ocean currents, changing the climate. Ice sheets waxed and waned in many cycles, and sea levels rose and fell.

History of the continents The continents are moving today, just as they have in the past. Data from orbiting satellites shows that rates of separation or of convergence between two plates are on the order of 5 to 10 cm (about 2 to 4 in.) per year, or 50 to 100 km (about 30 to 60 mi) per million years. At that rate, global geography must have been very different in past geologic eras than it is today (**Figure 8.25**).

Topography of the Earth 265

PROCESS DIAGRAM

Wegener was largely correct about the supercontinent Pangea, but today we know that there were even earlier supercontinents than Pangea. An earlier supercontinent dubbed Rodinia was fully formed about 700 million years ago. Some interesting evidence suggests that there may have been yet another supercontinent before Rodinia. In fact, over the billions of years of the Earth's geologic history, the union of the continents and their subsequent breakup is actually a repeating process that has occurred half a dozen times or more.

By studying the changing continental arrangements and locations over time, we can recognize that as the continents changed latitude, their climates, soils, and vegetation changed.

CONCEPT CHECK   STOP

1. **What** are the six major types of continental landforms?

2. **How** is plate motion related to the formation of midoceanic ridges and ocean trenches?

3. **How** did scientists confirm that the seafloor is spreading?

THE PLANNER ✓

Summary

1 The Changing Earth 240

- **Geomorphology** is the study of how the Earth's features have changed over time. **Uniformitarianism** expresses the idea that the processes that shape the Earth today have been operating throughout the Earth's history.

- The slow pace of geologic change is measured on the **geologic timescale**, which divides the Earth's history into epochs, periods, and eras. Precambrian time includes the Earth's earliest history, followed by the Paleozoic, Mesozoic, and Cenozoic eras.

- Through **radiometric dating**, geographers apply what they know about the rate at which certain atoms decay to determine the age of ancient rocks.

- **Endogenic processes** such as volcanic and tectonic activity create **initial landforms**, which are then sculpted into **sequential landforms** by **exogenic processes** such as wind and water, as shown here.

Initial and sequential landforms • Figure 8.3

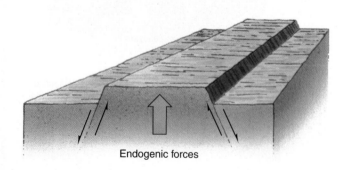

Endogenic forces

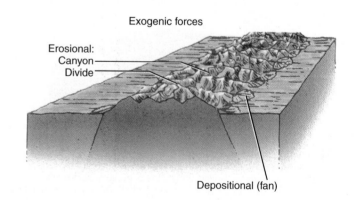

Exogenic forces

Erosional:
Canyon
Divide

Depositional (fan)

2 The Structure of the Earth 244

- The Earth is composed of three main layers: the **core**, **mantle**, and **crust**, shown in this diagram.

The structure of the Earth • Figure 8.4

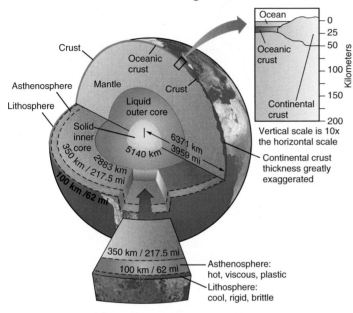

- The Earth's core is a dense mass of liquid iron and nickel that is solid at the very center.

- The mantle surrounds the core and includes the soft, plastic **asthenosphere**, as well as more rigid regions. The boundary between the crust and the upper mantle is called the **Moho**.

- The outermost layer is the crust. Continental crust is thicker and less dense than oceanic crust. The **lithosphere** includes the crust and the rigid portion of the mantle below it. The lithosphere is divided into **lithospheric plates** that slide over the underlying soft asthenosphere. **Isostasy** describes how the crust sinks and rises as weight is added and removed.

3 Earth Materials and the Cycle of Rock Change 247

- **Igneous rocks** are made from cooled **magma**. Such rocks are either **extrusive igneous rocks** that emerge in the form of lava (see photo) or **intrusive igneous rocks** that intrude into the crust, where they cool to form **plutons**. The largest plutons are **batholiths**, which become exposed at the surface through erosion.

Lava • Figure 8.10

- **Sedimentary rocks** are formed in layers, or **strata**, as fragments of **rocks** and **minerals** are compressed and cemented together. Sedimentary rocks include clastic sediments, formed from small pieces of rock, chemically precipitated sediments from solution, and organic sediments made from plant and animal remains.

- **Metamorphic rocks** are formed when igneous or sedimentary rocks are exposed to extreme heat and pressure.

- The cycle of rock change is responsible for constantly transforming rocks over millions of years.

4 Topography of the Earth 258

- Continental masses consist of active mountain-making belts and inactive regions of old, stable rock. Mountain-building occurs through volcanism and tectonic activity. Continental shields are large regions of low-lying igneous and metamorphic rocks.

- The motion of lithospheric plates is explained in part by the process of seafloor spreading, when new ocean crust wells up at midoceanic ridges and then is recycled back into the mantle at oceanic trenches.

- As a result of plate motion, a single large supercontinent—**Pangea**, shown here—broke apart into the present arrangement of continents and ocean basins.

Wegener's Pangea • Figure 8.23

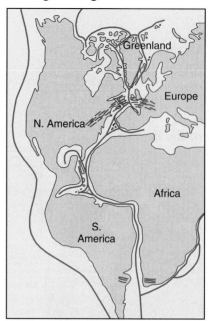

Key Terms

- asthenosphere 244
- batholith 251
- core 244
- crust 244
- endogenic process 243
- exogenic process 243
- extrusive igneous rock 249
- geologic timescale 241

- geomorphology 240
- igneous rock 249
- initial landform 243
- intrusive igneous rock 249
- isostasy 246
- landform 243
- lithosphere 245

- lithospheric plate 246
- magma 249
- mantle 244
- metamorphic rock 256
- mineral 247
- Moho 244
- Pangea 263

- pluton 251
- radiometric dating 241
- rock 247
- sedimentary rock 252
- sequential landform 243
- strata 252
- uniformitarianism 240

Critical and Creative Thinking Questions

1. Sketch the cycle of rock change and describe how each of three major rock types is formed and re-formed.

2. The rock cycle described in the accompanying figure is really a network of interlocking cycles. The most common rock cycle is sedimentary to metamorphic to igneous and back to sedimentary. Describe the situations that would give (1) igneous to metamorphic to sedimentary and (2) sedimentary to metamorphic to sedimentary.

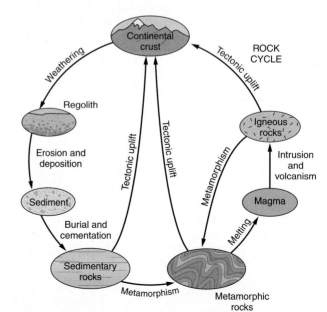

3. Snow-covered mountain belts provide our most striking landforms on all continents. Describe two processes that are believed to form these landscapes.

4. Looking at a modern world map, one can easily be convinced that continents now across the Atlantic Ocean from one another once fit together. When Wegener's theory about continental drift was presented in the early 1900s, why was it not accepted by the science community?

What is happening in this picture?

This 2.5-billion-year-old sedimentary rock strata in the Hamersley Range in Australia indicates a record of changing oxygen levels in the atmosphere. When iron reacts with oxygen (oxidation), it forms compounds that don't dissolve in seawater. During periods when there was less oxygen in the atmosphere, more iron dissolved in seawater and precipitated as chemical sediment, forming darker color bands in this rock formation. During periods when there was more oxygen in the atmosphere, less iron dissolved in seawater, and chemical precipitation formed lighter bands.

Scientists believe the red layers were deposited during hot climate periods when organic processes generated oxygen and created iron oxide. Geologically younger and thicker red strata are primarily sedimentary deposits containing iron oxides in sandstone and shale.

Black layers are rich in precipitated iron, low in O_2

Red layers are rich in oxidized iron (rust), high O_2

White layers are low in precipitated iron, rich in silica sediment from microorganisms.

Think Critically

1. In which eon did this sedimentary rock form?
2. What do the dark bands say about life on the planet in this eon?
3. Considering that photosynthesis by microorganisms releases oxygen, what do the alternating dark and light layers indicate about the abundance of these microorganisms?

Self-Test

(Check your answers in the Appendix.)

1. All time older than 570 million years before the present is _____ time.

   a. Cambrian

   b. Mesozoic

   c. Triassic

   d. Precambrian

2. Extrapolating geologic time to a single 24-hour day, the proliferation of life on planet Earth began only _____ before midnight.

   a. 18 minutes

   b. 2 hours and 50 minutes

   c. 45 minutes

   d. 1 hour

3. This cutaway diagram of the Earth's interior shows several layers. Label the mantle, outer core, inner core, asthenosphere, and lithosphere. Indicate whether each layer is solid, plastic, or liquid.

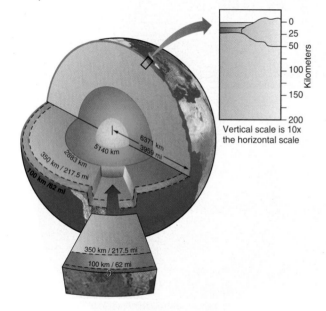

4. The _____ is a soft, plastic layer that underlies the lithosphere.

   a. mantle

   b. outer core

   c. inner core

   d. asthenosphere

5. _____ describes the rise of floating mountains over the soft asthenosphere as they erode.

   a. Subduction

   b. Isostacy

   c. Erosion

   d. Exogenic motion

6. A(n) _____ is a naturally occurring, inorganic chemical compound, often with a characteristic crystal shape.

   a. isotope

   b. mineral

   c. rock

   d. metamorphic rock

7. To determine the age of rocks, scientists _____.

   a. calculate magnetic pole reversal

   b. examine microscopic components

   c. determine radiometric timescales

   d. All of the above statements are correct.

8. Intrusive igneous rocks are noted for their _____.

   a. large mineral crystals

   b. organic mineral composition

   c. small mineral crystals

   d. darker colors

9. The completely metamorphosed equivalent of sandstone is _____, which is formed by the addition of silica to completely fill the open spaces between the grains.

   a. slate

   b. quartzite

   c. schist

   d. marble

10. The fossil shown in this photo is found between layered accumulations of mineral particles derived mostly through weathering and erosion of preexisting rocks. Which rock class do these layered accumulations belong to?

a. sedimentary rocks

b. igneous rocks

c. metamorphic rocks

d. volcanic rocks

11. The diagram shows a batholith with portions of the batholiths exposed. How are batholiths formed? How do they become exposed at the surface?

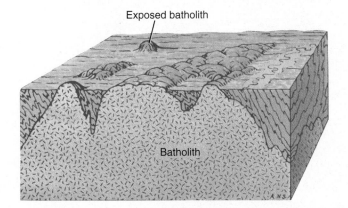

Exposed batholith

Batholith

12. Where do fossil fuels come from? How is the burning of fossil fuels related to levels of atmospheric CO_2?

13. The theory describing the motions and changes through time of the continents and ocean basins and the processes that fracture and fuse them is called _____.

a. continental drift

b. Earth dynamics

c. plate tectonics

d. the Pangea cycle

14. Ocean basins are characterized by a central _____ consisting of submarine hills that rise gradually to a rugged central zone delineated by a narrow, trench-like feature called a(n) _____.

a. cordillera, rift

b. subduction zone, rift

c. midoceanic ridge, rift

d. batholith, abyss

15. Alfred Wegener, a German meteorologist and geophysicist, suggested in 1915 that the continents had once been joined together as a supercontinent, which he named _____.

a. Rodina

b. Gondwanaland

c. Pangea

d. Laurasia

THE PLANNER ✓

Review your Chapter Planner on the chapter opener and check off your completed work.

Plate Tectonics, Earthquakes, and Volcanoes

The Earth trembled off the northeast seacoast of Honshu, Japan, on March 11, 2011, from the largest earthquake in Japan's history. The great Tohoku earthquake registered a magnitude of 9.0; the world's fourth-largest quake since 1900. The quake resulted from a rupture along a 300-kilometer (186-mile) subduction boundary in the Japan Trench, located on the Pacific Ring of Fire. More than 700 aftershocks continued for several days as the Pacific plate finally settled under the Eurasian plate.

Thousands perished in the initial moments after the quake, despite the fact that Japan has the best earthquake preparation in the world, regularly testing state-of-the-art early-warning sirens and practicing evacuation drills. Minutes after the quake, a 9.2-meter (30-foot) tsunami swept away tens of thousands of people, literally wiping towns off the map along the coast of Sendai and Fukushima. Without the early warning and constant training, hundreds of thousands more would have perished. Adding to the crisis, the tsunami damaged nuclear reactors located on the coast at Fukushima Dai-ichi, which led to the release of radioactive material into the local environment, further complicating rescue and recovery operations. In the wake of this natural disaster, seismic experts humbly acknowledged that no one had predicted such a massive earthquake and that science still could not foresee the timing, location, or magnitude of the next devastating earthquake.

The great Tohoku earthquake and subsequent tsunami destroyed coastal Fukushima, Japan.

CHAPTER OUTLINE

CHAPTER PLANNER ✓

- ❏ Study the picture and read the opening story.
- ❏ Scan the Learning Objectives in each section:
  p. 274 ❏   p. 282 ❏   p. 293 ❏
- ❏ Read the text and study all figures and visuals. Answer any questions.

Analyze key features

- ❏ Process Diagram p. 275 ❏   p. 281 ❏   p. 295 ❏   p. 301 ❏   p. 302 ❏
- ❏ Geography InSight p. 276 ❏   p. 298 ❏
- ❏ What a Geographer Sees, p. 284
- ❏ Video Explorations, p. 285
- ❏ Where Geographers Click, p. 297
- ❏ Stop: Answer the Concept Checks before you go on.
  p. 281 ❏   p. 293 ❏   p. 303 ❏

End of chapter

- ❏ Review the Summary and Key Terms.
- ❏ Answer the Critical and Creative Thinking Questions.
- ❏ Answer What is happening in this picture?
- ❏ Complete the Self-Test and check your answers.

Plate Tectonics

LEARNING OBJECTIVES

1. **Differentiate** the three main types of plate boundaries.
2. **Explain** the process of subduction.
3. **Describe** the process of continent–continent collision.

Today we know that the Earth's lithosphere is fractured into more than 50 separate tectonic plates ranging from very large to very small. Seven of these plates are significantly larger than the others: the Pacific, North American, Eurasian, African, Antarctic, Australian, and South American plates (**Figure 9.1**). The motions of lithospheric plates are responsible for creating the shape of our planet, from the tops of mountains to the trenches of the sea bottom. They also trigger some of our deadliest natural disasters. Over geologic time, changes in the configuration of oceans and continents produced by the movement of lithospheric plates have changed the Earth's climates and influenced the distribution and abundance of plants and animals across the Earth's landscapes. The body of knowledge about lithospheric plates and their motions is referred to as **plate tectonics**.

Types of Plate Boundaries

As the lithospheric plates move, they can move away from each other, at a **spreading boundary**, also called a divergent boundary, or they can collide at a **convergent boundary**. Plates may also simply slide past each other at a **transform boundary**.

Each type of boundary is associated with one of two basic forms of tectonic activity: compression or extension (**Figure 9.2**). Compression occurs when lithospheric plates are squeezed together along convergent lithospheric plate boundaries, and extension happens along spreading boundaries, where plates are being pulled apart.

> **plate tectonics** The theory of tectonic activity, which deals with lithospheric plates and their motions.

> **spreading boundary** A boundary where two lithospheric plates move apart.

> **convergent boundary** A boundary where two lithospheric plates come together.

> **transform boundary** A boundary where two lithospheric plates slide past each other.

Major and minor lithospheric plates of the globe • Figure 9.1

The Earth's surface crust is divided into a set of lithospheric plates that are in slow but steady motion. Movement of these plates is driven by heat from the Earth's core.

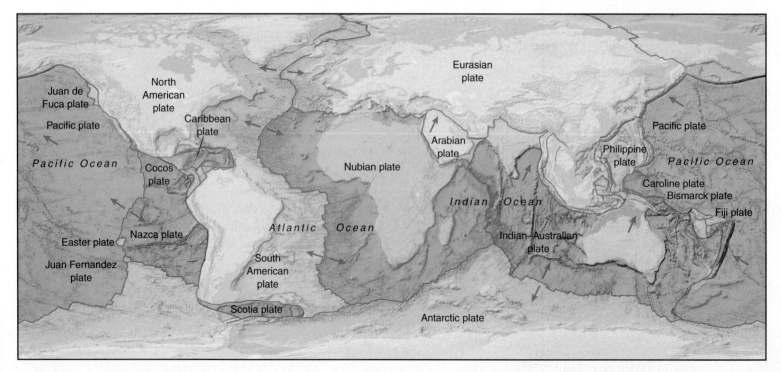

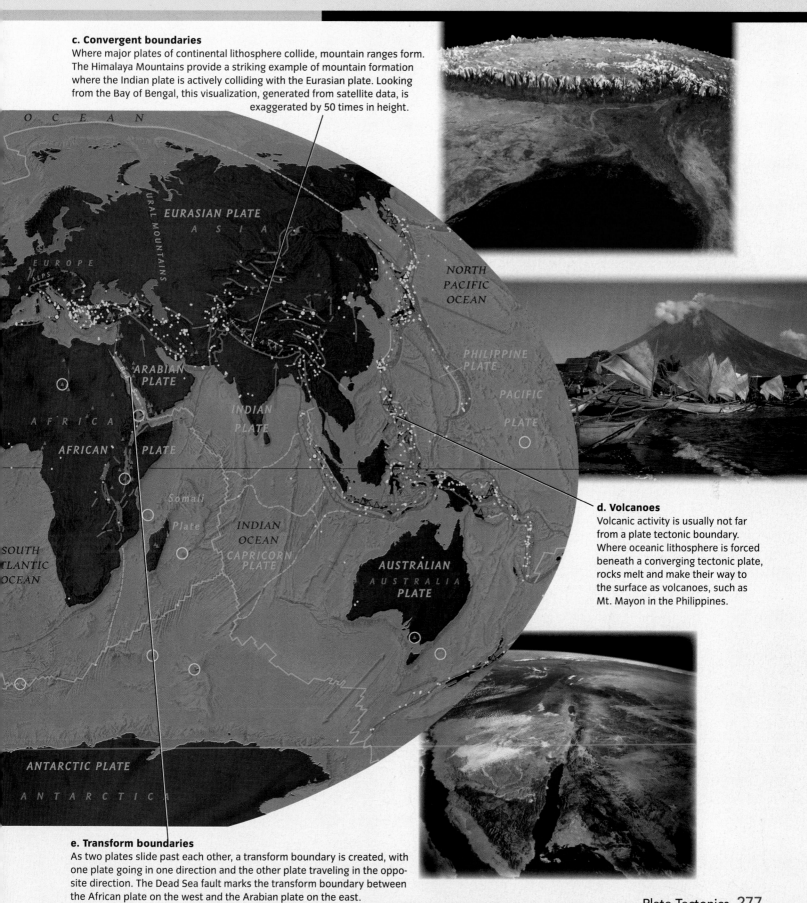

 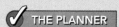

c. Convergent boundaries

Where major plates of continental lithosphere collide, mountain ranges form. The Himalaya Mountains provide a striking example of mountain formation where the Indian plate is actively colliding with the Eurasian plate. Looking from the Bay of Bengal, this visualization, generated from satellite data, is exaggerated by 50 times in height.

d. Volcanoes

Volcanic activity is usually not far from a plate tectonic boundary. Where oceanic lithosphere is forced beneath a converging tectonic plate, rocks melt and make their way to the surface as volcanoes, such as Mt. Mayon in the Philippines.

e. Transform boundaries

As two plates slide past each other, a transform boundary is created, with one plate going in one direction and the other plate traveling in the opposite direction. The Dead Sea fault marks the transform boundary between the African plate on the west and the Arabian plate on the east.

Plate Tectonics 277

Subduction • Figure 9.6

Where an oceanic plate collides with a continental plate, the oceanic plate is subducted into the asthenosphere.

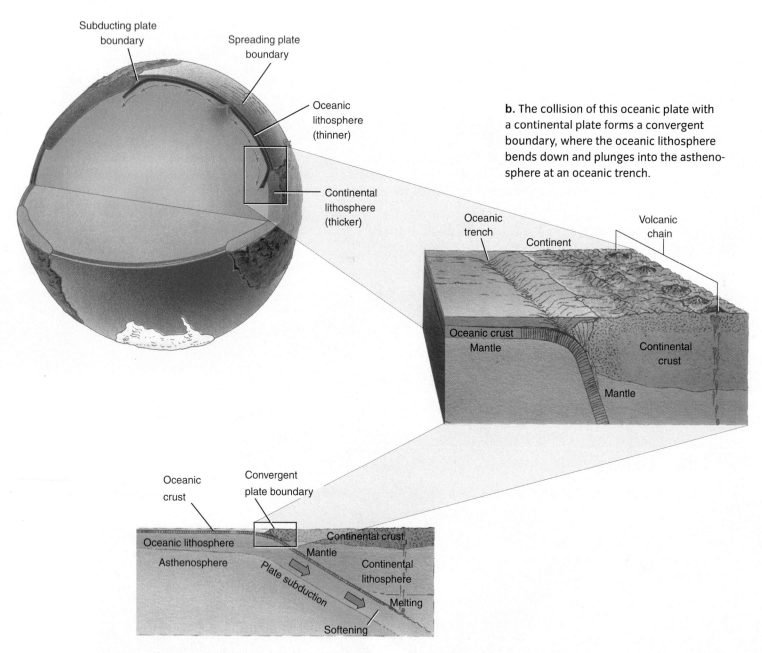

a. Plates of oceanic lithosphere, moving away from a spreading center, collide with other plates.

Subducting plate boundary

Spreading plate boundary

Oceanic lithosphere (thinner)

Continental lithosphere (thicker)

b. The collision of this oceanic plate with a continental plate forms a convergent boundary, where the oceanic lithosphere bends down and plunges into the asthenosphere at an oceanic trench.

Oceanic trench

Continent

Volcanic chain

Oceanic crust

Mantle

Continental crust

Mantle

Oceanic crust

Convergent plate boundary

Oceanic lithosphere

Continental crust

Mantle

Asthenosphere

Plate subduction

Continental lithosphere

Melting

Softening

c. This much broader view of the same area, seen from the side, shows the subducted descending plate gradually reverting to mantle rock as it softens.

Subduction Zones

Where two plates push against each other at a convergent boundary, the way they interact is determined by

subduction Descent of the edge of a lithospheric plate under an adjoining plate and into the asthenosphere.

the thickness and density of each of the plates. When one plate is denser and thinner than the other, it is forced to slide beneath the less dense, thicker plate in a process called **subduction** (**Figure 9.6**). As a result of this plate motion, subduction zones are associated with both volcanoes and earthquakes.

Subduction zones create a variety of features as one plate sinks and the other rises. Deep **oceanic trenches** result when the descending plate pulls material into the asthenosphere, such as the Mariana Trench, the deepest location in the ocean (about 11 km, or 6.85 mi, below sea level) and the Japan Trench, responsible for the Great Tohoku earthquake mentioned in the chapter opening. Mountain ranges, such as the Andes Mountains or the Cascade Range of mountains along the northwest Pacific coast in North America, are created by volcanic activity and crustal uplift at the collision boundary. Chains of volcanoes also occur along subduction zones, where magma, formed by the melting of ocean sediments carried into the asthenosphere, rises to the surface.

Collision Zones

When two continental plates collide at a convergent boundary, they are both too thick and too buoyant for either plate to slip under the other, so they meet head-on. As the two plates collide, various kinds of crustal rocks are first crumpled into tightly compressed wave-like structures called **folds**.

folds Corrugations of strata caused by crustal compression.

Orogeny The folding of crust where two continental plates meet contributes to a mountain-building process called **orogeny**. You can re-create the mountain-building process using an ordinary towel laid on a smooth tabletop. Using both hands, palms down, bring the ends of the towel together slowly. First, a simple upfold will develop and grow, until it overturns to one side or the other. Then more folds will form, grow, and overturn. The same thing happens when plates of continental lithosphere collide.

The great Himalaya Range is a result of such a collision between continents. This collision occurred in the Cenozoic Era, along a great tectonic line that marks the southern boundary of the Eurasian plate. The line begins with the Atlas Mountains of North Africa and runs, with a few gaps, to the great Himalaya Range, where it is still active. Each segment of this collision zone represents the collision of a different north-moving plate against the single Eurasian plate.

The folds create a set of alternating upbends, called **anticlines**, and troughs, called **synclines**. Anticlines usually form broadly rounded mountain ridges, and synclines create open valleys (**Figure 9.7**). Accompanying the folding is a form of faulting in which slices of rock move over the underlying rock on fault surfaces with gentle inclination angles. These are called **overthrust faults**.

anticlines The arch in folds of the Earth's crust.

synclines The U-shaped base of folds in the Earth's crust.

Folding • Figure 9.7

Compression of the Earth's crust as a result of plate motion raises mountain ranges in a process called folding. The initial landform associated with an anticline is a broadly rounded mountain ridge, and the landform corresponding to a syncline is an elongate, open valley.

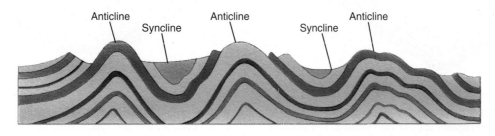

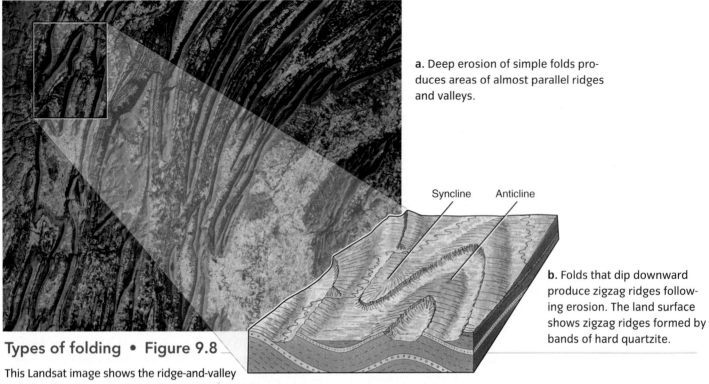

a. Deep erosion of simple folds produces areas of almost parallel ridges and valleys.

Syncline   Anticline

b. Folds that dip downward produce zigzag ridges following erosion. The land surface shows zigzag ridges formed by bands of hard quartzite.

Types of folding • Figure 9.8

This Landsat image shows the ridge-and-valley country of south-central Pennsylvania in color infrared. The strata were crumpled during a continental collision that took place over 200 million years ago.

As fold areas erode, they create different patterns, based on the type of folding. Erosion, and the distinct rates of erosion on various Earth materials from water, wind, and gravity, are discussed in Chapter 10. Simple folds erode to create a landscape of continuous and even-crested ridges that are almost parallel. Where folds rise up in some places and dip down in others, they erode to form a pattern of zigzag ridges. The Appalachian Mountains show examples of many different types of folding under various erosion regimes (**Figure 9.8**).

Sequential landforms are also determined by the different strengths or resistances of rock layers. Differential erosion of layers leads to gaps in the ridges and can even lead to anticlinal valleys and synclinal mountains (**Figure 9.9**).

Differential erosion of strata • Figure 9.9

The ways fold areas erode to form sequential landforms depends on the strengths of the rocks in the different strata.

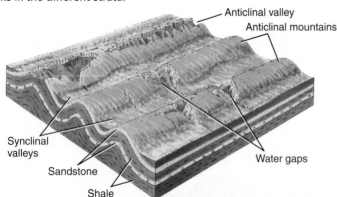

Anticlinal valley
Anticlinal mountains

Synclinal valleys

Sandstone

Shale

Water gaps

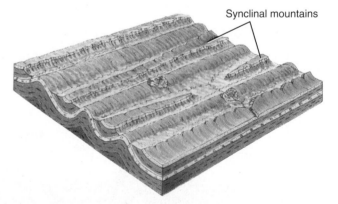

Synclinal mountains

a. Weaker formations such as shale and limestone erode away to leave long, narrow ridges of hard strata, such as sandstone or quartzite. Sometimes, the resistant rock at the center of the anticline is eroded through to reveal softer rocks underneath, creating an anticlinal valley. In a few places, a larger stream cuts across a ridge as the ridge is forming, creating a deep, steep-walled water gap.

b. Synclinal mountains can also be formed after continued erosion, when resistant rock at the center of a syncline is exposed. The resistant rock stands up as a ridge.

Continental suture • Figure 9.10

Where plates of continental lithosphere collide, the rock formations along the boundaries may be transformed under extreme pressure into newly formed metamorphic rocks. The resulting continental suture binds the two plates into a single, larger plate.

1 The thinner oceanic crust of the left-hand plate is subducted under the continental crust on the right.

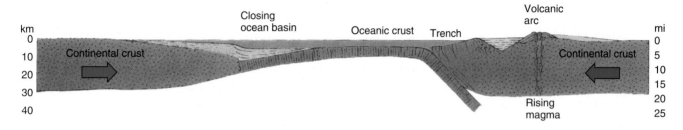

2 As the continents move closer, the ocean between the converging continents is eliminated. The oceanic crust is compressed, creating a succession of overlapping thrust faults, which ride up, one over the other.

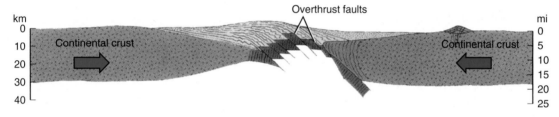

3 As the slices become more and more tightly squeezed, they are forced upward. A mass of metamorphic rock forms between the joined continental plates, welding them together. This new rock mass is the continental suture.

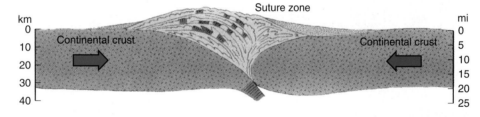

Continental suture Over long periods of time, two continents can converge to eliminate an ocean between them. A continental suture permanently unites the two plates, so that there is no further tectonic activity along that collision zone (**Figure 9.10**). Geologists have identified several ancient sutures in the continental shields. One example is the Ural Mountains, which divide Europe from Asia and were formed near the end of the Paleozoic Era. Others are the Appalachian Mountains of eastern North America and the Caledonides of Norway, which date from mid-Paleozoic time.

CONCEPT CHECK STOP

1. **How** are the three major plate boundaries formed?

2. **What** distinctive landforms are associated with subduction zones?

3. **What** distinctive landforms are associated with continent–continent collisions?

Tectonic Activity and Earthquakes

LEARNING OBJECTIVES

1. **Differentiate** the four main types of faults.
2. **Explain** how earthquakes are triggered.
3. **Identify** two tectonic environments associated with major earthquakes.

So far we have focused on the incremental motions of plates that build mountain ranges and create ocean basins over the course of millennia. But some plate motions can be sudden and dramatic. The sudden breaking of the Earth's crust, due to stresses created from the plate motion, can result in one of the most potentially devastating natural hazards that humankind faces: an earthquake.

Faults and Fault Landforms

As lithospheric plates move in different directions, the brittle rocks of the Earth's surface experience stress and strain. When these rocks finally break, they form a fracture in the crust called a **fault**. Major fractures in the crust can be followed along the ground for many kilometers and extend down into the crust for at least several kilometers.

> **fault** A sharp break in rock associated with a slippage of the crustal block on one side of a tectonic plate with respect to the block on the other.

A single fault movement can cause slippage of as little as a centimeter or as much as 15 m (about 50 ft). Fault movements can happen many years or decades apart, even several centuries apart. But when we add up all these

Types of faults • Figure 9.11

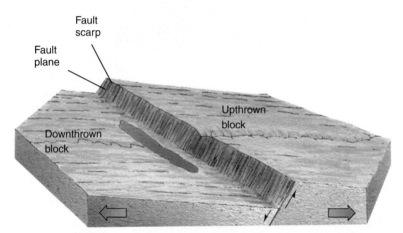

a. Normal fault

As a result of extensional activity, the crust on one side of a normal fault drops down relative to the other side. This creates a steep, straight fault scarp.

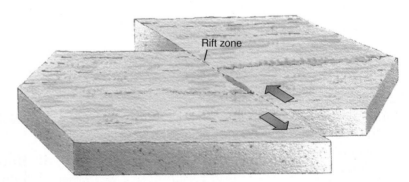

b. Strike-slip fault

Movement along a strike-slip fault is mostly horizontal, so we don't see a scarp, or at most we see a very low one. We can usually only trace a thin fault line across the surface, although in some places the fault is marked by a narrow trench, or rift.

small motions over long geologic time spans, they can amount to tens or hundreds of kilometers of displacement.

There are four main types of faults, characterized by how the crustal blocks move: **normal**, **strike-slip** or transcurrent, **reverse**, and the previously mentioned overthrust faults. The type of fault can usually be identified by the relationship between the moving blocks after they move (**Figure 9.11**). The presence of a newly exposed cliff face, called a **fault scarp**, can indicate how much the fault moved (**Figure 9.12**).

Fault scarp • Figure 9.12

This fault scarp was formed during the Hebgen Lake, Montana, earthquake of 1959. In a few moments, a displacement of 6 m (20 ft) took place on a normal fault.

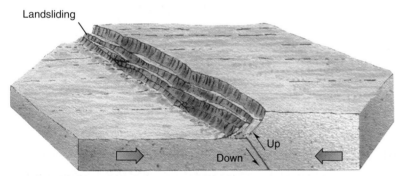

c. Reverse fault

As a result of compressional activity, the fault plane along a reverse fault is inclined such that one side rides up over the other. Reverse faults produce fault scarps similar to those of normal faults. But because the scarp tends to be overhanging there's a much greater risk of a landslide. The San Fernando, California, earthquake of 1971 was generated by slippage on a reverse fault.

d. Overthrust fault

Overthrust faults involve mostly horizontal movement. One slice of rock rides over the adjacent ground surface. A thrust slice may be up to 50 km (30 mi) wide.

Put It Together

Review the section Types of Plate Boundaries and answer this question. Which type of plate boundary is most associated with overthrust faults?

a. divergent boundary
b. transform boundary
c. spreading center
d. convergent boundary

Landforms created by parallel faults • Figure 9.13

a. Graben

A narrow block dropped down between two normal faults creates a graben—a trench with straight, parallel walls.

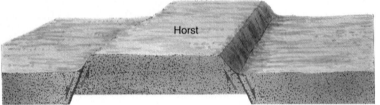

b. Horst

A narrow block elevated between two normal faults is a horst—making block-like plateaus or mountains, often with a flat top but steep, straight sides.

---

graben A crustal block dropped down between two normal faults.

Just as with other tectonic activity, fault movements give rise to distinctive landforms. For example, parallel normal faults can drop a block of crust down into a graben, or push a block up, into a **horst (Figure 9.13)**. Systems of parallel faulting are typical of rift valleys where continental spreading is occurring (see *What a Geographer Sees*).

horst A crustal block pushed up between two normal faults.

WHAT A GEOGRAPHER SEES

The Basin and Range Province of the Southwestern United States

A hiker traversing the northern portion of Nevada would enjoy a spectacular landscape of valley basins and mountain ranges (Figure a). From the distinctive landforms in this satellite image, however, a geographer would recognize a classic example of the basin and range, or horst and graben, landscape. As shown in this satellite image, this landscape covers much of Nevada and southern Utah (Figure b).

This striking landscape is a result of horizontal stress or stretching of the crust. Based on analysis of landforms and fault systems, geographers know that the basin and range is actually dropping down along a massive system of faults as the underlying lithosphere thins. The horsts and grabens are like keystone blocks of a masonry arch that have slipped down between neighboring blocks as the arch has spread apart. Each separate basin varies in width from about 30 to 60 km (20 to 40 mi).

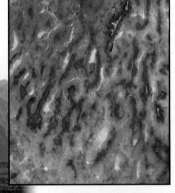

a.

b.

Think Critically

1. If the basins in this landscape continue to spread horizontally, what do you expect will happen?
2. Along which pair of compass directions does the basin and range system appear to be stretched?

Fault scarp evolution • Figure 9.14

When vertical displacement at a fault occurs, the rising block can create a steep fault scarp. Once it is formed, the geology and weathering patterns influence the form and erosion characteristics of the fault scarp.

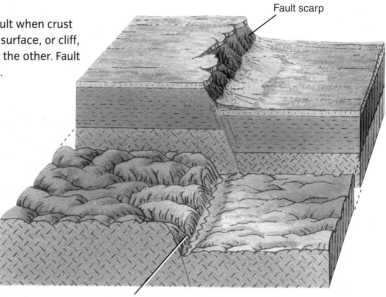

a. A recently formed fault scarp may result when crust stretched by tension leaves an exposed surface, or cliff, from the shifting of one block relative to the other. Fault scarps create distinctive and sheer cliffs.

b. Even though the cover of sedimentary strata has been completely removed, exposing the ancient shield rock, the fault continues to produce a landform, known as a **fault-line scarp**.

Fault scarp

Fault-line scarp

Repeated faulting can produce a great rock cliff hundreds of meters high. Because fault planes extend hundreds of meters down into the bedrock, their landforms can persist even after millions of years of erosion. **Figure 9.14** diagrams the effect of erosion on a fault scarp.

Earthquakes

As rock on both sides of the fault is slowly bent over many years by tectonic forces, energy builds up in the

> **earthquake** A trembling or shaking of the ground produced by movements along a fault.

bent rock, just as it does in a bent archer's bow. When that tension reaches a critical point, the fault slips, relieving the strain in the form of an **earthquake** (see *Video Explorations*).

When a fault slips, rocks on opposite sides of the fault move in different directions, instantaneously releasing a large amount of energy, in the form of seismic waves. These earthquake waves radiate out, traveling through the Earth's surface layer and shaking the ground. Like ripples produced when a pebble is thrown into a quiet pond, these waves gradually lose energy as they travel outward in all directions.

Video Explorations
When the Earth Shakes

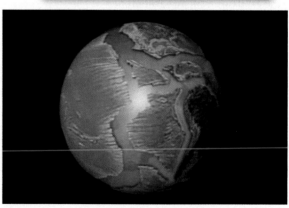

Every day, the Earth is shaken by hundreds of little earthquakes, most of which pass unnoticed. Earthquakes usually happen along lithospheric plate boundaries. This video explains how earthquakes are triggered.

Earthquake focus and epicenter • Figure 9.15

Seismic waves emanate in all directions from the earthquake site of origin, called the focus, located along a fault line. The epicenter is the surface location directly above the subterranean focus. The earthquake focus can originate from various types of fault zones.

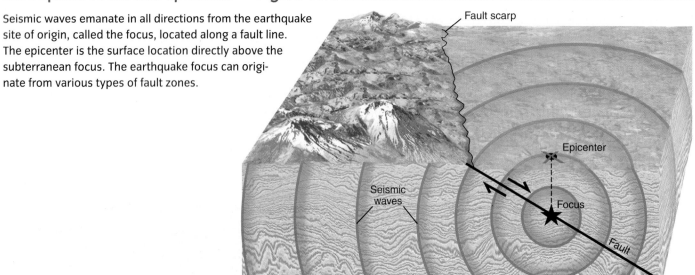

Epicenter and focus

In an earthquake, the location where the fault slipped, called the **focus**, can be near the surface or deep underground. The depth of an earthquake's focus in part determines the amount of shaking felt on the ground; shallow-focus earthquakes are likely to cause more damage than deep-focus earthquakes. The point on the Earth's surface that is directly above the focus is called the **epicenter** of the earthquake (**Figure 9.15**).

> **epicenter** The location on the Earth's surface directly above where a fault slipped to produce an earthquake.

To determine where the epicenter is, and therefore where to send emergency services, scientists can measure how long it takes different types of earthquake waves to reach sensors, called seismometers, positioned in different locations. The first waves to reach the seismometer are the faster-moving P waves, or primary waves, which shake the ground from side to side. Next come the slower-moving S waves, or secondary waves, which are characterized by up-and-down shaking. The amount of time it takes between the arrival of P and S waves at nearby seismograph stations is used to calculate the point of origin of the earthquake (**Figure 9.16**).

Determining earthquake epicenters • Figure 9.16

Epicenters are calculated by triangulating the readings from three different seismometer reading centers (**A**, **B**, and **C**), using the difference in travel times for P and S waves.

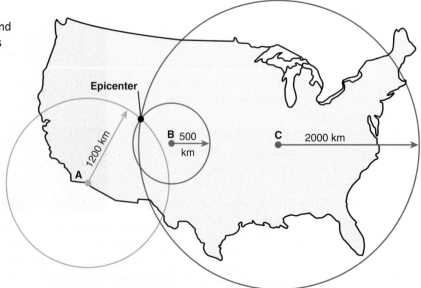

Magnitude The amount of energy released by an earthquake, called its **magnitude**, can be measured by the amplitude of the seismic waves produced. In 1935, the distinguished seismologist Charles F. Richter devised the scale of earthquake magnitudes that bears his name. **Figure 9.17** shows how the Richter scale relates to the energy released by some of the largest earthquakes on record.

Earthquake magnitude • Figure 9.17

Earthquake magnitude is a measure of how much energy is released by an earthquake and is not a direct indicator of the severity of damages.

a. For each whole unit of increase on the Richter scale (say, from 5.0 to 6.0), the quantity of energy released increases by a factor of 32. The most severe earthquake ever measured is the magnitude 9.5 Valdivia, or great Chilean, earthquake of 1960.

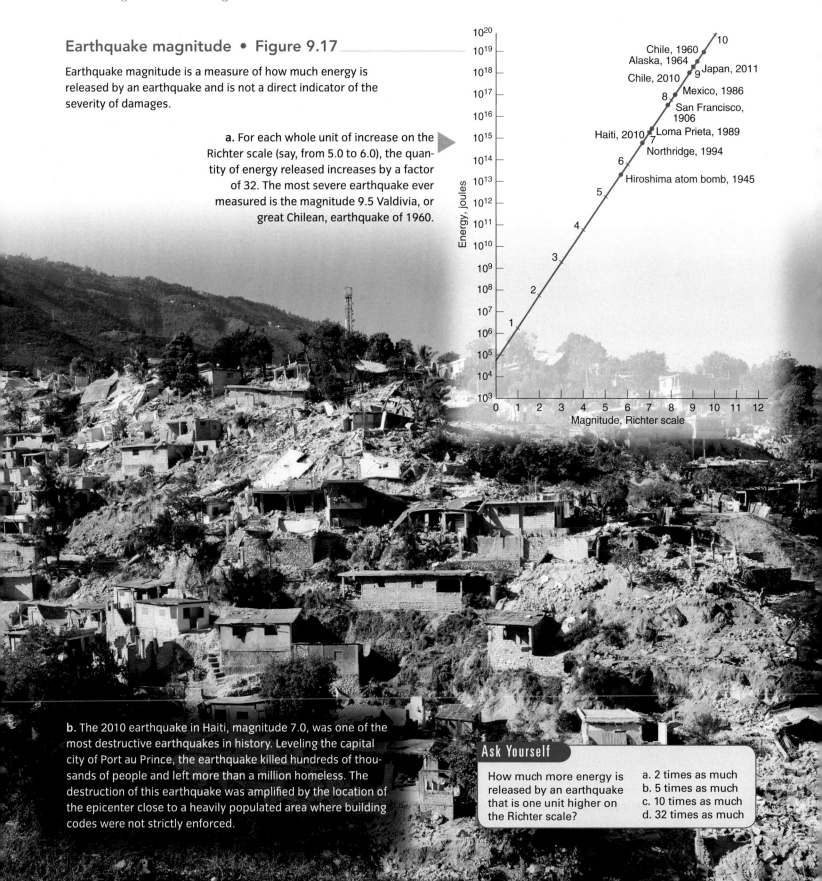

b. The 2010 earthquake in Haiti, magnitude 7.0, was one of the most destructive earthquakes in history. Leveling the capital city of Port au Prince, the earthquake killed hundreds of thousands of people and left more than a million homeless. The destruction of this earthquake was amplified by the location of the epicenter close to a heavily populated area where building codes were not strictly enforced.

Ask Yourself

How much more energy is released by an earthquake that is one unit higher on the Richter scale?

a. 2 times as much
b. 5 times as much
c. 10 times as much
d. 32 times as much

Earthquakes and plate boundaries • Figure 9.18

If you compare the map of earthquake locations with the tectonic features, you can see that seismic activity is primarily associated with lithospheric plate motion. This world map plots hundreds of thousands of earthquake epicenters recorded over a 20-year period by international earthquake information centers. The greatest-magnitude earthquakes are located where the yellow denotes the deepest earthquakes. According to geophysicists, about half of all moderate or strong quakes occur on previously unknown, or blind, faults.

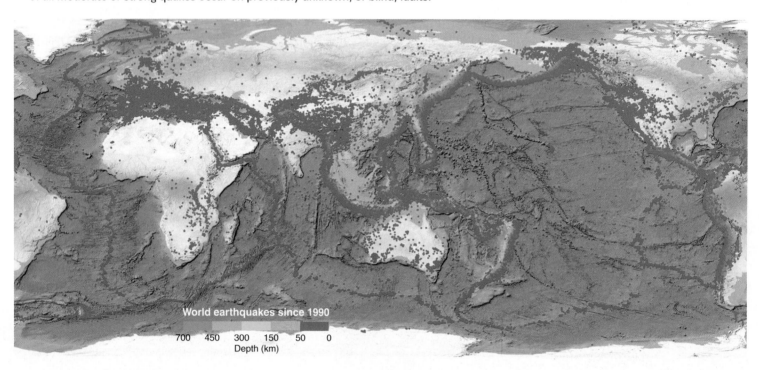

World earthquakes since 1990

700   450   300   150   50   0
Depth (km)

The magnitude of an earthquake is related to the length of the section of fault that broke. This means that longer faults have the potential to produce larger earthquakes. Keep in mind that earthquake magnitude is not a direct indicator of earthquake damages. Earthquake damages are also related to proximity of the epicenter to populated areas, the depth of focus, the type of ground materials, and whether local buildings meet earthquake safety codes. The magnitude 9.0 Great Tohoku earthquake occurred 129 km (80 mi) off the coast of Japan, 40 km (24 mi) deep, and spread across a 300-km (186-mi) fault, causing the Earth to move 30 meters (100 ft). Japan's strict building codes and public education program reduced the damages from this massive quake.

Tectonic Environments of Earthquakes

Earthquake activity is primarily (although not exclusively) associated with plate boundaries (**Figure 9.18**).

Subduction zone earthquakes Intense seismic activity occurs along convergent lithospheric plate boundaries where oceanic plates are undergoing subduction. This mechanism is responsible for the greatest earthquakes, including those in Japan, Alaska, North America, Central America, and Chile, and other narrow zones close to the trenches and volcanic arcs of the Pacific Ocean Basin.

On the Pacific coast of Mexico and Central America, the subduction boundary of the Cocos plate lies close to the shoreline. The great earthquake that devastated Mexico City in 1986 was centered in the deep trench offshore. Two great shocks in close succession, the first of magnitude 8.1 and the second of 7.5, damaged cities along the coasts of the Mexican states of Michoacoán and Guerrero. Although Mexico City lies inland about 300 km (about 185 mi) distant from the earthquake epicenter, it experienced intense ground shaking, killing some 10,000 people.

Undersea subduction zone earthquakes pose an additional hazard to coastal residents. When the subducted plate suddenly snaps back during an earthquake, it displaces a large volume of water, generating long-wavelength water waves that can travel hundreds of miles across the open ocean at high speeds. As these waves reach the shore, they slow and build into a wall of water that can be many meters high, called a tsunami. In 2004, a subduction zone earthquake of magnitude 9.0 in the Java Trench generated a tsunami that devastated populations around the Indian Ocean, killing hundreds of thousands of people. An international response to this event has led to the creation of an ocean buoy network for tsunami early warning systems. Such systems saved hundreds of thousands of lives along the Honshu, Japan, coast on March 11, 2011.

Tsunamis can affect coastal populations far from the earthquake's epicenter, but they are most hazardous when the earthquake epicenter is close to shore, leaving little time for warning between the earthquake and the arrival of the waves. Coastal residents who feel an earthquake should seek higher ground immediately, without waiting for a tsunami warning. We will discuss tsunamis in more detail in Chapter 13.

Transform boundaries Strike-slip faults on transform boundaries that cut through the continental lithosphere cause moderate to strong earthquakes. The San Andreas fault, which runs about 1000 km (about 600 mi) from the Gulf of California to Cape Mendocino, is an active strike-slip fault (**Figure 9.19**).

Movement on the San Andreas fault generated the great San Francisco earthquake of 1906, one of the worst natural disasters in the history of the United States. Since then, this sector of the fault has been locked—that is, rocks on the two sides of the fault have been held together without sudden slippage. In the meantime, the two lithospheric plates that meet along the fault have been moving steadily with respect to each other. This means that a huge amount of unrelieved strain (potential energy) has already accumulated in the crustal rock on either side of the fault.

San Andreas fault in Southern California • Figure 9.19

Throughout many kilometers of its length, the San Andreas fault looks like a straight, narrow scar. In some places this scar is a trench-like feature, and elsewhere it is a low scarp. It marks the active boundary between the Pacific plate and the North American plate. The Pacific plate is moving toward the northwest, carrying a great portion of the state of California and all of Lower (Baja) California with it.

On October 17, 1989, the San Francisco Bay area was severely jolted by an earthquake with a magnitude of 7.1 on the Richter scale. The earthquake's epicenter was located near Loma Prieta peak, about 80 km (50 mi) southeast of San Francisco, at a point only 12 km (7 mi) from the city of Santa Cruz, on Monterey Bay. Altogether, 62 lives were lost in this earthquake, and the damage was estimated to be about $6 billion. In comparison, the 1906 earthquake took a toll of 700 lives and caused property damage equivalent to about $30 billion in present-day dollars. The Loma Prieta slippage, though near the San Andreas fault, probably has not relieved more than a small portion of the strain on this fault.

In Southern California, three severe earthquakes occurred in close succession in 1992, along active local faults a short distance north of the San Andreas fault in the southern Mojave Desert. The second of these, the Landers earthquake, a powerful 7.5 on the Richter scale, occurred on a transcurrent fault trending north–northwest. It caused an 80-km (50-mi) rupture across the desert landscape.

For residents of the Los Angeles area, an additional serious threat lies in the large number of active faults nearby (**Figure 9.20**). Movements on these local faults have produced more than 40 damaging earthquakes since 1800, including the Long Beach earthquakes of the 1930s and the San Fernando earthquake of 1971. The San Fernando earthquake measured 6.6 on the Richter scale and severely damaged structures near the earthquake epicenter. In 1987 an earthquake of magnitude 6.1 struck the vicinity of Pasadena and Whittier, located within about

Earthquake activity in Southern California • Figure 9.20

California earthquakes, identified as circles with size proportional to the earthquake magnitude, are overlaid on top of delineated and mapped fault lines. The correlation of earthquakes and fault lines, as shown in this figure, is found around the globe.

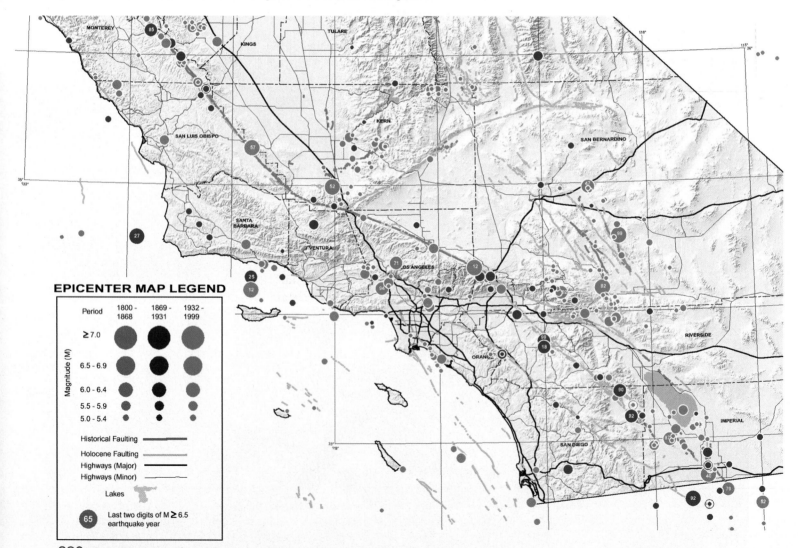

USGS earthquake hazard maps depict the New Madrid seismic zone, far from any
active plate boundaries, as one of the highest-hazard areas in the nation.

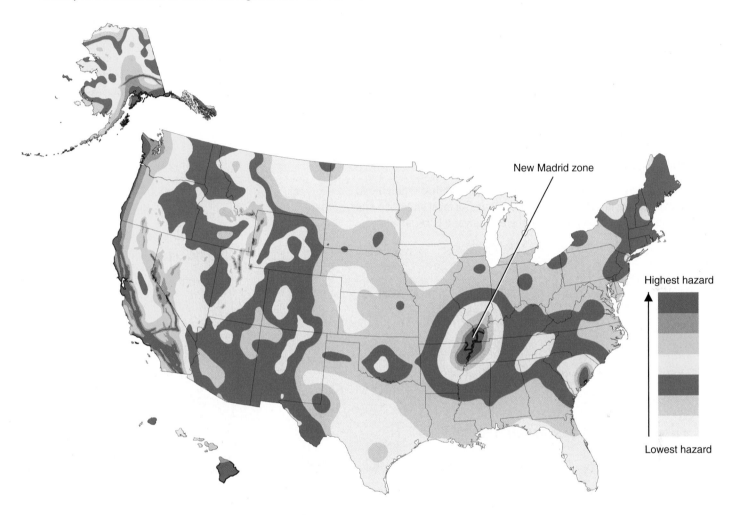

New Madrid zone

Highest hazard

Lowest hazard

20 km (12 mi) of downtown Los Angeles. Known as the
Whittier Narrows earthquake, it was generated along a local fault system that had not previously shown significant
seismic activity. The Northridge earthquake of 1994, at 6.7
on the Richter scale, produced the strongest ground motions ever recorded in an urban setting in North America
and the greatest financial losses from an earthquake in
the United States since the San Francisco earthquake of
1906. Sections of three freeways were closed, including the
busiest highway in the country, Interstate 5.

Spreading center earthquakes Spreading plate
boundaries also produce seismic activity. Most of these
boundaries are identified with the midoceanic ridge and its
branches, and the earthquakes are moderate. Close exami-
nation of the midoceanic ridge displays numerous strike-slip
faults that form perpendicular to the spreading plate line.

Earthquakes away from plate boundaries A
few earthquake epicenters are scattered over the continental plates, far from active plate boundaries. Scientists are uncertain why these occur but suggest that
spreading continental plates may be a factor. In many
cases, no active fault is visible. For example, the great
New Madrid earthquake of 1811 was centered in the
Mississippi River floodplain in Missouri. It produced
three great shocks in close succession, rated from 8.1 to
8.3 on the Richter scale. The Earth movements caused
the Mississippi River to change its course and even run
backward in a few stretches for a short time. Large areas
of land dropped, and new lakes were formed. The U.S.
Geological Survey (USGS) considers the present populated areas located over the New Madrid seismic zone
to be in the high-risk category (**Figure 9.21**). Researchers conducting underwater seismic surveys have recently

mapped two new unknown faults along the Mississippi River north of the city of Memphis, demonstrating an increased probability of a major earthquake of magnitude 7 on the Richter scale.

Earthquakes along blind faults Although scientists understand the role that faults play in earthquake risk, they don't always know where fault systems are located. Some faults, called blind faults, are not apparent on the Earth's surface. Although seismographic research can help scientists identify blind faults, areas of political unrest or geographic inaccessibility may contain fault systems that have never been mapped.

In 2003, a major earthquake on a blind strike-slip fault, previously unknown to seismologists and geoscientists, destroyed the ancient city of Bam in southern Iran (**Figure 9.22**). Approximately 30,000 people died and an equal number were injured and made homeless by the event, which was rated a 6.6 on the Richter scale.

Bam, Iran, 2003 earthquake • Figure 9.22

Bam, Iran, is located on a previously unknown blind fault that experienced a 6.6 Richter magnitude earthquake in 2003. The fault near Bam is a local strike-slip fault, not far from the convergent boundary between the Arabian and Eurasian plates.

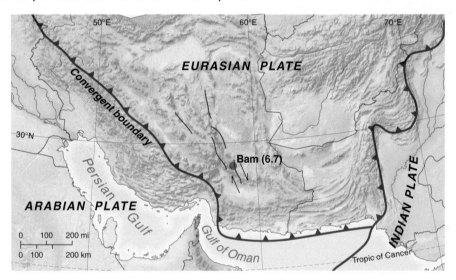

Before

After

Bam Citadel, a UNESCO Heritage Site,
before and after the 2003 earthquake.

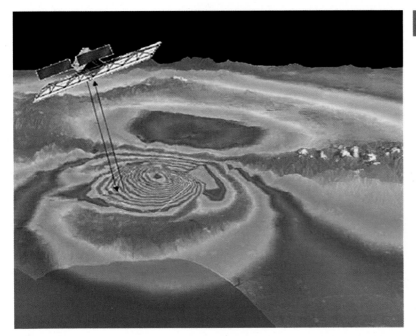

Space-based instruments imaging Earth movements • Figure 9.23

Scientists at NASA's Jet Propulsion Laboratory used satellite measurements using the Interferometric synthetic aperture radar (InSAR) taken on separate dates to detect landform deformation of tectonic stresses after the Northridge earthquake north of Los Angeles. Active faults building up strain were seen as colored contours radiating from the points of distortion. An artist's rendition of the satellite sits over the InSAR calibrated distortion rings.

This devastating earthquake highlights the lesson that while the major plate boundaries and most seismic areas have been mapped by geologists, as human populations grow, they will be increasingly exposed to unknown seismic events from blind, or unmapped, faults. This is why increasing research and seismic measurements are needed to improve the performance of early warning systems. Research scientists are using satellites to take measurements of seismic movement to within fractions of an inch. These precise measurements are identifying the slow buildup of deformation along faults and helping to map ground deformation after an earthquake (**Figure 9.23**).

CONCEPT CHECK STOP

1. **What** does a fault scarp indicate about fault movement?

2. **How** is the location of an earthquake determined?

3. **How** do scientists know that there will be another earthquake on the San Andreas fault?

Volcanic Activity and Landforms

LEARNING OBJECTIVES

1. **Identify** three tectonic environments that give rise to volcanoes.

2. **Distinguish** between the characteristics of felsic and basaltic lavas.

3. **Explain** how stratovolcanoes and shield volcanoes form and evolve.

Like earthquakes and plate tectonics, volcanic activity creates initial landforms. Underground molten mineral matter, called **magma**, is extruded through constricted vents and fissures in the Earth's surface. When the magma reaches the surface, where it is called **lava**, it cools and hardens, building a landform, called a **volcano**, that is typically conical or dome shaped.

> **volcano** A conical or dome-shaped structure built by accumulation of lava flows and volcanic ash.

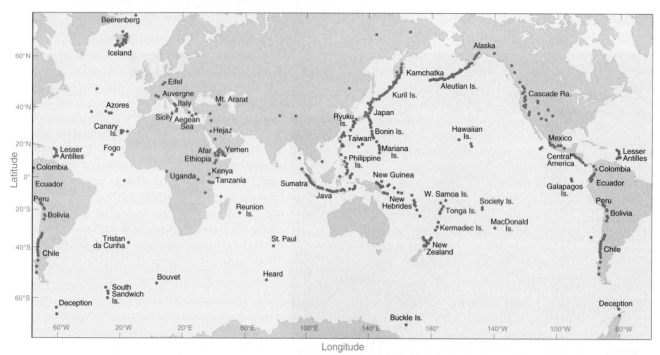

a. Dots show the locations of volcanoes known or believed to have erupted within the past 12,000 years. Each dot represents a single volcano or cluster of volcanoes.

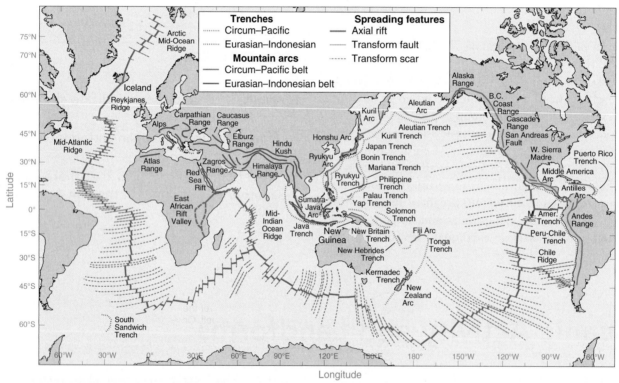

b. This map shows the principal tectonic features of the world such as midoceanic ridges, trenches, and subduction boundaries. Many volcanoes occur along these features—for example, the Ring of Fire around the Pacific Rim occurs on subduction boundaries. Iceland's volcanoes are produced by spreading on the Mid-Atlantic Ridge, and spreading in East Africa has also produced volcanoes.

Volcanic activity of the Earth • Figure 9.24

Tectonic Environments of Volcanoes

Volcanic activity is related to plate movement. If you compare the maps in **Figure 9.24**, you can see that many volcanoes are located on plate boundaries. Subduction zones around the Pacific Rim perpetuate significant volcanic activity, producing the Pacific Ring of Fire.

Volcanoes also occur along seafloor spreading centers. Iceland, in the North Atlantic Ocean, provides an outstanding example (**Figure 9.25**). Other islands of volcanoes located along or close to the axis of the Mid-Atlantic Ridge are the Azores, Ascension, and Tristan da Cunha.

Chains of island volcanoes can also appear away from plate boundaries, where a plume of magma, called a **hotspot**, rises through the crust (**Figure 9.26**). The chain of Hawaiian volcanoes, from the newest islands to the oldest seamounts, marks the motion of the Pacific plate over the stationary hotspot.

> **hotspot** A center of volcanic activity thought to be located over a rising mantle plume.

Helgafell   Lava from Eldfell   Eldfell

Spreading center volcano • Figure 9.25

Iceland is constructed entirely of eruptions of basalt emerging from the Mid-Atlantic Ridge. The volcanic cone on the right, christened Eldfell, is surrounded by lava flows extending into the ocean. Note the tongue of fresh lava from Eldfell invading the right side of the town. Eldfell's older sister, Helgafell, is on the left.

Hotspot volcano formation • Figure 9.26

✓ THE PLANNER

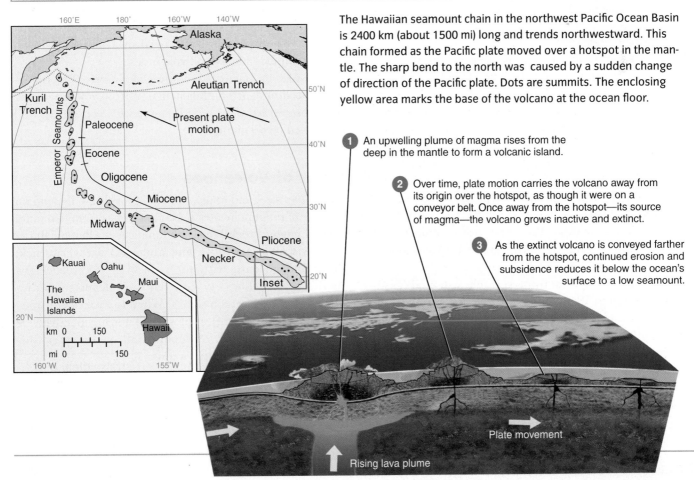

The Hawaiian seamount chain in the northwest Pacific Ocean Basin is 2400 km (about 1500 mi) long and trends northwestward. This chain formed as the Pacific plate moved over a hotspot in the mantle. The sharp bend to the north was caused by a sudden change of direction of the Pacific plate. Dots are summits. The enclosing yellow area marks the base of the volcano at the ocean floor.

1. An upwelling plume of magma rises from the deep in the mantle to form a volcanic island.

2. Over time, plate motion carries the volcano away from its origin over the hotspot, as though it were on a conveyor belt. Once away from the hotspot—its source of magma—the volcano grows inactive and extinct.

3. As the extinct volcano is conveyed farther from the hotspot, continued erosion and subsidence reduces it below the ocean's surface to a low seamount.

Plate movement

Rising lava plume

PROCESS DIAGRAM

Atmospheric CO₂ measurements • Figure 9.27

Levels of CO_2 in the atmosphere have been recorded from the top of Hawaii's Mauna Loa observatory since 1958. These measurements show that atmospheric levels of CO_2 have been steadily increasing over the past 50 years. The eruption of two major volcanoes during this period, marked on the graph, had virtually no impact on this trend.

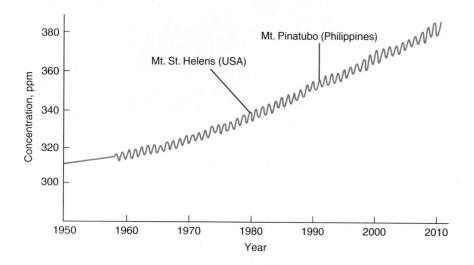

Think Critically

Based on this graph, what can you conclude about the role of volcanoes in global warming?

Volcanic Eruptions

When lava is forced out of a volcano violently, in the form of a volcanic eruption, it presents one of the most severe environmental hazards on the planet. During the Mt. Pelée disaster on the Caribbean island of Martinique in 1902, for example, thousands of lives were snuffed out in seconds. The destruction associated with volcanoes is wrought by a number of related effects. Sweeping clouds of incandescent gases descend the volcano slopes like great avalanches. Relentless lava flows can engulf whole cities. And showers of ash, cinders, and volcanic bombs (solidified masses of lava) cause terrible devastation. When hot ash combines with melting snow, it can race down the sides of a volcano in a hot mudflow called a lahar.

Atmospheric impacts from volcanic ash and gases have been shown to affect global weather and surface temperatures for months or years after an event. Following a major eruption, volcanic gases and ash in the atmosphere can reflect enough incoming sunlight to cause a global reduction in temperatures. For example, the year following the eruption of Mt. Tambora in 1815 is known as "the year without a summer" because unusually low temperatures around the world led to crop failures and famine.

On the other hand, volcanoes release gases into the atmosphere, including CO_2 and other greenhouse gases, which have a warming effect. Recall from Chapter 2 that greenhouse gases in the atmosphere absorb and reradiate longwave radiation back toward the Earth in a process called the greenhouse effect. An increase in greenhouse gases in the atmosphere increases the greenhouse effect and leads to more warming.

Some climate change critics have argued that volcanoes may be responsible for the documented rise in atmospheric levels of greenhouse gases over the past 150 years. In fact, volcanic sources have generated less than 1% of the greenhouse gases produced by human activities, as noted in the Keeling curve documenting monthly atmospheric levels of CO_2 (**Figure 9.27**).

Violent earthquakes are also associated with volcanic activity. Multiple seismic tremors or small earthquakes can inform scientists of volcanic behavior and can be used to warn citizens prior to major eruptions. Finally, for habitations along low-lying coasts, there is the peril of a tsunami generated by undersea or island volcanoes.

Thanks to scientific monitoring, we're now reducing the toll of death and destruction from volcanoes (see *Where Geographers Click*). Still, not every volcano is well monitored or predictable.

Types of Volcanoes

The shape, size, and explosiveness of a volcano depend on the type of magma involved. Magma comes from two main types of igneous rocks: felsic and mafic (see Chapter 8).

Felsic lavas (rhyolite and andesite) are very thick, gassy, and gummy, with high viscosity. So, felsic lava doesn't usually flow very far from the volcano's vent; instead, it builds up steep slopes. When the volcano erupts, ejected particles of different sizes, known collectively as **pyroclastic material**, or tephra, fall on the area surrounding the crater, creating a cone shape. The most common type of volcano associated with felsic magma is the **stratovolcano**, sometimes called a composite volcano.

> **stratovolcano** A volcano constructed of multiple layers of lava and volcanic ash.

Where Geographers CLICK

The Alaska Volcano Observatory

www.avo.alaska.edu

a.

Based in Anchorage and Fairbanks, the Alaska Volcano Observatory (AVO) uses a network of continuously recording seismometers to predict volcanic activity. Before an eruption, magma moves through the Earth's crust. This activity can be measured by seismometers—often months in advance. **Figure a** shows a seismometer at Mt. Spurr.

The observatory now monitors more than 20 volcanoes to help predict hazards to population centers and also to planes. Geographers also monitor satellite data from 80 volcanoes in the north Pacific for thermal activity and ash plumes. Thermal anomalies at volcanic vents have been detected up to several weeks before large eruptions. The AVO works with the National Weather Service to figure out where winds will carry the ash, which could be hazardous to aircraft. Such monitoring was a key factor in regulating air traffic after the Icelandic volcano that spread an ash cloud over Europe in 2010, grounding aircraft for days. Volcano updates include a volcano alert level for ground-based communities and aviation-specific color code for planes. (Green is for the least hazardous, and red is for the most hazardous, as shown in **Figure b**.)

b.

In contrast to felsic lava, mafic lava (basalt) is not very viscous and holds little gas. As you might expect, this means that eruptions of basaltic lava are usually quiet, and the lava can travel long distances to spread out in thin layers. Typically, then, large basaltic volcanoes are broadly rounded domes with gentle slopes. The most common type of volcano associated with mafic lava is the **shield volcano**.

The differences between felsic and mafic volcanoes are summarized in **Figure 9.28** on the next page.

> **shield volcano**
> A low, often large, dome-like accumulation of basalt lava flows emerging from long, radial fissures on flanks.

Shield volcanoes and stratovol-canoes are the two most common types of volcanoes. The key difference between them is their magma source, which controls their shape, size, and explosiveness.

a. Mafic magma has low viscosity, so it flows smoothly and easily. It emerges under low pressure, causing mostly gentle eruptions. Repeated eruptions build broad mountains that resemble a warrior's shield, giving shield volcanoes their name. They can become so massive that they bend the Earth's crust under their weight.

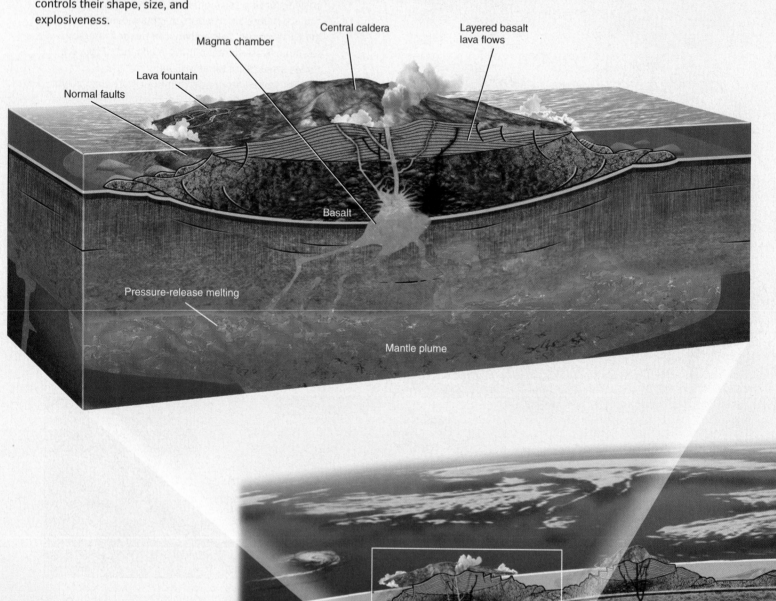

Central caldera

Magma chamber

Layered basalt lava flows

Lava fountain

Normal faults

Basalt

Pressure-release melting

Mantle plume

Rising lava plume

b. The mafic magma that forms shield volcanoes originates in the mantle. Mafic magmas are usually produced at spreading centers or hotspots and composed of dark material rich in magnesium and iron.

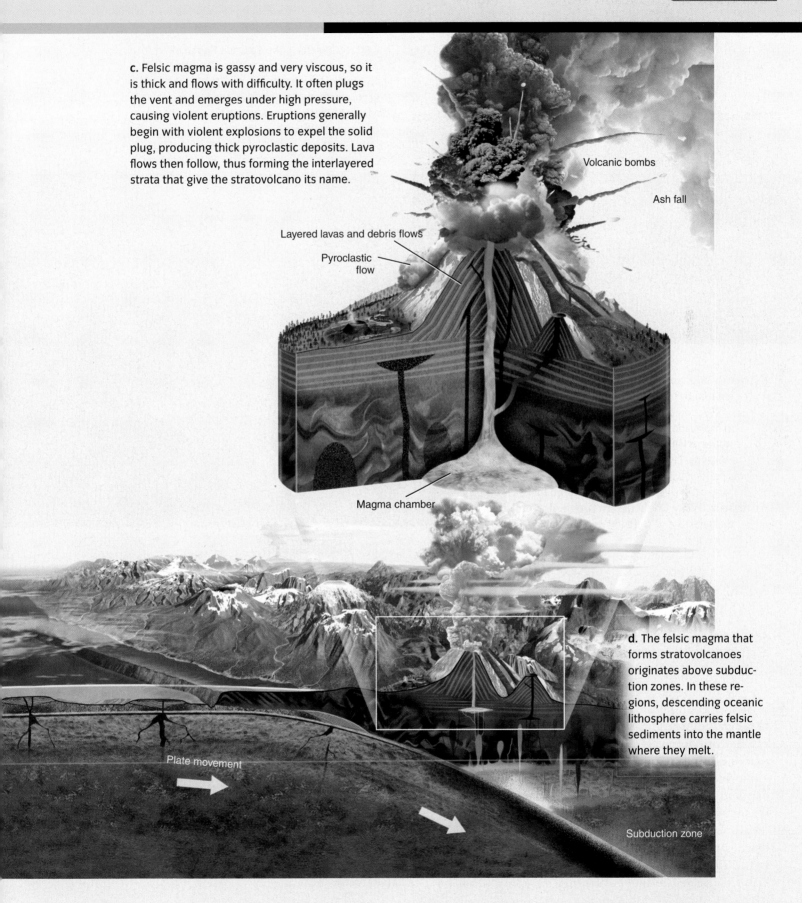

c. Felsic magma is gassy and very viscous, so it is thick and flows with difficulty. It often plugs the vent and emerges under high pressure, causing violent eruptions. Eruptions generally begin with violent explosions to expel the solid plug, producing thick pyroclastic deposits. Lava flows then follow, thus forming the interlayered strata that give the stratovolcano its name.

Volcanic bombs

Ash fall

Layered lavas and debris flows

Pyroclastic flow

Magma chamber

d. The felsic magma that forms stratovolcanoes originates above subduction zones. In these regions, descending oceanic lithosphere carries felsic sediments into the mantle where they melt.

Plate movement

Subduction zone

Volcanic Activity and Landforms **299**

Explosive stratovolcano eruptions • Figure 9.29

Because felsic lava holds gas under high pressure, it can cause explosive eruptions.

a. Mt. St. Helens in the Cascade Range in southwestern Washington State is a classic stratovolcano. It erupted violently on the morning of May 18, 1980, emitting a great cloud of heated gases and ash from the summit crater. Within a few minutes, the plume had risen to a height of 20 km (12 mi).

b. A cloud of hot, dense volcanic ash, emitted by the Soufrière Hills volcano in 1997, coursed down this narrow valley on the island of Montserrat in the Lesser Antilles. It killed about 20 people in small villages. Plymouth, the island's capital, was flooded with hot ash and debris, causing extensive fires that devastated the evacuated city. The southern two-thirds of the small island were left uninhabitable.

Ask Yourself

Which type of volcano is likely to have the most impact on human activities and/or the atmosphere?

a. a shield volcano
b. a stratovolcano

Stratovolcanoes Most of the world's active stratovolcanoes lie within the circum-Pacific mountain belt, where there is active subduction of the Pacific, Nazca, Cocos, and Juan de Fuca plates. One good example is the volcanic arc of Sumatra and Java, which lies over the subduction zone between the Australian and Eurasian plates.

Because felsic lavas from stratovolcanoes hold large amounts of gas under high pressure, they can produce explosive eruptions (**Figure 9.29**). Vast quantities of ash and dust fill the atmosphere for many hundreds of square kilometers around the volcano. Clouds of white-hot gases and fine ash sometimes travel rapidly down the flank of the volcanic cone, searing everything in their path. A glowing cloud from the 1902 Mt. Pelée eruption issued without warning, sweeping down on the city of St. Pierre and killing all but 2 of its 30,000 inhabitants.

These explosive eruptions can dramatically change the form of a stratovolcano. During an eruption, some of the upper part of the volcano is blown outward in fragments, although most of it settles back into the cavity formed beneath the former volcano by the explosion. Sometimes the explosion of a stratovolcano is so violent, from the pressure built up by gases captured in the felsic material, that it actually destroys the entire central portion of the volcano. The only thing remaining after the explosion is a great central depression, called a **caldera**. Over time, exogenic processes erode stratovolcanoes, creating new landscapes. **Figure 9.30** summarizes the stages in the building and eroding of stratovolcanoes.

Erosion of stratovolcanoes • Figure 9.30

1 These active volcanoes are in the process of building. They are initial landforms. Lava flows from the volcanoes, spreading down into a stream valley and forming a lake behind the lava dam.

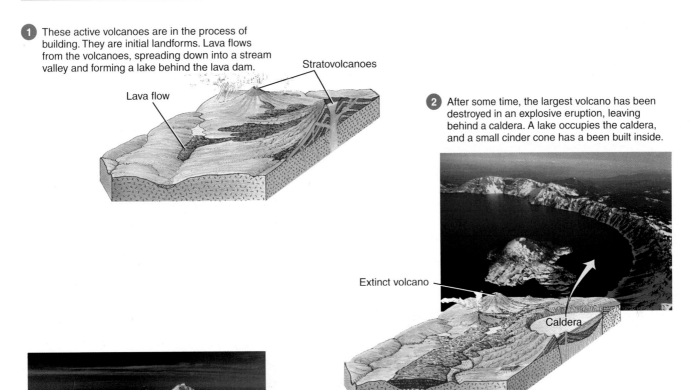

Stratovolcanoes

Lava flow

2 After some time, the largest volcano has been destroyed in an explosive eruption, leaving behind a caldera. A lake occupies the caldera, and a small cinder cone has a been built inside.

Extinct volcano

Caldera

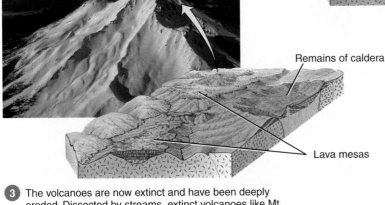

Remains of caldera

Lava mesas

3 The volcanoes are now extinct and have been deeply eroded. Dissected by streams, extinct volcanoes like Mt. Shasta, in the Cascade Range of northern California, lose their smooth, conical form. The caldera lake has been drained, and the rim has been worn to a low, circular ridge. The lava flows have resisted erosion far better than the rock of the surrounding area. They now stand high above the general level of the region, as mesas.

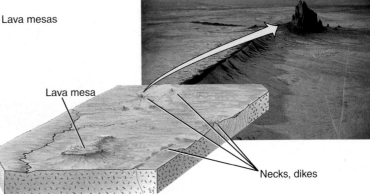

Lava mesa

Necks, dikes

4 All that remains now of each volcano is a small, sharp peak, called a volcanic neck. This is the remains of lava that solidified in the pipe of the volcano. Perhaps the finest illustration of a volcanic neck with radial dikes is Ship Rock, New Mexico.

Put It Together

Review Figure 9.10 and complete this statement.
Near a continental suture, you would expect to find _____ .

a. active stratovolcanoes
b. extinct stratovolcanoes
c. active shield volcanoes
d. extinct shield volcanoes

Volcanic Activity and Landforms 301

Erosion of shield volcanoes • Figure 9.31

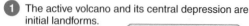

1 The active volcano and its central depression are initial landforms.

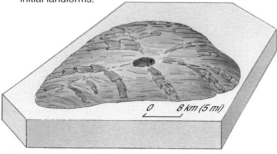

0   8 km (5 mi)

2 In the early stage of erosion, radial streams cut deep canyons into the flanks of the extinct shield volcano. These canyons are opened out into deep, steep-walled amphitheaters.

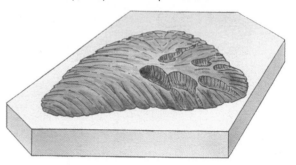

3 In the last stages, the original surface of the shield volcano is entirely obliterated, leaving a rugged mountain mass made up of sharp-crested divides and deep canyons.

This Landsat image shows the island of Kauai, the oldest of the Hawaiian shield volcanoes. You can see the radial pattern of streams and ridge crests leading away from the central summit. The intense red colors in this color infrared image are lush vegetation.

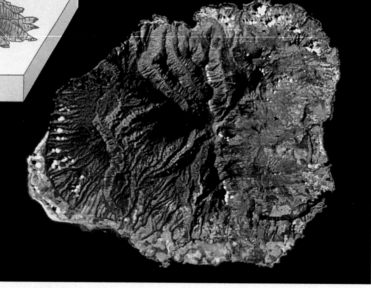

Shield volcanoes In contrast to the explosive eruptions of felsic stratovolcanoes, eruptions of shield volcanoes are usually quiet. Basaltic lava flows easily over long distances and spreads out in thin layers. Most of the lava flows from fissures (long, gaping cracks) on the flanks of the volcano. Shield volcanoes show erosion features that are quite different from those of stratovolcanoes (**Figure 9.31**). Mauna Loa in Hawaii is a distinctive shield volcano (**Figure 9.32**).

Other volcanic landforms Sometimes extruded lava creates a broad plateau rather than a cone. If a hotspot lies beneath a continental lithospheric plate, it can generate enormous volumes of basaltic lava that emerge from numerous vents and fissures and accumulate layer upon layer. These basalt layers, called **flood basalts**, can become thousands of meters thick and cover thousands of square kilometers (**Figure 9.33**).

Shield volcanoes • Figure 9.32

The shield volcanoes of the Hawaiian Islands have gently rising, smooth slopes that flatten near the top, producing a broad-topped volcano. Their domes rise to about 4000 m (about 13,000 ft) above sea level, but if you include the basal portion below sea level, they are more than twice that high. In width they range from 16 to 80 km (10 to 50 mi) at sea level and up to 160 km (about 100 mi) at the submerged base. On the distant skyline is the summit of Mauna Loa volcano. In the foreground is the summit cone of Mauna Kea, a smaller volcano on the north flank of Mauna Loa.

Cinder cones are small volcanoes that form when frothy magma is ejected under high pressure from a narrow vent, producing tephra. The rain of tephra accumulates around the vent to form a roughly circular hill with a central crater. Cinder cones rarely grow more than a few hundred meters high. An exceptionally fine example of a cinder cone is Wizard Island, Oregon, which was built on the floor of Crater Lake long after the caldera was formed (see Figure 9.30).

CONCEPT CHECK STOP

1. **Why** are so many volcanoes found along the Ring of Fire?

2. **How** do the properties of felsic versus mafic lavas determine the shape and explosiveness of a volcano?

3. **How** are the distinct landforms associated with stratovolcanoes and shield volcanoes produced?

Flood basalts • Figure 9.33

An important American example of flood basalts is found in the Columbia Plateau region of southeastern Washington, northeastern Oregon, and westernmost Idaho. Here, basalts of Cenozoic age cover an area of about 130,000 km² (about 50,000 mi²)—nearly the same area as the state of New York. Each set of cliffs in the photo is a major lava flow. Vertical cracks form in the lava as it cools, creating tall columns.

Summary

1 Plate Tectonics 274

- The motion of lithospheric plates is described through **plate tectonics**. Where plates move apart, there is a **spreading boundary**. At **converging boundaries**, plates collide. At **transform boundaries**, plates move past one another.

- At spreading boundaries, **lava** emerges from the **rifting** crust to form a **midoceanic ridge**.

- When oceanic lithosphere and continental lithosphere collide, the denser oceanic lithosphere plunges beneath the continental lithospheric plate at an **oceanic trench**, in a process called **subduction**.

- Compression occurs at lithospheric plate collisions. At first, the compression produces folding—**anticlines** (upfolds) and **synclines** (downfolds)—as is characteristic of the Appalachian region, shown here. If compression continues, **folds** may be overturned, creating **overthrust faults**. The mountain-building process that occurs where plates collide is called **orogeny**.

Types of folding • Figure 9.8

2 Tectonic Activity and Earthquakes 282

- In response to the built up strain as plates move, brittle rocks in the Earth's crust break into **faults**. The four main types of faults are **normal**, **strike-slip**, **reverse**, and overthrust faults.

- Fault movement sometimes produces an exposed cliff face called a **fault scarp**. Parallel normal faults can drop a block of crust down, into a **graben**, or push a block up, into a **horst**.

- **Earthquakes** occur when rock layers, bent by tectonic activity, suddenly fracture and move, initiating a series of underground waves that can cause shaking at the surface. Earthquake **magnitude** is a measure of how much the ground shakes. The location of the earthquake, its **epicenter**, can be determined by recording the arrival time of P waves and then S waves with a seismometer.

- Most severe earthquakes occur near plate collision boundaries. Transform boundaries, such as the one along the San Andreas fault, shown here, can also produce major earthquakes. Undersea subduction zone earthquakes can produce dangerous tsunamis.

San Andreas fault in Southern California • Figure 9.19

Explosive stratovolcano eruptions • Figure 9.29

3 Volcanic Activity and Landforms 293

- **Volcanoes** are landforms that mark the eruption of lava at the Earth's surface. Volcanic activity primarily occurs near plate boundaries. Chains of island volcanoes can also arise where plates move over a **hotspot**.

- Volcanic eruptions produce hazards such as volcanic gases, **pyroclastic material**, lava flows, and lahars. Eruptions can also trigger earthquakes and tsunamis.

- Volcanic eruptions and landforms are related to magma type. Thick, gassy felsic lava produces steep-sided **stratovolcanoes**, which erupt explosively, as shown in the photo. In contrast, basaltic lava holds less gas and flows easily, forming broadly rounded **shield volcanoes**, which erupt quietly.

- Volcanic activity produces **cinder cones**, small volcanoes formed from piles of accumulated ash, and **flood basalts**, thick lava accumulations over large areas. **Calderas** are basins that form after the collapse of a volcano.

Key Terms

- anticline 279
- caldera 300
- cinder cone 303
- convergent boundary 274
- earthquake 285
- epicenter 286
- fault 282
- fault-line scarp 285
- fault scarp 283
- flood basalt 302
- focus 286
- fold 279
- graben 284
- horst 284
- hotspot 295
- lava 293
- magma 293
- magnitude 287
- midoceanic ridge 275
- normal fault 283
- oceanic trench 279
- orogeny 279
- overthrust fault 279
- plate tectonics 274
- pyroclastic material 296
- reverse fault 283
- rifting 275
- shield volcano 297
- spreading boundary 274
- stratovolcano 296
- strike-slip fault 283
- subduction 279
- syncline 279
- transform boundary 274
- volcano 293

Critical and Creative Thinking Questions

1. Sketch a cross section showing a collision between an oceanic plate and a continental lithospheric plate. Label the following features: oceanic crust, continental crust, mantle, oceanic trench, and rising magma. Indicate where subduction is occurring.

2. Sketch the collision between continents and describe the formation of a continental suture. Provide a present-day example where a continental suture is being formed and give an example of an ancient continental suture.

3. Sketch the classic diagrams for a stratovolcano and shield volcano. Describe the impacts to nature and humans that can result from volcanic activity and provide recent examples.

4. Look at the photo. Describe the tectonic processes that shaped the rock layers exposed in this road cut located in western Maryland, in the Appalachian Mountains.

What is happening in this picture?

This spectacular photo, taken by astronauts on the Gemini XI mission, shows the southern tip of the Red Sea and the southern tip of the Arabian Peninsula.

Arabian Peninsula

Red Sea

Think Critically

1. How do you explain the fact that these two land masses almost seem like jigsaw puzzle pieces that could fit back together?
2. What plate boundary would you expect to find here?
3. What would you expect to happen in this region in the future?

Self-Test

(Check your answers in the Appendix.)

1. The diagram shows two basic forms of tectonic activity. Identify and briefly describe both processes. Which would you expect to see between separating oceanic plates? Which occurs between convergent plate boundaries?

   _____

   _____

2. The process in which one plate is carried beneath another is called _____.

   a. advection   c. convection

   b. liposuction   d. subduction

3. _____ plate boundaries with _____ in progress are zones of intense tectonic and volcanic activity.

   a. Transform, subduction

   b. Subduction, orogeny

   c. Convergent, subduction

   d. Spreading, axial motion

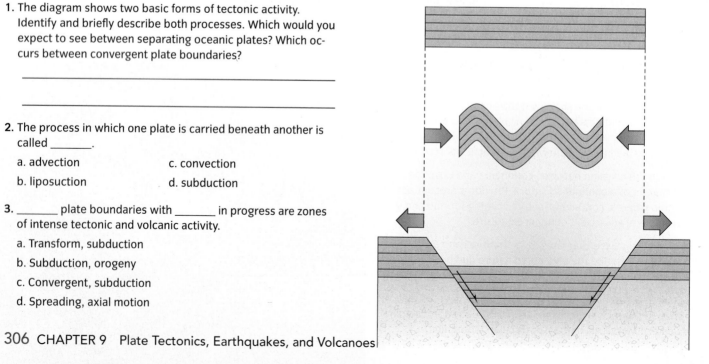

4. There are basically two different forms of tectonic activity: _____.

   a. compressional and extensional

   b. stressful and decompressional

   c. decompression and extensional

   d. compressional and stressful

5. Compression leads to the folding of the crust, which results in the formation of _____.

   a. anticlines and synclines

   b. synclines and troughs

   c. upfolds and anticlines

   d. transform and subduction zones

6. Orogeny may be caused by _____.

   a. strike-slip faults

   b. differential erosion

   c. subsidence

   d. subduction

7. Midoceanic ridges are _____.

   a. only a theory as they are underwater and can't be seen

   b. located where subduction creates rising ridges

   c. underwater horst and graben landforms

   d. visibly creating spreading centers in Iceland

8. Identify the type of fault shown in this diagram.

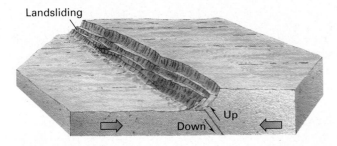

9. The diagram shows two initial landforms that can be created by normal faults. Identify each type.

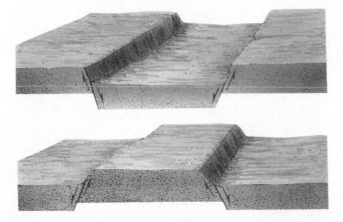

10. _____ result from vertical rise in one block relative to another, creating spectacular mountain cliffs.

    a. Rift valleys

    b. Horst and graben

    c. Fault scarps

    d. Reverse faults

11. The emanation point of the seismic waves released during an earthquake is called the _____.

    a. epicenter

    b. fault surface

    c. seismic front

    d. focus

12. The _____ of the magma present within a volcano primarily determines whether the volcano will erupt explosively or quietly.

    a. temperature

    b. viscosity

    c. gas pressure

    d. chemistry

13. Occasionally, stratovolcanoes erupt so violently that the entire central portion of the volcano collapses into its empty magma chamber to form a large depression called a _____.

    a. caldera

    b. dike

    c. cinder cone

    d. batholith

14. Rapid mixing of volcanic ash and the water produced by the flash melting of ice and snow that has accumulated at the tops of erupting stratovolcanoes produces a deadly mud avalanche known as a _____.

    a. debris flow

    b. rock fall

    c. lahar

    d. mudslide

15. Landforms created by tectonic activity include _____.

    a. basin and range

    b. fault scarps

    c. mountain range

    d. all of the above

THE PLANNER ✓

Review your Chapter Planner on the chapter opener and check off your completed work.

10 Weathering and Mass Wasting

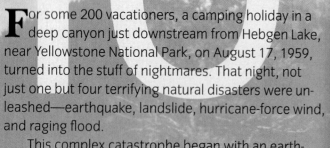

For some 200 vacationers, a camping holiday in a deep canyon just downstream from Hebgen Lake, near Yellowstone National Park, on August 17, 1959, turned into the stuff of nightmares. That night, not just one but four terrifying natural disasters were unleashed—earthquake, landslide, hurricane-force wind, and raging flood.

This complex catastrophe began with an earthquake, measuring 7.1 on the Richter scale, which triggered a landslide of 28 million cubic meters (37 million cubic yards) of rock over 600 meters (about 2000 feet) long and 300 m (about 1000 ft) thick. The mass of rock descended at 160 kilometers (100 miles) per hour. Vacationers reported that the mountain seemed to move across the canyon, with trees flying from its surface like toothpicks. This moving mountain pushed a vicious blast of wind that tumbled travel trailers. As the landslide mass hit the Madison River, it generated a wall of water that surged over the campsite. At least 26 people died beneath the Madison Slide.

Mass movements accounted for 12,733 deaths between 1991 and 2005, according to the United Nations. Those triggered by floods, earthquakes, and volcanoes raise the toll much higher. These mass movements begin with weathering processes that break rocks apart and upset slope stability.

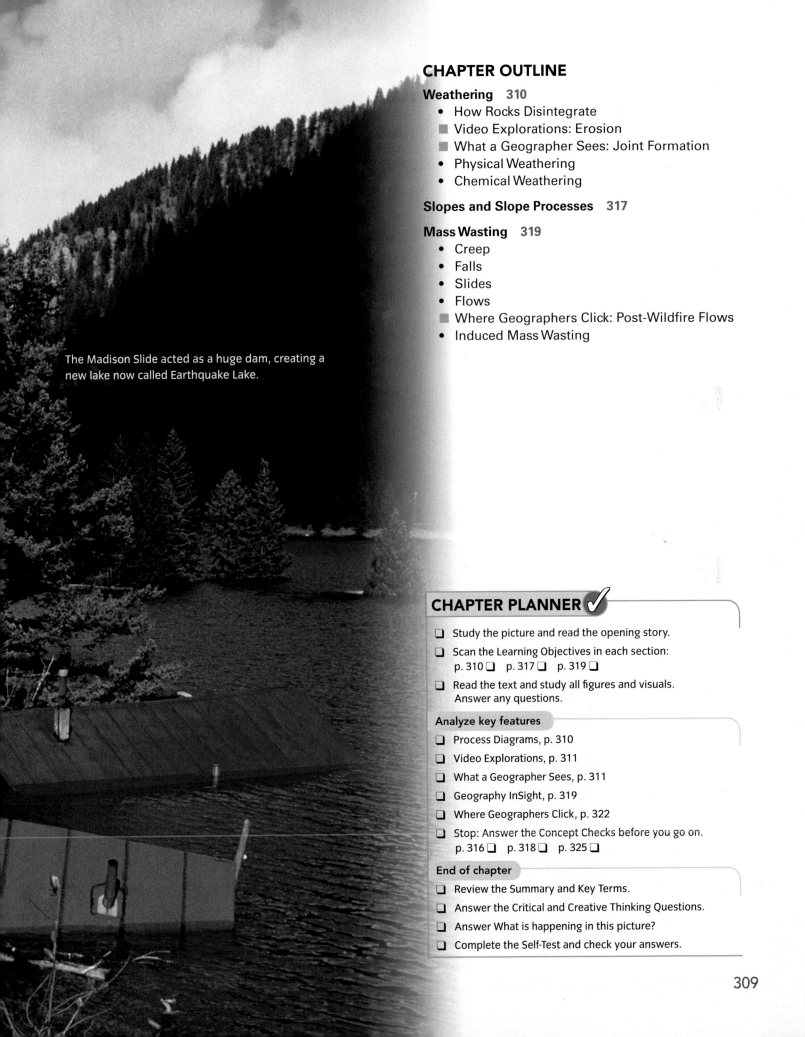

The Madison Slide acted as a huge dam, creating a
new lake now called Earthquake Lake.

CHAPTER OUTLINE

CHAPTER PLANNER ✓

- ❏ Study the picture and read the opening story.
- ❏ Scan the Learning Objectives in each section:
  p. 310 ❏   p. 317 ❏   p. 319 ❏
- ❏ Read the text and study all figures and visuals.
  Answer any questions.

Analyze key features

- ❏ Process Diagrams, p. 310
- ❏ Video Explorations, p. 311
- ❏ What a Geographer Sees, p. 311
- ❏ Geography InSight, p. 319
- ❏ Where Geographers Click, p. 322
- ❏ Stop: Answer the Concept Checks before you go on.
  p. 316 ❏   p. 318 ❏   p. 325 ❏

End of chapter

- ❏ Review the Summary and Key Terms.
- ❏ Answer the Critical and Creative Thinking Questions.
- ❏ Answer What is happening in this picture?
- ❏ Complete the Self-Test and check your answers.

Weathering

LEARNING OBJECTIVES

1. **Explain** how rocks break down to form soil.
2. **Explain** how physical weathering creates new landforms.
3. **Describe** the processes of chemical weathering.

As a part of the lithosphere's cycle, new crust is formed and gradually breaks down through the long-term process of denudation. Denudation involves first **weathering**, by exposure of rocks to surface elements, then **mass wasting**, as gravity pulls broken rock downhill, and finally **erosion** by wind and water, which transport and redeposit weathered materials. We will address weathering and mass wasting in this chapter and return to erosion and deposition by wind and water in Chapters 12 and 13.

> **weathering** All the processes that physically disintegrate or chemically decompose a rock at or near the Earth's surface.

> **mass wasting** The abrupt or incremental downhill movement of soil, regolith, and bedrock under the influence of gravity.

How Rocks Disintegrate

Weathering describes the combined action of all processes that cause rock to disintegrate physically and decompose chemically because of exposure near the Earth's surface. Weathering produces **regolith**—a surface layer of weathered rock particles that lies above solid, unaltered rock, called **bedrock**. Through further weathering and in combination with biological processes, the upper portion of regolith forms soil (**Figure 10.1**). Soil formation processes will be covered in Chapter 15.

> **regolith** The layer of rock and mineral particles that lies above bedrock.

> **bedrock** The solid rock layer under soil and regolith, which is relatively unchanged by weathering.

PROCESS DIAGRAM

From rock to soil • Figure 10.1

These exposed layers of rock in South Africa show how soil has formed through disintegration of a layer of sedimentary rock.

3 In the top layer, rock fragments and particles undergo a complex set of processes that continue the weathering process and form soil.

2 Nearer to the surface, the regolith has been exposed to weathering agents longer and is more weathered. Wetting, drying, freezing, thawing, and plant root growth have disturbed the regolith and begun to move and mix fine and coarse particles.

1 Air and water invade fractures in the rock, carrying gases, acid solutions, and microorganisms from the soil above. Weathering and the formation of regolith begin.

Video Explorations

Erosion

This video shows how the processes of erosion from water and wind have shaped the landscape. Differential rates of resistance in rock types leads to the variation seen in highly eroded rock formations.

The vulnerability of rock surfaces to weathering depends on how much of the rock is exposed, existing climate conditions, and the ability of water, air, and microscopic organisms to penetrate the exposed surface (see *Video Explorations*). The main passageways through which these weathering agents enter a rock are fractures called **joints** and small spaces between grains called **pores** (see *What a Geographer Sees*).

WHAT A GEOGRAPHER SEES

Joint Formation

A visitor to the Joshua Tree National Monument in California (Figure a) might think this rock formation was a work of modern sculpture. A geographer would see that the dramatic shape of the formation is a result of long-term weathering of the rock materials along joint planes (Figure b). The rock originally formed underground, where it was subject to great pressure from the overlying and surrounding rocks. Squeezing and twisting by tectonic forces caused the rock to fracture and form joints. Joints typically occur in parallel and intersecting planes, creating surfaces along which water, microorganisms, and tree roots can penetrate.

As the overlying rock was eroded, the pressure decreased, causing the rock to expand and crack along joint planes. Later, weathering rounded off and widened the joints even further, into the dramatic diagonal lines seen in the photo. Weathering has widened some of the joints to the point of detaching stones from the main rock body.

a.

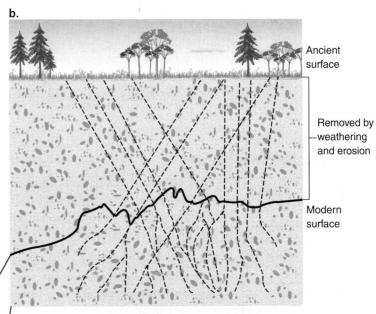

b.

Ancient surface

Removed by weathering and erosion

Modern surface

Think Critically

1. Why are some granite bodies extensively jointed, while others are essentially joint free?
2. How might the presence of joints affect the rate of weathering?

Weathering agents break up rocks in different ways, depending on how they penetrate the rocks.

a. Joint-block disintegration

When weathering agents penetrate along joints in the rock, joint-block disintegration breaks away large boulders.

b. Granular disintegration

When weathering agents penetrate microscopic openings between rock crystals or grains, granular disintegration breaks rocks into smaller pieces.

Ask Yourself

When rocks with right angles appear in a landscape, this is a sign of probable _____.

a. granular disintegration
b. joint-block disintegration
c. lack of regolith
d. slow weathering processes

Weathering agents can also actively penetrate rocks through microscopic openings that are not visible to the eye. Igneous and metamorphic rocks contain tiny spaces between crystals, whereas sedimentary rocks have pore space between grains. The types of openings a rock has, along with the chemical composition of the minerals exposed by the openings, determines how a rock will break down as a result of weathering (**Figure 10.2**).

Physical Weathering

physical weathering The breakup of massive rock (bedrock) by physical forces at or near the Earth's surface.

Physical weathering, or mechanical weathering, fractures rock into smaller pieces, without chemical alteration of the minerals.

Frost action In cold climates, an important physical weathering process is **frost cracking**, which occurs when water penetrates joints or pores in the bedrock and then weakens or breaks the rock through freezing and expansion (**Figure 10.3**). Once thought to be a widespread process, frost cracking only occurs when there is slow but steady cooling in the range –4°C (25°F) to –15°C (5°F). Once rocks are cracked,

however, the growth of ground ice in soils and regolith can move, churn, and sort rock fragments, forming fields of shattered rock rubble above the tree line in high-altitude and high-latitude environments.

Frost wedging • Figure 10.3

Unlike most other liquids, water expands when it freezes. Under proper circumstances, the expansion of freezing water can fragment rocks in a frost action process called *frost cracking*.

Niche formation in sandstone cliffs • Figure 10.4

In dry climates, salt-crystal growth and weathering lead to the formation of distinctive hollowed-out niches near the bases of sandstone cliffs.

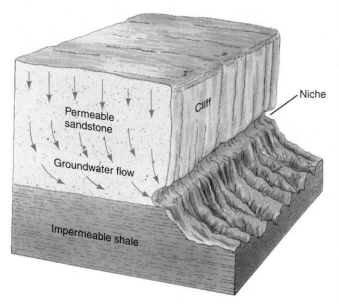

a. Ground water flows through the porous sandstone at the top of the cliff until it reaches a layer of impermeable shale at its base and is forced to seep out and evaporate. Here salt-crystal growth separates the grains of sandstone, breaking them loose and creating a niche.

b. Many of the deep niches or cave-like recesses formed in this way in the southwestern United States were occupied by Native Americans. Their cliff dwellings gave them protection from the elements and safety from armed attack. This is the White House Ruin, a large niche in sandstone in the lower wall of Canyon de Chelly, Arizona.

Salt-crystal growth A weathering process similar to frost cracking occurs in dry climates, where **salt-crystal growth** in crevices and pores creates pressure. Salt-crystal growth in rock pores can disintegrate rock, carving out many small and large niches, shallow caves, rock arches, and pits, as seen in sandstones of arid regions. During long drought periods, water is drawn up through fine openings and passages in the rock. The water then evaporates when reaching the rock surface, leaving behind tiny salt crystals. Over time, the force of these growing crystals breaks the rock apart, grain by grain.

Rock at the bases of cliffs where groundwater seepage occurs is especially susceptible to salt-crystal growth (**Figure 10.4**). Salt crystallization also damages masonry buildings, concrete sidewalks, and streets. Salt-crystal growth occurs naturally in arid and semiarid regions. In humid climates, salt-crystal growth is not a factor in weathering because rainfall dissolves salts and carries them downward to ground water.

Exfoliation • Figure 10.5

Exfoliation occurs when underlying rock is released from the pressure of overlying rock. The rock expands, creating joint planes that follow the rock surface. In this photo of North Dome, Yosemite National Park, California, the curving joint pattern above the treetops is created by exfoliation. Weathering has loosened the rock along the joints, and several shells of rock have fallen away.

Exfoliation Rocks are also weathered through **exfoliation**, in which layers of rock are peeled away. Rock that forms deep beneath the Earth's surface is compressed by the rock above. As the upper rock is slowly worn away, the pressure is reduced, and the rock below expands slightly. This expansion makes the rock crack in layers that are more or less parallel to the surface, creating a type of jointing called sheeting structure. In massive rocks such as granite, thick curved layers or shells of rock peel free from the parent mass below, much like the layers of an onion. When sheeting structure forms over the top of a single large knob or hill of massive rock, an exfoliation dome is produced (**Figure 10.5**).

Thermal action Most rock-forming minerals expand when heated and contract when cooled. This process is called **thermal action**. Because different minerals expand at different rates, internal stresses can crack rocks between mineral crystals. Intense heating of rock surfaces by the Sun during the day, followed by nightly cooling, enhances thermal action. Heat flows slowly inside rocks, increasing the build-up of stress.

Biological action Plant roots can produce pressure strong enough to break up rock as active root growth wedges joint blocks apart. You have probably observed concrete sidewalk blocks that have been fractured and uplifted by the growth of tree roots. This same process occurs when roots grow between rock layers or joint blocks. Even plant seedlings on cliff surfaces can affect the physical weathering of rocks.

Chemical Weathering

Chemical weathering occurs when rocks are broken down by chemical transformation. Chemical reactions can turn rock minerals into new minerals that are softer and bulkier and therefore easier to erode. Naturally occurring acids, produced by microorganisms digesting organic matter in soils, can dissolve some types of minerals, washing them away in runoff. Physical weathering and mechanical weathering often work in conjunction,

> **chemical weathering** The decomposition or decay of minerals in rocks by chemical reactions with water, oxygen, and acids.

when physical weathering exposes more rock surface area to the effects of chemical weathering.

Chemical reactions proceed more rapidly at warmer temperatures, so chemical weathering is most effective in warm, moist climates of the equatorial, tropical, and subtropical zones. Chemical weathering in these climates, working over thousands of years, has decayed igneous and metamorphic rocks down to depths as great as 100 m (about 330 ft). Geologists conducting exploratory digs have discovered the decayed rock material to be soft, clay rich, and easily eroded.

Chemical weathering includes three important processes: hydrolysis, oxidation, and carbonation.

Hydrolysis

In **hydrolysis**, minerals react chemically with water. For example, the mineral feldspar, a component of granite, reacts to form a soft clay mineral and silica residue that are readily washed or blown away. Through this process, granite can weather to produce interesting boulder and pinnacle forms (**Figure 10.6**). Thermal action or other weathering processes can then further degrade the rock.

Oxidation

In **oxidation**, oxygen and water react with metallic elements in minerals, making the minerals unstable and causing the rock to degrade in strength and eventually crumble into smaller particles. Eventually, the elements form oxides—pure compounds of elements with oxygen. Iron-bearing rocks, for example, yield rusty-red iron oxide (Fe_2O_3), which is formed by this process. Because chemical oxidation produces distinctive colors in rock, geologists can use the color of ancient rock layers to identify periods when less oxygen existed in the atmosphere.

Disintegration of granite through hydrolysis • Figure 10.6

Although there is not much rainfall in dry climates, water still penetrates the granite along planes between crystals of quartz and feldspar. Hydrolysis attacks the sides and edges of the feldspar crystals, breaking feldspar grains away from the main mass of rock and creating the rounded forms shown here in Owens Valley, California.

Chemical weathering of tombstones • Figure 10.7

These two tombstones from a burial ground in Massachusetts have weathered differently because of the susceptibility of different rock materials to carbonation.

a. This marker is carved in marble. It has been strongly weathered, weakening the lettering.

b. This marker is made of slate and is much more resistant to weathering. The engraved letters are clear and sharp, even though the tombstone is almost a century older.

Put It Together

Review Figure 10.6 and answer this question.
What would be the result if granite were used for a tombstone?
a. Granite is very resistant to chemical weathering, so the tombstone would last a long time.
b. Granite is susceptible to hydrolysis, so the tombstone would last a long time.
c. Granite is susceptible to hydrolysis, so the tombstone would weather more quickly.
d. Granite is susceptible to carbonation, so the tombstone would weather more quickly.

Carbonation Carbonic acid—a weak acid that is formed when carbon dioxide dissolves in water—also weathers rocks through a process known as **carbonation**. Carbonic acid is often found in rainwater, soil water, and stream water, and it slowly dissolves some types of minerals. Carbonate sedimentary rocks, such as limestone and marble, are particularly susceptible to carbonation (**Figure 10.7**). Carbonic acid in ground water dissolves limestone, creating underground caverns and distinctive landscapes that form when these caverns collapse.

CONCEPT CHECK	

1. **How** do joints contribute to the disintegration of rocks?

2. **What** is the role of climate in physical weathering?

3. **How** do physical and chemical weathering work in conjunction with one another?

Slopes and Slope Processes

LEARNING OBJECTIVES

1. **Describe** the forces at work on a slope.
2. **Explain** how the angle of repose determines the steepness of a slope.

Once rock fragments have been loosened from parent rock through physical or chemical weathering, they are subjected to gravity, running water, waves, wind, and the flow of glacial ice. In this chapter, we concentrate on how Earth materials are moved by gravity. We will return to the other effects in the following chapters.

The way that Earth materials move downhill is governed by the mechanics of slopes—patches of the land surface that are inclined from the horizontal. Slopes guide the flow of surface water downhill and fit together to form stream channels. Nearly all natural surfaces slope to some degree. Slopes have a characteristic form and distribution of materials when in equilibrium (**Figure 10.8**).

Gravity universally and continuously pulls down on all materials across the Earth's surface. Slopes are constantly reduced by weathering and the transport of weathered debris by gravity. Weathered rock moves down the slope into a collection of degraded debris at the base called **colluvium**. If water transports and deposits the debris farther away from the slope base, it becomes alluvium.

Counterbalancing the downward force of gravity are the resisting force of the cohesiveness of the rock material and the internal friction holding it in place. Slopes maintain a condition of dynamic equilibrium, which means that material on a slope stays in place until one counterbalancing factor or another causes the slope system to readjust into a new state of equilibrium. Where and when materials on a slope respond to the constant forces of gravity by moving downhill is determined by the angle of the slope, the slope material, and the combination of weathering processes we have just discussed.

If a slope is too steep, the downward force of gravity will overcome the resisting force of friction, and the slope materials will move until the slope is less steep and equilibrium is restored. The **angle of repose** is the maximum

Typical distribution of materials on a slope • Figure 10.8

On a typical slope, soil and regolith blanket the bedrock, except in a few places where the bedrock is particularly hard and projects in the form of outcrops. Regolith that breaks off from the rock beneath moves gradually down the slope toward the stream and accumulates at the foot of the slope as colluvium.

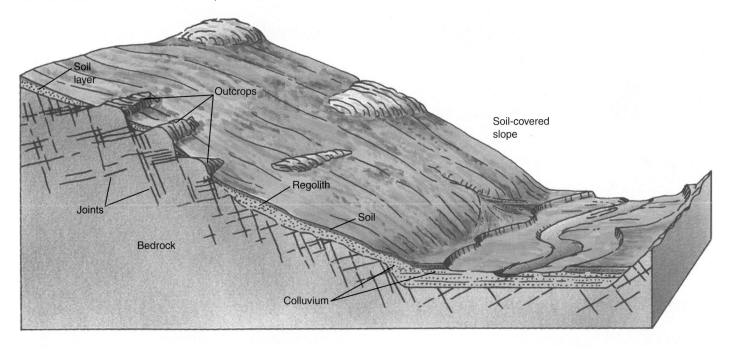

Angle of repose • Figure 10.9

The angle of repose is the maximum angle at which debris, sand, or rock of a given size can rest before the force of gravity causes it to slide downslope. Coarser materials have a steeper angle of repose.

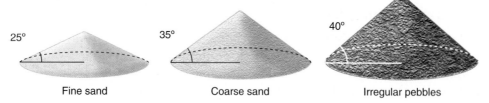

25°   35°   40°

Fine sand   Coarse sand   Irregular pebbles

slope angle that allows rock, sand, clay, and debris to stay in place without movement. The angle of repose depends on the slope material (**Figure 10.9**) as well as its moisture content.

Water, vibration, and additional falling debris also shift the dynamic equilibrium, overcoming friction and allowing the slope materials to overcome the frictional resistance and move downslope. This is why landslides are often triggered by weather events such as heavy rain or by natural disasters such as earthquakes.

The stability of slopes is important for transportation routes. Transportation engineers must account for differences in earth materials, weights and pressures of loads, and vibrational forces in designing and maintaining railway and railroad cuts, fills, and grades. Geographic information systems can identify areas with high risk of slope movement, using vegetation cover, topography, drainage patterns, and rock type, and other factors. At a global scale, high-resolution satellite data can identify locations that are at risk for mass movements of slopes (**Figure 10.10**).

SCIENCE TOOLS

Mapping slope stability • Figure 10.10

NASA scientists using data from the Tropical Rain Measuring Mission satellite have mapped out areas of the globe where slope stability can be undermined through rainwater saturation. Mountainous areas are delineated as having a high probability of landslides due to the preponderance of steep slopes.

Global landslide hotspots

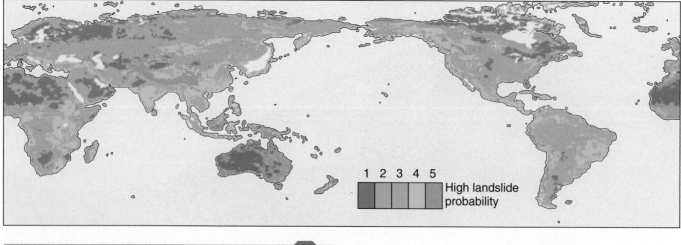

1 2 3 4 5
High landslide probability

CONCEPT CHECK   STOP

1. **What** determines the resisting force of a slope?

2. **How** is the angle of repose related to the coarseness of the sediments that make up a slope?

Mass Wasting

LEARNING OBJECTIVES

1. **Describe** soil creep, rockfalls, landslides, and flows.

2. **Explain** induced mass wasting.

When Earth materials move down a slope in response to gravity, we call it mass wasting. Factors such as the steepness of the slope, the type of material, and its water content determine whether the material creeps, falls, slides, or flows downhill (**Figure 10.11**).

Creep

The slowest form of mass wasting is **soil creep**. On almost every soil-covered slope, soil and regolith are slowly and imperceptibly moving downhill. This common movement

> **soil creep** The extremely slow downhill movement of soil and regolith.

Geography InSight   Types of mass wasting •   ✓ THE PLANNER

Figure 10.11

Different types of mass wasting are extremely varied and grade into one another. We've selected only a few of the most important forms of mass wasting here, categorized according to the water content of the material and the speed of motion that occurs.

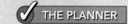

Dry ⟶ Wet

Slow

Fast

a. Creep is the slowest form of mass wasting, as Earth materials move at an imperceptible rate downslope. Here creep has carried surface blocks of nearly vertical rock beds leftward, making the beds appear to bend.

d. Flows are the wettest forms of mass wasting and range from slow-moving earthflows to dangerously rapid mudflows.

When soils were saturated by rain after Hurricane Mitch in 1998, an entire neighborhood in Tegucigalpa, Honduras, flowed downhill, blocking the Choluteca River.

b. Falls of rock and debris occur often along steep or sheer cliffs or road cuts, where rock falls can be triggered by earthquakes and road vibration.

c. Landslides can be triggered by earthquakes or heavy rains. In 2001, a 7.6 earthquake triggered this landslide in Santa Tecla, El Salvador.

The slow downhill creep of soil and regolith shows up in many ways on a hillside.

a. Where soil creep is active, fence posts, monuments, and utility poles lean downhill. Retaining walls are broken or pushed over.

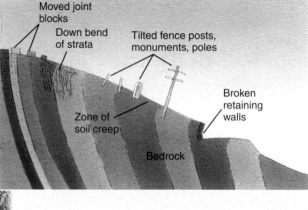

b. This photo of a roadcut near Downieville, California, shows how soil creep moves joint blocks. Although the strata appear to bend to the left, the motion is actually that of detached bedrock fragments, moving downhill along with the soil.

appears in older neighborhoods and farms, where one can see fence posts and retaining walls leaning in a downhill direction. The process is triggered when soil and regolith are disturbed by alternate drying and wetting, growth of ice needles and lenses, heating and cooling, trampling and burrowing by animals, and shaking by earthquakes. Gravity pulls on every such rearrangement, and the particles very gradually work their way downslope. In some layered rocks, such as shales or slates, edges of the strata can appear to bend in the downhill direction (**Figure 10.12**).

Solifluction is a type of mass wasting that occurs when soils become thoroughly saturated with water and slowly flow downslope by gravity. It occurs where the ground underlying the soil is impermeable—a clay layer, for example—and soil water cannot drain downward. Movement is slow, perhaps a few millimeters or centimeters per day or per year. Tropical rainforest soils on slopes and tundra soils underlain by frozen ground often show this form of mass movement.

Falls

The most visible form of mass wasting we are likely to encounter is a **rockfall**, in which rocks fall down

Talus slopes • Figure 10.13

In high mountains, rock fragments falling from a rock face or cliff can form an accumulation called talus. As shown here at Bow Lake, in Alberta, Canada, the fragments form steep slopes or cone-shaped piles at the base of steep gullies. Fresh talus slopes are unstable. Walking across the slope or dropping a large rock fragment from the cliff above will disturb the talus and cause the surface fragments to slide or roll down.

Put It Together

Review Figure 10.9 and complete this statement.
If the size of the rocks that make up a talus slope increases, the angle of the slope with respect to the ground will _____.

a. increase
b. decrease
c. stay the same
d. impossible to determine

steep slopes or cliff sides, often bringing soil or regolith along for the ride. The typical site of a rockfall is a cliff face where a sandstone layer is underlain by shale. Sandstone blocks, separated along joint planes, slowly creep outward, moving on the weak and impermeable shale layer underneath, until the block falls or slides down the cliff face. Places where a road has been cut into a steep mountainside are also particularly vulnerable to falls of rocks and rocky debris from the oversteepened slopes. In mountainous regions, rockfalls can be extremely dangerous to those living below steep hillsides.

Loose, fallen rocks called **talus**, or scree, collect at the bottom of a slope or cliff in a cone-shaped pile (**Figure 10.13**). Talus slopes provide picturesque boundaries along the base of many majestic mountain escarpments.

landslide A rapid sliding of large masses of bedrock on a steep mountain slope or from a high cliff.

Slides

A large mass of bedrock or regolith sliding downhill is known as a **landslide**. Landslides can be set off by earthquakes or sudden rock failures. Landslides can also result when the base of a slope is made too steep by excavation or river erosion. Large, disastrous landslides are possible wherever mountain slopes are steep. In China, Switzerland, Norway, and the Canadian Rockies, for example, villages built on the floors of steep-sided valleys have been destroyed when millions of cubic meters of rock descended without warning.

During a landslide, rubble travels down a mountainside at amazing speed. Geologists think this speed is possible because somehow the particles at the bottom of the slide translate the energy of the slide into an ultrasonic vibration mode that supports the moving mass on a fluid bed without mixing it.

Types of slides • Figure 10.14

Landslides and flows can be classified as rotational or translational.

a. Rotational slide

In a rotational slide, the mass rotates as it slides downslope, resulting in displacement of soil from the top of the rupture. Slumping describes the rippling effect toward the base.

No gap

Surface of rupture

Rotational displacement

b. Translational slide

In a translational slide, the mass slides directly down the slope, resulting in a distinct displacement, or gap, at the surface of rupture in hillside materials. Slippage is often along an impermeable layer or permafrost.

Visible gap

Directional displacement

Impermeable layer

Toe

Ask Yourself

If an underlying shale bed has been exposed, this indicates that a _____ landslide has occurred.

Flows

Mass wasting of Earth materials with high water content results in flows. Flows occur when precipitation or snowmelt is greater than absorption into the underlying sediment or rock. In deserts, for example, thunderstorms produce rain much faster than the rain can be absorbed by the soil. After a wildfire, hillsides are similarly vulnerable to flows (see *Where Geographers Click*).

Flows and landslides can take different forms, depending on the way a slide moves. In a **rotational slide**, the material rotates along an axis that is perpendicular to the direction of the slope. This occurs when there is a downward and outward movement of the soil mass on top of a concave failure in the surface below (**Figure 10.14a**). The rupture of a rotational slide is curved upward like a spoon, and there is no distinct gap or scarp at the surface rupture but rather a slumping of the soil that creates a small to large cliff. When a series of curved ridges are created in the mass movement of a rotational slide, it is called a slump.

In a **translational slide**, the mass moves out and down the slope with little or no rotation (**Figure 10.14b**). Translational slides can move considerable distances, often originating along geologic faults, joints, or discontinuities between rock and soil. In climates with permafrost, slides can occur along the top of the frozen ground layer.

Where Geographers CLICK

Post-Wildfire Flows

http://landslides.usgs.gov/research/wildfire/

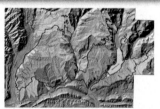

This Web site is a resource center for the U.S. Geological Survey's (USGS's) efforts to monitor and assess landslide risk around the United States. These data help forecast when storms will trigger mass movements. Areas that have been struck by wildfires are often susceptible to flows because burned hillsides, stripped of vegetation, have a reduced ability to absorb heavy rainfall. The USGS is developing tools to predict where post-wildfire flows are most likely to occur and how big these events might be. Federal, state, and local land-management agencies can use these tools to mitigate the effects of flows in these areas.

Earthflows When fine-grained or clay-rich materials are saturated with water, an **earthflow** can occur. The saturated soil and regolith turns into a thick liquid that flows downhill, forming a bowl or depression at the head. The earthflow often has an hourglass shape with a spreading toe (**Figure 10.15**). Earthflows can range in size from a few dozen square meters to massive flows in which millions of metric tons of clay-rich or weak bedrock move like a great mass of thick mud.

> **earthflow** The moderately rapid downhill flow of water-saturated soil, regolith, or weak shale.

During heavy rains, earthflows can block highways and railroad lines. Flows aren't usually a threat to life because they move quite slowly, but they can often severely damage buildings, pavement, and utility lines that have been constructed on unstable slopes.

Mudflows A **mudflow** is a rapid flow of water and soil or regolith that pours swiftly down canyons in mountainous regions. In arid regions with little or no vegetation, thunderstorm runoff picks up fine particles to form a thin mud that flows down to the canyon floors and then follows the stream courses. As it flows, it picks up additional sediment and becomes thicker and thicker until it is too thick to flow farther. Roads, bridges, and houses on the canyon floor are engulfed and destroyed. A mudflow can severely damage property and even cause death as it emerges from a canyon and spreads out.

> **mudflow** A flowing mixture of water and soil or regolith that flows rapidly downhill.

Earthflow • Figure 10.15

In an earthflow, saturated soil, regolith, or clay-rich bedrock moves downhill as a mass of thick, flowing mud. A bowl or depression occurs at the head, and a spreading toe marks the depositional area.

Source area

Main track

Depositional area

Mudflows on the slopes of erupting volcanoes are called **lahars**. Lava and hot ash rapidly melt snow atop the volcano, creating flows of rock, soil, volcanic ash, and water that accelerate down the steep slopes and flow great distances (**Figure 10.16**). Herculaneum, a city at the base of Mt. Vesuvius, was destroyed by a lahar during the CE 79 eruption. At the same time, the neighboring city of Pompeii was buried under volcanic ash.

Mudflow • Figure 10.16

More than 20,000 lives were lost when a volcanic mudflow—known as a lahar—swept through the town of Armero, Colombia. The mudflow was caused by a minor volcanic eruption of Nevado del Ruiz, which melted the snow and ice on its summit and created a cascade of mud and debris that engulfed the town. Many homes were simply swept away.

Debris flow • Figure 10.17

Heavy rains saturate the soil and regolith of steep slopes, which then flow rapidly down narrow creek valleys and canyons, carrying mud, rocks, and trees. This debris flow, also known as an alpine debris avalanche, struck the village of Schlans, Switzerland, in November 2002.

When mudflows pick up rocks and boulders, they are called **debris flows (Figure 10.17)**. These fast-moving flows can carry anything from fine particles and boulders to tree trunks and limbs down steep mountain slopes.

Induced Mass Wasting

Human activities can induce mass wasting processes by creating unstable piles of waste soil and rock and by removing the underlying support of natural masses of soil, regolith, and bedrock. Mass movements produced by human activities are called *induced mass wasting*.

For example, mass wasting is induced where roads and home sites have been bulldozed out of the deep regolith on very steep hillsides and mountainsides, steepening the slopes. The excavated regolith is piled up to form nearby embankments, leading to further steepening. Soils receive excess water from septic tanks and lawn irrigation. This combination of raising the water content and steepening makes slopes unstable and leads to mass wasting **(Figure 10.18)**.

Artificial forms of mass wasting are distinguished from natural forms because they use machinery to raise earth materials against the force of gravity. Explosives produce disruptive forces many times more powerful than the natural forces of physical weathering. Industrial societies now move great masses of regolith and bedrock from one place to another using such technology. We do this to extract mineral resources or to move earth when constructing highway grades, airfields, building foundations, dams, canals, and various other large structures. These activities destroy the preexisting ecosystems and plant and animal habitats. When the removed materials are then used to build up new land on adjacent surfaces, ecosystems and habitats are buried.

Scarification is a general term for excavations and other land disturbances produced to extract mineral resources. Strip mining is a particularly destructive scarification activity

Induced earthflow • Figure 10.18

Infiltrating water from septic tanks and irrigation water applied to lawns and gardens is thought to have weakened the clay layer enough to cause these houses in Palos Verdes Hills, California, to slide downward toward the ocean.

(Figure 10.19). Even though strip mining is under strict control in most locations, scarification is on the increase, driven by the ever-increasing human population and the rising demand for coal and other industrial minerals.

Vocal communities throughout the Appalachian region have recently raised concerns about the negative social and environmental consequences of mountaintop removal to reach coal beds. High mountain valleys near the mines are filled with rock waste and tailings, creating toxic runoff and groundwater contamination that have caused many communities to migrate away. As people move out of the area, the local economy degrades further, causing additional social stress.

Mass wasting is often triggered by abnormally heavy rains, which are predicted to become more frequent as a result of climate change caused by rising global carbon emissions. The predicted increase in frequency and severity of storms should then increase the frequency and severity of mass wasting events endured by humans and their settlements around the globe. This unexpected linkage demonstrates the complexity of global and environmental change that human society is likely to face in the future.

Strip mining • Figure 10.19

This aerial view shows the impact of strip mining on the landscape at Samples Mine in West Virginia. Reclamation is costly and difficult. Mining companies have buried approximately 2,000 miles of Appalachian streams beneath piles of toxic waste and debris, initiating decades of controversy and legal wrangling.

CONCEPT CHECK STOP

1. **What** factors determine whether mass wasting occurs by creep, fall, slide, or flow?

2. **How** can development of hillside communities induce mass wasting?

Summary

1 Weathering 310

• **Denudation** is the disintegration of continental surfaces through **weathering**, **mass wasting**, and **erosion**.

• As shown in the diagram, **bedrock** is broken into smaller pieces to form **regolith**, and ultimately soil, when weathering agents such as air, water, and microorganisms penetrate rocks through **joints** and **pores**.

Bedrock disintegration • Figure 10.2

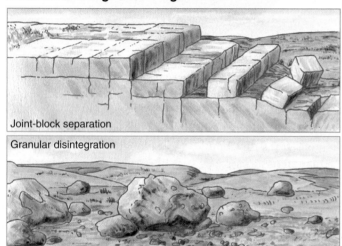

Joint-block separation

Granular disintegration

• **Physical weathering** processes include frost action, **thermal action**, and **salt-crystal growth**. **Exfoliation** occurs when layers of rock peel away as the weight of overlying rock layers is reduced.

• **Chemical weathering** results from mineral alteration. **Hydrolysis** and **oxidation** produce a regolith that is often rich in clay minerals and oxides. **Carbonation** dissolves limestone.

2 Slopes and Slope Processes 317

• On a slope at equilibrium, the downward force of gravity on slope materials is balanced by the resisting force of friction.

• The **angle of repose** is the maximum angle at which materials can remain in place on a slope. As shown in the diagram, the angle of repose is greater for coarser materials than for finer materials.

Angle of repose • Figure 10.9

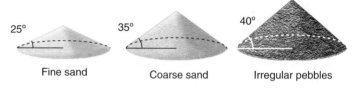

25°   35°   40°

Fine sand   Coarse sand   Irregular pebbles

• If slope angle increases or water or debris is added, slope materials will move downhill in order to reestablish equilibrium.

3 Mass Wasting 319

• **Mass wasting** is the spontaneous downhill motion of soil, regolith, or rock under gravity.

• **Soil creep** is the almost imperceptible movement of regolith down a slope, resulting in evidence such as tilted fence posts and poles, as shown in the diagram.

Soil creep • Figure 10.12

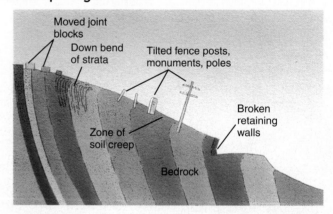

Moved joint blocks
Down bend of strata
Tilted fence posts, monuments, poles
Broken retaining walls
Zone of soil creep
Bedrock

• **Rockfalls** occur when rocks fall directly down a steep slope or cliff. Rock fragments, called **talus**, accumulate in unstable cone-shaped piles at the bases of cliffs.

- A **landslide** is a rapid sliding of large masses of bedrock, sometimes triggered by an earthquake. Flows are slower motions of wet or saturated materials. **Translational slides** move directly down a slope, whereas **rotational slides** rotate as they move.

- In an **earthflow**, water-saturated soil or regolith slowly flows downhill. A **mudflow** is much swifter than an earthflow.

- Mass wasting can be induced by heaping up soil and regolith into unstable piles and by removing layers that naturally support soil and regolith.

Key Terms

- angle of repose 318
- bedrock 310
- carbonation 316
- chemical weathering 314
- colluvium 317
- debris flow 324
- earthflow 323
- erosion 310
- exfoliation 314
- frost cracking 312
- hydrolysis 315
- joint 311
- lahar 323
- landslide 321
- mass wasting 310
- mudflow 323
- oxidation 315
- physical weathering 312
- pore 311
- regolith 310
- rockfall 320
- rotational slide 322
- salt-crystal growth 313
- scarification 320
- soil creep 319
- solifluction 320
- talus 321
- thermal action 314
- translational slide 322
- weathering 310

Critical and Creative Thinking Questions

1. Consider the region in which you live for evidence of physical and chemical weathering. Which weathering process is most dominant in your area?

2. When astronauts first landed on the Moon, they discovered that it has a deep regolith. The Moon lacks an atmosphere, a hydrosphere, and a biosphere. How might this regolith have been formed?

3. If a landscape that was previously dry experiences more precipitation due to climate change, how will this affect the types of weathering and mass wasting that occur? Give specific examples of processes that will become less prevalent and others that will become more prevalent.

4. Identify the type of mass movement associated with each of the following locations and plot each one on the diagram: Palos Verdes Hills, Madison River, and Herculaneum.

5. Imagine yourself as the newly appointed director of public safety and disaster planning for your state. One of your first jobs is to identify locations where human populations are threatened by potential disasters, including those of mass wasting. Where would you look for mass wasting hazards and why? In preparing your answer, you may want to consult maps of your state.

6. Give two reasons that support the prediction that the property damage from mass wasting events will increase in the future. For each reason, propose a strategy that could minimize this projected increase.

What is happening in this picture?

The California town of La Conchita is situated at the base of this steep cliff, at the top of which is an avocado orchard. In January 2005, 400,000 tons of earth flowed down on the town, killing 10 people.

Think Critically

1. Why was this a bad location for a town?
2. What mass wasting factors do you think might have caused the mudslide?

Self-Test

(Check your answers in the Appendix.)

1. The process in which rocks are fractured, broken, and/or transformed to softer and weaker forms is known as _____.

   a. erosion

   b. mass wasting

   c. weathering

   d. soil creep

2. Which is the correct order, from bottom to top, in the accompanying photo?

   a. soil, regolith, bedrock

   b. soil, bedrock, regolith

   c. regolith, soil, bedrock

   d. bedrock, regolith, soil

3. _____ is the spontaneous movement of soil, regolith, and rock under the influence of gravity.

   a. Erosion

   b. Mass wasting

   c. Weathering

   d. Soil creep

4. _____ is the source for sediment that may be transformed into soil.

   a. Bedrock

   b. Granite rock

   c. Alluvium

   d. Regolith

5. What do we call openings that form at the base of cliffs in arid climates, such as the one pictured in this photo?

a. rock arch   c. niche

b. pit   d. talus cone

6. Rocks can be weathered through exfoliation, when _____.

a. layers of rock are peeled away

b. temperature changes break up surface layers of rock

c. water penetrates joints or pores in the bedrock

d. salt-crystal growth in crevices and pores creates pressure

7. When carbon dioxide naturally dissolves in rain, soil, and ground waters, it produces _____, which can easily dissolve _____.

a. carbonic acid; carbonate sedimentary rocks

b. carbonic acid; igneous rocks

c. carbonic acid; clastic sedimentary rocks

d. sodium bicarbonate; carbonate sedimentary rocks

8. In _____, minerals react chemically with water.

a. carbonation   c. oxidation

b. hydrolysis   d. talus

9. Which of the following is the correct order of materials in the diagram, from left to right?

a. coarse sand, fine sand, and irregular pebbles

b. fine sand, irregular pebbles, and coarse sand

c. fine sand, coarse sand, and irregular pebbles

d. irregular pebbles, coarse sand, and fine sand

 25°    35°    40°

10. The extremely slow downhill movement of soil and regolith is called _____.

a. earthflow   c. debris flow

b. mudflow   d. soil creep

11. A(n) _____ is a mass of water-saturated soil that slowly moves downhill.

a. earthflow   c. landslide

b. mudflow   d. slump

12. A(n) _____ is the rapid sliding of large masses of bedrock that is often triggered by a(n) _____.

a. landslide; earthquake

b. landslide; thunderstorm

c. mudflow; thunderstorm

d. earthflow; earthquake

13. Mudflows that occur on the flanks of erupting volcanoes are called _____.

a. pyroclastic flows   c. lahars

b. debris floods   d. mudflows

14. This photo shows examples of _____.

a. lahars   c. mudflows

b. talus slopes   d. earthflows

15. What is the general term for excavations and other land disturbances produced to extract mineral resources?

a. talus   c. scarification

b. lahar   d. debris flood

THE PLANNER

Review your Chapter Planner on the chapter opener and check off your completed work.

Fresh Water of the Continents

Famed for its canals, Venice was built in the 11th century CE, on low-lying islands in a coastal lagoon. The city is now dropping about 1 millimeter (a quarter inch) per year due to the natural movement of the lithospheric plates underneath. But Venice's plight was greatly exacerbated after World War II, when industries began pumping ground water from below the city. That caused Venice to sink a foot (one-third of a meter) in two decades, leaving St. Mark's Square, the center of Venetian social life, just five centimeters (2 inches) above the normal high-tide level.

Although groundwater removal has now stopped in Venice, flooding is still a huge problem. December 2008 saw the worst sustained flooding in 22 years, with water standing more than 150 cm (59 in) deep before receding. Venice could be permanently flooded within a century. The city's fate may now depend on the construction of floodgates on the barrier beach that lies between Venice and the open ocean.

Venice is not the only city in peril. Groundwater withdrawal affects the Thai city of Bangkok—now the world's most rapidly sinking city—and regions in California and Texas, as well as Mexico City. Venice provides a dramatic reminder of the importance of understanding the processes that remove and replenish our water supply.

A gondola passes through high-water flooding near the Rialto Bridge on December 2, 2008, in Venice, Italy.

CHAPTER OUTLINE

CHAPTER PLANNER ✔

- ❑ Study the picture and read the opening story.
- ❑ Scan the Learning Objectives in each section:
  p. 332 ❑   p. 335 ❑   p. 342 ❑   p. 351 ❑   p. 353 ❑
- ❑ Read the text and study all figures and visuals. Answer any questions.

Analyze key features

- ❑ Process Diagram   p. 334 ❑   p. 338 ❑   p. 350 ❑
- ❑ What a Geographer Sees, p. 336
- ❑ Where Geographers Click   p. 342 ❑   p. 349 ❑
- ❑ Geography InSight, p. 354
- ❑ Video Explorations, p. 357
- ❑ Stop: Answer the Concept Checks before you go on.
  p. 334 ❑   p. 342 ❑   p. 351 ❑   p. 353 ❑   p. 357 ❑

End of chapter

- ❑ Review the Summary and Key Terms.
- ❑ Answer the Critical and Creative Thinking Questions.
- ❑ Answer What is happening in this picture?
- ❑ Complete the Self-Test and check your answers.

Fresh Water and the Hydrologic Cycle

LEARNING OBJECTIVES

1. **Describe** the global distribution of fresh water.

2. **Describe** the processes by which fresh water moves through the hydrologic cycle.

Water is essential to life. Nearly all organisms require a constant supply of water for survival. In particular, humans depend on fresh water, without salt content, for our survival. Much of this fresh water is stored in soils, regolith, and pores in bedrock, and a small amount flows in streams and rivers. In this chapter, we will focus on water on the land surface and underground.

Distribution of Fresh Water

Recall from Chapter 4 that more than 97% of the Earth's water is salt water contained in the oceans. Fresh water on the continents in surface and subsurface water makes up only about 2.5% of the hydrosphere's total water (**Figure 11.1**). Almost 69% of this fresh water is locked in ice sheets and mountain glaciers. Of the fresh water that is not frozen, 30% is stored underground

Distribution of fresh water • Figure 11.1

Fresh water makes up only 2.5% of the Earth's water, with the rest being salt water. Of this fresh water, most is contained in glaciers and stored underground as ground water. Rivers and streams, such as this Appalachian stream in Maryland, account for only a tiny fraction of global fresh water.

Fresh water

- Glaciers 68.7%
- Permafrost 0.8%
- Ground water 30.1%
- Surface and atmospheric water 0.4%

Surface and atmospheric water

- Freshwater lakes 67.4%
- Plants and animals 0.8%
- Rivers 1.6%
- Wetlands 8.5%
- Atmosphere 9.5%
- Soil moisture 12.2%

Ask Yourself

What percentage of the total water on Earth is in freshwater lakes?

a. 0.0067%
b. 0.27%
c. 0.4%
d. 2.5%

The hydrologic cycle • Figure 11.2

The hydrologic cycle traces the paths of water as it moves from oceans, through the atmosphere, to land, and returns to the ocean once more.

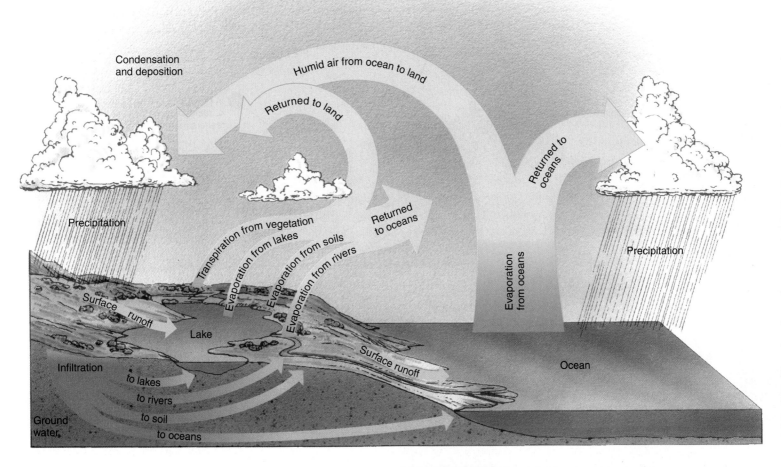

Think Critically

How will a global temperature rise affect the hydrologic cycle?

ground water The subsurface water in the saturated zone that can move as part of the hydrologic cycle.

as **ground water**. Ground water can be found at almost every location on land that receives rainfall, but it accounts for only a little more than half of 1% of the hydrosphere. This small fraction is still many times larger than the amount of fresh water in lakes, streams, and rivers, which only amount to about three-hundredths of 1% of the total water (0.03%). Fresh surface water varies widely around the globe, and many arid regions do not have permanent streams or rivers.

Movement of Fresh Water Through the Hydrologic Cycle

Recall from Chapter 4 that all of the Earth's water is a finite resource that is constantly cycling through evaporation, condensation, and precipitation as a part of the hydrologic cycle (**Figure 11.2**).

Fresh water in the hydrologic cycle • Figure 11.3

Fresh water moves from the atmosphere to land and then returns to streams and lakes.

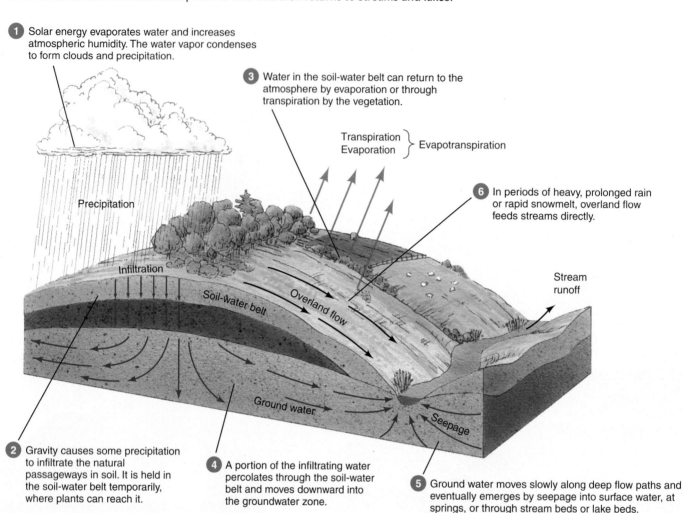

1 Solar energy evaporates water and increases atmospheric humidity. The water vapor condenses to form clouds and precipitation.

3 Water in the soil-water belt can return to the atmosphere by evaporation or through transpiration by the vegetation.

Transpiration
Evaporation } Evapotranspiration

6 In periods of heavy, prolonged rain or rapid snowmelt, overland flow feeds streams directly.

Precipitation

Infiltration

Soil-Water belt

Overland flow

Stream runoff

Ground water

Seepage

2 Gravity causes some precipitation to infiltrate the natural passageways in soil. It is held in the soil-water belt temporarily, where plants can reach it.

4 A portion of the infiltrating water percolates through the soil-water belt and moves downward into the groundwater zone.

5 Ground water moves slowly along deep flow paths and eventually emerges by seepage into surface water, at springs, or through stream beds or lake beds.

In this chapter, we look in greater detail at the portion of the hydrologic cycle that specifically involves fresh water (**Figure 11.3**). During light or moderate rain, most natural soil surfaces absorb rain water through **infiltration**. Infiltrating rain water is stored temporarily in a region called the **soil-water belt**, where it is available to plants. Some water from the soil-water belt returns to the atmosphere through **evapotranspiration**, the combined effect of evaporation and transpiration through plants. Other water moves downward into ground water, where it eventually feeds streams and marshes.

Water that drains from an area by surface, subsurface, or groundwater paths is called **runoff**. By supplying water to streams and rivers, runoff allows rivers to carve out canyons and gorges and to carry sediment to the ocean. During periods of heavy rainfall, some runoff travels over the ground surface rather than being absorbed into the ground. This surface runoff is called **overland flow**. Overland flow moves sediment from hills to valleys, and by doing so it helps shape landforms.

> **infiltration** Absorption and downward movement of precipitation into the soil and regolith.

> **runoff** The flow of water leaving an area through surface, subsurface, or groundwater flow.

CONCEPT CHECK STOP

1. **What** is the largest reservoir of fresh water on Earth?

2. **When** does overland flow occur?

Ground Water

LEARNING OBJECTIVES

1. **Describe** the different zones of underground water.

2. **Describe** the evolution of landforms that result from carbonic acid solution.

3. **Explain** the problems associated with managing ground water.

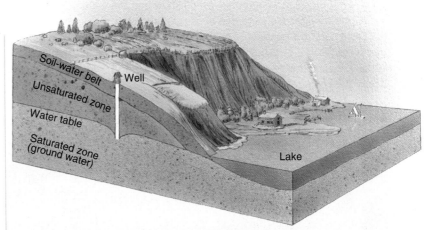

Zones of subsurface water • Figure 11.4

Water in the soil-water belt is available to plants. Water in the unsaturated zone percolates downward to the saturated zone of ground water, where all pores and spaces are filled with water.

Now let's take a closer look at how precipitation feeds ground water. Water from precipitation sinks into the soil and flows downward through the soil-water belt under the force of gravity. This process is called **percolation**. Eventually, the percolating water fully saturates the pore spaces in bedrock, regolith, or soil, at which point it is called ground water (**Figure 11.4**). The **water table** represents the upper limit of this **saturated zone**.

> **water table** The upper limit of the body of ground water, marking the boundary between the saturated and unsaturated zones.

The amount of ground water that can be held in the saturated zone depends on the porosity of the sediments that make up this layer. If the layer is porous and permeable enough to hold and conduct a usable quantity of water and the water can be easily pumped from the material, then it is called an **aquifer**. A bed of sand or sandstone is often a good aquifer because clean, well-sorted sand—such as that found in beaches, dunes, or stream deposits—can hold an amount of ground water equal to about one-third of its bulk volume. Sandy

> **aquifer** A layer of rock or sediment that contains abundant freely flowing ground water.

materials have large pore spaces that allow the water to move through the sediment and be readily pumped from wells.

By contrast, layers that are relatively impermeable to ground water are known as **aquicludes**. Clay and shale beds are examples of materials that do not conduct water in usable amounts. When an aquifer is sandwiched between two impermeable aquicludes, pressure may force the water to rise to the surface as a self-flowing **artesian well** (**Figure 11.5**). A fault can serve as a natural conduit for ground water, producing artesian springs.

Artesian well • Figure 11.5

Precipitation provides water that saturates the porous sandstone aquifer. The water cannot penetrate farther through the impervious aquicludes. Because the elevation of the well that taps the aquifer is below that of the range of hills feeding the aquifer, pressure forces water to rise in the well.

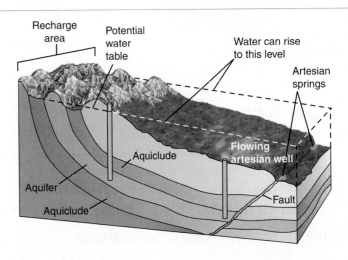

WHAT A GEOGRAPHER SEES

Hot Springs and Geysers

Visitors to Yellowstone National Park may be stunned to see hot water emerging from the Earth, where Mammoth Hot Springs features pools of steaming water (Figure a) and where Old Faithful Geyser shoots hot water dramatically up from the ground (Figure b). A geographer would know that the features in the photographs were created because hot rock material near the Earth's surface heated the nearby ground water to high temperatures (Figure c). When this heated ground water reaches the surface, it can emerge as hot springs, with temperatures not far below the boiling point of water.

When underground crevices and fissures create just the right circulation pattern, or plumbing system, super-heated water in contact with hot subterranean rocks flashes into steam that blasts the hot water up pipe-like crevices to the surface in the form of a geyser. This process repeats as water seeps in from surrounding ground water and replaces the ejected geyser water. Lacking the special underground circulation pattern, pools of hot water or hot springs form when the water reaches the surface.

a.

Hot spring

Geyser

Water heated by contact with hot rocks

Super-heated water rising to surface

Water heated under pressure

Heated water moves upward

b.

c.

Think Critically

1. Geysers eject water intermittently. What factors might influence the length of the intervals between eruptions?
2. How might an earthquake affect a region of geysers and hot springs?

The Water Table

The height and position of the water table controls the flow of ground water. In places where the water table is at or above the land surface, ground water emerges by seepage into streams, ponds, lakes, and marshes. The level of the water table thus corresponds to the water height in these surface water features. Under special geologic conditions, ground water can be dramatically ejected into the air in geysers, as illustrated in *What a Geographer Sees*.

The level of the water table changes, depending on the amount of precipitation that feeds the ground water compared to the amount of water that is withdrawn, either

Ground water and the water table • Figure 11.6

In places where the water table is at or above ground level, ground water feeds streams and rivers through seepage.

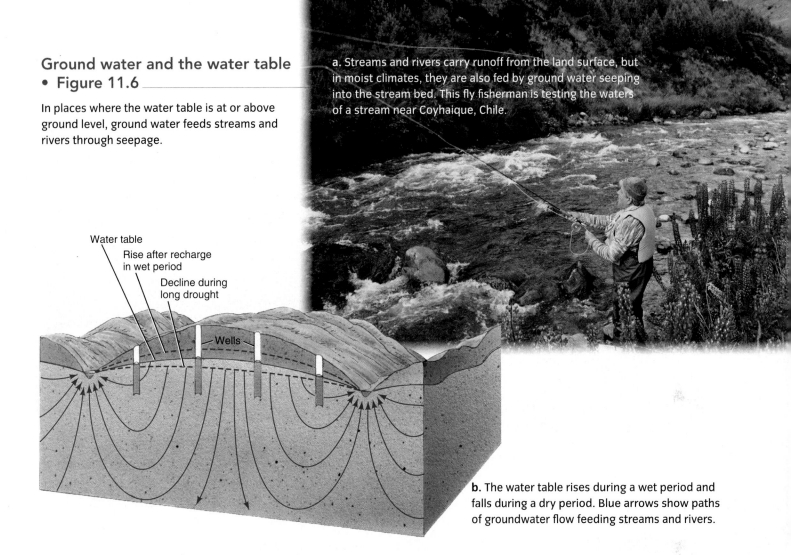

Water table
Rise after recharge in wet period
Decline during long drought
Wells

a. Streams and rivers carry runoff from the land surface, but in moist climates, they are also fed by ground water seeping into the stream bed. This fly fisherman is testing the waters of a stream near Coyhaique, Chile.

b. The water table rises during a wet period and falls during a dry period. Blue arrows show paths of groundwater flow feeding streams and rivers.

from seepage or by human extraction (**Figure 11.6**). Water percolating down through the unsaturated zone tends to raise the water table, while seepage into lakes, streams, and marshes draws off ground water and tends to lower its level. When there is a large amount of precipitation, the water table rises under hilltops or divide areas. During droughts, the water table falls. Over the long term, however, the average position of the water table tends to remain stable, and the flow of water released to streams and lakes balances the flow of water percolating down into the water table.

Limestone Solution by Ground Water

The flow of ground water can carve out immense caverns underground. Carbonic acid (H_2CO_3)—a weak acid produced from carbon dioxide dissolved in water—slowly erodes limestone in moist climates. As we saw in Chapter 10, this chemical weathering process is called *carbonation*.

Below the surface, carbonation can dissolve limestone to form deep underground caverns (**Figure 11.7**). Inside caverns, water containing dissolved minerals

drips through the cave ceiling and onto the floor. Where the water drips from the ceiling, calcium carbonate and other minerals are deposited to form **stalactites**,

Cavern development • Figure 11.7

Rainfall absorbs carbon dioxide in the atmosphere and soil and becomes carbonic acid. Carbonic acid dissolves limestone to form underground caverns.

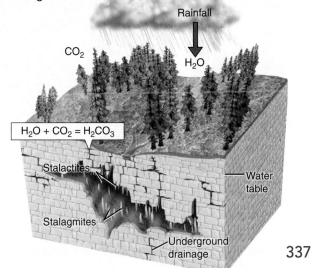

Rainfall
CO_2
H_2O
$H_2O + CO_2 = H_2CO_3$
Stalactites
Water table
Stalagmites
Underground drainage

Evolution of a karst landscape • Figure 11.8

✓ THE PLANNER

1 Over time, rainwater dissolves limestone, producing caverns and sinkholes, such as this one in Belize, known as Nohoch Ch'en.

Sinkholes

Cavern

Global Locator

North America

Belize

NATIONAL GEOGRAPHIC

2 Eventually, the caverns collapse, leaving open, flat-floored valleys. In warm, humid climates, the valley may be surrounded by limestone towers, such as these near Guilin in southern China.

Valley produced by cavern collapse

Cavern

Global Locator

ASIA

CHINA

Guilin

NATIONAL GEOGRAPHIC

shaped like icicles hanging from the ceiling. Where the water hits the cavern floor, the deposits form an upward-pointing formation called a **stalagmite**. If the process continues for a long time, stalactites and stalagmites meet to form a column. The thickness and chemical composition of these formations gives climate scientists information about historical rainfall data to re-create past climates.

Underground caverns near the surface can collapse to form surface depressions, or **sinkholes**. Some sinkholes are filled with soil washed from nearby hillsides, whereas others are steep-sided, deep holes. Over time, solution of limestone by ground water forms distinctive **karst topography**, noted for caverns, sinkholes, karst towers, and disappearing streams (**Figure 11.8**). Important regions of karst or karst-like topography are the Mammoth Cave region of Kentucky, the Yucatan Peninsula, the Dalmatian coastal area of Croatia, and parts of Cuba and Puerto Rico. The karst landscapes of southern China and west Malaysia are dominated by steep-sided, conical limestone hills and towers. In arid or semiarid climates, however, limestone is a resistant rock that forms ridges. In these regions, precipitation is insufficient to dissolve the rock by carbonation.

Groundwater Use and Management

Ground water is a substantial source of fresh water for human use. However, as you have also seen, ground water is a finite resource that can be replenished by precipitation. Withdrawal and pollution of ground water can have serious and long-lasting environmental consequences.

Groundwater withdrawal
Our human civilizations have long relied on springs and wells to tap ground water (**Figure 11.9**). Today, vast underground reservoirs of ground water provide whole regions with water for domestic, industrial, and agricultural uses. The yield of a single drilled well ranges from as low as a few hundred liters or gallons per day in a domestic well, which is enough to support an average American household, to many millions of liters or gallons per day for a large industrial or irrigation well.

Dug well • Figure 11.9

This old-fashioned dug well supplies water for household needs in Uttar Pradesh, India. The well is lined with bricks, and ground water seeps in around them to fill the well.

As the well draws water, the water table is depressed in a cone shape centered on the well. The cone of depression may extend out as far as 20 km (12 mi) or more from a well where heavy pumping is continued.

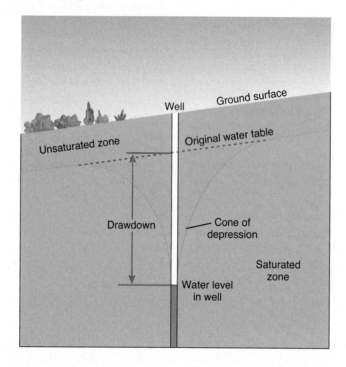

As water is pumped from a well, the level of water in the well drops. At the same time, the surrounding water table is lowered in the shape of a downward-pointing cone, which is called the **cone of depression** (Figure 11.10). The difference in height between the cone tip and the original water table is known as the **drawdown**. Where many wells are in operation, their intersecting cones will lower the water table.

Ground water is often depleted far faster than it can be replaced by water infiltrating downward to the saturated zone. Vast numbers of wells use powerful pumps to draw huge volumes of ground water to the surface—greatly altering nature's balance of groundwater input and output. As a result, we are exhausting a natural resource that is not renewable except over very long periods of time.

In many arid and semiarid regions, the ground water pumped from wells today was accumulated thousands of years ago and is not replenished by present rainfall. Large areas of the southwestern United States supported large lakes when the climate was cooler and wetter between 28,000 and 7000 years ago. When the climate changed to our present conditions, the lakes dried up, leaving behind vast groundwater reservoirs. Today, growing populations and farming communities in these regions depend upon this nonrenewable ground water as part of a nonsustainable system of development and agriculture.

Removing water from underlying sediments can also cause **subsidence**, or sinking of the landscape. Ground water residing in the pore spaces of the subsurface sediment serves to support the overlying material. When this water is withdrawn, the empty pore spaces collapse from the soil pressure above. Venice, Italy, shown in the chapter-opening feature, provides a classic example. This phenomenon has also lowered the landscape at many locations throughout the American West and Southwest, as farming communities have depleted the ground water in order to grow crops (**Figure 11.11**). The sinking land undermines homes and road infrastructure.

When aquifers are over-pumped in coastal regions, salt water can contaminate ground water through a process called saltwater intrusion. Because fresh water is less dense than salt water, a layer of salt water from the ocean can lie below a coastal aquifer. As the aquifer is depleted, the level of salt water rises and eventually reaches the well from below, making the well unusable. The contamination of ground water by saltwater intrusion affects salt-intolerant ecological systems, as native vegetation is replaced by salt-tolerant species. The coastal agricultural economy of the southern United States is also affected when yields are reduced and the soils are made unsuitable for future farming.

Land subsidence mapping • Figure 11.11

These shaded relief maps portray areas of significant subsidence over a period of 7 months in 1997 in the Santa Clara Valley, California. The deep central subsidence (red) is a result of seasonal groundwater pumping. Areas of uplift show where groundwater supplies are at higher levels. Nationwide, subsidence has directly affected more than 44,000 km² (17,000 mi²) in 45 states, costing farmers and governments over $125 million annually. Scientists at the USGS are measuring small deformations of the Earth's crust to map and monitor these areas using the Interferometric Synthetic Aperture Radar (In-SAR) satellite sensor. The InSAR sensor allows for high-resolution imaging (called an interferogram) of the ground surface to determine subtle changes in the surface elevation of subsiding regions.

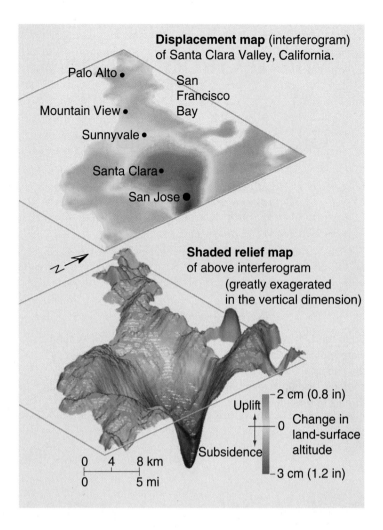

Displacement map (interferogram) of Santa Clara Valley, California.

Palo Alto
San Francisco Bay
Mountain View
Sunnyvale
Santa Clara
San Jose

Shaded relief map of above interferogram (greatly exaggerated in the vertical dimension)

Uplift
−2 cm (0.8 in)
0
Change in land-surface altitude
Subsidence
−3 cm (1.2 in)

0 4 8 km
0 5 mi

Pollution of ground water

Another major environmental problem is the contamination of ground water by pollutants such as agricultural runoff, industrial waste, and acid rain, among other sources. For example, as rainwater filters through buried landfill waste, it picks up pollutants and carries them down to the water table (**Figure 11.12**). Flow of water over the land surface can carry excess fertilizers, agriculture pesticides, or toxic road runoff into

Groundwater contamination • Figure 11.12

Polluted water, leached from a waste disposal site, such as the one shown in the photo, moves toward a stream (left) and a supply well (right).

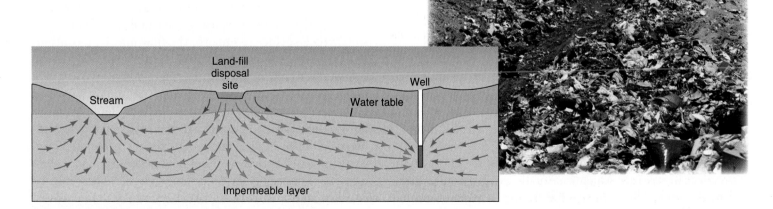

Stream
Land-fill disposal site
Well
Water table
Impermeable layer

streams and lakes, from where it gradually infiltrates into ground water. Although waves of chemical pollution in streams and rivers can move rapidly downstream, pollutants remain in relatively slow-moving ground water for long periods of time. Chemical pollution of ground water makes it necessary for well owners to regularly test their water for contaminants. The U.S. Environmental Protection Agency (EPA) tracks sources of water pollution around the country (see *Where Geographers Click*). Contamination of large urban aquifers has caused the EPA to close down wells in many locations and introduce costly filtration and decontamination technologies.

CONCEPT CHECK | STOP

1. **What** factors affect the height of the water table?

2. **What** are the key features of a karst landscape, and how do they form?

3. **How** is ground water contaminated?

Surface Water and Fluvial Systems

LEARNING OBJECTIVES

1. **Diagram** a drainage system and explain how it works.

2. **Explain** the relationship between stream channels and discharge.

3. **Describe** flooding processes.

Recall from our discussion of the hydrologic cycle at the beginning of this chapter that some precipitation filters through the soil to become ground water and some travels over the surface to streams, which eventually carry the water back to the ocean. We have already seen what happens to water that infiltrates the soil. Now we take a closer look at water that travels over the surface through fluvial systems.

Drainage Systems

Water naturally travels downhill by the force of gravity to collect in streams, which eventually carry the water back to the ocean. Streams are fed by the seepage of ground water, melting of ice and snow, lakes, and surface runoff from precipitation.

Overland flow When soils are saturated or rain falls too quickly to be absorbed into the ground, water travels directly over the surface as overland flow. Where the soil or rock surface is smooth, the flow may be a continuous thin film, called **sheet flow** (**Figure 11.13**). If the ground is rough or pitted, overland flow may be made up of a series of tiny rivulets connecting one water-filled hollow with another. On a grass-covered slope, overland flow is divided into countless tiny threads of water, passing around the stems. Even in a heavy and prolonged rain, you might not notice overland flow in progress on a sloping lawn. On heavily forested slopes, overland flow may be entirely concealed beneath a thick mat of decaying leaves.

Overland flow • Figure 11.13

During heavy rains, water runs off the land surface in a thin layer. It converges in lower areas, forming wide temporary streams that move water and sediment into creeks and rivers. This torrent followed a severe thunderstorm near Las Vegas, Nevada.

Drainage basins As runoff moves to lower and lower levels and eventually to the sea, it becomes organized into a branched network of stream channels. This network and the sloping ground surfaces next to the channels that contribute overland flow to the streams are together called a **drainage system**. **Drainage divides** mark the boundary between slopes

> **drainage system**
> A branched network of stream channels and adjacent land slopes that converge to a single channel at the outlet.

that contribute water to different streams or drainage systems. The entire system is bounded by an outer drainage divide that outlines a more-or-less pear-shaped **drainage basin**, or **watershed** (Figure 11.14). Drainage systems funnel overland flow into streams and smaller streams into larger ones.

Channel network of a stream • Figure 11.14

Smaller and larger streams merge in a network that carries runoff downstream. Each small tributary has its own small drainage basin, bounded by drainage divides. An outer drainage divide marks the stream's watershed at any point on the stream.

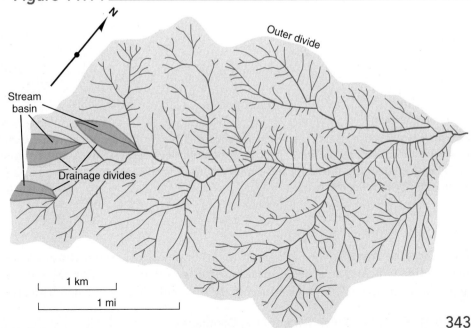

Stream basin

Drainage divides

Outer divide

1 km

1 mi

343

The arrangement of streams that develops on a landscape often depends on the underlying rock structure.

a. Dendritic pattern

A drainage pattern much like a branching tree, known as a *dendritic pattern*, usually develops on uniform rocks or on flat-lying rock layers. This example shows a stream pattern developed on a part of the Idaho batholith, a large area of relatively uniform granitic rocks.

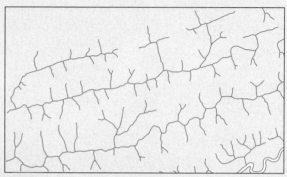

b. Trellis pattern

As streams erode their valleys, they cut into the softer rock and leave the harder rocks behind to form the drainage divides. Where rocks are folded into anticlines and synclines, streams follow the long, parallel valleys, creating a trellis pattern. This pattern is taken from a folded belt in the central Appalachians.

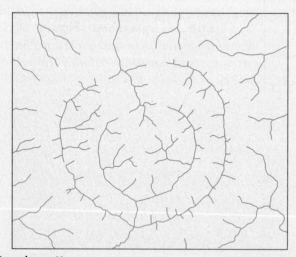

c. Annular pattern

If tectonic forces cause layers of sedimentary rock to bulge upward and form a dome, erosion can produce circular valleys as the dome is eroded. The streams in these valleys then follow an annular, or ring-shaped, pattern.

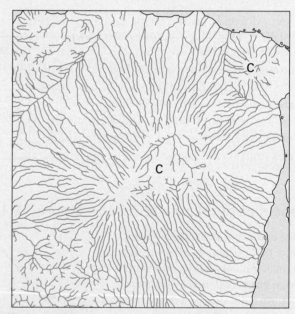

d. Radial pattern

Stratovolcanoes often develop drainage patterns of streams leading away from the central crater, marked "C" in the map. This example is from a volcanic island in the East Indies. Most of the radial streams drain the large volcano in the middle. At the upper right is a smaller cone with its own central crater.

Drainage patterns The outline of streams draining a region often falls into a particular **drainage pattern** that depends on the underlying rock structure. For example, a steep stratovolcano will have a radial pattern of streams leading away from the central crater. Other patterns include dendritic, trellis, and annular (**Figure 11.15**).

Stream Channels and Discharge

stream A long, narrow body of flowing water moving along a channel to lower levels under the force of gravity.

We can define a **stream** as a long, narrow body of water flowing through a channel and moving to lower levels under the force of gravity. A stream may be so narrow that you can easily jump across it, or, in the case of the Mississippi River, it may be as wide as 1.5 km (about 1 mi) or more.

Scientists who study streams measure them by their cross-sectional area (A), which depends on the width (w) and depth (d) of their channels. Two other key characteristics are stream velocity (V), which measures how rapidly the water in the stream flows, and **discharge** (Q), which is the volume of water per unit of time passing through a cross section of the stream at that location. The slope (S) of the stream, also called the **stream gradient**, is another important characteristic. Streams are also measured with attention to their potential floodplain, an extended, relatively flat area where the stream spreads out during flooding.

Stream velocity

The velocity of water flowing through a stream depends on the shape and slope of the channel (**Figure 11.16a**). The water meets resistance as it flows because of friction with the channel walls. So, water close to the bed and banks moves more slowly than water in the central part of the flow. If the channel is straight and symmetrical, the line of maximum velocity is located in midstream. If the stream curves, the maximum velocity shifts toward the bank on the outside of the curve.

Similarly, when the stream gradient is steep, water velocity will increase as the force of gravity has a greater effect on water in the stream. When the slope is gentler, stream velocity will decrease. Water meets frictional resistance, or shear, as it drags along the stream bed. Therefore, velocity is greatest at the surface of the stream (**Figure 11.16b**).

In all but the most sluggish streams, the movement of water is affected by turbulence. If we could follow a particular water molecule, we would see it travel a highly irregular, corkscrew path as it is swept downstream. If we measure the water velocity at a certain fixed point for several minutes, however, we see that the average motion is downstream.

Stream discharge

The amount of water that travels through a stream, its discharge, is a product of the stream's cross-sectional area and the velocity of its flow. As stream discharge changes, such as during periods of flooding or drought, the stream's cross-sectional area and

Stream velocity • Figure 11.16

Stream velocity is determined by the shape and slope of a stream.

a. In straight stretches, velocity is greatest in the middle of the stream, where the water meets less frictional resistance from the stream banks.

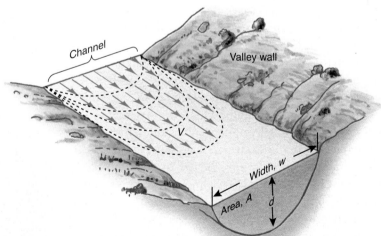

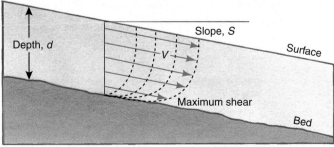

b. Velocity is more rapid where the stream gradient, measured by the slope, S, is steep. Velocity is greatest at the surface of the stream, where there is less frictional resistance, or shear, from the stream bed.

Discharge within a drainage system • Figure 11.17

The amount of water flowing in the Mississippi watershed increases as the water moves downstream because of the addition of tributaries.

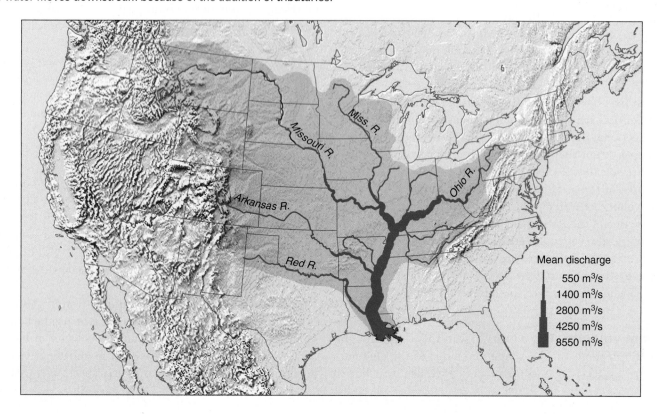

Mean discharge
- 550 m³/s
- 1400 m³/s
- 2800 m³/s
- 4250 m³/s
- 8550 m³/s

velocity will change accordingly. Within a drainage system, stream discharge increases downstream as streams and rivers combine to deliver runoff and sediment to the oceans (**Figure 11.17**).

The discharge of a given stream at a given time remains nearly constant over a short stretch, even as the velocity of the stream changes with the stream gradient. This means that in stretches of rapids, where the stream flows swiftly, the stream channel will be shallow and narrow. In contrast, in pools, where the stream flows more slowly, the stream channel will be wider and deeper to maintain the same discharge (**Figure 11.18**). Sequences of pools and rapids can be found along streams of all sizes.

The discharge of a stream varies over the course of the year as a result of annual cycles of precipitation. The relationship between stream discharge and precipitation is shown by a hydrograph, which plots the discharge of a stream with time at a particular location (**Figure 11.19**). Water supplied to a stream by ground water is known as **base flow**. It increases during wet periods when the water table is high and decreases during dry periods when the water table is low. When it rains, overland flow combines with base flow to further increase the stream's discharge.

Pools and rapids • Figure 11.18

In steep areas where the velocity of the flow increases, the channel is shallow, forming rapids, whereas in flatter areas where the stream flow slows, the channel is deep, forming pools.

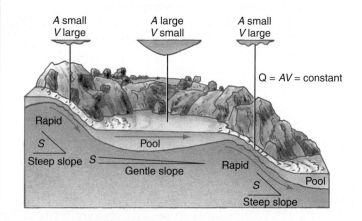

A small V large A large V small A small V large

$Q = AV = $ constant

Rapid
S
Steep slope S
Pool
Gentle slope   Rapid
S
Pool
Steep slope

Ask Yourself

As the slope increases, the _____ of the stream is constant and the _____ decreases.

a. velocity; area
b. area; velocity
c. area; discharge
d. discharge; area

Discharge related to base flow and overland flow • Figure 11.19

This hydrograph shows the fluctuating discharge of the Chattahoochee River, Georgia, throughout a typical year. The sharp, abrupt fluctuations in discharge are produced by overland flow after rainfall periods lasting 1 to 3 days.

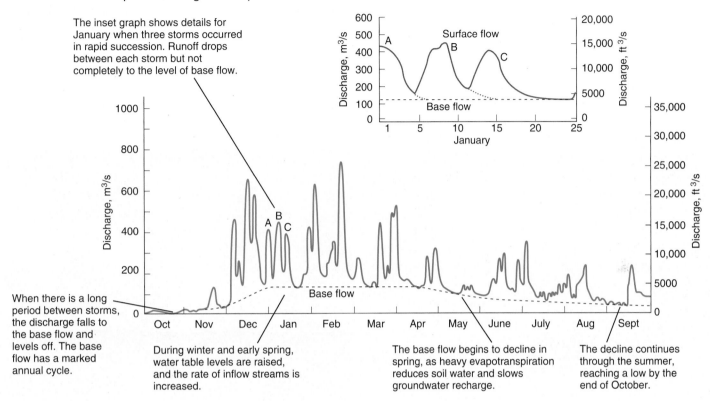

The inset graph shows details for January when three storms occurred in rapid succession. Runoff drops between each storm but not completely to the level of base flow.

When there is a long period between storms, the discharge falls to the base flow and levels off. The base flow has a marked annual cycle.

During winter and early spring, water table levels are raised, and the rate of inflow streams is increased.

The base flow begins to decline in spring, as heavy evapotranspiration reduces soil water and slows groundwater recharge.

The decline continues through the summer, reaching a low by the end of October.

Flooding

When soil is saturated by snowmelt or precipitation, runoff fills streams and rivers. When the discharge of a river cannot be accommodated within the normal channel, the water spreads over the adjacent ground, called the floodplain.

> **floodplain** A broad belt of low, flat ground bordering a river channel that floods regularly.

Floodplains Flooding is a regular occurrence in stream systems. Most low-gradient rivers of humid climates have a **floodplain**—a flat area bordering the channel on one or both sides that fills with water during a flood (**Figure 11.20**). As we will discuss in Chapter 13, floodplains are carved out over time as a river migrates across the

Floodplain • Figure 11.20

In humid climates, most rivers are bordered by a plain of flat ground called a floodplain.

a. Migration of a river through a valley over time produces a wide floodplain.

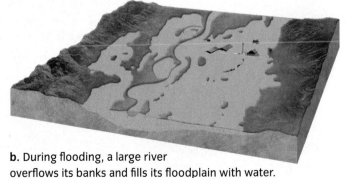

b. During flooding, a large river overflows its banks and fills its floodplain with water.

Mississippi flood of 2011 • Figure 11.21

The spring of 2011 saw epic flooding along the Mississippi River. The discharge of the swollen river, about 65,000 m³ (2 million ft³) per second, matched the most destructive river flood in U.S. history, which occurred on the Mississippi in 1927. Many stream gauges recorded record or near-record values. At Memphis, the river peaked at a stage of 47.9 ft, only 0.8 ft below the record, set in 1937. In Mississippi, record crests were recorded at Vicksburg and Natchez. In the city of New Orleans, the river came 3 ft (1 m) from overtopping the levees. Early damage estimates for the total economic cost ranged as high as $7–9 billion. The flood was triggered by four major storms in April, which triggered extensive melting of winter snow.

a. This satellite image, acquired by NASA's MODIS imager, shows the Mississippi Valley between Cairo, Illinois, and Memphis, Tennessee, on May 6, 2010. The brown areas are fertile agricultural lands in tributary valleys that drain into the Mississippi below Memphis.

b. On May 6, 2011, the Mississippi left its banks and inundated broad areas of its floodplain. Flooding occurred as far west as the Black and White rivers in Arkansas. Two levels of clouds are visible in the upper-left part of the image.

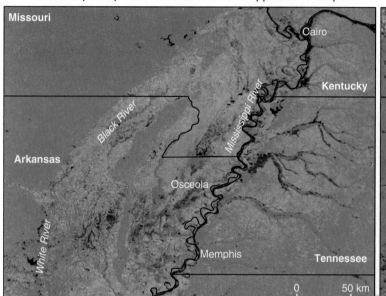

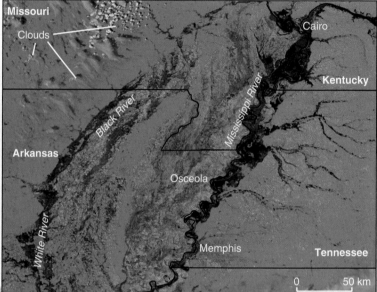

c. Although much of Memphis escaped flooding, some areas of the city and nearby areas were immersed in flood waters. Pictured here is a flooded U.S. Navy facility in Millington, just north of Memphis. The circular feature is a recreational park with a sail-like canopy at its center.

> **Put It Together**
>
> *Review Figure 11.15 and answer this question.* Which drainage pattern do the streams in these images most closely follow?
> a. dendritic
> b. trellis
> c. annular
> d. radial

valley floor. Periodic inundation of floodplains is expected and does not prevent either crop cultivation after the flood has subsided or the growth of dense forests, which are widely distributed over low, marshy floodplains in all humid regions of the world.

Flooding on a floodplain becomes a human catastrophe only when cities and homes are situated there (**Figure 11.21**). Historically, these areas were developed because they were easily accessible to river transportation and because the floodplains provided fertile agricultural ground. Today, people often build homes near a river to enjoy a picturesque view or easy access to the water for recreational activities.

The National Weather Service designates a particular water surface level as the **flood stage** for a particular river at a given place. If water rises above this critical level, the floodplain will be inundated. Over time, we see examples where even higher discharges cause rare and disastrous floods that inundate land well above the floodplain. The National Weather Service operates a River and Flood Forecasting Service through 85 offices located at strategic points along major river systems of the United States (see *Where Geographers Click*). This effort is augmented by hundreds of stations maintained by each state. When flooding potential is high, forecasters analyze precipitation patterns and the progress of high waters moving downstream. They develop specific flood forecasts after examining the flood history of the rivers and streams concerned. The forecasts are then delivered to communities that could potentially be affected.

Where Geographers CLICK

Flood Forecasting

http://water.weather.gov/ahps/

NOAA's Advanced Hydrologic Prediction Service is designed to provide improved river and flood forecasting and water information over the Web to citizens, community leaders, and emergency managers to assist in making decisions about evacuations and movement of property before flooding occurs.

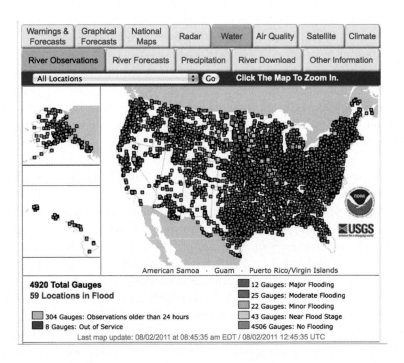

Stream discharge during flooding • Figure 11.22

The graph shows the discharge of Sugar Creek (blue line), a tributary of the Muskingum River, Ohio, during a 4-day period that included a heavy rainstorm.

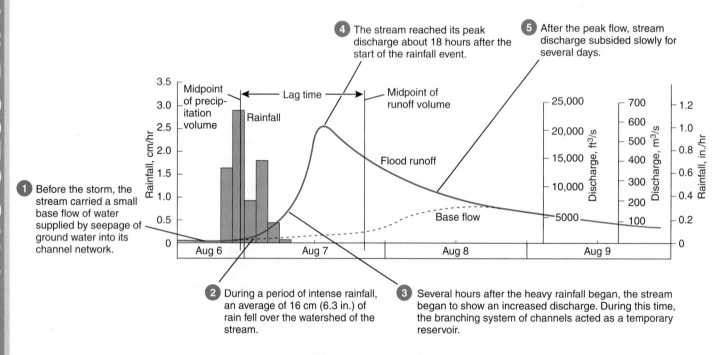

4 The stream reached its peak discharge about 18 hours after the start of the rainfall event.

5 After the peak flow, stream discharge subsided slowly for several days.

1 Before the storm, the stream carried a small base flow of water supplied by seepage of ground water into its channel network.

2 During a period of intense rainfall, an average of 16 cm (6.3 in.) of rain fell over the watershed of the stream.

3 Several hours after the heavy rainfall began, the stream began to show an increased discharge. During this time, the branching system of channels acted as a temporary reservoir.

Flood profiles Following a period of increased precipitation or snowmelt, stream discharge rises and then falls over the following days or weeks (**Figure 11.22**). There is a delay, or lag time, between the precipitation event and the peak flow of a flood because it takes time for the water to move into stream channels. The length of this delay depends on several factors, including the size of the drainage basin feeding the stream. Larger drainage basins show a longer delay.

Under some conditions, the lag time between precipitation and flooding can be significantly less, increasing the danger to local populations. These **flash floods** are characteristic of streams draining small watersheds with steep slopes. These streams have short lag times of only 1 or 2 hours, so when there's intense rainfall, the stream quickly rises to a high level. The flood arrives as a swiftly moving wall of turbulent water, sweeping away buildings and vehicles in its path. In arid western watersheds, great quantities of coarse rock debris are swept into the main channel and travel with the flood water, producing debris floods. In forested landscapes, tree limbs and trunks, soil, rocks, and boulders are swept downstream in the flood waters. Flash floods often occur too quickly to warn people, so they can cause significant loss of life.

Urbanization and flooding The growth of cities and suburbs affects the flow of small streams in two ways (**Figure 11.23**). First, it is far more difficult for water to infiltrate the ground, which is increasingly covered by impervious surfaces such as roads, buildings, driveways, walks, pavement, and parking lots. In a closely built-up residential area, 80% of the surface may be impervious to water. This in turn increases overland flow, making flash flooding more common during heavy storms for small watersheds lying largely within the urbanized area. It's also harder to refill the groundwater body beneath. The reduction in ground water decreases the base flow to channels in the same area. So, in dry periods, stream discharges will tend to be lower in urban areas, while in wet periods, the chance and amount of flooding rise.

A second change caused by urbanization comes from the introduction of storm sewers. This system of large underground pipes quickly carries storm runoff from paved areas directly to stream channels for discharge.

This shortens the time it takes runoff to travel to channels, while the proportion of runoff is increased by the expansion in impervious surfaces. Together, these changes reduce the lag time of urban streams and increase their peak discharge levels. Many rapidly expanding suburban communities are finding that low-lying, formerly flood-free residential areas now experience periodic flooding as a result of urbanization.

Urban water flow • Figure 11.23

Impervious surfaces like this street in Bangkok increase runoff and hasten the flow of water into streams and rivers draining urban environments.

Lakes

LEARNING OBJECTIVES

1. **Define** lakes and describe their features.
2. **Describe** how lakes gain and lose water.

 lake is a body of standing water with an upper surface that is exposed to the atmosphere and does not have an appreciable gradient or current. Ponds, marshes, and swamps with standing water can all be included under the definition

> **lake** A body of standing water without an appreciable gradient or current.

of a lake (**Figure 11.24**). Lakes are quite important for humans as sources of fresh water and food, such as fish. Artificial lakes created by damming rivers can also be used to generate hydroelectric power.

Lakes receive water from streams, overland flow, and ground water, and so they form part of drainage systems. Unlike ground water, which represents a large water storage body, fresh surface water in the liquid state is stored only in small quantities. (An exception is the Great Lakes system.) About 20 times as much ground water is available

Lakes and ponds • Figure 11.24

Standing water collects on the surface in lakes, ponds, marshes, and swamps.

a. Lake

Lakes, such as Lake Louise in Alberta, Canada, are usually larger and deeper than ponds. This example was formed when a rockslide dammed a stream in a narrow valley.

b. Pond

The shallow waters of this pond in Nicolet National Forest, Wisconsin, support an almost continuous cover of grasses and sedges.

Hydroelectric dam • Figure 11.25

Many lakes are artificial, dammed to provide water supplies or hydroelectric power. The Picote Dam, shown here, is on the Douro River, in Portugal. It is about 100 m (328 ft) high, providing ample water pressure at its base to drive its power-generating turbines.

globally as water held in freshwater lakes. Many lakes lose water at an outlet that becomes an outflowing stream. Lakes also lose water by evaporation and extraction for human use.

Lakes, like streams, are landscape features but are not usually considered to be landforms. By contrast, lake basins, like stream channels, are true landforms, created by a number of geologic processes and ranging widely in size. For example, the tectonic process of crustal faulting creates many large, deep lakes. Lava flows often form a dam in a river valley, causing water to back up as a lake. Landslides can also suddenly create lakes by blocking the flow of a stream.

Many of the world's lakes are artificially constructed or enhanced by placing dams across stream channels (**Figure 11.25**). Artificial lakes range from small ponds built to serve ranches and farms to vast lakes that cover hundreds of square kilometers. In some areas, the number of artificial lakes is large enough to have a significant effect on the region's hydrologic cycle.

On a geologic timescale, lakes are short-lived features. Lakes disappear by one of two processes or a combination of both. First, lakes that have stream outlets will be gradually drained as the outlets are eroded to lower levels. Even when the outlet lies above strong bedrock, erosion will still occur slowly over time. Second, inorganic

Water level and the water table • Figure 11.26

The water level of lakes and ponds is close to the level of the water table. In this example, retreating glaciers have left an irregular landscape of sand and gravel deposits, with depressions that are occupied by lakes.

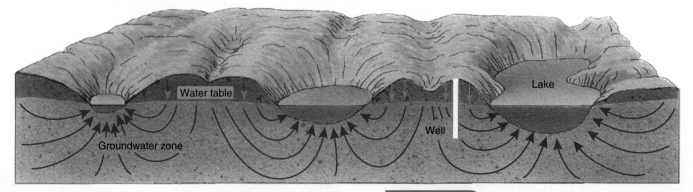

Put It Together

Review Figure 11.5 and complete this statement.
If the water level of a lake is high above the level of the water table, the lake is likely on top of a(n) _____.

sediment carried by streams enters the lake and builds up, along with organic matter produced by plants and animals within the lake. Eventually, the lake fills, forming a boggy wetland with little or no free water surface. Many former freshwater ponds have become partially or entirely filled by organic matter from the growth and decay of water-loving plants.

Lakes can also disappear when the climate changes. Shifts in the climate as the planet warms may increase or reduce regional precipitation, thereby affecting the water table. The water level of lakes and ponds in moist climates closely coincides with the surrounding water table (**Figure 11.26**). The water surface is maintained at this level as ground water seeps into the lake and as precipitation runs off. If precipitation is reduced, or if temperatures and net radiation increase, evaporation can exceed input by ground water and runoff, and the lake will dry up.

Many former lakes of the southwestern United States that flourished during the Ice Age have now shrunk greatly or have disappeared entirely. Lake Bonneville was a huge lake covering Utah and the Great Basin from 32,000 years ago until about 14,500 years ago, when it dried up as a result of changes in the climate at the end of the last glacial period. The Bonneville Salt Flats, famous for the land speed records set there, is the remnant of that lake.

Most lakes contain fresh water. However, streams bring dissolved solids into a lake. In lakes without a surface outlet or sufficient flow rate to flush out mineral buildup, salts remain behind after pure water is evaporated. The salinity, or saltiness, of the water slowly increases over time, especially in warm, dry climates with high evaporation rates. Some of the world's largest lakes are saline.

CONCEPT CHECK

1. **How** and why do humans create artificial lakes?
2. **What** is the relationship between the level of lakes and the water table?

Water as a Natural Resource

LEARNING OBJECTIVES

1. **Explain** how water is used as a natural resource.
2. **Describe** how surface water can be polluted.

Fresh water is a basic natural resource that is essential to human agricultural and industrial activities, but it is also a limited resource. Population growth and increased demand have placed a strain on fresh water supplies around the world. The supply of usable fresh water is also threatened by water pollution.

Water Access and Supply

We rely on clean water for food, recreation, and countless other uses (**Figure 11.27** on the next page). However, despite the abundance of rainfall in many areas, fresh water access is becoming a chronic issue.

Our heavily industrialized society requires enormous supplies of fresh water to sustain it, and demand is increasing. Urban dwellers in developed nations consume 150 to 400 L (50 to 100 gal) of water per person per day in their homes. We use large quantities of water in air conditioning units and power plants, and much of this water is obtained from surface water.

To increase the availability of fresh water, we build dams to trap surface water and provide water supplies for great urban centers, such as New York City and Los Angeles, and we build reservoirs to store precipitation and runoff that would otherwise escape to the sea. However, dams bring a host of negative environmental ramifications. These include disruption of native fisheries, drowning of the river and creek ecosystems, displacement of local people, and loss of nutrients that would otherwise flow into the river floodplains below the dams. Furthermore, dams have finite lifetimes due to silt buildup and structural decay.

Global water is a finite resource, and in the long term, we can use only as much water as is supplied in precipitation. **Desalination**, a process that separates fresh water from sea water, offers an additional source of fresh water in locations lacking sufficient precipitation. The high energy cost of operating desalination plants, however, precludes the use of this technology in most of the developing world.

Water is essential to support human populations. Because watersheds are not limited to single countries or political regions, water conflicts arise, as shown on the map by red triangles.

b. Power

In many parts of the world, water is valued as a source of hydroelectric power. Massive water pipes funnel water from the Euphrates River to turbines in Turkey's Ataturk Dam, the centerpiece of a controversial plan to irrigate southeastern Turkey.

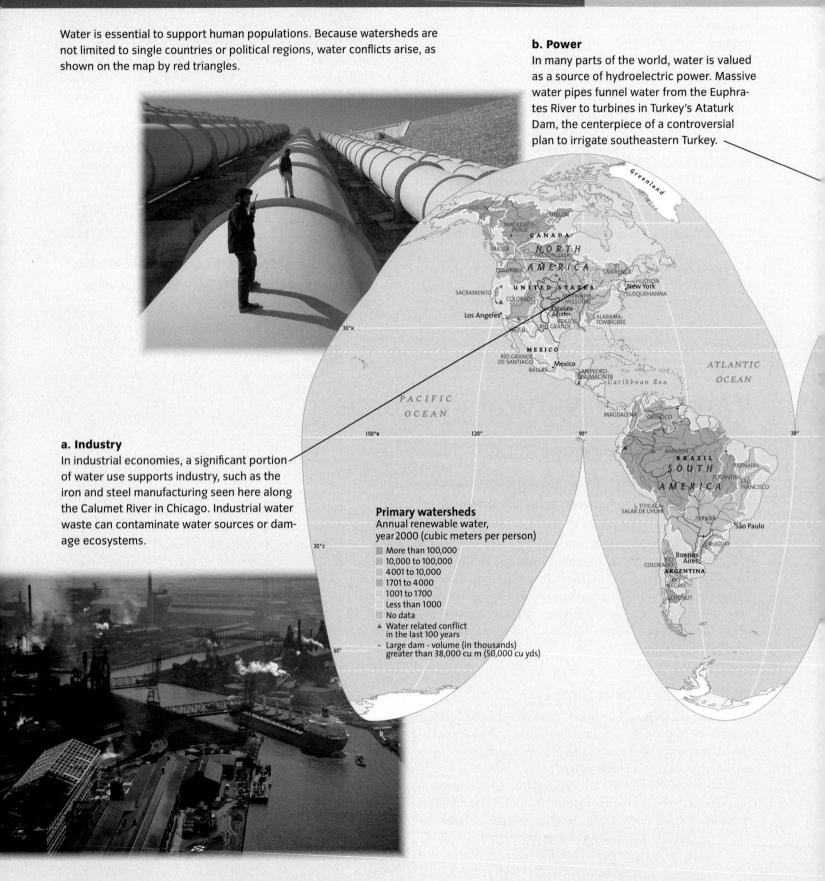

a. Industry

In industrial economies, a significant portion of water use supports industry, such as the iron and steel manufacturing seen here along the Calumet River in Chicago. Industrial water waste can contaminate water sources or damage ecosystems.

Primary watersheds
Annual renewable water, year 2000 (cubic meters per person)

- More than 100,000
- 10,000 to 100,000
- 4001 to 10,000
- 1701 to 4000
- 1001 to 1700
- Less than 1000
- No data
- ▲ Water related conflict in the last 100 years
- – Large dam - volume (in thousands) greater than 38,000 cu m (50,000 cu yds)

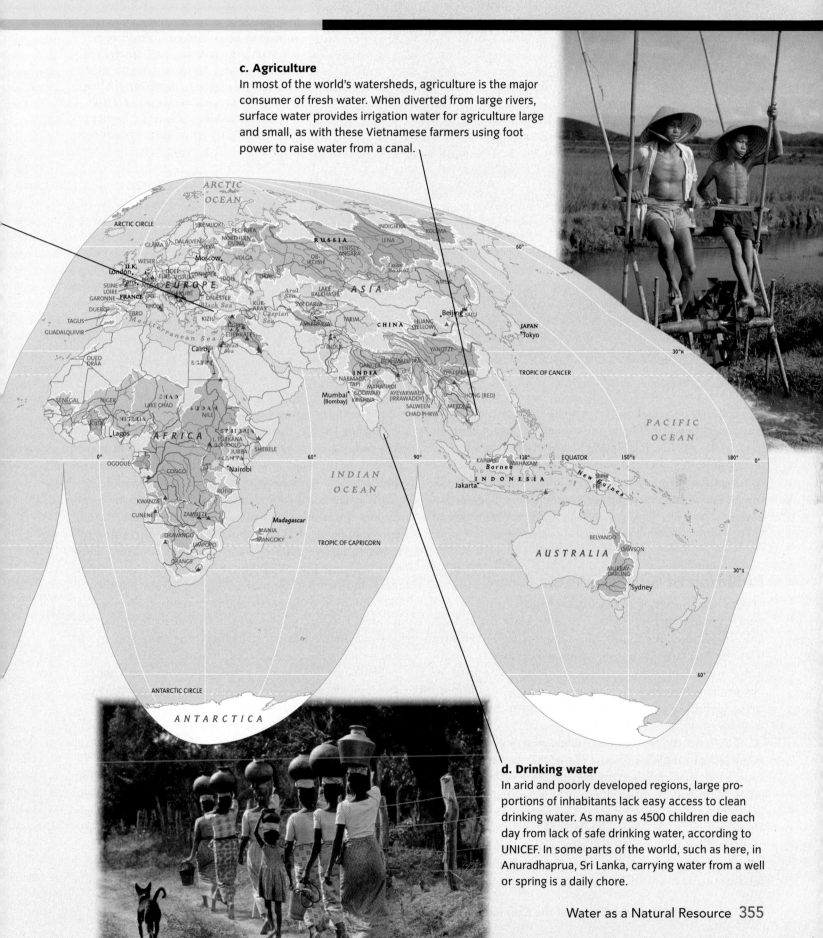

c. Agriculture

In most of the world's watersheds, agriculture is the major consumer of fresh water. When diverted from large rivers, surface water provides irrigation water for agriculture large and small, as with these Vietnamese farmers using foot power to raise water from a canal.

d. Drinking water

In arid and poorly developed regions, large proportions of inhabitants lack easy access to clean drinking water. As many as 4500 children die each day from lack of safe drinking water, according to UNICEF. In some parts of the world, such as here, in Anuradhaprua, Sri Lanka, carrying water from a well or spring is a daily chore.

Water as a Natural Resource 355

In Appalachia, water percolating through strip-mine waste banks and abandoned mines often finds its way into streams and rivers. This water contains sulfuric acid and various salts of metals, particularly of iron. In sufficient concentrations, the acid from these sources is lethal to certain species of fish. Over 2000 miles of streams have been destroyed by strip-mining fill and toxic runoff from the mines.

Water resources have risen to the top of the United Nations Millennium Goals and other international agendas, as the world's population shares dwindling supplies of this finite resource. Climate change scientists predict a continuation of the current trends toward increased flooding and droughts scattered geographically throughout all nations. More efficient use of fresh water will be required to cope with increased scarcity and competition.

Pollution of Surface Water

Our modern society makes radical changes to the flow of water by constructing dams, irrigation systems, and canals. In addition, it pollutes and contaminates our surface waters with a large variety of wastes and chemicals. Chemicals entering our water systems are linked to a variety of health concerns. Blood tests reveal the presence of more than 200 industrial chemicals in the blood stream of newborn babies in the United States.

There are many different sources of water pollutants. Some industrial plants dispose of toxic metals and organic compounds directly into streams and lakes through outlet pipes. Many communities still discharge untreated or partly treated sewage wastes into surface waters. In urban and suburban areas, de-icing salt and lawn conditioners (lime and fertilizers) enter and pollute streams and lakes, and also contaminate ground water. In agricultural regions, fertilizers and livestock wastes are significant pollutants.

Mining and processing of mineral deposits also pollute water (**Figure 11.28**). Surface water can also be contaminated by radioactive substances released from nuclear power and processing plants. *Video Explorations* documents the pollution of the Volga River by industrial accidents.

Many chemical compounds dissolve in water by forming ions—charged forms of molecules or atoms. Among the common chemical pollutants of both surface water and ground water are sulfate, chloride, sodium, nitrate, phosphate, and calcium ions. Sulfate ions enter runoff from both polluted urban air and sewage. Chloride and sodium ions come from polluted air and from deicing salts used on highways. Fertilizers and sewage also contribute nitrate and phosphate ions. Nitrates can be highly toxic in large concentrations and are difficult and expensive to remove.

Phosphate and nitrates are plant nutrients and can encourage algae and other aquatic plants to grow to excessive amounts in streams and lakes. In lakes, this process is known as **eutrophication** and is often described as the aging of a lake. Nutrients stimulate plant growth, which in turn produces a large supply of dead organic matter in the lake. Microorganisms break down this organic matter but require oxygen in the process. However, oxygen is normally present only in low concentrations because it dissolves only slightly in water. The microorganisms use up oxygen to the point called **hypoxia**, at which other organisms, including desirable types of fish, cannot survive. After a few years of nutrient pollution, the lake takes on the characteristics of a shallow pond that has been slowly filled with sediment and organic matter over thousands of years.

Plants and animals have also been killed by toxic metals, including mercury, pesticides, and a host of other industrial chemicals introduced into streams and lakes. Sewage introduces live bacteria and viruses—classed as biological pollutants—that can harm humans and animals alike. Another form of pollution is thermal pollution, which is created when heat, generated from fuel combustion and from the conversion of nuclear fuel to electricity, is discharged into the environment. Heated water put into streams, estuaries, and lakes can have drastic effects on local aquatic life. The impact can be quite large in a small area.

CONCEPT CHECK   **STOP**

1. **What** factors limit our use of surface water as a natural resource?

2. **How** do the sources and effects of biological pollutants and chemical pollutants in surface water differ?

Video Explorations

Volga: The Soul of Russia

WILEY PLUS   NATIONAL GEOGRAPHIC   Video

Unchecked industrial processes, power production, and nuclear technologies can all pollute water supplies. Some of the most extreme examples of this are found in the former Soviet Union, where industries severely polluted the environment with toxic wastes.

The Volga River, which flows through Moscow, Russia, is the largest river in Europe. Local people believe that water in the river was polluted by underground nuclear explosions set off by a nearby industrial plant. This video demonstrates the environmental effects of the industrial plant's pollution of the Volga River.

THE PLANNER ✓

Summary

1   Fresh Water and the Hydrologic Cycle 332

• Most of the Earth's fresh water is in ice caps or mountain glaciers and **ground water**, as shown in the chart.

Distribution of fresh water • Figure 11.1

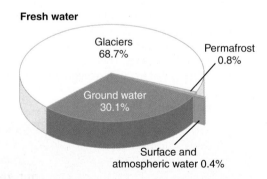

Fresh water

Glaciers 68.7%

Permafrost 0.8%

Ground water 30.1%

Surface and atmospheric water 0.4%

• As a part of the hydrologic cycle, the **soil-water belt** absorbs precipitation through **infiltration**. From there, some water is diverted to the atmosphere through **evapotranspiration** and some filters down into ground water, where it feeds springs and streams through seepage.

• Water from streams and rivers that returns to the ocean is **runoff**. During periods of heavy precipitation, runoff travels over the surface as **overland flow**.

2   Ground Water 335

• Rainwater percolates through the soil-water belt until it reaches the **saturated zone**, where it fully occupies the pore spaces in rock and regolith. The top of the saturated zone is the **water table**, which rises during wet periods and falls during dry periods, as shown in the diagram.

Ground water and the water table • Figure 11.6

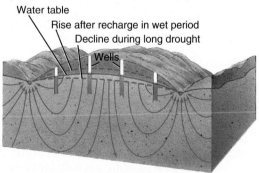

Water table

Rise after recharge in wet period

Decline during long drought

Wells

• Ground water flows underground through layers of porous sediment called **aquifers**. Impermeable layers are **aquicludes**.

(continued on next page)

- **Karst topography**, characterized by caverns, **sinkholes**, and limestone towers, forms when ground water dissolves limestone.

- Removing too much ground water can deplete groundwater supplies and lead to **subsidence**. Ground water can be polluted by contaminants as a result of rainwater seeping through landfills, as well as industrial runoff and byproducts of mining.

3 Surface Water and Fluvial Systems 342

- **Streams** are organized into a **drainage system** that moves runoff from slopes into channels and from smaller channels into larger ones.

- The **drainage pattern** of streams is often related to underlying rock structures.

- Water moves faster at the middle and top of streams and where the **stream gradient** is steeper.

- There is a fixed relationship between the cross-sectional area of a stream channel, the velocity of the stream flow, and the **discharge** of the stream, as shown in the diagram. As a result, stream channels often develop a series of shallow rapids and deeper pools.

Pools and rapids • Figure 11.18

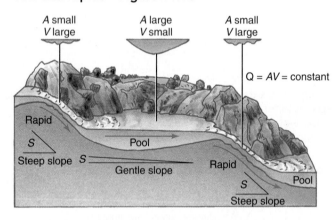

- A hydrograph plots the discharge of a stream at a location through time.

- Floods occur when river discharge increases and the flow can't be contained within the river's usual channel and it flows to fill a **floodplain**.

- In small, steep watersheds, flash floods can occur. Urban areas are prone to flash flooding because precipitation cannot infiltrate the impermeable surfaces of a city and runs off quickly into streams.

4 Lakes 351

- **Lakes** are fed by streams, runoff, and ground water, as shown in the diagram. Lakes lose water to evaporation and stream outlets.

Water level and the water table • Figure 11.26

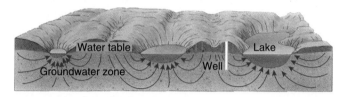

- A lake can be formed by natural processes or can be artificially dammed as a water reservoir or source of hydroelectric power.

- Lakes can disappear as stream outlets are eroded to lower levels or when the climate changes.

5 Water as a Natural Resource 353

- Human populations rely on surface water for drinking water, as well as for industrial and agricultural use.

- Supplies of fresh water are a limited global resource. **Desalination** can provide additional freshwater resources by removing salt from sea water but has high energy costs.

- Water pollution arises from many sources, including industrial sites, sewage treatment plants, agricultural activities, and mining, shown in the photo. **Eutrophication** occurs when the buildup of pollutants in lakes leads to excessive growth of algae and other plants. This can result in **hypoxia**, a point at which there is not enough oxygen for fish to survive.

Water pollution • Figure 11.28

Key Terms

- aquiclude 335
- aquifer 335
- artesian well 335
- base flow 346
- cone of depression 340
- desalination 353
- discharge 345
- drainage basin (watershed) 343
- drainage divide 343
- drainage pattern 344
- drainage system 343
- drawdown 340

- eutrophication 356
- evapotranspiration 334
- flash flood 350
- flood stage 349
- floodplain 347
- ground water 333
- hypoxia 356
- infiltration 334
- karst topography 339
- lake 351
- overland flow 334
- percolation 335

- runoff 334
- saturated zone 335
- sheet flow 342
- sinkhole 339
- soil-water belt 334
- stalactite 337
- stalagmite 339
- stream 345
- stream gradient 345
- subsidence 340
- water table 335

Critical and Creative Thinking Questions

1. A thundershower causes heavy rain to fall in a small region near the headwaters of a major river system. Describe the flow paths of that water as it returns to the atmosphere and ocean. What human structures might influence the flows?

2. What happens to precipitation falling on soil in your area? Discuss how the plants, drainage pattern, and soil in your area affect the ratio of water that becomes ground water as opposed to entering fluvial systems directly.

3. Consider the flowing artesian well in the diagram. Explain how a landfill far away could have a larger impact on the water quality of the well than could a pollution source much closer to the well.

4. How is groundwater contamination different from surface water contamination? How do these differences affect efforts to reduce water pollution?

5. What type of drainage pattern is most common in your area, and what does that tell you about the underlying geology?

6. What characteristics of lakes affect how susceptible they are to pollution? What type of lake is most likely to become polluted, and what types of lakes can quickly recover from pollution?

7. Imagine yourself a recently elected mayor of a small city located on the banks of a large river. What issues might you be concerned with that involve the river? In developing your answer, choose and specify some characteristics for this city—such as its population, its industries, its sewage systems, and the present uses of the river for water supply or recreation.

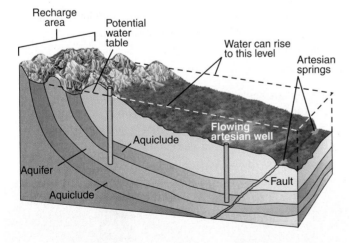

Recharge area · Potential water table · Water can rise to this level · Artesian springs · Aquiclude · Flowing artesian well · Aquifer · Fault · Aquiclude

What is happening in this picture?

A view deep within Luray Caverns, Luray, Virginia.

Think Critically

1. What rock is this cavern made from?
2. How did these features form?

Self-Test

(Check your answers in the Appendix.)

1. Label the following regions on this diagram showing the zones of subsurface water: (a) the saturated zone, (b) the soil-water belt, (c) the unsaturated zone, and (d) the water table.

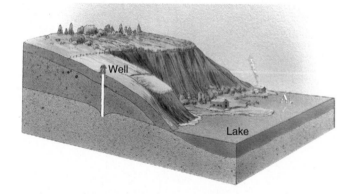

Well

Lake

2. The water table is at its highest _____.

a. under the lowest areas of land surface

b. under the highest areas of land surface

c. adjacent to perennial streams

d. in lakes

3. A layer of rock or sediment that contains abundant, freely flowing ground water is known as a(n) _____.

a. aquiclude   c. aquifer

b. infiltration   d. water table

4. Geographers apply the term _____ to the topography of any limestone area where sinkholes, as shown in this photograph, are numerous and small surface streams are nonexistent.

a. travertine   c. carbonate

b. karst   d. dolomite

5. A source of groundwater contamination in coastal wells is _____.

    a. saltwater intrusion

    b. solid-waste disposal

    c. high-temperature incineration

    d. air pollution

6. The _____ of a stream is a narrow trough, shaped by the forces of flowing water.

    a. course

    b. fall

    c. channel

    d. mouth

7. On this diagram of the channel network of a stream, label the outer divide, stream basin, and drainage divides.

1 km

1 mi

8. The volume of water per unit of time passing through a cross section of a stream at a given location is measured by its _____.

    a. gradient

    b. volume

    c. water velocity

    d. discharge

9. A _____ consists of a branched network of stream channels and adjacent slopes that feed the channels.

    a. drainage system

    b. drainage boundary

    c. drainage divide

    d. drainage pattern

10. The most important factor determining the lag time between a period of heavy rainfall or snowmelt and a stream's increased discharge response is the _____.

    a. size of the drainage basin feeding the stream

    b. number of drainage systems involved

    c. amount of drainage basin rainfall or snowmelt

    d. steepness of the gradient of the drainage basin

11. A _____ is a particular river surface height at a particular location above which floodplain inundation will occur.

    a. lag-time stage

    b. floodplain stage

    c. center of mass of runoff

    d. flood stage

12. Flash floods are characteristic of streams draining _____ watersheds with _____ slopes.

    a. large; gentle

    b. small; steep

    c. small; gentle

    d. large; steep

13. An important point about _____ is that they are short-lived features on the geologic time scale.

    a. rivers

    b. floodplains

    c. lakes

    d. drainage basins

14. Lakes without outlets other than evaporation often show _____.

    a. salt buildup

    b. a lesser surface area

    c. silty bottoms

    d. reduced volumes

15. What process offers an additional source of fresh water but requires very high energy costs?

    a. eutrophication

    b. desalination

    c. hypoxia

    d. infiltration

THE PLANNER ✓

Review your Chapter Planner on the chapter opener and check off your completed work.

Landforms Made by Running Water

The Victoria Falls are a series of waterfalls formed as the Zambezi River that borders Zambia and Zimbabwe plummets into a narrow chasm about 120 meters (400 feet) wide, carved by its waters along a fracture zone in the Earth's crust. At over 100 m (328 ft) tall and over 1.6 kilometers (1 mile) wide, this is the largest single vertical sheet of water in the world.

The waterfall converts the calm river to a ferocious torrent. The loud roar produced as 9.1 million m³ (321 million ft³) of water per minute thunder over the edge of the basalt cliff can be heard from 40 kilometers (25 mi) away.

Humans have learned to harness the power of waterfalls, here and elsewhere around the world. The first small power station was set up to generate hydroelectric power in the third gorge below Victoria Falls in 1938.

It's easy to imagine how the immense force of these spectacular waterfalls can mold the land below. Over the course of centuries, even gentle flows of running water have sculpted many of the distinctive landforms we see around us by eroding land and depositing sediment.

Victoria Falls, on the Zambezi River between Zambia and Zimbabwe, is one of the world's largest waterfalls.

CHAPTER OUTLINE

CHAPTER PLANNER ✓

- ❑ Study the picture and read the opening story.
- ❑ Scan the Learning Objectives in each section:
  p. 364 ❑   p. 370 ❑   p. 378 ❑
- ❑ Read the text and study all figures and visuals. Answer any questions.

Analyze key features

- ❑ Geography InSight, p. 364
- ❑ Where Geographers Click, p. 366
- ❑ Video Explorations, p. 370
- ❑ Process Diagram p. 373 ❑   p. 377 ❑
- ❑ What a Geographer Sees, p. 382
- ❑ Stop: Answer the Concept Checks before you go on.
  p. 370 ❑   p. 378 ❑   p. 384 ❑

End of chapter

- ❑ Review the Summary and Key Terms.
- ❑ Answer the Critical and Creative Thinking Questions.
- ❑ Answer What is happening in this picture?
- ❑ Complete the Self-Test and check your answers.

Erosion, Transportation, and Deposition

LEARNING OBJECTIVES

1. **Explain** how slopes erode.
2. **Identify** three mechanisms by which streams erode.
3. **Explain** how streams transport sediment.
4. **Describe** the situations when deposition occurs.

Most of the world's land surface has been sculpted by running water. Water is involved in three closely related processes—erosion, transportation, and deposition. Mineral materials, from bedrock or regolith, are removed from slopes and stream channels by erosion, carving out gorges and valleys. These

Geography InSight

Erosional and depositional landforms
• Figure 12.1

After a crustal block is uplifted by plate tectonic activity, it is shaped by running water, which erodes sediments in some places and deposits them in others. Valleys form as rock is weathered and then eroded away by fluvial agents.

a. Peaks and ravines
Erosion by water, coupled with mass wasting, has carved out this ravine in a biosphere reserve on the Kamchatka Peninsula in eastern Russia.

b. Fans
Deposition of sediment by a stream has formed these alluvial fans in Wrangell-Saint Elias National Park in Alaska.

particles are then transported by water, either in solution as ions or as sediment of many sizes. The sediments are finally deposited downstream, where they build up, forming plains, levees, fans, and deltas. Waves, glacial ice, and wind also carve out landforms, but these processes are restricted to certain areas on the globe, as we will see in later chapters.

The landforms shaped by the progressive removal of bedrock are erosional landforms. Fragments of soil, regolith, and bedrock that are removed from the parent rock mass are transported and deposited elsewhere, making an entirely different set of surface features—the depositional landforms (**Figure 12.1**).

c. Canyons
Plunging down the slope of South Africa's great Eastern Escarpment, the Blyde River eroded this steep, colorful canyon in flat-lying sedimentary rocks.

Peak

Erosional landforms

Ravine

Canyon

Fan

Floodplain

Depositional landforms

d. Floodplain
The Kustatan River, Alaska, carrying sediment-laden water from nearby glaciers, deposited this floodplain on its way to Cook Inlet, near Anchorage.

Slope Erosion

Fluvial erosion starts on the uplands as soil erosion (see *Where Geographers Click*). When falling raindrops hit bare soil, their force lifts soil particles, which fall back into new positions, creating splash erosion (**Figure 12.2**). A torrential rainstorm can disturb as much as 225 metric tons of soil per hectare (about 100 U.S. tons per acre). On a sloping ground surface, splash erosion shifts the soil slowly downhill. The soil surface also becomes much less able to absorb water. This important effect occurs because the natural soil openings become sealed by particles shifted by raindrop splash. Because water cannot infiltrate the soil as easily, a much greater depth of overland flow can be triggered from a smaller amount of rain. This intensifies the rate of soil erosion.

Recall from Chapter 11 that some precipitation infiltrates the surface. When rainfall is heavy or the ground is saturated, however, water begins to flow across the surface as overland flow. Overland flow removes the soil in thin uniform layers through sheet erosion. Where land slopes are steep, runoff from torrential rains is even more destructive. Rill erosion scores many closely spaced channels, or **rills**, into the soil and regolith. Over time, these rills can join together to make still larger channels. These deepen rapidly, turning into **gullies**—steep-walled, canyon-like trenches whose upper ends grow progressively upslope (**Figure 12.3**).

Soil erosion by rain splash • Figure 12.2

A large raindrop lands on a wet soil surface, producing a miniature crater (right). Grains of clay and silt are thrown into the air, and the soil surface is disturbed.

Rills and gullies • Figure 12.3

Rill erosion represents the starting point for gullies.

a. Rills begin with small channels or rivulets of water following a gravitational path downslope.

Global Locator

Katmandu, Nepal

NATIONAL GEOGRAPHIC

b. Deforestation on these mountain slopes near Katmandu, Nepal, has led to rapid erosion and gullying.

Soil particles picked up by overland flow are carried downslope until the surface slope meets the valley bottom. The deposited particles accumulate in a thickening layer known as colluvium. Because this deposit is built by overland flow, it is distributed in sheets, making it difficult to notice unless it eventually buries fence posts or tree trunks.

Overland flow picks up mineral particles ranging in size from fine clay to coarse sand or even gravel. The size of particles removed depends on how fast the flow moves and how tightly plant roots and leaves hold down the soil. Under stable natural conditions in a humid climate, the erosion rate is slow enough to allow soil to develop normally. Each year, a small amount of soil is washed away, and a small amount of solid rock material is turned into regolith and soil.

In arid or semiarid climates, or where land has been cleared for cultivation, erosion can be rapid and severe. If there is no foliage to intercept rain and no ground cover from fallen leaves and stems, raindrops fall directly on the mineral soil, eroding it much faster than it can be formed. Some natural events, such as forest fires, also speed up soil erosion.

Destroying vegetation also reduces the ground surface's resistance to erosion under overland flow. Even deep layers of overland flow can cause only a little soil erosion on slopes that have a protective covering of grass sod. This is because the grass stems are tough and elastic, creating friction with the moving water and taking up the water's gravitational energy. Without such a cover, the water can easily dislodge soil grains, sweeping them downslope. Ultimately, accelerated soil erosion creates a rugged, barren topography.

Stream Erosion

Once overland flow is channeled into a stream, it can erode in various ways, depending on the nature of the channel materials, the speed and volume of the flow, and what kind of debris the stream flow is carrying. First, the flowing water drags on the bed and banks and also forces particles to hit the bed and banks. This form of erosion, called **hydraulic action**, can rapidly erode loose alluvial materials, such as gravel, sand, silt, and clay, when river flow is high. Erosion then undermines the banks of the stream, causing large masses of loose material to slump into the river, where the particles are quickly separated and carried downstream.

Second, as the water picks up debris, it further breaks down the stream bed through mechanical erosion. Where rock fragments carried by the swift current strike against bedrock channel walls and floors, they knock off chips of rock. This process of mechanical wear is called **abrasion**. In bedrock that is too strong to be eroded by simple hydraulic action, abrasion is the main method of erosion. A striking example of abrasion is the erosion of a pothole (**Figure 12.4**). As the rock fragments move downstream, they become rounded. Cobbles and boulders roll along the stream bed, crushing and grinding the smaller grains, producing a wide assortment of grain sizes.

Third, chemical weathering can remove rock from the stream channel by acid reactions and solution, a process called **corrosion**. We see this type of chemical weathering in limestone, in particular, which often develops cupped and fluted surfaces where it is being actively dissolved in the stream bed. This is the same solution process that occurs when ground water dissolves limestone by carbonation, as described in Chapter 10.

Potholes • Figure 12.4

These potholes in river bedrock were created by abrasion on the bed of a swift mountain stream. Potholes form when grinding stones are caught in shallow depressions in the stream bed. Flowing water spins the stones around, carving holes in the rock.

Streams carry their load as dissolved, suspended, and bed load.

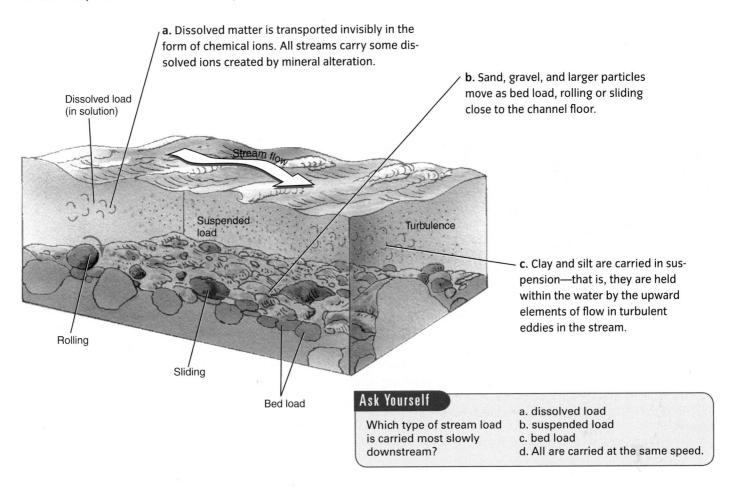

a. Dissolved matter is transported invisibly in the form of chemical ions. All streams carry some dissolved ions created by mineral alteration.

Dissolved load (in solution)

b. Sand, gravel, and larger particles move as bed load, rolling or sliding close to the channel floor.

Stream flow

Suspended load

Turbulence

c. Clay and silt are carried in suspension—that is, they are held within the water by the upward elements of flow in turbulent eddies in the stream.

Rolling

Sliding

Bed load

Ask Yourself

Which type of stream load is carried most slowly downstream?

a. dissolved load
b. suspended load
c. bed load
d. All are carried at the same speed.

Stream Transportation

stream load The solid matter carried by a stream in dissolved form, in suspension, and as bed load.

The solid matter carried by a stream is the **stream load**. Stream load is carried in three ways: as dissolved load, suspended load, or as bed load (**Figure 12.5**). Generally, most of the stream load is carried in suspension. A large river such as the Mississippi, for example, carries as much as 90% of its load in suspension.

Stream capacity measures the maximum solid load of debris—including bed load and suspended load—that a stream can carry. Capacity is given in units of metric tons per day passing downstream at a given location. A stream's capacity increases sharply as its velocity and discharge rise, such as during a flood. This is because there is more water volume to hold sediment and because swifter currents are more turbulent and have more energy to hold sediment in suspension.

The capacity to move the larger particles of the bed load also increases with velocity because faster-moving water drags against the bed harder. In fact, the capacity to move bed load increases according to the third to fourth power of the velocity. In other words, if a stream's velocity is doubled in times of flood, its ability to transport bed load will increase from 8 to 16 times.

Deposition

A stream carries its load downslope toward a valley, a lake, or an ocean. Along that journey, whenever a stream's load exceeds its capacity, it deposits some of its load. This deposited sediment is called **alluvium**.

alluvium Sediment laid by a stream that is found in a stream channel or in low parts of a stream valley subject to flooding.

Deposition typically occurs where the velocity of stream flow decreases. For example, deposition occurs along stream banks when the stream flow slows down on the inside of a bend in the channel. During flooding, fast-moving flood waters slow down and spread out over floodplains, depositing alluvium in layers

Video Explorations

The Power of Water

Harnessing the power of water using dams could be viewed as an environmentally safe, pollution-free, and renewable energy source. However, these systems can have many negative side effects.

As shown in this video, the Columbia River system in the Pacific Northwest has been heavily modified through the construction of numerous dams. These dams have already reduced the populations of several salmon species almost to the point of extinction, and they have caused environmental damage to the river ecosystem.

on the valley floor. Fine sediment, rich in organic matter, can increase soil fertility, but sometimes flooding leaves behind sterile layers of sand or gravel.

Dams represent a special case for sediment deposition and can have a variety of negative influences on fluvial systems (see *Video Explorations*). Where earthen or concrete barriers block the stream flow, transported sediment quickly settles at the base of the dam. Sediment that would otherwise continue downstream and settle out in floodplains continuously fills in the reservoir bottom. Eventually, the sediment may displace enough water to render the original dam inoperable for water storage or electrical generation. Costly dredging is the only remedy for removing depositional sediment.

CONCEPT CHECK STOP

1. **How** and why does deforestation affect slope erosion?
2. **What** factors determine how rapidly a stream can erode its channel?
3. **How** and why does stream capacity change during flooding?
4. **Why** does deposition occur along the banks at river bends?

Stream Gradation and Evolution

LEARNING OBJECTIVES

1. **Define** stream gradation.
2. **Describe** how a stream valley evolves over time.
3. **Describe** the landscapes associated with stream rejuvenation.
4. **Explain** the equilibrium model of landscape evolution.

Most major stream systems have experienced thousands of years of runoff, erosion, and deposition. Now that you have seen how these processes occur, let us take a closer look at how they shape the evolution of streams over time.

Stream Gradation

The downhill flow of water and sediment in stream channels is governed by basic physical laws. The water falling on a mountain landscape, high above the ocean, has potential energy induced by gravity, and as it flows downhill in a stream, it expends the potential energy as work. The work is done by friction on the bed and banks, which erodes sediment. Stones and boulders of bed load, pushed by the force of the water, collide, fracture, and break apart, doing more work. Energy is also expended in the internal friction of the turbulent flow of sediment-laden water.

The amount of work that a stream does within a particular length is determined largely by the potential energy available—that is, the drop in elevation within that length. But exactly how that work is done depends on a lot of different factors. For example, in a stretch with a deep and narrow channel of swiftly moving, clear water, a lot of energy may be expended in moving bed load. In a stretch with a wide, shallow channel of slowly moving sediment-laden water, the stream may expend most of its energy in internal friction and by eroding fine sediment from its bed and banks. Moreover, the amount of flowing water in the stream and the amount and type of sediment it carries are always changing, as precipitation brings flooding or as mass wasting and overland flow provide new inputs of sediment of different sizes. Although the laws of physics govern the motion of every particle of water or sediment, there are many possibilities for the shapes of channels and the slopes they maintain.

Through a number of feedback mechanisms, a stream can develop a more-or-less stable state, in which its long-term capacity to transport sediment is matched by the long-term rate at which it receives sediment. Channel shapes and slopes, which determine the transport rate, remain about the same over time. In this equilibrium situation, the stream is known as a **graded stream**. Streams do not reach a graded condition until they are flowing on alluvium—that is, material that the stream has already moved at least once and can move again, under the proper conditions (**Figure 12.6**).

> **graded stream** A stream with a gradient adjusted to carry its average sediment load.

Channel slope and load • Figure 12.6

Over time, a stream channel adjusts its gradient to transport its particular sediment load.

a. A graded stream carrying coarse sediment
Riley Creek, near the entrance to Denali National Park, Alaska, has a steep slope adjusted to carrying cobbles and larger stones, which are visible on stream banks and gravel bars.

b. A graded stream carrying fine sediment
The Avon River, located in southern England, has a shallow slope adjusted to carrying sand and silt, which can be readily seen on the river banks.

These profiles of elevation and distance along two rivers show the slope of each river at various points along its course. High in the Rocky Mountains, the slopes of the Arkansas River and its tributary, the Canadian River, are relatively steep and somewhat irregular. In the middle and lower portions, where the rivers are graded, the slopes are shallower and smoother. At the junction of the Arkansas with the Mississippi River, the slope is nearly flat.

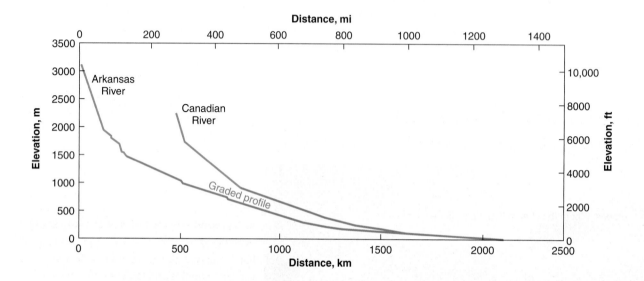

Put It Together

Review Figure 12.6 and answer this question.
Which stream will transport coarser sediment between an elevation of 1000 and 1500 m?

a. Arkansas River
b. Canadian River
c. We cannot determine this based on the information given.
d. Both rivers will be equally effective at transporting coarse sediment.

A key characteristic of a graded stream is that its slope decreases fairly smoothly in a downstream direction (**Figure 12.7**). As the river gets larger, it transports sediment more efficiently and so requires a lesser slope to move the sediment. In some cases, the average size of bed load particles decreases downstream, and less gradient is required to move the load.

meanders The sinuous bends of a graded stream flowing in the alluvial deposit of a floodplain.

Streams flowing on alluvium often develop characteristic channel patterns. For example, wide and shallow streams often develop a channel of broad, sweeping curves, called **meanders**. In other situations, stream flow is divided into multiple threads, providing a **braided channel**. We will return to these channel forms later in the chapter.

Evolution of Stream Valleys

Through the process of gradation, streams and their valleys evolve in a predictable way (**Figure 12.8**). In the early stages of stream evolution, streams include sudden changes in gradient, where faulting and tectonic uplift lead to a sudden drop-off, or where the resistance of stream bed materials suddenly changes. Points where the gradient of the stream changes abruptly, such as waterfalls and rapids, are called **nickpoints**. Stream flow is particularly fast and turbulent at nickpoints, so these segments of the stream are more rapidly eroded back to the average gradient of the stream, in a process called **downcutting**.

In some situations, nickpoints can migrate upstream as a result of the erosive action of the fast-moving water. For example, as water plunges at high velocity over the

Evolution of a graded stream and its valley • Figure 12.8

 THE PLANNER

Pictured here is a situation in which rapid tectonic uplift raises a highland landscape and then slows or stops. The newly created landscape is then eroded by streams and mass wasting. The process may take millions of years to create a broad valley in a landscape of low relief.

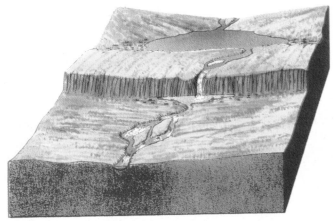

1 The flow of this ungraded stream is faster at the waterfalls and rapids, so these steeper segments of the stream are rapidly eroded. Where the stream flow slows (for example, at ponds and lakes), sediment is deposited.

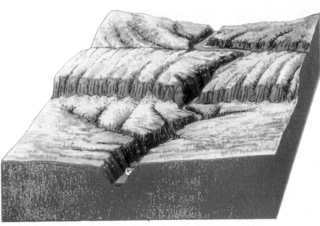

2 Waterfalls and rapids have eroded until their gradient is closer to the stream's average gradient. The lakes and ponds have drained and disappeared.

3 The river begins to wander sideways, eroding the side slopes, creating a curving path. Alluvium accumulates on the inside of each bend.

4 As the stream flow continues to erode the banks, the channel develops sweeping meanders. Over time, the migration of these curves downstream creates a floodplain of flat land between steep bluffs. The stream has achieved a graded condition.

Direction of migration

Nickpoint migration • Figure 12.9

Niagara Falls has migrated more than 11 km (7 mi) upstream from its original position of about 11,000 years ago, leaving behind a steep gorge. At the present time, the river's flow is divided into two channels. The near channel in the photo provides the horseshoe-shaped Canadian Falls, and the far channel provides the straighter American Falls.

edge of a waterfall, it exerts tremendous energy at the base of the falls, carving out a deep pool. The churning, oxygenated water may also enhance the chemical weathering of this rock. The pool gradually undercuts the cliff base of the waterfall, and eventually the top edge of the cliff collapses. This collapse moves the waterfall farther upstream (**Figure 12.9**). Debris from the collapsed cliff accumulates at the base of the waterfall, reducing the gradient.

A stream may erode its streambed rapidly through downcutting in early stages of its evolution, but eventually the balance between erosion and deposition stabilizes.

> **base level** The lower limit to how far a stream can erode its bed.

There is a lower limit to how far a stream can erode its bed, called its **base level**. The hypothetical base level for all streams is sea level, because once stream flow reaches the sea, gravity can carry it no farther downhill. However, some streams have base levels much higher than sea level, such as the level of a lake or reservoir into which the stream flows and is temporarily stored.

Stream Rejuvenation

Tectonic uplift sometimes raises the gradient of portions of a graded stream. For example, faulting may push up blocks of crust to form waterfalls and rapids. With new segments that are steeper, the stream begins a new cycle of downcutting. Such streams are said to be **rejuvenated streams**.

Entrenched meanders When a broadly meandering stream is uplifted by rapid tectonic activity, the uplift increases the river's gradient and, in turn, its velocity, so that it cuts downward into the bedrock below. This forms a steep-walled inner gorge. On either side of the gorge there will be the former floodplain, now a flat terrace high above river level. Any river deposits left on the terrace are rapidly stripped off by runoff because floods no longer reach the terraces to restore eroded sediment.

The meanders become impressed into the bedrock, passing on their meandering pattern to the inner gorge. We call these sinuous bends **entrenched meanders** (**Figure 12.10**) to distinguish them from the floodplain meanders of an alluvial river.

Entrenched meanders • Figure 12.10

Tectonic uplift of a meandering stream produces entrenched meanders.

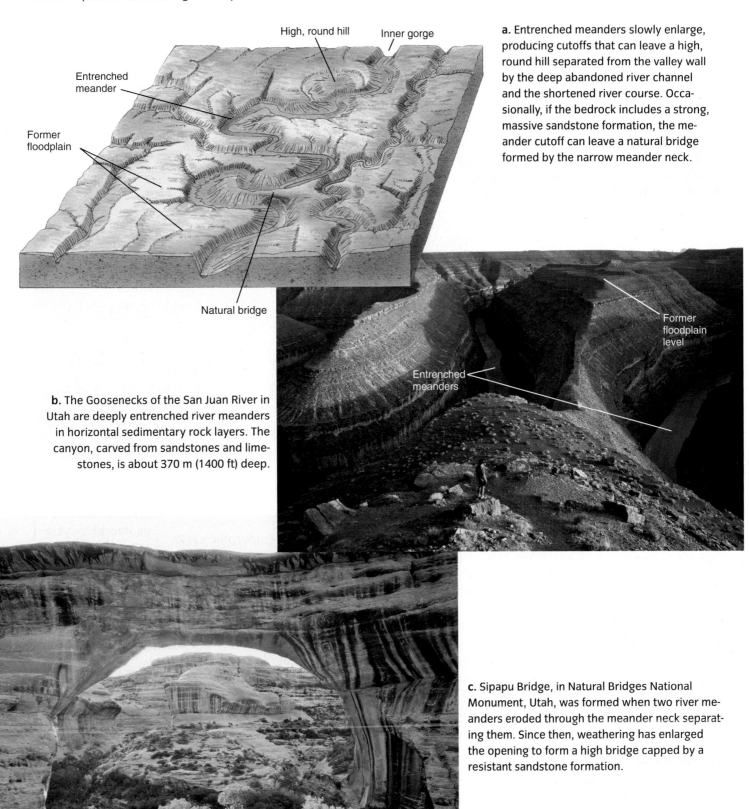

High, round hill

Inner gorge

Entrenched meander

Former floodplain

Natural bridge

a. Entrenched meanders slowly enlarge, producing cutoffs that can leave a high, round hill separated from the valley wall by the deep abandoned river channel and the shortened river course. Occasionally, if the bedrock includes a strong, massive sandstone formation, the meander cutoff can leave a natural bridge formed by the narrow meander neck.

Former floodplain level

Entrenched meanders

b. The Goosenecks of the San Juan River in Utah are deeply entrenched river meanders in horizontal sedimentary rock layers. The canyon, carved from sandstones and limestones, is about 370 m (1400 ft) deep.

c. Sipapu Bridge, in Natural Bridges National Monument, Utah, was formed when two river meanders eroded through the meander neck separating them. Since then, weathering has enlarged the opening to form a high bridge capped by a resistant sandstone formation.

Alluvial terraces • Figure 12.11

Alluvial terraces are the remnants of abandoned floodplains that are left behind as a river carves out a new, lower floodplain.

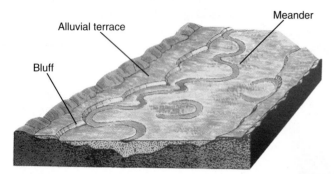

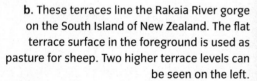

a. The stream excavates alluvium from the floodplain and carries it downstream. This leaves step-like alluvial surfaces on both sides of the valley. The treads of these steps are called alluvial terraces.

b. These terraces line the Rakaia River gorge on the South Island of New Zealand. The flat terrace surface in the foreground is used as pasture for sheep. Two higher terrace levels can be seen on the left.

alluvial terrace
A bench-like landform carved by a stream during degradation.

Alluvial terraces As a river lowers its bed through time, it can leave behind parts of its floodplain at a higher level as a series of **alluvial terraces** (Figure 12.11). Alluvial terraces attract human settlement because—unlike valley-bottom floodplains—they aren't subject to annual flooding. The terraces are easily tilled and make prime agricultural land. Downcutting and terrace formation can result from a period of uplift or from a change in climate affecting the amount of runoff and sediment the river receives, among other reasons.

Theories of Landscape Evolution

As we have seen, the landscapes formed by flowing water range from mountain regions of steep slopes and rugged peaks, to regions of gentle hills and valleys, to nearly flat plains that stretch from horizon to horizon. We can think of these constantly changing landscapes as stages of evolution in a cycle, often called the **geomorphic cycle**, that begins with rapid uplift by plate tectonic activity and follows with lengthy erosion by streams in a graded condition until again uplifted by tectonic activity (**Figure 12.12**). This cycle, originally called the geographic cycle, was first proposed by William Morris Davis, a prominent geographer and geomorphologist of the late 19th and early 20th centuries.

The geomorphic cycle is useful for describing landscape evolution over very long periods of time, but it does little to explain the diversity of the features observed in real landscapes. Most geomorphologists think of landforms and landscapes in terms of equilibrium. This approach explains a landform as the product of forces acting upon it, including both forces of uplift and denudation activities that wear down rocks.

geomorphic cycle
A theory which suggests that landscapes go through stages of development and that the rejuvenation of landscapes arises from tectonic uplift of the land.

Idealized stages of the geomorphic cycle • Figure 12.12

The cycle of the evolution of a land mass after uplift shows how a nearly flat, eroded plain is uplifted and eroded over geologic time until it returns to a similar nearly flat surface.

Uplift

1 Uplift
The cycle begins with the rapid uplift of a low, nearly flat erosion surface known as a peneplain to an elevation well above base level. (The prefix *pene-* means "almost.")

Base level

Base level

2 Youthful stage
The streams draining the uplifted plain, now with steeper slopes, erode canyons and gorges. Remnants of the original surface remain on the divides between streams.

5 Old age
The gradients of streams and valley-side slopes gradually lower. The land surface slowly approaches sea level. After millions of years, the land surface is reduced to a low, gently rolling peneplain, which is very close to the base level. Eventually the surface is uplifted again, rejuvenating the landscape.

4 Late mature stage
As time passes, the streams develop broad floodplains. Mass wasting reduces slopes and ridges to rounded forms.

3 Early mature stage
The landscape is now extremely rugged. Remnants of the former peneplain are gone. Rivers are now graded and just starting to build alluvial floodplains.

Base level

Ask Yourself

Which stage of the geomorphic cycle would have the broadest floodplains?

a. uplift
b. youthful stage
c. early mature stage
d. late mature stage
e. old age

One strength of this viewpoint is that we can take into account the characteristics of the rock material. Thus, we find steep slopes and high relief where the underlying rock is strong and highly resistant to erosion. Even a rugged landscape that appears to be quite "young" can be in a long-lived equilibrium state in which hill slopes and stream gradients remain steep in order to maintain a graded condition while eroding the strong rocks of the region.

Another problem with Davis's geomorphic cycle is that it applies only where the land surface is stable over long periods of time after the initial uplift. But we know from our study of plate tectonics that crustal movements are frequent on the geologic timescale. Few regions of the land surface remain untouched by tectonic forces in the long run. Recall also that continental lithosphere floats on a soft asthenosphere. As layer upon layer of rock is stripped from a land mass by erosion, the land

mass becomes lighter and is buoyed upward. A better model, then, is one of uplift as an ongoing process to which erosional processes are constantly adjusting rather than as a sudden event followed by denudation.

CONCEPT CHECK

1. **What** are the differences between a meandering stream and a braided stream?

2. **How** and why do nickpoints migrate upstream?

3. **Why** are alluvial terraces good places for settlement?

4. **What** is one limitation of the idealized geomorphic cycle?

Fluvial Landforms

LEARNING OBJECTIVES

1. **Describe** meandering streams and floodplains.

2. **Explain** how braided streams are formed.

3. **Describe** the formation of arid climate fluvial landforms.

4. **Discuss** the formation and features of deltas.

ow that you've seen how landscapes are shaped by water over time, let's take a closer look at several of the landforms associated with them, called **fluvial landforms**.

Meandering Streams and Floodplains

Meandering is a characteristic of graded rivers carrying substantial loads of sediment of varying sizes. In a meander bend, water moves with greater velocity on the outside of the bend, and the channel is deeper. The greater velocity erodes the floodplain sediment, creating a **cut bank** and causing the bend to grow outward or in a downstream direction (**Figure 12.13**). On the inside of the bend, the flow is slower, and sediment accumulates in a **point bar**. Sometimes the stream flow cuts off a meander loop by eroding through the narrow portion of the meander neck. After the cutoff, silt and sand

Growth and movement of meanders
• Figure 12.13 _____

A meander grows outward and migrates downstream as the river bank is eroded on the outside of the meander, forming a cut bank, while deposition occurs at the inside bank, creating a point bar. When meanders touch, the river quickly takes the shortcut, leaving a meander scar or an ox-bow lake.

Old ox-bow lake

Neck

Old channel filled in with sediment and vegetation

Deposition

Erosion

Ox-bow lakes

Floodplains • Figure 12.14

Meandering streams leave their banks every few years, flooding the surrounding floodplain.

a. Periodic flooding creates characteristic landforms on a floodplain.

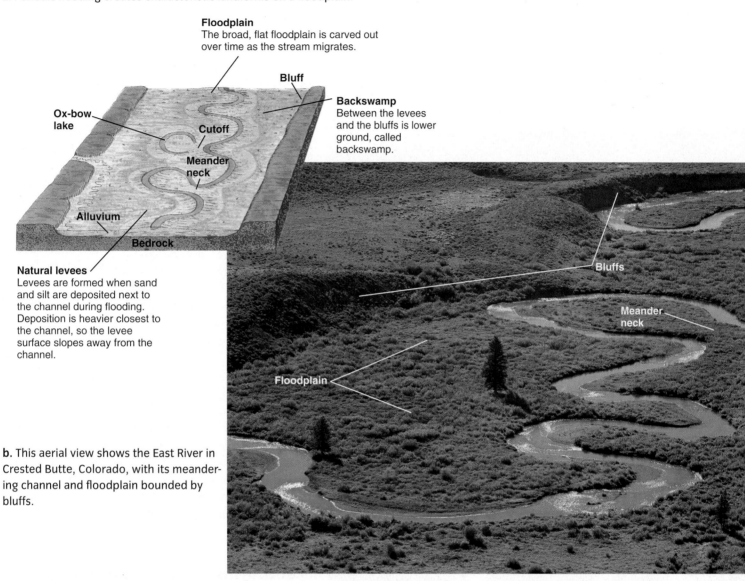

Floodplain
The broad, flat floodplain is carved out over time as the stream migrates.

Bluff

Backswamp
Between the levees and the bluffs is lower ground, called backswamp.

Ox-bow lake

Cutoff

Meander neck

Alluvium

Bedrock

Natural levees
Levees are formed when sand and silt are deposited next to the channel during flooding. Deposition is heavier closest to the channel, so the levee surface slopes away from the channel.

Bluffs

Meander neck

Floodplain

b. This aerial view shows the East River in Crested Butte, Colorado, with its meandering channel and floodplain bounded by bluffs.

---

ox-bow lake A crescent-shaped lake representing the abandoned channel left by the cutoff of a meander.

are deposited across the ends of the former channel, producing an **ox-bow lake**. Eventually the lake fills in with sediment and becomes first a swamp and then a **meander scar**. Cutoffs are sometimes human induced to straighten the stream channel and make it more convenient for transport.

Over time, the growth and movement of meanders form a broad, flat floodplain of alluvium. The floodplain is normally flooded each year or two, when stream flow increases and the river leaves its banks. As the velocity of the water flowing away from the bank and across the floodplain decreases, the sediment begins to settle out. Sand and silt accumulate first, building up **natural levees** along the channel (**Figure 12.14**). Farther away, fine sediment settles out of the nearly stagnant water, accumulating between the levees and the bluffs that bound the floodplain—an area known as the **backswamp**.

natural levee The belt of higher ground paralleling a meandering river on both sides of a stream channel and built up by deposition of sediment during flooding.

Periodic flooding not only adds silt, clay, and organic matter to the floodplain soils but also infuses the floodplain with dissolved minerals. The resupply of nutrients keeps floodplain soils fertile and productive, and the floodplains of some alluvial rivers—such as the Nile, Tigris, Euphrates, and Yellow rivers—have been farmed for thousands of years.

Braided Streams

In streams carrying a large sediment load, including a large proportion of bed load, the channel can take a braided form, with many individual threads of flow separated by shallow bars of sand or gravel. We refer to a stream with this type of channel as a **braided stream** (**Figure 12.15**). Braided channels are more likely to form in streams with large fluctuations in discharge. When the river is high, bed load is set in motion. Bars are eroded, and bed load moves downstream. As the water level falls and velocity decreases, new sand and gravel bars form. The low flow following the flood cannot move the bars, so the water flows in channels around them.

Braided channels are often found where fluvial sediments are accumulating and floodplains are being built up. This process is referred to as **aggradation**. It can occur in a number of ways. When a stream receives a large input of sediment—too much to handle, given its current capacity—it will accumulate sediment in the floodplain. Streams that drain glaciers, which receive abundant coarse sediment coupled with high flows during the season of ice melt, are typical examples. Another cause of aggradation is change of base level. When base level rises, it reduces the slope of the river, also reducing the river's capacity to transport sediment. As a result, sediment accumulates in the floodplain.

> **aggradation** The raising of stream channel altitude by continued deposition of bed load.

Braided stream • Figure 12.15

The braided channel of the Chitina River, Wrangell Mountains, Alaska, shows many channels separating and converging on a floodplain filled with coarse rock debris left by a modern valley glacier at the head of the valley.

Landforms of Flat-Lying Rocks in Arid Climates

Running water, along with winds, shapes landscapes even in regions with low moisture and precipitation. In arid climates, there isn't much vegetation to protect the ground, so overland flow is especially effective at sculpting landforms. Deserts feature many interesting landforms, but here we focus on some of the spectacular landforms of the American Southwest, where we find flat-lying rock strata of varying resistance to erosion (**Figure 12.16**).

When flat-lying rocks are eroded in an arid climate, successive rock layers are stripped or washed away, leaving behind a broad platform, or plateau.

Landforms of flat-lying rocks in arid climates • Figure 12.16

In arid climates, where there is little vegetative cover, mass wasting and overland flow carve out distinctive landforms.

a. Erosion by wind and water create plateaus, mesas, and buttes from resistant rock layers.

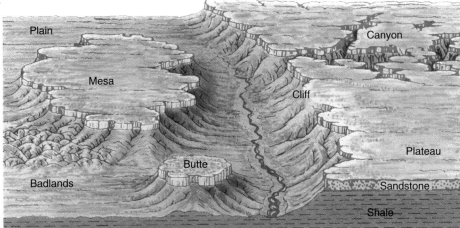

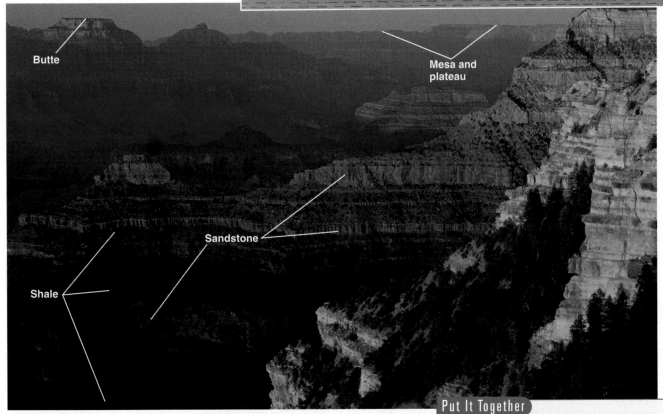

b. Arizona's Grand Canyon was carved by the Colorado River and its tributaries into a plateau of flat-lying sedimentary rocks.

Put It Together

Review Figure 12.10 and answer this question. Which of the landforms in Figure 12.16a could include an entrenched meander?

a. plain
b. badlands
c. butte
d. canyon

Such plateaus may be capped by hard rock layers, creating a sheer rock wall, or cliff, at the edge of the resistant rock layer. At the base of the cliff is an inclined slope, which flattens out into a plain beyond. As the weak clay or shale formations exposed at the cliff base are washed away, the rock in the upper cliff face repeatedly breaks away along vertical fractures. Cliff retreat produces a *mesa*—a table-topped plateau bordered on all sides by cliffs. As a mesa is reduced in area by retreat of the rimming cliffs, it maintains its flat top. Eventually, it becomes a small, steep-sided hill known as a *butte*. Further erosion may produce a single, tall column, called a *pinnacle*, before the landform is totally consumed.

WHAT A GEOGRAPHER SEES

Badlands

A traveler would have a hard time making his way through a landscape such as the one shown in this striking photo at Zabriskie Point, Death Valley National Monument, California (Figure a). These regions are known as badlands precisely because of the difficult terrain they presented to early travelers.

The Native American Lakota tribe called them *mako sica*—literally "bad lands"—and French trappers called them *les mauvaises terres à traverser*, or "the bad lands to cross."

A geographer would know that the deep gullies are a result of rapid erosion of the sparsely vegetated and clay-rich sediment. When it rains, overland flow from slopes sweeps clay particles into the stream channels (Figure b).

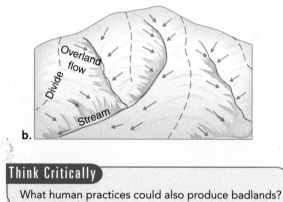

b.

Think Critically

What human practices could also produce badlands?

a.

Where clay is exposed at the surface, erosion is very rapid, and unstable slopes are soon dissected into **badlands** (see *What a Geographer Sees*). Erosion occurs too rapidly for plants to take hold, so no soil develops. A maze of small stream channels with steep slopes results.

Many deserts have closed, lowland basins of broad, arid plains. Often we find salt-rich lakes or former lake deposits in these basins. Streams heading in adjacent hills or mountains drain into these depressions, and when they are active, they contribute sediment, water, and dissolved salts to the valley bottom. But the water soon evaporates in the arid climate, leaving behind the salt and sediment. The result is a very flat surface called a **playa** (**Figure 12.17**). In the southwestern United States, many playa

---

Nearby uplands are rugged mountain masses dissected into canyons with steep, rocky walls.

This great uplifted fault block is the eastern face of the Inyo Mountains.

Desert playa
• Figure 12.17

Racetrack Playa, a flat, white plain, is surrounded by rugged mountains. This desert valley lies in the northern part of the Panamint Range, west of Death Valley, California.

The playa occupies the floor of the central part of the basin.

deposits are the residue of large lakes that formed during cooler, wetter climate periods but are now completely evaporated.

alluvial fan A gently sloping, conical accumulation of alluvium deposited by a stream emerging from a rock gorge or rock notch.

Another very common arid climate landform is the **alluvial fan** (**Figure 12.18**). Alluvial fans are built by streams carrying heavy loads of coarse rock waste from a mountain or an upland region. Constrained by the hard rock of the mountain, the stream flows in a narrow gorge until it emerges on the plain. The channel abruptly widens, and its ability to carry sediment rapidly decreases due to loss of depth, infiltration of water into the fan sediments, and evaporation. In many cases, the stream breaks into a few smaller channels, called **distributaries**, that spread the water and sediment over the fan surface. Alluvial fans are also found in other climates where a constrained stream emerges onto the floodplain of a major river or lowland plain.

Deltas

Where a stream enters a body of standing water, such as a lake or an ocean, it first drops its bed load, forming a bar across the mouth of the river. Coarse suspended sediment, carried near the channel bottom, then begins to settle out. The flow commonly breaks into several distributaries as it crosses the bar and the newly deposited sediment. The flow of water and fine sediment is carried into the open water, where it maintains its velocity for a short distance. Eventually the flow mixes into the water body, and fine sediment settles out as well. If the freshwater flow mixes with salt water, the salts cause even the finest clay particles to clot

Alluvial fans • Figure 12.18 _____

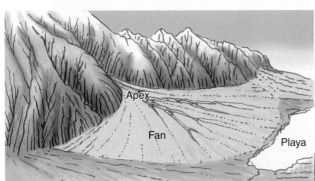

a. The apex of the fan is firmly fixed at the canyon mouth. The stream flow breaks up into distributaries that deposit sediment, forming the cone shape.

b. This vast alluvial fan extends out onto the floor of Death Valley. This fan was built by streams carrying rock waste from a great uplifted fault block, the Panamint Range, located to the west of the valley.

Delta dynamics • Figure 12.19

Earth scientists used the data from the Shuttle Radar Topography Mission (SRTM) to map the elevation of 33 deltas, including the Irrawaddy River Delta in Burma (Myanmar), to learn more about delta dynamics. The SRTM data combined with historic topographic maps show that the Irrawaddy River Delta is sinking, as a result of river diversions from upriver dams and irrigation. Erosion from farming and development is also cited as a cause of land loss in the delta, along with subsidence from oil, gas, and water drilling.

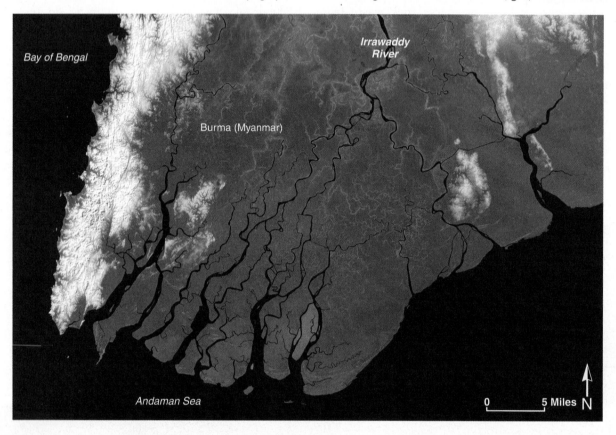

together and fall to the seafloor. The deposit built by this process is known as a **delta**.

> **delta** A sediment deposit built by a stream entering a body of standing water.

Deltas provide extensive flat and fertile lands, and most major river deltas have been cultivated since the earliest times of human history. The distributaries provide transportation corridors that link ocean and inland river traffic, and major port cities such as New Orleans on the Mississippi, Alexandria on the Nile, and Kolkata on the Ganges-Brahmaputra handle the trade.

Deltas can grow rapidly, at rates ranging from 3 m (about 10 ft) per year for the Nile River to 60 m (about 200 ft) per year for the Mississippi. Some cities and towns that were at river mouths several hundred years ago are today several kilometers inland.

Human activities have recently been linked to degradation of delta environments (**Figure 12.19**). In many locations, delta land is sinking due to compaction of sediment and removal of water, oil, and gas. Agricultural activities further accelerate loss of land with soil runoff. Upstream river water diversion projects remove from the river sediment that would otherwise replenish the delta landscape. Millions of people living in sinking deltas will need to develop adaptations as climate change increases the threat of flooding from storms and sea-level rise.

CONCEPT CHECK STOP

1. **How** is an ox-bow lake formed?
2. **Where** do braided streams typically form?
3. **What** causes the typical cone shape of an alluvial fan?
4. **How** is human activity affecting deltas?

Summary

1 Erosion, Transportation, and Deposition 364

• As shown in the diagram, running water erodes, transports, and deposits rock material, forming both erosional and depositional landforms.

Erosional and depositional landforms • Figure 12.1

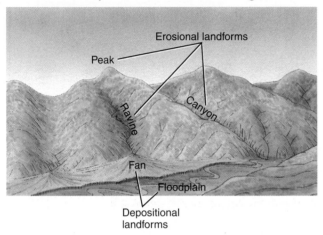

• Slope erosion occurs when overland flow sweeps away particles, forming **rills** and **gullies**. This erosion is accelerated when vegetation is removed.

• Stream channels carve into soft materials by **hydraulic action**. Where stream channels flow on bedrock, channels are deepened by **abrasion**. Bedrock in stream channels can also be lost by **corrosion**.

• **Stream load** consists of the sediment carried by a stream. **Stream capacity** to carry load increases greatly as water volume and velocity increase. In addition, velocity increases as stream gradient increases.

• When a stream's load is greater than its capacity, the stream deposits sediment in the form of **alluvium**. Deposition often occurs when stream velocity slows.

2 Stream Gradation and Evolution 370

• Streams gain the energy for their work when precipitation falls into the watershed above the stream and then flows downhill under gravity.

• **Graded streams** have developed channel shapes and slopes that are just able, over time, to transport the average sediment load the stream receives. The slope of a graded stream decreases smoothly in a downstream direction. Depending on their slope and sediment load, streams may form a channel with broad sweeping curves, called **meanders**, or a channel with multiple threads, called a **braided channel**.

• Ungraded streams rapidly erode steeper **nickpoints**, such as rapids and waterfalls, through **downcutting** until they are closer to the average gradient of the stream. Over time, a stream carves out a broad valley surrounded by steep bluffs. The **base level** of a stream is the lowest level to which it can erode its bed.

• A period of uplift or climate change leads to **rejuvenated streams**, where formerly graded streams resume downcutting, forming features such as **entrenched meanders** and **alluvial terraces**, as shown in the diagram.

Entrenched meanders • Figure 12.10

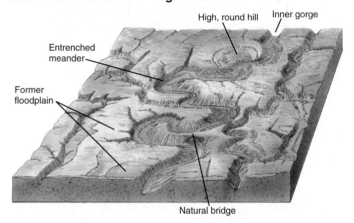

• The **geomorphic cycle** organizes fluvial landscapes in a cycle of uplift and subsequent erosion. The equilibrium approach views uplift and erosion as continuous processes.

3 Fluvial Landforms 378

- As meandering streams migrate through a valley, they carve out a broad floodplain that fills during periods of high water. **Fluvial landforms** associated with meandering streams include **cut banks**, **point bars**, **natural levees**, **ox-bow lakes**, **backswamps**, and **meander scars**.

- **Braided streams** form when a stream carries a large sediment load with abundant bed load. **Aggradation** is the process in which floodplains and other deposits are built up by deposition.

- Fluvial erosion carves arid landscapes of flat-lying rocks into plateaus, canyons, mesas, and buttes. **Playas** become lakes during wet periods and then evaporate to form flat basins of sedimentary and salt deposits.

- As shown in the diagram, **alluvial fans** occur when streams traveling downslope in a canyon or gorge emerge into an open plain and deposit sediment in a cone-shaped fan.

Alluvial fans • Figure 12.18

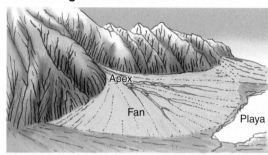

- **Deltas** are formed where streams empty into comparatively still water, such as a lake or an ocean. Human activities are linked with the subsidence of deltas.

Key Terms

- abrasion 368
- aggradation 380
- alluvial fan 383
- alluvial terrace 376
- alluvium 369
- backswamp 379
- badlands 382
- base level 374

- braided channel 372
- braided stream 380
- corrosion 368
- cut bank 378
- delta 384
- downcutting 372
- distributary 383
- entrenched meanders 374

- fluvial landform 378
- geomorphic cycle 376
- graded stream 371
- gully 366
- hydraulic action 368
- meander scar 379
- meanders 372
- natural levee 379

- nickpoint 372
- ox-bow lake 379
- playa 382
- point bar 378
- rejuvenated stream 374
- rill 366
- stream capacity 369
- stream load 369

Critical and Creative Thinking Questions

1. Compare erosional and depositional landforms. Sketch an example of each type.

2. Describe the importance of vegetation in controlling slope erosion.

3. Choose a river in your area. Research its course by using Google Earth or other aerial or satellite photos. Identify in this river at least two processes or landforms discussed in the chapter.

4. In the western United States, badlands got their name because they are difficult to traverse because of severe erosion by water. How can it be difficult to find water in a landscape completely shaped by running water?

5. Imagine that you are digging through the alluvial deposits of a stream that has dried up. What characteristics of the stream could you infer from knowing only the size and shape of the alluvium?

6. A river originates high in the Rocky Mountains, crosses the High Plains, flows through the agricultural regions of the Midwest, and finally reaches the sea. Describe the fluvial processes and landforms you might expect to find on a journey along the river from its headwaters to the ocean.

What is happening in this picture?

This aerial photo of the floodplain of the meandering Mississippi River was taken using infrared film, which displays green vegetation and agricultural crops as red patterns.

Think Critically

1. What are two examples of a curved characteristic landform of an alluvial floodplain in this image?
2. How do these features develop?
3. Describe where erosion and deposition are most likely to occur next in this aerial photo.

Self-Test

(Check your answers in the Appendix.)

1. Landforms shaped by _____ are described as fluvial landforms.
   a. glacial ice
   b. wave action
   c. denudation
   d. running water

2. Landforms that are shaped by progressive removal of the bedrock mass are _____ landforms.
   a. initial
   b. erosional
   c. fluvial
   d. ablation

3. In steeply sloped landscapes, a destructive form of soil erosion called _____ results in many closely spaced channels due to torrential rain episodes.
   a. rill erosion
   b. sheet erosion
   c. furrow erosion
   d. gully erosion

4. The term _____ is used to describe any stream-laid sediment deposit.
   a. colluvium
   b. alluvium
   c. fluvial
   d. sediment

5. The process of mechanical wear by the rolling of cobbles and boulders along the beds of streams is called _____.
   a. deflation
   b. ablation
   c. abrasion
   d. grinding

6. Chemical rock weathering processes such as acid reactions and solution collectively refer to a mechanism known as _____.
   a. erosion
   b. demineralization
   c. salinization
   d. corrosion

7. Streams carry sediment as they flow. On the diagram, label (a) suspended load and (b) bed load. What is the third type of load that streams carry?

8. The maximum solid load of debris that a stream can carry at a given discharge is a measure of the _____.
   a. stream suspension pattern
   b. deposition rate
   c. stream capacity
   d. stream velocity

9. In a graded stream, the capacity of a stream _____ the sediment load supplied to it.

a. exceeds

b. increases

c. erodes

d. equals

10. A gradual reduction in the channel gradient of a stream leads to _____.

a. a greater capacity to carry suspended load

b. an increase in dissolved load

c. a reduced ability of the stream to carry bed load

d. greater erosion at the mouth of the stream

11. An equilibrium condition in which the slopes of all stream channels form a coordinated network that is just able to carry the sediment load contributed by the drainage basin is referred to as a(n) _____.

a. braided stream

b. meandering stream

c. graded stream

d. entrenched stream

12. The diagram shows a variety of landforms created by a meandering stream. Label the following features: (a) a bluff, (b) the floodplain, (c) a meander neck, and (d) an ox-bow lake.

13. The points where the gradient of a stream changes abruptly are called _____.

a. braided channels

b. nickpoints

c. meanders

d. alluvial terraces

14. When clay is exposed at the surface, erosion is very rapid, and unstable slopes are soon dissected into _____.

a. badlands

b. playas

c. alluvial fans

d. deltas

15. The diagram shows a variety of landforms in an arid area. Label the following features: (a) a mesa, (b) badlands, (c) a butte, and (d) a canyon.

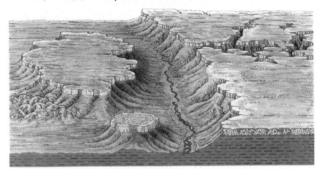

THE PLANNER ✓

Review your Chapter Planner on the chapter opener and check off your completed work.

Landforms Made by Wind and Waves

Singing sand dunes, as they are known, have been a part of desert folklore for centuries, but they aren't just fantasy. Some dunes are said to squeak, while others boom. The dunes are reported to sound like drum rolls or galloping horses, and even to be melodic at times. Dune songs have been reported to last for up to 15 minutes and can sound as loud as a low-flying airplane.

What causes these strange noises? A sand dune is a hill of loose sand shaped by the wind. Active dunes constantly change form under wind currents. Scientists believe that the songs are created by avalanches of sand on the dune. As the sand grains bounce down, they begin to move with the same frequency, setting up a hum. But the precise mechanism for generating such distinct sounds remains controversial. Not all sand dunes sing, and it's not clear why some dunes can sing while others can't—adding another layer of mystery.

Wind action is responsible for creating the immense and beautiful sand dunes—singing or not—that we find in deserts. Together with waves, the wind has also shaped coastlines around the world.

A guide leads a line of camels across the famous singing sands of Dunhuang, China.

CHAPTER PLANNER ✓

Wind Action

LEARNING OBJECTIVES

1. **Describe** wind deflation and abrasion and their effects.

2. **Discuss** how wind transports particles of different sizes.

This chapter brings together two different agents of erosion: wind and waves. Wind is the movement of air, a fluid of low density, and waves involve the movement of water, a fluid of much higher density. Because of this difference, the power of moving air to create landforms is generally much lower than the power of moving water as waves. On the other hand, there are vast desert areas of our planet with landforms shaped by wind, and waves act only on coastlines of oceans and large lakes. Yet the two agents are related because waves are created by wind blowing across a water surface. We will first examine how wind creates landforms.

Erosion by Wind

Wind performs two kinds of erosional work: deflation and abrasion. **Deflation** is the removal of particles, largely silt and clay, from the ground by wind. It acts on loose soil or sediment. Dry river courses, beaches, farmlands, and areas of recently formed glacial deposits are especially susceptible to deflation. In dry climates, much of the ground surface can be deflated because the soil or sediment is not held in place by vegetation.

> **deflation** The lifting and transport in turbulent suspension by wind of loose particles of soil or regolith from dry ground surfaces.

The ability of the wind to remove particles by deflation depends on the size of the particles. If there is a mixture of particles of different sizes on the ground, deflation removes the fine particles and leaves the coarser particles behind. As a result, rock fragments ranging in size from pebbles to small boulders become concentrated into a surface layer known as a **desert pavement** (**Figure 13.1**). The large fragments become closely fitted together,

Desert pavement • Figure 13.1

 THE PLANNER

After finer materials have been removed by deflation, the remaining coarser materials form a desert pavement.

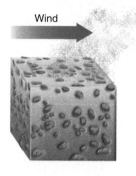

Wind

1 Newly deposited sediment contains a mixture of coarse and fine particles. Wind carries away the silt and clay and slowly reduces the volume of the deposit.

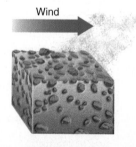

Wind

2 As the deflation continues, the surface lowers. The coarser particles—gravel, pebbles, and larger stones—are left behind and accumulate on the surface. Rare rainfall events may also remove finer particles by sheetflow.

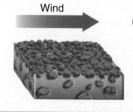

Wind

3 Eventually, the surface is covered by coarse particles, forming a desert pavement that resists further deflation or erosion.

Ask Yourself

If the size of the fine particles increases, the formation of desert pavement will _____.
a. occur more rapidly
b. occur more slowly
c. increase the depth of the pavement
d. lead to deflation more quickly

Wind-sculpted rocks • Figure 13.2

Wind abrasion blasts sand against exposed rocks to create rock sculptures of various sizes.

a. Wind abrasion can carve stones on a desert plain into shapes with flattened or grooved sides, creating a ventifact. In this example from the Mojave Desert, California, winds from two directions have created two facets with a sharp ridge between them.

b. In some types of layered rocks, wind abrasion can produce tongue- or teardrop-shaped ridges, called yardangs, oriented parallel to the prevailing wind direction. Although rare, they are quite striking. This example is near Gilf Kebir, in the western desert of Egypt.

sheltering the smaller particles—grains of sand, silt, and clay—that remain beneath. The pavement acts as an armor that protects the remaining fine particles from rapid removal by deflation or overland flow. This pavement is easily disturbed, however, by wheeled or tracked vehicles, which expose the finer particles underneath, resulting in renewed deflation and water erosion.

In drier plains regions, deflation can form a shallow basin called a *blowout*, especially where the grass cover has been broken or disturbed by grazing animals, for example. The size of the depression may range from a few meters (10 to 20 feet) to a kilometer (0.6 mile) or more in diameter, although it is usually only a few meters deep. A blowout can also form in a shallow depression in the plain where rains have filled the depression and created a shallow pond or lake. As the water evaporates, the mud bottom dries out and cracks, leaving small scales or pellets of dried mud. These particles are then lifted out by the wind.

The second process of wind erosion, **abrasion**, drives sand-sized particles against an exposed rock or soil surface, wearing down the surface by the impact of the particles. Wind abrasion is most active in a layer of about 10–40 centimeters (4–16 inches) above the surface. The weight of the sand grains prevents them from being lifted much higher into the air. Wooden utility poles on windswept plains often have protective metal sheathing or heaps of large stones placed around the bases. Without such protection, they would quickly be cut through at the base by wind abrasion.

Wind abrasion produces pits, grooves, and hollows in rock. Such wind-sculpted rocks are called *ventifacts* (**Figure 13.2a**). Common examples include rocks on desert plains that are given two or three smooth or grooved sides by wind abrasion. Another interesting form produced by wind abrasion is the *yardang*, a low tongue- or teardrop-shaped ridge (**Figure 13.2b**).

Transportation by Wind

The ability of winds to transport sediment depends on the strength and turbulence of the wind as well as the size of the sediment grains. The finest particles, those of clay and silt sizes, are lifted and raised into the air—sometimes to a height of 1000 m (about 3300 ft) or more, where they are carried in suspension by turbulence.

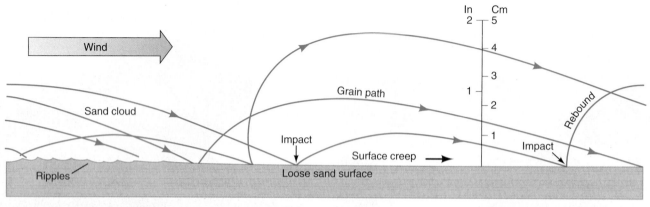

In | Cm
2 — 5
— 4
Wind
— 3
Grain path 1 —
— 2
Sand cloud
Rebound
— 1
Impact Impact
Surface creep →
Ripples Loose sand surface

Saltation • Figure 13.3

Wind picks up individual grains of sand and carries them downwind on a low arc. Most grains travel within a few centimeters of the surface, but some reach heights of a meter or so. Each collision of a grain with the surface creates a tiny crater, ejecting particles upward into the wind stream, which moves them downwind. At higher wind speeds, surface sand particles are rolled and pushed along as surface creep.

Sand grains are moved by **saltation**, a process in which sand grains fly in low arcs, perhaps a meter long, from one point to another (**Figure 13.3**). When they hit the surface, each collision forces other particles into the air, which are also carried downwind. At higher wind speeds, sand grains are also dragged along the ground, in the same way that a stream carries bedload. This movement is called *surface creep*.

Strong, turbulent winds blowing over barren surfaces can lift great quantities of fine dust into the air, forming a dense, high cloud called a *dust storm*. Vigorous saltation of coarse sand grains, driven by the strong wind, adds to the dust. Collisions disturb the soil, breaking down soil particles and releasing fine dust that can be carried upward by the turbulent winds.

A dust storm approaches as a dark cloud extending from the ground surface to heights of several thousand meters (**Figure 13.4**). Typically, the advancing cloud wall occurs along a rapidly moving cold front. Standing within

Dust storm • Figure 13.4

In a front reaching up to 1000 m (3300 ft) in altitude, a dust storm approaches an American facility in Iraq. It passed over in about 45 minutes, leaving a thick layer of dust in its wake.

Dust storm over Beijing • Figure 13.5

Springtime brings increasingly frequent dust storms into Beijing from loose silt blown off the growing desert regions of China's interior. Global warming is contributing to desertification and increasing seasonal winds, creating health-threatening conditions over the nation's capital city. This severe dust storm struck the city in March 2002.

the dust cloud, you are shrouded in deep gloom or even total darkness. Visibility is cut to a few meters, and a choking fine-grained dust penetrates everywhere. These small particles are a health hazard for the very young and the old, as well as for persons with sensitive lungs.

Severe deflation resulting from overgrazing or poor agricultural practices can cause significant dust storms, such as those that occurred during the American Dust Bowl of the 1930s. Cultivating short-grass prairie in a semiarid region makes the ground much more susceptible to deflation. Plowing disturbs the natural soil surface and grass cover, and in drought years, when vegetation dies out, the unprotected soil is easily eroded by wind action. That is why much of the Great Plains region of the United States has suffered dust storms generated by turbulent winds. Strong cold fronts frequently sweep over this area and lift dust high into the troposphere at times when soil moisture is low.

China's capital in Beijing continues to experience severe dust storms from loose silt blown from the Gobi Desert and Yellow River Basin (**Figure 13.5**). These dust storms normally occur five to seven times each spring. China has planted more than 40 billion trees in an attempt to help bind the soil and reduce wind strength at the ground surface. However, the combination of poor land use practices and climate change is increasing the frequency of these seasonal dust storms.

Scientists calculate that today's atmosphere contains the highest levels of dust particles in human history due to deforestation, modern agricultural practices, and the increasing number of droughts that result from the warming climate. Human activities in very dry, hot deserts have also significantly helped raise high dust clouds. As grazing animals and humans trample the fine-textured soils of the desert of northwest India and Pakistan (the Thar Desert bordering the Indus River), they produce a dust cloud. Such clouds hang over the region for long periods and can extend to heights of 9 km (about 30,000 ft).

CONCEPT CHECK **STOP**

1. **How** do ventifacts form?
2. **What** land use practices contribute to dust storms?

Eolian Landforms

LEARNING OBJECTIVES

1. **Explain** how sand dunes form and move.

2. **Describe** the formation of loess.

Eolian landforms are wind-generated landforms, named after the Greek god of wind, Aeolus. Ordinarily, wind is not strong enough to dislodge mineral matter from the surfaces of unweathered rock or from moist, clay-rich soils, or from soils bound by dense plant cover. Wind works best as a geomorphic agent when wind velocity is high and moisture and vegetation cover are low—typically in deserts and in semiarid lands or steppes. In coastal environments, beaches also provide a supply of loose sand to be shaped by the wind, even where the climate is humid and the land surface inland from the coast is well protected by plant cover. Eolian landforms are also found in continental interiors where layers of silt were deposited during the retreat of glaciers at the end of the Ice Age.

> **sand dune** A hill or ridge of loose, well-sorted sand shaped by wind and usually capable of downwind motion.

Sand Dunes

A **sand dune** is any hill of loose sand shaped by wind. Dunes are one of the most common types of eolian landforms. They form where there is a source of sand—for example, a sandstone formation that weathers easily and releases individual grains, or perhaps a beach supplied with abundant sand from a nearby river mouth, or a closed interior basin in an arid region that receives abundant river sand. Active dunes constantly change form under wind currents, but they must be nearly free of vegetation in order to move. They become inactive when stabilized by vegetation cover or when patterns of wind or sand sources change.

Dune formation A dune typically begins as a sand drift in the lee of some obstacle, such as a small hill, rock, or clump of brush that reduces wind speed, causing saltating sand to stop moving. Once a sufficient mass of sand has gathered, it begins to move downwind (**Figure 13.6**). Pushed by the prevailing wind, sand blows up the windward dune slope, passes over the dune crest, and slides down the steep leeward side of the dune, called the *slip face*, moving the dune forward. The slip face maintains a more-or-less constant angle from the horizontal, known as the *angle of repose*. For loose sand, this angle is about 33–34°.

Formation and movement of sand dunes • Figure 13.6

This cross section through a sand dune shows the typical gentle windward slope (facing the wind) and the steep slip face (facing away from the wind). The solid lines inside the dune show former slip faces before the dune moved forward into its current position.

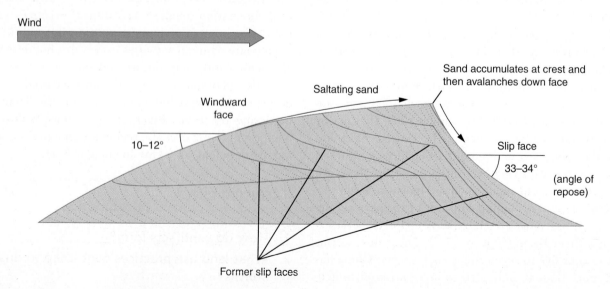

Wind

Saltating sand

Sand accumulates at crest and then avalanches down face

Windward face

10–12°

Slip face

33–34°

(angle of repose)

Former slip faces

Transformation of the Sahara • Figure 13.7

Spaceborne Imaging Radar (SIR) data along the Sudan–Egypt border allows scientists to see through up to 6 m (20 ft) of dry sand in the Sahara Desert. The black-and-white radar images, seen as a band against the Landsat view of the tan desert sand, reveal subsurface channels that hosted flowing rivers as recently as 10,000 years ago.

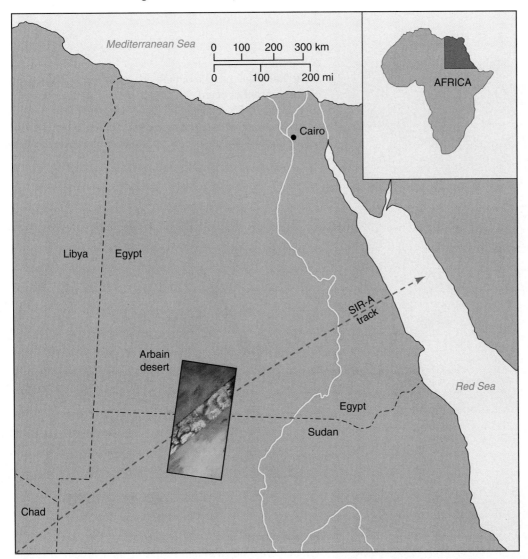

A vast expanse of sand dunes makes up a great desert landscape, called an **erg**, after the Arabic word for dune field. In the Sahara Desert, enormous quantities of dune sand have been weathered from sandstone to create extensive regions of erg desert. Also common are regions covered with an aggregate of pebbles that form a vast and semi-impermeable flat surface. This desert pavement surface is called a *reg*. These two major desert landscapes can be found throughout the globe.

We tend to think of the Sahara Desert as an ideal place for sand dunes, but it was not always so. During a wetter climate of the past, parts of the Sahara were drained by rivers carving fluvial landscapes. In 1981, a NASA radar instrument aboard the Space Shuttle generated images that penetrated the upper layer of dry dune sand to reveal this ancient topography of stream patterns hidden underneath (**Figure 13.7**).

Types of dunes Sand dunes take a variety of forms, depending on the supply of sand, the amount and direction of wind, and the amount of vegetation cover (see **Figure 13.8** on the next page).

The shape, size, and behavior of a dune depend on sand supply, wind, and amount of vegetative cover. To read the triangular diagram, project the location of a dune type onto the sides of the triangle. For example, longitudinal dunes (e) have a low sand supply, strong winds, and a varying amount of vegetation.

a. Barchan

A **barchan dune** has the outline of a crescent, with the points of the crescent directed downwind. Mobile barchan dunes, such as these in Namibia, form when the wind blows predominantly in one direction. They are often outliers moving away from a larger body of sand, such as a field of transverse dunes.

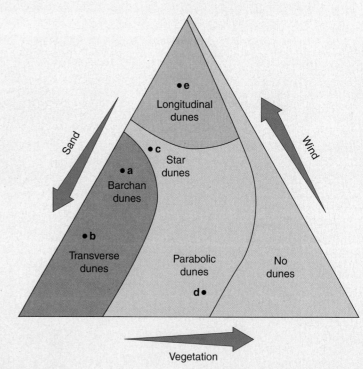

b. Transverse

When there is an abundant supply of sand, little or no vegetation, and a single dominant wind direction, **transverse dunes**, such as these in Baja California, Mexico, often form. Their crests are at right angles to the wind direction.

c. Star

Star dunes are elaborately shaped pyramidal mounds formed by winds blowing from several prevailing directions. Star dunes, such as these at Rub' al Khali on the Arabian Peninsula, are stable, almost-permanent features of the desert landscape.

d. Parabolic

Coastlines often provide abundant sand, along with a moist wind off the ocean that allows some vegetation to grow. In this environment, **parabolic dunes** often develop. The arms of these dunes, anchored by vegetation, point upwind.

e. Longitudinal

Longitudinal dunes consist of long, narrow ridges oriented parallel with the direction of the prevailing wind. Formed in deserts with meager sand supply, they are often only a few meters high but may be several kilometers long, such as these dunes in South Australia.

Put It Together

Review Figure 13.6 and complete this statement. For the transverse dune, the angle of the upwind side should be approximately _____ and the angle of the downwind side approximately _____.

a. 11°; 11°   c. 33°; 11°
b. 11°; 33°   d. 33°; 33°

Loess • Figure 13.9 _____

This thick layer of wind-transported silt in New Zealand was deposited during the Ice Age.
Loess has excellent cohesion and often forms vertical faces as it wastes away.

Loess

In several large midlatitude areas of the world, the surface is covered by deposits of wind-transported silt that has settled out from dust storms over many thousands of

> **loess** A surface deposit of wind-transported silt.

years. This material is known as **loess**. (The pronunciation of this German word is somewhere between "lerse" and "luss.") Loess is usually a uniform yellowish to buff color and lacks any visible layering. The dust is composed of fine silt and clay particles, which are often shaped like tiny flakes or plates. As they settle out in layers, the flat particles overlap each other, giving the deposit a weak cohesive structure. Loess often contains calcium carbonate dust, which can dissolve and recrystallize as rainwater penetrates the loess deposit, adding to the cohesiveness. As a result, loess tends to form vertical cliffs wherever it is exposed by the cutting of a stream or grading of a roadway (**Figure 13.9**). It is also very easily eroded by running water, and when the vegetation cover that protects it is broken, it is rapidly carved into gullies. Thick loess is a stable, easy-to-dig material, and it has been used for cave dwellings in some regions, including China and Central Europe.

The thickest deposits of loess are in northern China, where layers over 30 m (about 100 ft) thick are common and a maximum thickness of 100 m (about 330 ft) has been measured. This layer covers many hundreds of square kilometers and appears to have been brought as dust from the interior of Asia. Loess deposits are also common in the United States, Central Europe, Central Asia, and Argentina.

In the United States, there are thick loess deposits in the Missouri–Mississippi Valley (**Figure 13.10**). Eastern Washington State's famous Palouse farmland is derived from thick and fertile loess soils. American and European loess deposits are directly related to the continental glaciers of the Ice Age. At the time when the ice covered much of North America and Europe, the winter climate was generally dry bordering the ice sheets. Strong winds blew southward and eastward over the bare ground, picking up silt from the floodplains of braided streams that discharged the meltwater from the ice. This dust settled on the ground between streams, gradually building up a smooth, level ground surface. The loess is particularly thick along the eastern sides of the valleys because of prevailing westerly winds that carried the loess eastward. It is well exposed along the bluffs of most streams flowing through these regions today.

Loess distribution in the United States
• Figure 13.10

Large areas of the prairie plains region of Indiana, Illinois, Iowa, Missouri, Nebraska, and Kansas are underlain by loess ranging in thickness from 1 to 30 m (about 3 to 100 ft). Extensive deposits also occur in Tennessee and Mississippi in areas bordering the lower Mississippi River floodplain. Still other loess deposits are in the Palouse region of northeast Washington and western Idaho.

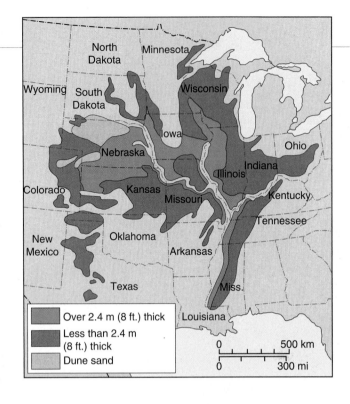

Loess is an important agricultural resource. It forms rich black soils that are especially suited to cultivation of grains. The highly productive plains of southern Russia, the Argentine Pampas, and the rich grain region of north China are underlain by loess. In the United States, corn is extensively cultivated on the loess plains in Kansas, Iowa, and Illinois. Wheat is grown farther west on the loess plains of Kansas and Nebraska and in the Palouse region of eastern Washington.

CONCEPT CHECK

1. **What** are the similarities and differences between the different types of sand dunes?
2. **Why** is loess particularly suited for cave dwellings?

The Work of Waves and Tides

LEARNING OBJECTIVES

1. **Describe** how waves form and break.
2. **Explain** the process of littoral drift.
3. **Explain** the effect of wave refraction on coastlines.
4. **Describe** the cause of tides.

ind plays an indirect role in shaping coastal landforms by generating waves. Breaking on ocean shores, they erode and transport sand and other sediments to new locations.

Waves

When winds blow over broad expanses of water, they generate waves. Waves are the most important agents in shaping coastal landforms.

Wind waves Both friction between moving air and the water surface and direct wind pressure on waves transfer energy to water, creating a train of waves. The height of waves depends on the wind speed and also on the wind duration and the *fetch*—the length of the water surface exposed to the wind. Dependence of wave height on fetch explains why waves on lakes never reach the heights of ocean waves.

WHAT A GEOGRAPHER SEES

Breaking Waves

Orienting his board for the best ride, this surfer skims the inside of a breaking wave off Maui (Figure a). A geographer would recognize the mechanism that produces such waves (Figure b). In deep water, a parcel of water makes circular loops as the waves pass by. As a wave reaches shallower water, it begins to drag against the bottom, and the circular loops become flatter. The wave's height increases and length decreases. Because the front of the wave is in shallower water, it is steeper and moves more slowly than the rear. Eventually, the front becomes too steep to support the advancing wave. As the rear part continues to move forward, the wave collapses, or breaks, forming turbulent surf that rushes onto shore.

a.

b.

The rhythmic rising and falling of the water surface causes the water molecules at or near the surface to move in a circle. The top of this circular motion represents the wave crest, and the bottom represents the wave base. The distance between two wave crests is the *wavelength*. Between two successive wave crests is a depression called the *wave trough*. You may have experienced this circular motion while floating in the waves at the beach. As a wave crest passes you, it pushes you forward, toward the beach. After it passes, the trough pulls you backward, toward the top of the next crest.

As a wave approaches the shore, when it encounters a depth of about half its wavelength, the circular motion begins to create friction on the bottom. The speed of the wave slows from the drag, making the wave steeper. The circular motion of the water particles changes shape, becoming more to-and-fro and finally back-and-forth. Eventually, the forward motion of the wave form becomes slower than the forward motion of the water, and the wave breaks. The breaker sends foaming, turbulent water forward in a sheet, washing up the beach slope as *surf* (see *What a Geographer Sees*).

As waves surge forward onto shore, they create a powerful *swash*. Once the force of the swash has been spent against the slope of the beach, the return flow, or *backwash*, pours down the beach (**Figure 13.11**). If the beach slope is steep and the waves are strong, this *undercurrent*, or *undertow*, as it is popularly called, can be strong enough to sweep unwary bathers off their feet and carry them seaward beneath the next oncoming breaker.

Tsunamis Although most waves are generated by wind, waves can also be generated by any sudden displacement of large amounts of water, such as by an earthquake or an undersea landslide. These waves are called **tsunamis**, or seismic sea waves. Although seismic sea waves are sometimes called *tidal waves*, they are unrelated to tides.

tsunami A train of sea waves triggered by an earthquake or another seafloor disturbance.

A tsunami originates when the sudden movement of the seafloor, resulting from an earthquake, for example, generates a train of water waves. These waves travel over the ocean at 700–800 km/hr (435–500 mi/hr), moving outward in all directions from their source. In deep ocean waters, the tsunami's wave motion is normally too gentle to be noticed, making tsunamis hard to detect in the open ocean. As the wave approaches land, however, it follows the same physics as wind-driven waves: As the wave drags along the bottom and slows, the wave shortens and steepens to a height of 15 m (50 ft) or more.

Ocean waters rush landward and surge far inland, destroying coastal structures and killing inhabitants. Ocean water flows inland at speeds of up to 15 m/s (34 mi/hr) for several minutes. Considering that each cubic meter of water weighs about 1000 kilograms, the power of the water surge is enormous. After some minutes, the waters retreat, continuing the devastation as people and debris are pulled back out to sea. Tsunamis typically consist of several surging waves following one after the other. The largest wave is not necessarily the first to arrive.

With their tremendous force, tsunamis can dramatically reshape coastal areas, inflicting damages in regions on both sides of the ocean. The deadliest tsunami recorded so far struck the Indian Ocean region in December 2004, following a massive undersea earthquake—measuring 9.0 on the Richter scale—in the Java Trench, west of Sumatra. About 1000 km (600 mi) of fault ruptured, causing the seafloor nearby to move upward about 5 m (16 ft). This rapid motion of a vast area of ocean bottom generated a giant tsunami that devastated populations in

Swash and backwash • Figure 13.11

As a wave breaks, swash rushes forward and up the slope of the beach. It loses momentum, stops, and is then drawn back by gravity as backwash.

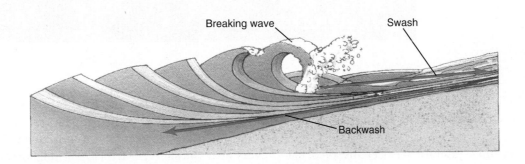

Breaking wave

Swash

Backwash

Tsunami destruction • Figure 13.12

Indonesia experienced the worst devastation of any country following the Indian Ocean tsunami of 2004.

a. Before the tsunami, Banda Aceh, Indonesia, seen from space on June 23, 2004, is a city of small buildings, parks, and trees.

b. Two days after the tsunami, the devastation caused by the earthquake and tsunami is complete. All that remains of most structures is their concrete floors. Trees are uprooted or stripped of leaves and branches.

Southeast Asia, India, and Africa, reshaping coastal landscapes, wiping out cities, and killing more than 265,000 people (**Figure 13.12**).

Where early-warning systems have been installed, tsunamis can usually be predicted far enough in advance to evacuate local populations. Monitoring of seismic activity in the seafloor triggers tsunami warnings for coastlines that might be affected. However, if an earthquake occurs very near the shore, or if local warning systems are not effective, tsunamis can take a devastating toll.

On March 11, 2011, the Pacific Tsunami Warning Center in Hawaii detected an undersea earthquake of magnitude 9.0, located 70 km (43 mi) off the east coast of Tohoku, the northeastern portion of Honshu, the largest island of Japan. The Great Tohoku earthquake, Japan's largest ever recorded, occurred along a 300 km (186 mi) subduction zone at a depth of 32 km (20 mi) under the ocean floor, with its epicenter 70 km (43 mi) east of the coastline. Japan's technological advances and regular emergency training could not stave off the onslaught of destruction set in motion by the earthquake and its huge tsunami.

Although many perished during the Great Tohoku earthquake, advanced tsunami early warning systems saved tens of thousands of lives minutes before the 9.2 m (30 ft) wave came ashore (**Figure 13.13**). Nuclear reactors built along the coastline were flooded, eventually producing partial core meltdowns that released harmful radioactivity. The final death toll exceeded 15,000.

Littoral Drift

Waves breaking along the shore provide tremendous energy to move sediment along the shoreline in a process called **littoral drift**. (*Littoral* means "pertaining to a coast or shore.") Littoral drift includes two transport processes. **Beach drift** is the transport of sediment along the beach. This occurs when waves reach the shore at an angle, however slight, and the swash from the breaking waves carries sand up the beach at a similar

> **littoral drift** The transport of sediment parallel with the shoreline by the combined action of beach drift and longshore current transport.

Japan tsunami of 2011 • Figure 13.13

Following Japan's largest underwater earthquake, tsunami warnings provided only a few minutes for hundreds of thousands to flee the coastal towns for higher ground along the Tohoku coastline of Japan. The disaster resulted in over $300 billion in damage, 15,000 deaths, and another 8000 missing persons.

Through the processes of littoral drift, individual rock particles are transported long distances along the beach and underwater.

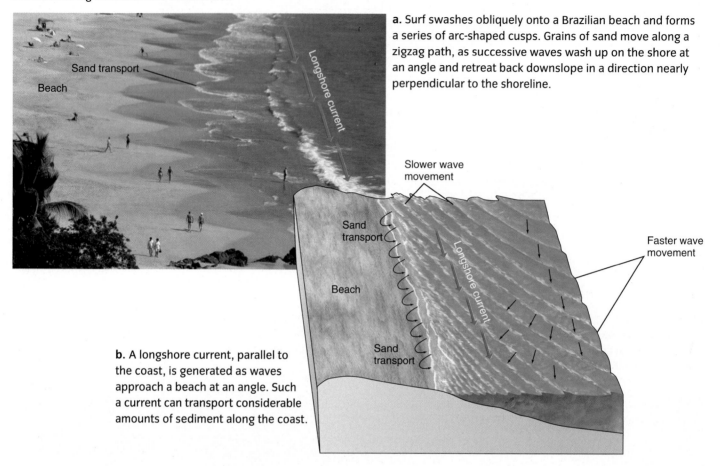

a. Surf swashes obliquely onto a Brazilian beach and forms a series of arc-shaped cusps. Grains of sand move along a zigzag path, as successive waves wash up on the shore at an angle and retreat back downslope in a direction nearly perpendicular to the shoreline.

b. A longshore current, parallel to the coast, is generated as waves approach a beach at an angle. Such a current can transport considerable amounts of sediment along the coast.

angle. The backwash, however, is drawn downward toward the breaker zone by gravity in a straighter path. The result is that sand is carried down the beach in a series of steps, creating a net movement of sand away from the angle of wave attack (**Figure 13.14**).

Out in the breaker zone, the waves pile up ocean water and push it along the shoreline, away from the direction of wave attack. This produces a longshore current, which moves sand along the bottom as **longshore drift**, the second component of littoral drift. If the wave direction remains more or less the same for long periods, littoral drift can transport sediment for long distances along the coast.

Wave Refraction

The process by which waves erode sediment along a shore depends on two things: how much energy the waves have and the resistance of the shore materials. Where a rocky coastline is struck by waves, waves erode areas of softer rock more quickly, carving out bays and coves and leaving behind jutting landforms of resistant rock, called *headlands*.

When the coastline has prominent headlands that project seaward and deep bays in between, approaching wave fronts slow when the water becomes shallow in front of the headlands. This slowing effect causes the wave front to wrap around the headland, in a process called *wave refraction* (**Figure 13.15**). This concentrates wave energy on the headland, enhancing erosion. The angle of the waves against the side of the headland also creates a longshore current that moves sediment from the headland into the surrounding bays, creating crescent-shaped *pocket beaches* in the bays. Over time, the action of waves against a coast tends to have the effect of straightening the coast as it erodes sediment from headlands and deposits it into bays.

Tides

Most marine coastlines are also influenced by the **ocean tide**—the rhythmic rise and fall of sea level under the influence of changing attractive forces of the Moon and Sun on the rotating Earth. In the tidal system, the Earth and Moon are coupled together by their mutual gravitational

Wave refraction • Figure 13.15

Wave energy is concentrated at the headlands. Sediment is eroded from cliffs on the headland and carried by littoral drift along the sides of the bay. The sand is deposited at the head of the bay, forming a pocket beach.

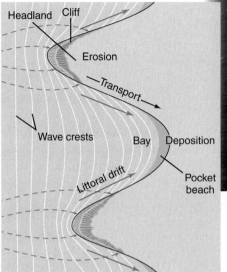

Think Critically

If you were going to build a house close to the shore shown in this photo, would the headland or the pocket beach be a more stable site? Why?

attraction, which is balanced by an inertial force created by their revolution around a common center of mass (**Figure 13.16**). While the inertial force is constant at all points on the globe, the gravitational attraction of the Moon is greater on the near side of the Earth. Sea water responds to this attraction, and the ocean "bulges" toward the Moon.

On the far side of the globe, the Moon's gravitational force is weaker. Because the inertial force is the same, the ocean water is pushed away from the Moon, creating a second bulge on the far side. The bulges remain essentially stationary while Earth rotates through them, creating two high tides and two low tides per day. The Sun also affects

The cause of tides • Figure 13.16

The balance between the Moon's gravitational attraction and the opposing inertial force created by the revolution of the Earth and Moon around a common center of mass causes the Earth's oceans to bulge on the sides nearest to and farthest from the Moon. As the Earth rotates through these bulges, local areas experience two daily high tides.

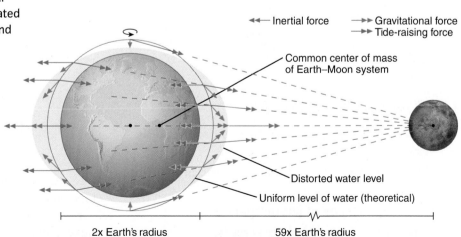

Tidal range • Figure 13.17 _____

The Bay of Fundy in Nova Scotia, Canada, is known for its exceptional tidal range, with a record of 17 m (55.8 ft) between high water and low water. The shape of the bay makes it naturally resonate with the rise and fall of the tide, reinforcing the tidal flood and ebb. Boats like the ones pictured here must allow for extra length in the mooring lines tied to the dock to account for the full tidal range twice each day. The lower boat shown here will be lifted to the height of the dock by the high tide.

the eggs to incubate without subsequent exposure to wave action. Hatchlings emerge at the next spring tide to be washed to the sea. These species' survival is based on close synchronization with these extreme tides.

Each day, coastal regions experience two high tides (semidiurnal), called **flood tides**, and two low tides, called **ebb tides**. Unique orbital geometry, however, provides exceptions in the Gulf of Mexico and South China Sea, where only one high tide and one low tide (diurnal) occur daily. The difference between the heights of successive high and low waters is known as the *tidal range* (**Figure 13.17**). If the tidal range is large, the changing water level can play an important role in shaping coastal landforms.

Tides contribute to the erosion, transportation, and deposition of sediment by ocean waters. As tides ebb and flood, ocean waters flow into and out of bays. Tidal currents pouring through narrow inlets that connect bays with the ocean are very swift and pick up sediment as they travel, eroding the inlets. This keeps the inlet open, despite littoral drift that would otherwise close the inlet with sand. Tidal currents also carry large amounts of fine silt and clay in suspension. Some of this sediment is deposited in bays. Over time, this sediment settles to the floors of the bays and builds up in layers, gradually filling in the bays.

CONCEPT CHECK	STOP

1. **How** does a tsunami wave form differently from a wind wave?
2. **What** is the difference between beach drift and longshore drift?
3. **Why** do waves erode headlands more vigorously than bays?
4. **What** are spring and neap tides and why do they occur?

tides in a similar way, but its tide-producing force is only about half as strong as the Moon's. When the Earth, Sun, and Moon are aligned, their gravitational forces combine to create a higher tide called a *spring tide*. In contrast, when the Moon and Sun are positioned at right angles to each other, this results in a lower tide, called a *neap tide*.

Some fish and marine animals time their egg laying on the beach to coincide with the spring tides, enabling

Coastal Landforms

LEARNING OBJECTIVES

1. **Describe** the erosional landforms produced by waves.
2. **Describe** the depositional landforms produced by waves.
3. **Define** coastlines of submergence.
4. **Explain** the features of coral reefs.

A s waves meet the shore, they expend tremendous amounts of energy. This energy, along with the coastal currents it produces, shapes coastlines, carving out bays and steep cliffs, and building beaches and barrier islands. The **coastline** (or coast) includes the shallow water

coastline (coast) The zone in which coastal processes operate or have a strong influence

shoreline The shifting line of contact between water and land.

zone in which waves perform their work, as well as the landforms they create. In contrast, the **shoreline** refers to the dynamic zone of contact between water and land.

The Earth's coastlines are enormously varied. For example, along most of the east coast of the United States, we find a coastal plain gently sloping toward the sea, with shallow lagoons and barrier islands at the coastline. On the west coast, we often find rocky shorelines, with dramatic sea cliffs, headlands, and pocket beaches.

Plate tectonics provides a way to explain these differences. The eastern coastline is a passive continental margin, without tectonic activity. Abundant sediment from rivers on the continent, accumulating over millions of years, provided the material to build the coastal plain, beaches, and islands. Here, the characteristic coastal landforms are depositional—built by sediment moved by waves and currents. The western coastline is the site of great tectonic activity, with rocks rising from the sea along subduction boundaries and transform faults. Here, the characteristic landforms are erosional, with waves and weathering slowly eroding the strong rock exposed at the shoreline.

World coastlines are varied for other reasons as well. During each glaciation of the recent ice age, sea level fell by as much as 125 m (410 ft). Rivers cut canyons to depths well below present sea level to reach the ocean. Waves broke against coastlines that are now well underwater. Now that sea level has risen, many coastal landscapes are coastlines of submergence. Still other coastlines are dominated by coral reefs, where ocean waters are warm enough for coral to grow.

Erosional Coastal Landforms

The breaking of waves against a shoreline provides a variety of distinctive features. Where resistant rocks meet the waves, **sea cliffs** often occur. At the base of a sea cliff is a notch, formed largely by physical weathering. Constant splashing by waves followed by evaporation causes salt crystals to grow in tiny crevices and fissures of the rock, breaking it apart, grain by grain. Hydraulic pressure of waves and abrasion by rock fragments thrust against the cliff also erode the notch. Undercut by the notch, blocks fall from the cliff face into the surf zone. As the cliff erodes, the shoreline gradually retreats shoreward. Sea cliff erosion forms a variety of erosional landforms, including sea caves, sea arches, and sea stacks (**Figure 13.18**).

sea cliff A rock cliff shaped and maintained by the weathering and erosion of breaking waves at its base.

Formation of erosional coastal features • Figure 13.18

THE PLANNER

Cliffs are active coastal features that change as a result of sculpting by waves.

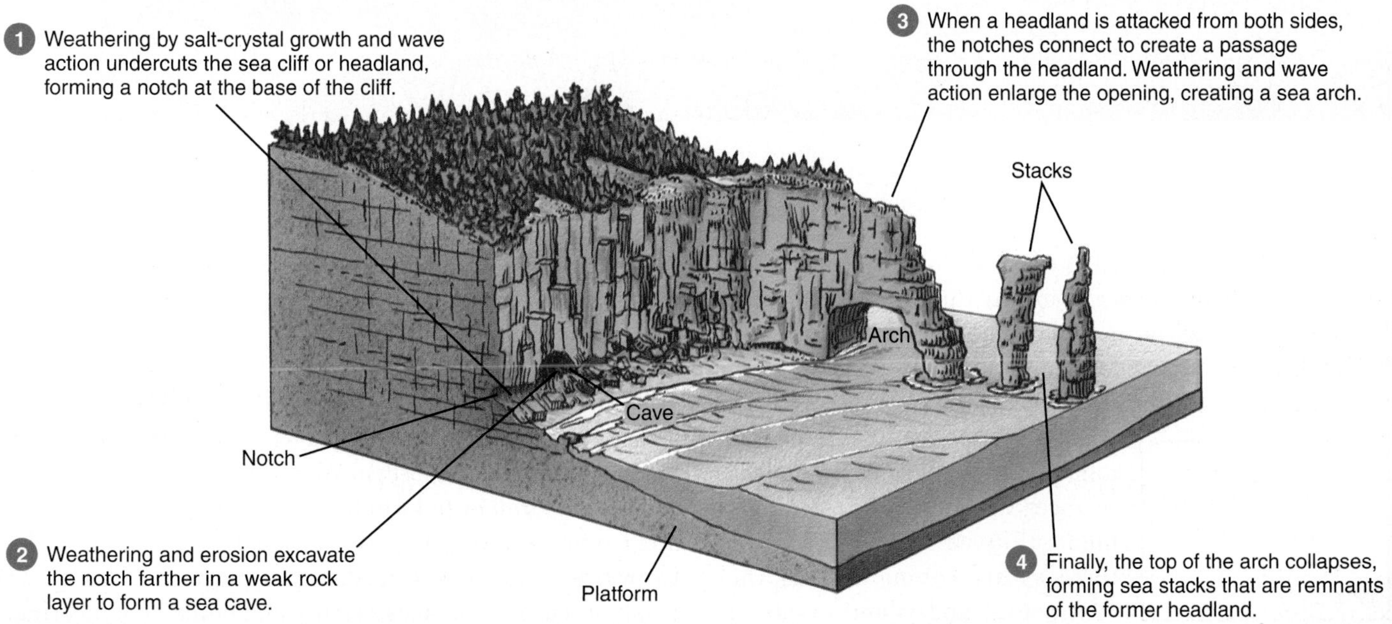

1 Weathering by salt-crystal growth and wave action undercuts the sea cliff or headland, forming a notch at the base of the cliff.

2 Weathering and erosion excavate the notch farther in a weak rock layer to form a sea cave.

3 When a headland is attacked from both sides, the notches connect to create a passage through the headland. Weathering and wave action enlarge the opening, creating a sea arch.

4 Finally, the top of the arch collapses, forming sea stacks that are remnants of the former headland.

Notch

Cave

Platform

Arch

Stacks

PROCESS DIAGRAM

Marine terraces provide a visual record of coastlines and sea levels from the Earth's past.

a. After a period of tectonic uplift or lowered sea level, the shore platform at the base of the cliffs is raised up to become a marine terrace.

Former sea cliff

Former sea level

Marine terrace

b. This series of marine terraces appears on the western slope of San Clemente Island, off the Southern California coast. More than 20 different terrace levels have been identified. The highest has an elevation of about 400 m (about 1300 ft).

The retreat of the cliffs creates a broad, gently sloping plane, called a *shore platform*, at the base of the cliffs. If tectonic activity pushes the shoreline up or if the sea level falls, the platform is abruptly lifted above the level of wave attack. The former shore platform is raised up to become a **marine terrace**. Repeated uplifts create a series of marine terraces with a step-like arrangement (**Figure 13.19**). Marine terraces are common along the continental and island coasts of the Pacific Ocean, where tectonic processes are active along the mountain and island arcs.

marine terrace A former shore platform elevated to become a step-like coastal landform.

Depositional Coastal Landforms

Most of the sediment we find along a coastline is provided by rivers reaching the ocean. Waves then transport and deposit this sediment to form features such as beaches, spits, and barrier islands. These depositional landforms are relatively transitory, appearing, disappearing, or migrating as a result of seasonal changes, storms, and human engineering.

Beaches A **beach** is a wedge-shaped sedimentary deposit, built and worked by wave action. The form of a beach varies over time as waves deposit or erode more. During short periods of storm activity, waves cut back the beach, forming a long, flat, sloping profile. The sand moves just offshore and along the shore by longshore drift. Gentler waves return the sand to the beach, building a steeper beach face and a bench of sand at the top of the beach (**Figure 13.20**).

When sand leaves a section of beach more rapidly than it is brought in, the beach is narrowed and the shoreline moves

> **beach** A thick, wedge-shaped deposit of sand, gravel, or larger stones in the zone of breaking waves.

landward. When sand arrives at a particular section of the beach more rapidly than it is carried away, the beach is widened and built oceanward. Many midlatitude beaches experience more active waves during the winter and are eroded to narrow strips. This change is called **retrogradation** (cutting back). In summer, the gentler wave climate allows sediment to accumulate and the beach is replenished. This change is called **progradation** (building out). Retrogradation and progradation can also happen on longer time cycles, related to changes in sediment input, climate, or human activity in the coastal zone.

Changing beach profile • Figure 13.20

As the wave climate changes with the seasons, so does the beach profile. With climate change, an increase in the frequency and intensity of storms could lead to greater erosion of beach sand.

a. In winter, larger waves and occasional storms create a relatively flat beach profile. Malibu Beach, on California's Pacific Coast Highway, provides an example.

b. The gentler waves of summer create a steeper beach face, shown in this photo of Malibu Beach in summer, taken not far from the winter example. Often the summer beach is wider, but not at this location.

Coastal foredunes • Figure 13.21

Beach grass growing on coastal foredunes traps drifting sand, producing a dune ridge. Nauset Beach, Cape Cod National Seashore, Massachusetts.

Coastal dunes Where ample sand is available, a narrow belt of dunes, called *foredunes*, often occurs in the region landward of beaches. These dunes are usually held in place by a cover of beach grass (**Figure 13.21**). Although these dunes are typically built by wind, they play an important role in maintaining a stable coastline by trapping sand blown landward from the adjacent beach.

As sand from the beach collects along the foredunes, the dune ridge builds upward, becoming a barrier several meters above high-tide level. This forms a protective barrier for tidal lands on the landward side of a beach ridge or barrier island. In a severe storm, the swash of storm waves cuts away the upper part of the beach. Although the foredune barrier may then be eroded by wave action and partly cut away, it will not usually yield. Between storms, the beach is rebuilt, and, in due time, wind action restores the dune ridge if a vegetative cover is maintained. If the plant cover of the dune ridge is reduced by human foot and vehicular traffic, however, inlets may open up and allow ocean water to wash in during storms or high tides. Many coastal communities now protect their dunes by building raised pathways over the dunes to restrict foot traffic or by planting protective vegetation.

Spits Where littoral drift moves the sand along the beach toward a bay, the sand is carried out into the open water as a long finger, or **spit**. As the spit grows, it forms a barrier, called a *baymouth bar*, across the mouth of the bay. Once the bay is isolated from the ocean, it is transformed into a **lagoon**. Where a spit grows to connect the mainland to a near-shore island, it forms a *tombolo* (**Figure 13.22**).

Barrier islands Much of the length of the Atlantic and Gulf coasts of North America is flanked by **barrier islands**—low ridges of sand with beaches and dunes and a landward lagoon

Depositional landforms resulting from littoral drift • Figure 13.22

Littoral drift deposits sand in a distinctive way, forming depositional landforms such as spits, baymouth bars, and tombolos.

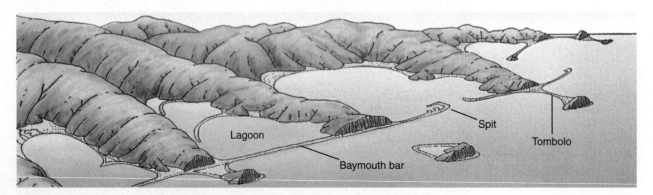

(Figure 13.23). These islands were formed as sea level rose after the last glaciation, causing the existing beaches to grow and migrate landward with the invading shoreline. Eventually, they became too big to move, and the sea flooded the area behind them to form the lagoons. Today barrier islands are still being worked by wind and waves, and they provide many fine examples of depositional landforms.

Gaps in the barrier islands, called *tidal inlets*, allow tidal waters to circulate in lagoons behind the islands. Strong currents flow back and forth though the inlet as the tide rises and falls. New inlets are formed in severe storms, and they are kept open by the tidal current. The inlet may later be closed later by longshore drift.

A lagoon is a shallow bay that accumulates fine sediments brought in by the tidal circulation. In time, tidal sediments can fill bays and produce *mud flats*, which are barren expanses of silt and clay. Mud flats are exposed at low tide but covered at high tide. Salt-tolerant plants start to grow on mud flats, and the plant stems trap more sediment, building up the flats to form a *salt marsh*. A thick layer of peat eventually forms at the salt marsh surface. Salt marshes are ecologically important, serving as nurseries for young ocean-going fishes.

Barrier islands and their lagoons serve to protect the mainland from hurricane waves and storm surge, although development on the islands bears the brunt of the storm damage. Climate change is increasing the risk to life and property in these coastal zones as a result of increased storm frequency and intensity as well as sea-level rise.

Coastlines of Submergence

Coastlines of submergence are formed when rising sea level partially drowns a coast or when part of the coast sinks. At the end of the last glaciation, sea level rose by about 120 m (394 ft), submerging coastal landscapes. Climate scientists expect sea levels to rise even higher over the next century as a result of global climate change. Today, many of the major river delta areas are experiencing submergence resulting from land use practices and sea level rise.

Barrier island coast • Figure 13.23

Where the gently sloping coastal plain of the Atlantic and Gulf coasts meets the ocean, sea-level rise and wave and wind action have built a coastline of barrier islands, beaches, and lagoons.

a. A barrier island is separated from the mainland by a wide lagoon. Sediments fill the lagoon, and dune ridges advance over the tidal flats. An inlet allows tidal flows to pass in and out of the lagoon.

b. Shown here is the Outer Banks of North Carolina. The barrier beach is a prime site for development, and the lagoon is used for fishing and boating.

Put It Together

Review Figure 13.16 and answer this question. If you have a sailboat in the lagoon and need to follow the ebb tide to get out to sea, how many opportunities do you have each day (24-hour period)?

a. none
b. one
c. two
d. four

A ria coastline • Figure 13.24

A ria coast is a river valley (a) that is flooded with ocean water (b) after the submergence of a land mass or a rise in sea level.

a. Before submergence, active streams have carved the valleys.

b. After submergence, the shoreline rises up the sides of the valleys, creating estuaries of mixed ocean and fresh water.

Island

Sandbar

Submerged valleys

c. This view of Ria Ortigueira, Coruña, Spain, shows many of the features of the ria coastline, including submerged valleys, islands, and sandbars.

A **ria coast** is formed when a rise of sea level or a crustal sinking (or both) brings the shoreline to rest against the sides of river valleys previously carved by streams (**Figure 13.24**). The shoreline rises up the sides of the stream-carved valleys, creating narrow bays. Streams that occupied the valleys add fresh water to the bays, making them estuaries of mixed fresh and salt waters, creating a unique habitat for many plants and animals.

A **fiord coast** is similar to a ria coast except that the bays are formed in valleys that have been scoured by glaciers (**Figure 13.25**). Glaciers eroded their walls and scraped away loose sediment and rock to form broad, steep-sided valleys that are now flooded with sea water. Fiords, the Scandinavian word for bays, are common features along the northern coastal countries and can be found where glaciers occupied coastlines during the Ice Age. These deep, glacially-scoured bays can be hundreds of meters deeper than the adjacent seas, and they often meander tens of kilometers inland.

Coral Reefs

So far we have been talking about coastal landforms built by sediment. **Coral reefs** are unique because the new land is of biological origin, made by living organisms. Growing together, corals and algae secrete rock-like deposits of carbonate minerals. As coral colonies die, new ones are built on them, accumulating as composite layers of limestone.

coral reef A rock-like accumulation of carbonates secreted by corals and algae in shallow water along a marine shoreline.

Fiord coast • Figure 13.25

Along a fiord coast, valleys that were once scoured by glaciers are now flooded with sea water. This fiord coast is in Fiordland National Park, South Island, New Zealand.

Coral fragments are torn free by wave attack, and the pulverized fragments accumulate as sand beaches.

Because water temperatures above 20°C (68°F) are needed for dense coral reefs to grow, coral reef coasts are usually found in warm tropical and equatorial waters between 30° N and 25° S. The sea water must be free of suspended sediment, and it must be well aerated for vigorous coral growth to take place. For this reason, corals live near the water surface and thrive in positions that are exposed to waves from the open sea. Because muddy water prevents coral growth, reefs are not found near the mouths of muddy streams. Sediment runoff from poor land-use practices and development has been identified as a primary factor in most coastal coral reef die-offs.

There are three distinctive types of coral reefs—fringing reefs, barrier reefs, and atolls. These reefs are often found around a hotspot volcano (Chapter 9) at different stages of submergence. *Fringing reefs* are built as platforms attached to shore. They are widest in front of headlands where the wave attack is strongest. *Barrier reefs* lie out from shore and are separated from the mainland by a lagoon (**Figure 13.26**). At intervals along barrier reefs, there are

Coral reef coast • Figure 13.26

A coral barrier reef with a lagoon rings the island of Moorea, Society Islands, South Pacific Ocean. The island is a deeply dissected volcano with a history of submergence.

narrow gaps through which excess water from breaking waves is returned from the lagoon to the open sea.

Atolls are more-or-less circular coral reefs enclosing a lagoon but have no land inside. Most atolls are rings of coral growing on top of old, sunken hotspot volcanoes. They begin as fringing reefs surrounding a volcanic island. Then, as the volcano sinks, the reef continues to grow, and eventually only the reef remains.

Reefs are highly productive ecosystems that support a diversity of marine life-forms. They also perform an important role in recycling nutrients in shallow coastal environments. They provide physical barriers that dissipate the force of waves, protecting ports, lagoons, and beaches that lie behind them, and they are an important aesthetic and economic resource. Unfortunately, because of their very specific environmental requirements, coral reefs are highly susceptible to damage from human activities as well as from natural causes such as tropical storms. Some 50% of coral reefs worldwide are currently threatened by human activities.

When corals are stressed by a change in their local environment, such as water temperature or sedimentation, they expel the algae that usually live symbiotically inside their structures. Because the corals become a white color without the algae, this phenomenon is known as **coral bleaching**. Without the algae, the corals die.

Major coral bleaching occurs during strong El Niños, in locations where water temperature increases by 1°C (1.8°F) or more. Thus, long-term global warming could devastate coral life. Geographers are also worried about the impact of rising carbon dioxide levels in the atmosphere on corals. As more CO_2 dissolves in sea water, the water becomes more acidic, making it harder for corals to build their calcium carbonate skeleton structure.

CONCEPT CHECK STOP

1. **How** are marine terraces formed?

2. **How** and why do some beaches change from season to season?

3. **What** might cause a coastline to be submerged?

4. **What** is coral bleaching and why does it occur?

Human Interactions with Coastal Processes

LEARNING OBJECTIVES

1. **Explain** what measures can be undertaken to control coastal erosion.

2. **Describe** the effect of sea-level rise on coastal populations.

A round the world, human populations have settled along coasts, drawn by sea-related industries such as shipping, fishing, and tourism, as well as scenic views and coastal climates. More than 70% of the world's population lives on coastal plains, and 11 of the world's 15 largest cities are on coasts or estuaries. Human efforts to control dynamic coastal processes in order to protect their property and investments are often futile and can disrupt the balance of sediment supply and erosion.

Coastal Engineering

Homes built on sea cliffs, beaches, or barrier islands are threatened by the erosive action of waves, especially during storms (**Figure 13.27**). Homeowners sometimes build seawalls or pile up mounds of boulders, called *rip-rap*, in order to absorb some of the destructive erosive energy of the waves and protect their property. However, these fortifications can reduce the diversity and abundance of fish, crabs, and other marine species, negatively impacting marine ecosystems.

Coastal property damage • Figure 13.27

Insurance companies are adjusting their policies and raising or canceling premiums for many coastal areas in response to the increased threat to these areas posed by global warming.

a. Storm waves from Hurricane Isabel breached the North Carolina barrier beach to create a new inlet, as shown in this aerial photo from September 2003. Notice also the widespread destruction of the shoreline in the foreground, with streaks of sand carried far inland by wind and wave action.

b. The storm surge from Isabel swept this beachfront home in Rodanthe, North Carolina, from its foundation piers.

Along stretches of shoreline affected by retrogradation, the beach may be seriously depleted or even entirely disappear, destroying valuable shore property. Property owners or governments sometimes try to build a broad, protective beach through a process called *artificial beach nourishment*. One method of beach nourishment is to simply pump sand from offshore onto the beach. These programs are costly and only temporarily solve the problem because waves immediately begin to carry away the newly added sand. Geographers are consistent in warning governments that these futile public works programs waste tax dollars.

Another way to protect beaches is to try to prevent the loss of sand by trapping it as it is transported down the shore by littoral drift. This is accomplished by installing walls or embankments called **groins** at close intervals along the beach. Groins are usually built at right angles to the shoreline and made from large rock masses, concrete, or wooden pilings. Although groins effectively keep sand

Groins • Figure 13.28

Groins such as these along Lake Michigan in Chicago trap sediment that is being transported by the longshore current. The result of these structures is that the beaches on the up-current side of the groins build up, while the beaches on the down-current side of the groins erode.

on a specific beach, they deprive other beaches farther down the shore of sand that they would normally receive through littoral drift (**Figure 13.28**). **Jetties**, which are structures built to keep navigation channels open, similarly disrupt littoral drift.

In some cases, human influences on beaches result from actions taken far from the coast. Most beach sand comes from sediment that is delivered to the coast by rivers. Dams constructed far upstream can drastically reduce the sediment load of the river, starving the beach of sediment and causing retrogradation on a long stretch of shoreline.

Global Warming and Sea-Level Rise

The sea-level rise that is resulting from the current global temperature rise is another serious threat to coastlines. As glaciers and snow packs melt, the sea level rises and the upper layers of the ocean expand as they get warmer. Sea level has risen 20 cm (7.9 in.) since 1900, at a rate of 2 mm (0.08 in.) per year. This rate has increased to 3 mm (0.12 in.) per year in the 21st century. At the current rates of glacier melt and thermal expansion, scientists estimate that global warming will cause the sea level to rise by about 21 to 47 cm (8.3 to 18.5 in.) between now and 2100.

Sea-level rise has a number of negative effects on coastal environments. Most coastal erosion occurs during severe storms, and increased storm frequency and intensity have been linked to global warming. As a result of this increased erosion, combined with sea-level rise, we can expect to see beaches, salt marshes, and estuaries pushed landward. Land subsidence will also increase, with recent estimates suggesting that as much as 22% of the world's coastal wetlands could be lost by the year 2100. This would have a major effect on commercially important fish and shellfish populations. *Video Explorations* shows what is in store for Great Britain's shoreline as sea level rises, and *Where Geographers Click* focuses on a tool you can download to visualize the effect of sea-level rise at most locations.

Video Explorations
Sinking England

Great Britain is an ever-shrinking island. Its coastline has been eroding for thousands of years. The video explores the decision that its government must make about whether it will continue to spend a great deal of money to defend its coastline or abandon large areas of Britain to the sea as global warming raises sea levels by up to half a meter (1.6 ft) in the next 100 years.

Where Geographers CLICK

Visualizing Global Sea-Level Rise

www.cresis.ku.edu/data/sea-level-rise-maps

The mission of the Center for Remote Sensing of Ice Sheets (CReSIS) is to develop new technologies and computer models to measure and predict how the melting of ice sheets in Greenland and Antarctica will affect global rises in sea level. Visit this Web site to explore an online tool that illustrates the effect of a sea-level rise of 1 to 6 m at different locations around the world.

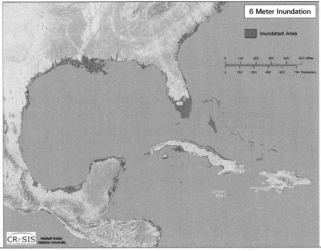

CONCEPT CHECK STOP

1. **What** are the drawbacks of constructing seawalls, groins, and other structures?

2. **How** will global warming affect coasts?

Summary

THE PLANNER ✓

1 Wind Action 392

- Wind action erodes landforms through **abrasion** and **deflation**.

- Human activities can hasten the action of deflation by breaking protective surface covers of vegetation and **desert pavement**, which is shown in the diagram.

Desert pavement • Figure 13.1

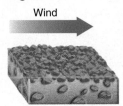

Wind

- Deflation creates blowouts in semidesert regions; in arid regions, deflation produces dust storms.

2 Eolian Landforms 396

- **Sand dunes** form when a source provides abundant sand that can be moved by wind action. Types of dunes include **barchan, transverse, parabolic, star,** and **longitudinal dunes,** as shown in the diagram.

Types of sand dunes • Figure 13.8

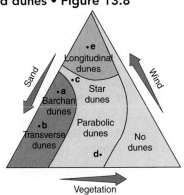

- **Loess** is a surface deposit of fine, wind-transported silt.

3 The Work of Waves and Tides 401

- Waves are usually formed as a result of surface winds. **Tsunamis** form as a result of the sudden movement of the ocean floor, as in an earthquake.

- Wave action produces **littoral drift**, which moves sediment parallel to the beach.

- Wave refraction focuses wave energy against resistant headlands. Sediment is then transported to bays, creating pocket beaches, as shown in the diagram.

Wave refraction • Figure 13.15

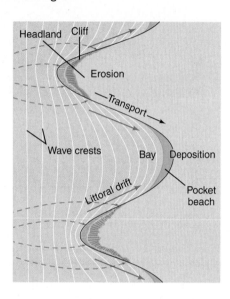

- **Ocean tides** cause sea level to rise and fall rhythmically as a result of gravitational attraction to the Sun and Moon.

4 Coastal Landforms 408

- A **shoreline** is the zone where water meets the land. **Coastlines** include the shallow water and nearby land.

- Waves weather and erode hard rock into **sea cliffs** and create horizontal shore platforms at the base of the cliffs. If the shore is raised by tectonic uplift or the sea level drops, these shore platforms are exposed as **marine terraces**.

- **Beaches**, usually formed of sand, are shaped by the swash and backwash of waves. Deposition of sand along coasts forms **spits** and **barrier islands**.

- Coastlines of submergence result when coastal lands sink below sea level or sea level rises rapidly. As shown in the diagram, **ria coasts** have flooded river valleys. **Fiord coasts** have flooded valleys scoured out by glaciers.

A ria coastline • Figure 13.24

- **Coral reefs** are landforms built from skeletons of corals. They typically occur in warm, sediment-free ocean environments.

5 Human Interactions with Coastal Processes 416

- Seawalls and rip-rap protect coastal properties from the impact of waves.

- Beach nourishment involves adding new sand to a beach.

- **Groins** and **jetties** prevent the transport of sediment away from a beach, but they also deprive beaches farther down the shoreline from their normal supply of sand, as shown.

Groins • Figure 13.28

- As glaciers melt as a result of global warming, sea levels will rise, flooding some coastal areas. Climate change is also expected to increase the severity of coastal storms.

Key Terms

- abrasion 393
- barchan dune 398
- barrier island 412
- beach 410
- beach drift 405
- coastline (coast) 408
- coral bleaching 416
- coral reef 414
- deflation 392

- desert pavement 392
- ebb tide 408
- eolian landform 396
- erg 397
- fiord coast 414
- flood tide 408
- groin 417
- jetty 418
- lagoon 412

- littoral drift 405
- loess 400
- longitudinal dune 399
- longshore drift 406
- marine terrace 410
- ocean tide 406
- parabolic dune 399
- progradation 411
- retrogradation 411

- ria coast 414
- saltation 394
- sand dune 396
- sea cliff 409
- shoreline 409
- spit 412
- star dune 399
- transverse dune 398
- tsunami 403

Critical and Creative Thinking Questions

1. How can agricultural practices be modified to minimize the occurrence of dust storms?

2. Sketch each of the following dunes and indicate the wind direction: barchan dunes, transverse dunes, parabolic dunes, and longitudinal dunes.

3. List three ways that a loess deposit differs from an erg.

4. Find the ocean coastline nearest where you live or study and describe its features. How would a 1-m rise in sea level affect this coastline?

5. What is the role of coastal dunes in beach preservation? How are coastal dunes influenced by human activity? What problems can result?

6. Consult an atlas or Google Earth to identify a good example of each of the following types of coastlines: ria coast, fiord coast, barrier island coast, and coral reef coast. For each example, provide a brief description of the key features you used to identify the coastline type.

7. Discuss how some of the human-built structures in the photograph could have increased the amount of beach lost during the storm.

What is happening in this picture?

Two of the most popular beach destinations along Maryland's Atlantic coast are Ocean City and Assateague Island, famous for its annual wild pony swim. Due to the high economic and tourism value of this coastal area, many engineering projects are maintained to preserve the features expected by the seasonal visitors. As discussed in the chapter, construction of jetties and groins in one area may directly affect beach erosion or replenishment in another.

Ocean City

Jetty

Assateague Island

Think Critically

1. What is the direction of the long-shore current along this coast? How can you tell?
2. Based on this photo, what effect has the construction of the jetty had on beaches in Ocean City and Assateague? In particular, comment on the relative depth and size of the beaches.

Self-Test

(Check your answers in the Appendix.)

1. _____ is formed when fine particles of silt and clay are removed from the surface by wind deflation.
   a. Desert bedrock
   b. Desert pavement
   c. Silt
   d. Loess

2. The hopping movement of small grains of sand, caused by wind is called _____.
   a. slip
   b. loess
   c. saltation
   d. ventifact

3. _____ sand dunes form in regions of abundant sand supply and have wave crests at right angles to the wind direction.
   a. Barchan
   b. Transverse
   c. Star
   d. Parabolic

4. A great sand sea, like the one found in the Sahara Desert, is called a(n) _____.
   a. beach
   b. desert pavement
   c. reg
   d. erg

5. The distinctive type of sand dune shown in the photograph is a _____ dune.
   a. barchan
   b. transverse
   c. star
   d. parabolic

6. The most important agent shaping coastal landforms is _____ action.

   a. storm

   b. stream

   c. salinization

   d. wave

7. Littoral drift includes _____.

   a. beach drift and ebb tide

   b. ebb tide and longshore drift

   c. beach drift and longshore drift

   d. flood tide and ebb tide

8. Tidal currents are made up of two opposing currents called _____ currents.

   a. longshore and littoral

   b. ebb and flood

   c. longshore and flood

   d. ebb and littoral

9. The shifting line of contact between water and land is referred to as a _____, and the broader term _____ refers to a zone in which coastal processes operate or have strong influence.

   a. coastline; shoreline

   b. beach; coastline

   c. shoreline; coastline

   d. seashore; coastline

10. Label the landforms in this diagram that have been created by weathering and wave action against a sea cliff.

11. When sand arrives at a particular section of the beach more rapidly than it is carried away, the beach is widened and built oceanward. This is called _____.

   a. retrogradation

   b. progradation

   c. propagation

   d. retreading

12. Broad expanses of enclosed shallow water called _____ are common features immediately adjacent to barrier islands.

   a. salt marshes

   b. marine terraces

   c. lagoons

   d. tidal inlets

13. _____ coasts are unique in that the addition of new land is made by organisms in warm oceans.

   a. Fiord

   b. Coral reef

   c. Ria

   d. Barrier island

14. In the accompanying photo, what is the name of the structures that are influencing the shape of the beach? Indicate the direction of the longshore drift.

15. Which of the following is *not* an anticipated impact of global warming on coastal environments?

   a. increased coastal erosion

   b. increased coastal wetlands

   c. estuaries pushed landward

   d. increased land subsidence

THE PLANNER ✓

Review your Chapter Planner on the chapter opener and check off your completed work.

Glacial and Periglacial Landforms

For about the past 2.5 million years, the Earth has experienced the Late-Cenozoic Ice Age. During this Ice Age, glaciers have advanced and retreated many times. As recently as 20,000 years ago, a huge ice sheet hundreds of meters thick ranged as far south as 38° N latitude in North America. But soon afterward, climate began to change, and the ice sheet began to retreat. By 11,000 years ago, the ice front was north of the Great Lakes, and by 6500 years ago, the ice sheet was completely gone.

This ice sheet and its predecessors have altered vast expanses of the North American landscape. Each time the ice advanced, it scraped and carved the Earth's surface, widening stream valleys and leveling soil and regolith into flattened plains. As the ice retreated, it littered the landscape with debris and left behind plains and hills of gravel and sand. Each glacial episode has been accompanied by climate change—cooling to start the glaciation and warming to end it.

Today's ice sheets are restricted to Greenland and Antarctica, but highland glaciers still survive. Here we can watch glaciers in action to learn how they move, erode, and deposit sediment to create their distinctive landforms.

The Margerie Glacier, a tidewater glacier in Glacier Bay National Park, Alaska, heads high in the Fairweather Range near Mount Quincy Adams and flows into an arm of Glacier Bay.

CHAPTER OUTLINE

CHAPTER PLANNER ✓

- ❏ Study the picture and read the opening story.
- ❏ Scan the Learning Objectives in each section:
  p. 426 ❏   p. 429 ❏   p. 433 ❏   p. 440 ❏   p. 445 ❏
- ❏ Read the text and study all figures and visuals. Answer any questions.

Analyze key features
- ❏ Geography InSight, p. 426
- ❏ Where Geographers Click, p. 428
- ❏ Process Diagram p. 434 ❏   p. 437 ❏
- ❏ What a Geographer Sees, p. 442
- ❏ Video Explorations, p. 449
- ❏ Stop: Answer the Concept Checks before you go on.
  p. 429 ❏   p. 433 ❏   p. 439 ❏   p. 444 ❏   p. 449 ❏

End of chapter
- ❏ Review the Summary and Key Terms.
- ❏ Answer the Critical and Creative Thinking Questions.
- ❏ Answer What is happening in this picture?
- ❏ Complete the Self-Test and check your answers.

Types of Glaciers

LEARNING OBJECTIVES

1. **Describe** the features and types of alpine glaciers.

2. **Explain** the features of ice sheets.

A lmost 70% of the Earth's fresh water is stored in the **cryosphere**, primarily in the form of **glaciers**. Glaciers form in regions that have low temperatures and sufficient snowfall. These conditions are found at both high elevations and high latitudes. In mountains, glacial ice can form even in tropical and equatorial zones if the elevation is high enough to keep average annual temperatures below freezing. Glaciers can take a variety of forms, but they can generally be divided into two broad categories: alpine glaciers and ice sheets, also called *continental glaciers*.

> **cryosphere** The portion of the hydrosphere in which water is stored as ice.
>
> **glacier** Any large natural accumulation of land ice affected by present or past motion.

Alpine Glaciers

> **alpine glacier** A glacier formed at high elevation, typically flowing down steep mountain slopes and filling valleys below.

Alpine glaciers, or mountain glaciers, form in high mountain ranges where snow accumulates and temperatures are cold enough to maintain year-round snow cover. Alpine glaciers take a variety of specific forms, depending on where they occur and the prevailing weather patterns (**Figure 14.1**). In high mountains, glaciers flow from high-elevation collecting grounds down to lower elevations, where temperatures are warmer. Here the ice melts and evaporates. Glacial meltwater provides freshwater sources for large portions of the globe. Himalayan glaciers, for example, feed rivers throughout China, Southeast Asia, and India. The recent decline in many alpine glaciers due to global climate change, however, presents a serious threat to the supply of fresh water for these regions.

Geography InSight

Alpine glaciers form in high mountain regions where there is sufficient snowfall and temperatures are cold enough to maintain year-round snow cover.

a. Cirque glacier
A cirque glacier, such as this one in Montana's Glacier National Park, occupies a steep, bowl-shaped depression near a mountain summit. This glacier, now shrunken in the present warm interglacial climate, would have been much larger during the last glacial period, flowing downhill and joining other glaciers to form a valley glacier below.

c. Tidewater glacier

A glacier that ends in a valley filled by an arm of the sea is called a tidewater glacier. By virtue of its bulk and weight, the glacier may stay grounded—resting on the valley floor—for some distance away from the shoreline before sea water begins to float the ice. Tidewater glaciers often produce icebergs that break off and float away.

Ocean

d. Piedmont glacier

d. When a mountain glacier flows out of the mountains and onto a surrounding lowland, it is called a piedmont glacier. This example is the Turnstone Glacier on Ellesmere Island, Canada, which is fed by a highland ice cap.

b. Valley glacier

Valley glaciers flow down steep-sided valleys once carved by streams. Several glaciers have coalesced to form this valley glacier in the Saint Elias National Park and Preserve, Alaska. Medial moraines of rock debris on the ice surface (see also Figure 14.8) parallel the downslope flow of the glacier.

Alpine glaciers originate from a snowfield that has collected in a bowl-shaped depression called a **cirque**. When alpine glaciers are contained within these basins, they are called cirque glaciers. However, most alpine glaciers flow out of these basins to become valley glaciers, which occupy sloping stream valleys between steep rock walls. When a valley glacier flows out onto a surrounding plain, it forms a piedmont glacier. When a valley glacier terminates in sea water, as a tidewater glacier, blocks of ice break off to form icebergs.

Large masses of ice often occur at the very tops of mountain ranges. **Ice caps** are continuous masses that cover mountaintops with distinctive domes. **Ice fields** consist of interconnected valley glaciers with protruding rock ridges or summits called *nunataks*. Ice caps and ice fields feed individual glaciers that travel down the mountainside to lower elevations.

Ice Sheets

In arctic and polar regions, temperatures are low enough year-round for snow to collect over broad areas, eventually forming a vast layer of glacial ice. Snow begins to accumulate on uplands, which are eventually buried under enormous volumes of ice. The layers of ice can reach a thickness of several thousand meters. The ice is thickest at the interior and thins toward the margins. The ice then spreads outward, over surrounding lowlands, and covers all landforms it encounters. We call this extensive type of ice mass an **ice sheet**, or a continental glacier.

> **ice sheet** A large, thick plate of glacial ice that moves outward in all directions.

At some locations, ice sheets extend long tongues to reach the sea, known as *outlet glaciers*. In other locations, glaciers meet the sea to form great plates of floating glacial ice, called *ice shelves*. Ice shelves are fed by the ice sheet, and they also accumulate new ice through the compaction of snow. Large masses of ice break off from outlet glaciers and the floating edges of the ice sheets. These masses drift out to open sea with tidal currents to become icebergs. As global climate warms and the rate of ice shelf melting increases, some ice shelves are at risk of breaking loose from their ice sheets (see *Where Geographers Click*). As ice shelves break up and drift away, the broken ice creates navigational hazards.

Where Geographers CLICK

Monitoring the Wilkins Ice Shelf

www.esa.int/esaEO/SEMWZS5DHNF_index_0.html

The Wilkins Ice Shelf was connected to the coastline of Antarctica by an ice bridge approximately 100 kilometers (62 miles) long and only a few kilometers wide. In April 2009, the bridge collapsed and destabilized the shelf, releasing icebergs from the northern front. Now that the Wilkins Ice Shelf is at risk of breaking away from the Antarctic peninsula, the European Space Agency's (ESA's) Envisat satellite is observing the area on a daily basis. These satellite images of the ice shelf are updated automatically on the ESA's Web site to monitor the developments as they occur.

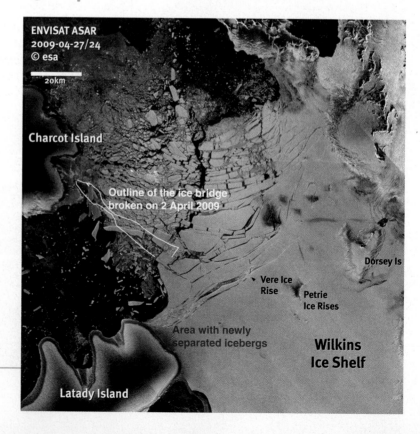

Greenland Ice Sheet • Figure 14.2

The Greenland Ice Sheet, seen here in a photo from space, occupies more than 1.7 million km² (about 670,000 mi²) and covers about seven-eighths of this vast island. The view here is of the southern tip, showing the snow-covered terrain of hills and fiords around the central mass of ice.

Although ice sheets covered parts of northern North America and Eurasia many times during the most recent ice age, today the only remaining ice sheets are those of Antarctica and Greenland. The Greenland Ice Sheet has the form of a broad, smooth dome (**Figure 14.2**). The only land exposed is a narrow, mountainous coastal strip. The Antarctic Ice Sheet covers 13 million km² (about 5 million mi²). It is as much as 4000 meters (about 13,000 ft) thick—about 1000 m (about 3200 ft) thicker than the Greenland Ice Sheet. Together, these ice sheets make up over 99% of the world's ice. No ice sheet exists near the North Pole, which lies in the middle of the vast Arctic Ocean. Ice there occurs only as floating sea ice.

CONCEPT CHECK

1. **How** do icebergs form?
2. **Why** is there no ice sheet at the North Pole?

Glacial Processes

LEARNING OBJECTIVES

1. **Explain** how glaciers form.
2. **Describe** how glaciers move.
3. **Explain** glacial erosion and deposition.

Although many parts of the world have snow and ice, only certain regions can produce glaciers. The ability of glaciers to survive and advance is also dependent on specific climatic conditions. Glaciologists study the formation and behavior of glaciers.

Formation of Glaciers

Glaciers form as snow accumulates from year to year. When the snowfall of the winter exceeds the loss of snow in the summer due to evaporation and melting, a new layer of snow is added to the surface of the glacier. Surface snow melts and refreezes on warm days and cold nights, compacting the snow and turning it into granular ice called **firn**. This intermediate ice is then compressed over a period of 25 to 100 years into hard, crystalline glacial ice by the weight of the layers above it. The **zone of accumulation** identifies the part of the glacier where snow is accumulating. Precipitation rates, temperatures, and the volume of snow layers dictate the rate of glacial ice formation. When the ice mass becomes so thick that it begins to move under the force of gravity, it becomes an active glacier. Flowing downhill, the glacier encounters less snowfall and warmer temperatures. Evaporation and melting increase, and eventually more ice is lost than is formed. This part of the glacier is the **zone of ablation**. Between the two zones is the **equilibrium line**, where the rate of snow accumulation balances the rate of evaporation and melting (**Figure 14.3**).

A glacier is an open system that has a dynamic balance between the input of accumulated snow and the output of melting or evaporating snow. The glacier's input and output, called its *mass balance*, changes each year as a result of the annual temperature and amount of snowfall. If the mass balance remains positive for some years, the glacier grows in bulk, and its end point, called the *terminus*, moves forward. If the mass balance remains negative, the glacier shrinks. Detailed field measurements allow scientists to monitor how the mass balance of glaciers changes over time—for example, as a result of climate change (**Figure 14.4**).

Glacial mass balance • Figure 14.3

Snowfall exceeds evaporation and melting in the zone of accumulation, where new glacial ice is formed. At lower elevations, increased evaporation and melting exceed snowfall. This is the zone of ablation. The equilibrium line marks the location where accumulation and ablation are about equal. Although the amounts of snow accumulation and ablation change from year to year, the glacier remains more or less in equilibrium until climate changes. Plus and minus signs indicate the relative degree of gain (+) or loss (–) to the glacier by each process in each zone.

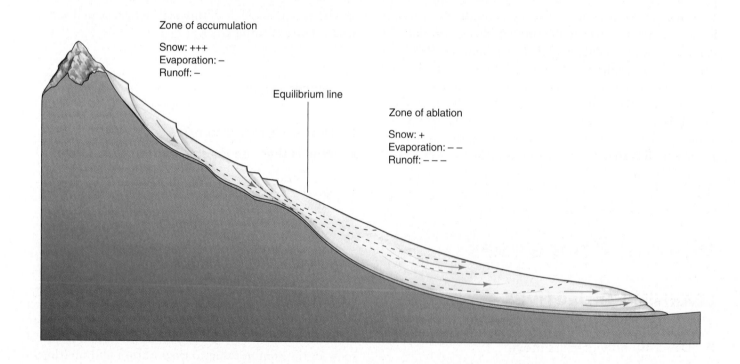

Zone of accumulation

Snow: +++
Evaporation: –
Runoff: –

Equilibrium line

Zone of ablation

Snow: +
Evaporation: – –
Runoff: – – –

Think Critically

1. Explain why volcanic ash might form a dark band crossing the lower portion of an alpine glacier.
2. Would you expect the depth to the ash layer to increase or decrease in an upvalley direction? Why?

Monitoring the effect of climate change on glaciers • Figure 14.4

McCall Glacier, located in what is now the Arctic National Wildlife Refuge in Alaska, has the longest history of scientific observation for any U.S. arctic glacier, with records beginning in 1958. Such long-term monitoring allows scientists to gauge the effect of climate change on a glacier's mass balance.

a. A comparison of photos taken in 1958 (left) and 2003 (right) indicates that the glacier is retreating.

b. Field researchers set instrumentation for making mass balance measurements, including surface mass balance, ice volume, ice temperature, ice velocities, bed properties, ice cores, subsurface ice accumulation, surface albedo, and local weather.

Think Critically

1. Melting glaciers are visually striking evidence of climate change. List, in order, the steps that would lead to a glacier's retreat due to climate change.
2. How might the loss of the glacier's ice volume affect downstream ecosystems?

Movement of Glaciers

When we think of ice, most of us picture a brittle, crystalline solid. But large bodies of ice, with a great thickness and weight, can flow in response to gravity. That's because the pressure on the ice at the bottom of an ice mass changes the physical properties of the ice, causing it to lose rigidity and become plastic. This means that a huge body of ice can move downhill or spread out over a large area. This motion typically happens at a rate of a few centimeters per day.

A glacier is always flowing outward or forward, even if the front of the glacier is retreating. Most of a glacier's movement occurs by slippage between the bottom and

Crevasses • Figure 14.5

Deep fissures in the ice, called crevasses, open up as a result of stresses in the brittle surface layer of a glacier. Crevasses form perpendicular to the direction of glacial flow, often where the glacier descends a steep slope or ledge.

Glacial flow →

sides of the glacier and the rock materials that confine it. In all but the coldest glaciers, high pressure at the base, coupled with the slow flow of geothermal heat and heat caused by friction, creates a layer of liquid water that allows the glacier to slide downhill. This process is called *basal sliding*. Some plastic flow of the glacier also occurs. It occurs most rapidly along the centerline of the glacier. In contrast to the plastic ice in the deeper parts of the glacier, the surface of the glacier bears less weight and is more brittle. As the glacier moves over a ridge or cliff, this brittle surface layer may crack as a result of the tension and form a **crevasse** (**Figure 14.5**).

Glacial Erosion and Deposition

Glacial ice normally contains rock and sediment that it has picked up along the way. These rock fragments range from large angular boulders to pulverized rock flour. Most of this material is loose rock debris and sediments found on the landscape as the ice overrides it. Alpine glaciers also carry rock debris that slides or falls from valley walls onto the surface of the ice.

Glaciers can create their own sediment by eroding underlying bedrock. Glaciers don't create enough pressure to fracture bedrock directly, but they can loosen and pick up bedrock blocks that are already fractured and easy to split off. Glaciers also push and drag rock particles against the bedrock, creating small fractures that release rock chips and particles. As a glacier flows over an irregular bed, small volumes of water at the base of the ice commonly refreeze, plucking loose rock fragments from the bedrock and carrying them into the glacial mass. Rock fragments carried on the bottom grind along the bedrock, sometimes smoothing and polishing the rock into grooves that mark the direction of glacial flow (**Figure 14.6**).

The glacier finally deposits the rock debris at its lower end, or terminus, where the ice melts. We use the term **glacial drift** to refer to all the varieties of rock debris that are deposited by glaciers. There are two types of drift. **Stratified drift** consists of layers of sorted and stratified clays, silts, sands, or gravels.

> **glacial drift** A general term for all varieties and forms of rock debris deposited by ice sheets.

Glacial abrasion • Figure 14.6

This grooved and polished surface marks the former path of glacial ice on the side of Tracy Arm Fiord in Alaska.

These materials were deposited by meltwater streams or in bodies of water adjacent to the ice. **Till** is an unstratified mixture of rock fragments, ranging in size from clay to boulders, that is deposited directly from the ice, without water transport. Glaciers deposit debris along their margins and where melting occurs, in the form of a **moraine**, as we will see in the following discussion.

> **moraine** The accumulation of rock debris carried by an alpine glacier of an ice sheet and deposited to become a depositional landform.

CONCEPT CHECK STOP

1. **What** climate characteristics are required for glacial ice to form?
2. **How** do crevasses form as a result of glacial movement?
3. **How** does glacial abrasion indicate the path of a glacier?

Glacial Landforms

LEARNING OBJECTIVES

1. **Describe** the landforms made by alpine glaciers.
2. **Describe** the landforms made by ice sheets.

G laciers create both erosional and depositional landforms in mountain ranges and in areas formerly covered by continental ice sheets. During the 2.5 million years of the present ice age, the Earth has experienced dozens of **glaciations**. Many parts of northern North America and Eurasia have been covered many times by massive sheets of glacial ice. As a result, glacial ice has shaped many landforms now visible in regions from the midlatitudes to subarctic zones.

> **glaciation** A single episode or time period in which ice sheets and alpine glaciers formed, spread, and disappeared.

Landforms produced by alpine glaciers • Figure 14.7

Alpine glaciers erode and shape mountains into distinctive landforms.

1 Before glaciation
This region has been sculptured entirely by weathering, mass wasting, and streams. The mountains look rugged, with fairly steep slopes and ridges. Small valleys are steep and V-shaped; larger valleys have narrow floodplains filled with alluvium and debris.

2 Snow accumulation
Climate change (cooling) increases snow buildup in the higher valley heads, forming cirques. Their cup shapes develop as regolith is stripped from slopes and bedrock is plucked out by ice. Slopes above the ice undergo rapid wasting from intense frost action.

3 Glaciation
As the cirque is eroded, steep bedrock walls replace the original slopes, forming jagged cols, horns, and arêtes. Glaciers descend from their cirques and flow together, joining the trunk glacier in the main valley. The glacial flow strips the valley slopes of regolith and flattens the valley floors by grinding and plucking out chunks of bedrock.

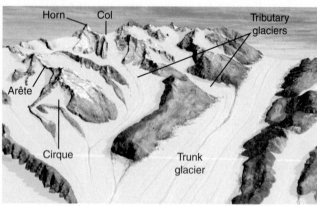

4 Melting glaciers
When the ice melts, a system of wide troughs with relatively flat floors is revealed. Smaller tributary valleys that join the main glacier valley may be left hanging at a higher elevation as the trunk glacier deepens the main valley. Some cirques fill with water to become lakes called tarns.

Ask Yourself

Picturesque mountain lakes called tarns are formed from what alpine glacial landscape feature?
a. col   c. cirque
b. arête   d. moraine

Landforms Made by Alpine Glaciers

Alpine glaciers typically erode existing valleys down to hard bedrock, stripping them of regolith formed by weathering and mass wasting. (**Figure 14.7**). Valley heads are enlarged and hollowed out by glaciers, producing bowl-shaped cirques. A cirque marks the origin of the glacier and is the first landform produced. When a glacier carves a depression into the bottom of a cirque and then melts away, a small lake called a **tarn** can be formed.

Over time, glacial erosion of bedrock and mass wasting of adjacent slopes steepen the sides of the cirque. Intersecting cirques carve away the mountain, leaving peaks called **horns** and sharp ridges called **arêtes**. Where opposed cirques have intersected deeply, a notch called a **col** forms. As the glacier descends from steeper slopes into the valley, smaller tributary glaciers join together to form larger trunk glaciers occupying valleys originally carved by streams. Glacial flow strips away loose regolith and then erodes and widens the glacial valley.

> **glacial trough**
> A deep, steep-sided valley shaped by the action of alpine glaciers.

After the ice has finally melted, a deep **glacial trough** remains. Tributary glaciers also produce troughs, but they are smaller in cross section and less deeply eroded by the smaller glaciers. When the floors of these troughs lie above the level of the main trough, they are called **hanging valleys**. When streams form later in these valleys, they create scenic waterfalls and rapids that cascade down steep slopes to the main trough below. Major troughs sometimes hold large, elongated trough lakes.

When the floor of a trough that is open to the sea lies below sea level, the sea water enters as the ice front recedes, creating a **fiord**. Fiords are opening up today along the Alaskan coast, where some glaciers are melting back rapidly and ocean waters are filling their troughs. Fiords are found largely along mountainous coasts between latitudes 50° and 70° N and S.

> **fiord** A narrow, deep ocean inlet that partially fills a glacial trough.

In addition to erosional features, alpine glaciers also leave behind distinctive depositional features. Rock waste slides or falls onto the glacier's edges, where it is worked into a ridge called a **lateral moraine** on either side of the trough (**Figure 14.8**). When two tributary glaciers come together to form a larger trunk glacier, their lateral moraines often combine and continue together down the center of the trough, forming a **medial moraine**.

Medial moraine • Figure 14.8

When two ice streams flow together, their separate lateral moraines join together to form a medial moraine. This aerial view shows medial moraines at the junction of two ice streams in the Saint Elias Range, Wrangell-St. Elias National Park, Alaska.

Lateral moraines

Medial moraines

435

In glacial landscapes, a variety of features mark both the expansion and retreat of ice sheets.

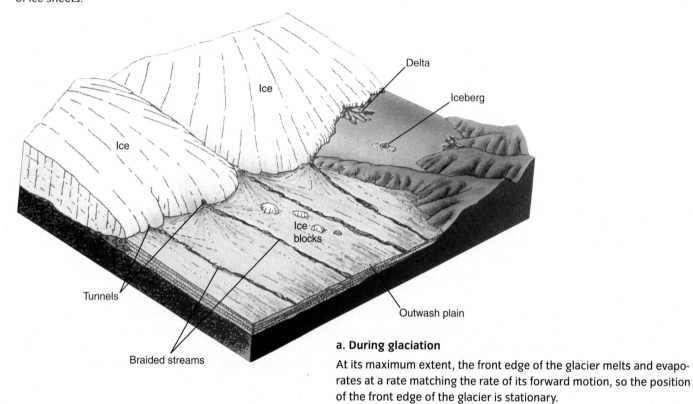

a. During glaciation

At its maximum extent, the front edge of the glacier melts and evaporates at a rate matching the rate of its forward motion, so the position of the front edge of the glacier is stationary.

b. After melting

When the glacier retreats, it leaves behind deposits in the form of moraines, eskers, drumlins, and kames.

Landforms Made by Ice Sheets

Like alpine glaciers, ice sheets strip away surface materials and erode bedrock. Repeated glaciations can excavate enormous amounts of rock at locations where the bedrock is weak and fractured and the flow of ice is channeled by a valley along the ice flow direction. Under these conditions, the ice sheet behaves like a valley glacier, scooping out a deep glacial trough.

Ice sheets are also responsible for depositional landforms. Like huge conveyor belts, traveling ice sheets transport and deposit debris. When the glacial margin is stationary over long periods of time, as was the case during the last glaciation, deposited debris accumulates in distinctive patterns. As the glaciers retreat, they leave behind several types of depositional landforms (**Figure 14.9**).

Moraines Sediment deposited in a ridge along the front or side edge of a glacier is known as a moraine (**Figure 14.10**). Glacial till and water-laid sediment that accumulates at the front edge of the ice forms an irregular jumbled heap of debris, sand, and gravel called the **terminal moraine**. When the ice front retreats, it may pause for some time along a number of positions, forming *recessional moraines*.

Formation of a moraine • Figure 14.10

A moraine forms at the end of an advancing glacier, which carries debris forward, like a conveyor belt, toward the glacial front. As the ice moves forward, melting and evaporation reduce the glacier's bulk and leave behind a deposit of glacial debris on the ice surface. Surface and internal debris then accumulate at the glacial front, along with some water-laid sediment, forming the moraine.

1 At the maximum extent of the glacier, debris accumulates at the ice front. Sediment under the glacier is sheared, compressed, and compacted by the weight and motion of the ice, forming lodgement till. Meltwater sorts and carries sediment beyond the moraine, forming the outwash plain.

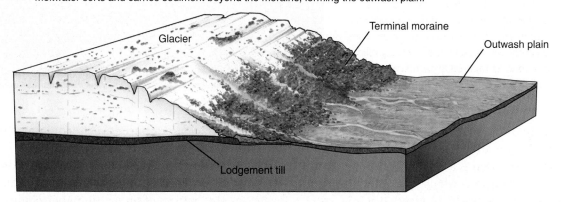

2 When the glacier retreats, the terminal moraine is left as a series of irregular piles of debris, often mixed with jumbled water-laid sediments. Debris in and on the glacier is left behind as a layer of melt-out till.

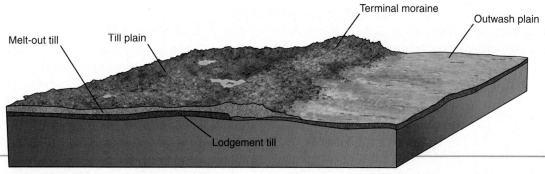

Landforms associated with till plains and outwash plains • Figure 14.11

Till and outwash plains often include several characteristic glacial landforms.

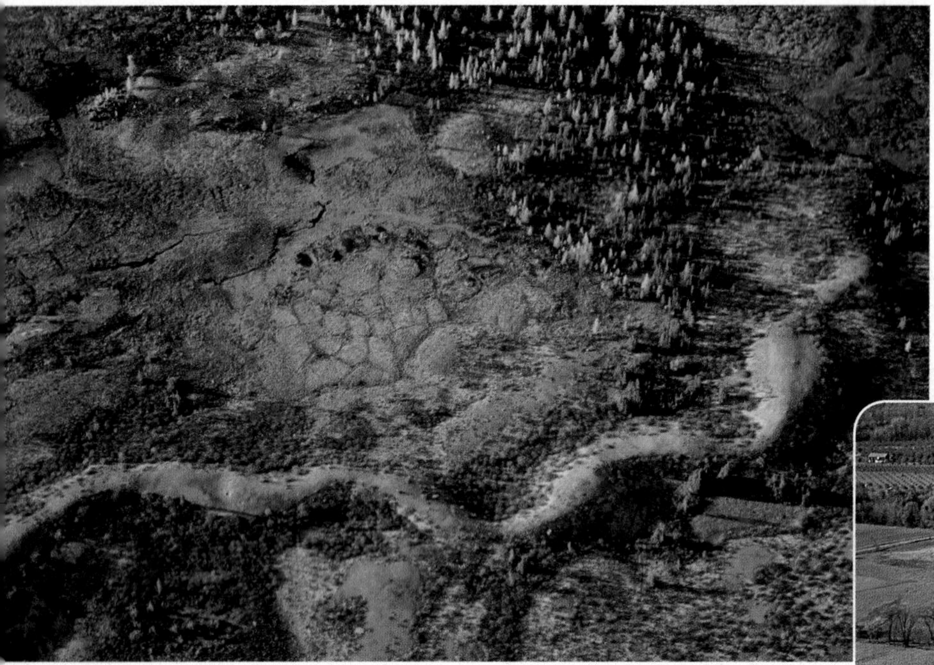

a. Esker

An esker is a curving ridge of sand and gravel found on the till plain. It is not till but water-laid sand and gravel marking the flow of water in an ice tunnel or channel at the bottom of the glacier. This example is from Kettle Moraine State Park, Wisconsin.

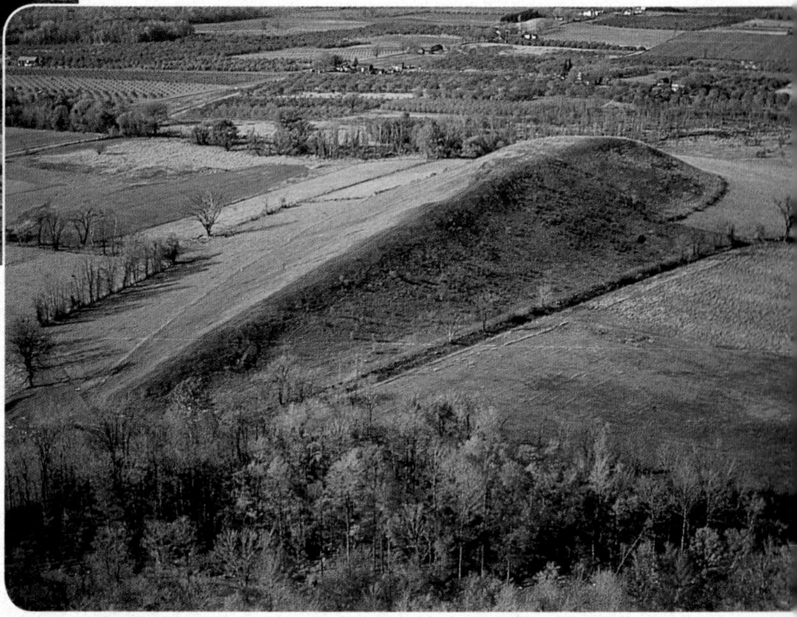

b. Drumlin

This small drumlin is constructed from lodgement till deposited beneath moving glacial ice. Drumlins are usually found on the till plain in groups, with the long axis of each drumlin parallel to the direction of ice movement. Drumlins often form at rock projections. As the glacier moves forward, till piles up in front of the projection and is also deposited behind the rock in a narrowing tail. However, some drumlins lack rock projections and are composed entirely of till. This drumlin is located south of Sodus, New York. Its tapered form indicates that it was deposited as ice moved from upper right to lower left.

Till plain As glacial ice moves forward, it compacts and compresses the sediment underneath it, creating *lodgement till*, a layer of particles of all sizes that is usually tough and dense. On top is a layer of *melt-out till*, which is let down as the ice retreats or stagnates and melts in place. It is a lighter and looser deposit that is easier to plow and work. Together, these two layers make up the **till plain** behind the terminal moraine.

The till plain can be thick and can bury the preexisting hills and valleys. Over parts of North America that were formerly covered by ice sheets, glacial drift thickness averages from 6 m (about 20 ft) over mountainous terrain, such as New England, to 15 m (about 50 ft) and more over the lowlands of the north-central United States. Over Iowa, drift thickness is from 45 to 60 m (about 150 to 200 ft), and over Illinois, it averages more than 30 m (about 100 ft). In some places where deep stream valleys already existed before the glaciers advanced, such as in parts of Ohio, drift is much thicker.

Two features of the till plain are eskers and drumlins (**Figure 14.11**). An **esker** is a sinuous ridge of sand and gravel, formed at the bottom of an ice tunnel or channel that drains the ice sheet. A **drumlin** is a smoothly rounded, oval hill of glacial till that resembles the bowl of an inverted teaspoon.

Outwash plains Beyond the moraine, an **outwash plain** is formed from water-laid sediment left by streams carrying water away from the ice front (Figure 14.10). The plain is built up from layers of sand and gravel. Although the plain is generally smooth, it is often marked by kettles and kames (Figure 14.11). A **kettle** is a depression, originally formed as outwash sand and gravel builds up around a block of stagnant ice. When the ice block melts, the kettle

c. Kettle

The outwash plain typically contains depressions, called kettles, that mark the position of ice blocks embedded in the outwash deposits. Often the kettle dips below the water table, creating a kettle pond. This example is near the headwaters of the Thelon River in Northwest Territories, Canada.

d. Kame

An isolated mound or hill of sand and gravel on the outwash plain is known as a kame. Some kames are formed as terraces between ice and a valley wall. Others are deltas built out into ice-front lakes dammed by ice. Still others accumulate when sediment is washed into openings within or between ice blocks. This type of kame is shown in the photo, taken in Antarctica.

Put It Together

Review Figure 14.9 and answer this question.
Which landform associated with outwash plains would be found beyond the farthest terminal moraine of a glacier?

a. kame
b. esker
c. drumlin
d. till plain

hole in the outwash plain remains. A **kame** is a hill or mound of water-laid sediment found on the outwash plain.

Human development of glacial landforms

Landforms associated with glaciers are of major environmental importance. Glaciation can have both positive and negative effects on the land for agricultural development. In hilly or mountainous regions, such as New England, the glacial till is thinly distributed and extremely stony, making cultivation difficult. Till deposits built up on steep mountains or hillslopes can absorb water from melting snows and spring rains and then become earthflows. Crop cultivation is also hindered along moraine belts because of the steep slopes, the irregularity of the topography, and the number of boulders. Moraine belts, however, are well suited to pastures.

Flat till plains, outwash plains, and lake plains, on the other hand, can sometimes provide very productive agricultural land. There are fertile soils on till plains and on exposed lakebeds bordering the Great Lakes. Their fertility is enhanced by the blanket of wind-deposited silt (loess) that covers these plains.

Stratified drift provides sand and gravel deposits from outwash plains, kames, and eskers that are used for road construction and in concrete. Thick, stratified drift deposits can also serve as aquifers, which are a major source of ground water.

CONCEPT CHECK STOP

1. **How** is a mountain landscape different before and after glaciation?

2. **How** does the formation of recessional moraines differ from the formation of terminal moraines?

Periglacial Processes and Landforms

LEARNING OBJECTIVES

1. **Analyze** the spatial distribution of permafrost.
2. **Explain** the formation of periglacial landforms.
3. **Outline** the causes and effects of permafrost thawing.

I n the arctic and alpine tundra regions, year-round cold temperatures keep the ground frozen and prevent the growth of substantial vegetation. These **periglacial** regions (*peri-* means "near") were formerly covered with ice and were shaped by glacial processes. Their present-day climate, which is exemplified by treeless, frozen ground, gives rise to a unique set of periglacial processes and landforms.

> **periglacial** In an environment located in cold climates or near the margins of alpine glaciers or large ice sheets.

Permafrost

In the tundra, mean annual temperatures are below freezing, and only the top layer of soil or regolith thaws during the warmest month or two. Ground and bedrock that lie below the freezing point of fresh water (0°C [32°F]) all year round are called **permafrost**. Permafrost includes clay, silt, sand, pebbles, and boulders, as well as solid bedrock perennially below freezing.

> **permafrost** Soil, regolith, and bedrock at a temperature that remains below 0°C (32°F) on a perennial basis.

Distribution of permafrost Permafrost is more or less extensive based on climate conditions in a particular region (**Figure 14.12**). In the coldest regions of the northern hemisphere, there is continuous permafrost, in which all ground surfaces are frozen except those beneath deep lakes. The zone of continuous permafrost coincides largely with the tundra climate, but it also includes a large part of the boreal forest climate in Siberia. Permafrost reaches to a depth of 300 to 450 m (about 1000 to 1500 ft) in the continuous zone near latitude 70° N. Discontinuous permafrost is defined as areas where frost-free zones exist under lakes and rivers, creating patches of frozen and non-frozen soil. These areas are common in much of the boreal forest climate zone of North America and Eurasia. Farther south, zones of alpine permafrost occur in high mountain areas where temperatures remain cold as a result of high elevation.

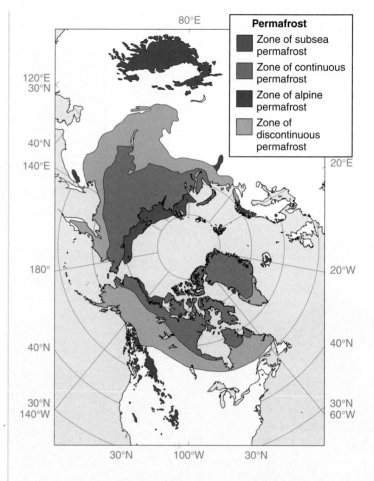

Distribution of permafrost • Figure 14.12

Most of the world's permafrost occurs in the northern hemisphere, where it can be divided into regions of continuous and discontinuous permafrost. Alpine permafrost occurs in high mountain regions, and subsea permafrost occurs along continental margins at the highest latitudes.

Today's climate change is increasing temperatures and bringing more precipitation to much of the North American Arctic region, which in turn will change the distribution of permafrost. Substantial warming is being recorded, and climate scientists predict that by the end of the century, arctic temperatures will have risen by 4°C to 8°C (7.2°F to 14.4°F), and annual precipitation will have increased by as much as 20%. As the climate warms, the border between continuous and discontinuous permafrost will shift poleward as increased snowfall insulates the ground under new forest. Discontinuous permafrost will be reduced at the southern boundary, and the isolated permafrost to the south will largely disappear.

Permafrost processes Permafrost terrains have a shallow surface layer, known as the *active layer*, which thaws with the changing seasons. The active layer ranges in thickness from about 15 cm (6 in.) to about 4 m (13 ft), depending on the latitude and nature of the ground. The thickness of the active layer changes as a result of climate changes, getting thinner during colder periods and thicker during warmer periods. As a result of current warming and increased temperatures in the arctic, the active layer over broad areas of permafrost is deepening.

Taliks are pockets of unfrozen ground that occur in a zone of discontinuous permafrost or underneath a deep lake or river in a zone of continuous permafrost. Taliks form connective links between ground water and the active layer.

Ground Ice and Periglacial Landforms

Water is commonly found in pore spaces in the ground, and in its frozen state it is known as *ground ice*. The amount of ground ice present within and above the permafrost layer varies greatly. In the active layer, it can take the form of a body of almost 100% ice. At increasing depth, the percentage of ice becomes lower, approaching zero at great depth.

During the short summer season, ground ice thaws, leaving the soil saturated with water and vulnerable to mass wasting and water erosion. With the return of cold temperatures, the freezing of soil water exerts a strong mechanical influence on the surface layer, creating a number of distinctive landforms. Recall from Chapter 10 that water expands when it freezes, and ice accumulation in the soil can cause the ground to shift horizontally through frost heaving or vertically through frost thrusting. During warm periods, melting of the ice in the soil forms a layer of mud that can slump down to form terraces and lobes through the process of *gelifluction* (**Figure 14.13**). Repeated cycles of melting and freezing can open up cracks in the ground, leading

Gelifluction • Figure 14.13

Bulging masses of water-saturated regolith, along with plants and soil, have slowly moved downslope, flowing atop an impermeable bed of permafrost to form gelifluction lobes and low terraces. These examples are found at Breiddalfjellet, Breivikeidet, Troms, Norway. A cloud shadow crosses the lower middle portion of the air photo.

WHAT A GEOGRAPHER SEES

a.

Ice Wedges

Exposed by river erosion, this great wedge of solid ice fills a vertical crack in organic-rich floodplain silt along the Yukon River, near Galena in western Alaska (**Figure a**). A tourist in the region might be amazed at the ice formation, but a geographer would know that this type of ice wedge is a typical periglacial phenomenon that occurs as a result of frost action.

During extreme winter cold, a crack forms when permafrost shrinks (**Figure b**). In the spring, surface meltwater enters the crack and freezes, expanding the crack. After several hundred winters, the ice wedge has grown and continues to grow as the same seasonal sequence is repeated. The ice wedge thickens until it becomes as wide as 3 m (about 10 ft) and as deep as 30 m (about 100 ft).

b.

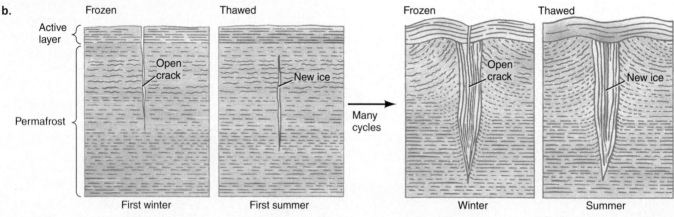

to the formation of ice wedges (see *What a Geographer Sees*).

Another remarkable ice-formed feature of the arctic tundra is a conspicuous conical mound called a *pingo* (**Figure 14.14**). Pingos form when pockets of unfrozen ground under a lakebed freeze after a lake is drained. As the unfrozen ground water is converted to ice, its volume expands, and it pushes upward into a dome shape. In extreme cases, pingos reach heights of 50 m (164 ft), with a base diameter of 600 m (about 2000 ft).

Pingo formation • Figure 14.14

This low hill, found on the tundra landscape of Northwest Territories, Canada, is a pingo. Formed by the freezing of water in a drained lake bed, it has a core of ice that is covered with soil and regolith.

Areas that contain coarse-textured regolith, consisting of rock particles in a range of sizes, can give rise to some very beautiful and distinctive features. The annual cycle of thawing and freezing sorts the coarsest fragments—pebbles and cobbles—from finer particles by moving them horizontally and vertically. This sorting produces rings and stripes of coarse fragments called *patterned ground* (**Figure 14.15**).

Patterned ground • Figure 14.15

Frost action sorts stones into distinctive patterns.

a. Polygon patterns

Sorted polygons of gravel form stone rings on this nearly flat land surface in Northeastland, Svalbard, Norway. The rings are 3 to 4 m (10 to 13 ft) across, and the gravel ridges are 20 to 30 cm (8 to 12 in) high.

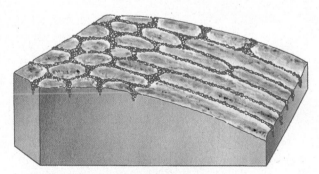

b. Stripe patterns

On steep slopes, soil creep elongates these polygons in the downslope direction, turning them into parallel stone stripes.

Periglacial Processes and Landforms **443**

Human Interactions with Periglacial Environments

Today increased permafrost thaw is affecting the periglacial environment by increasing the depth and extent of the melted soils. Localized thawing occurs as a result of construction and industry, while regional thawing is a result of human-induced climate change. Permafrost thaw further contributes to global warming by releasing greenhouse gases into the atmosphere. The frozen ground contains vast amounts of carbon in the form of peat, which decays to release CO_2 more rapidly under warming temperatures. Also abundant is methane, held within ice as methane hydrate. As the active layer warms and deepens, more of these gases are released into the atmosphere. This phenomenon has the potential to boost the total greenhouse gas volume in the atmosphere rapidly, and yet it is not accounted for in most current climate models.

The upper layers of permafrost are often rich in ice, and with climate warming, melting of these layers releases water. The ground collapses, producing water-filled depressions. Because the water conducts heat more readily than vegetation-covered soil, permafrost thawing expands, and the depressions form shallow lakes. We call the new

Permafrost thaw • Figure 14.17

These old buildings, located in Dawson City, Yukon Territory, Canada, have thawed the permafrost beneath them, causing their foundations to settle and the buildings to sag together.

Thermokarst • Figure 14.16

This aerial photo is used by the USGS to monitor thermokarst expansion and drainage along a section of Alaska's North Slope coastline. Thermokarst landscapes have compounded the challenges facing the oil resource extraction industry due to the unstable soils and poor drainage affecting roads and facilities.

terrain **thermokarst** because the landscape resembles limestone karst formations (**Figure 14.16**). Instead of being shaped by chemical solution, however, thermokarst terrain is shaped by heat flow (*thermo-* means "heat").

Permafrost can also thaw when people burn or scrape away layers of decaying matter and plants from the surface of the tundra or arctic forest. The presence of a heated building or an insulated pipe directly on the frozen ground leads to thawing. As the ice under the structure melts, the ground subsides, causing the structure to settle (**Figure 14.17**). In some cases, the disturbance can set off thermokarst formation.

Responsible building practices can reduce permafrost thaw. These practices include placing buildings on piles with an insulating air space below or, alternatively, insulating the surface with a thick pad of coarse gravel between the permafrost and the new building. In many arctic settlements, steam and hot-water lines are now placed above ground to prevent thaw of the permafrost layer.

CONCEPT CHECK

1. **How** will climate change affect the global distribution of permafrost?
2. **What** conditions give rise to patterned ground?
3. **What** steps can be taken during construction to prevent permafrost thaw?

Global Climate and Glaciation

LEARNING OBJECTIVES

1. **Explain** what an ice age is.
2. **Explain** the possible causes of glaciation cycles.
3. **Describe** the predicted effect of climate change on glaciation.

Glacial ice sheets have a significant impact on our global climate. Because glacial ice sheets reflect much of the solar radiation they receive, they influence the Earth's radiation and heat balance. The temperature difference between ice sheets and warm regions near the equator helps drive the system of global heat transport. When the volume of glacial ice increases, as it does during a glaciation, sea levels fall. When the ice sheets melt away, sea level rises. Today's coastal environments evolved during a rising sea level accompanying the melting of the last ice sheets of the Ice Age.

History of Glaciation

Glaciation occurs when temperatures fall in regions of ample snowfall, allowing ice to accumulate and build. *Deglaciation* occurs when the ice melts at the beginning of a period of milder climate called an *interglaciation*. An **ice age** is a period of millions of years of generally cold climate consisting of many alternating glaciations and interglaciations. Throughout its long history, the Earth has cycled through a number of ice ages.

The Ice Age The Cenozoic Era, which began about 66 million years ago, has seen a gradually cooling climate, culminating in the recent Ice Age. By about 40 million years ago, the Antarctic Ice Sheet began to form, and by about 5 million years ago, the Arctic Ocean was ice covered. Throughout the past 2.5 million years or so, the Earth has been experiencing the Late-Cenozoic Ice Age (or, simply, the Ice Age). This ice age consists of alternating glacial and interglacial periods; five major glaciations have occurred in the past 500,000 years.

Scientists have established a record of climate cycles during this period by analyzing layers of sediment deep under the sea. The Earth's magnetic field experienced many sudden reversals of polarity in Cenozoic time, and the absolute ages of these reversals are known with certainty. By observing these reversals in the sediments, scientists were able to date the layers accurately. They also studied the composition and chemistry of the core layers, which provide a record of ancient temperature cycles in the air and ocean. Deep-sea cores reveal a long history of alternating glaciations and interglaciations going back at least 2.5 million years, with climatic oscillations beginning even earlier.

The last major glaciation, called the Wisconsin Glaciation, started about 120,000 years ago. During this glaciation, ice sheets covered much of North America and Europe, as well as parts of northern Asia and southern South America (**Figure 14.18**). All high mountain areas of the world developed alpine glaciers. The white surface

Maximum glaciation during the Ice Age • Figure 14.18

During the last glacial period, ice sheets covered large portions of North America, Europe, and the Andes. Global sea level was about 120 m (about 400 ft) lower than present. Most of Canada was engulfed by the vast Laurentide ice sheet, which then spread south into the United States.

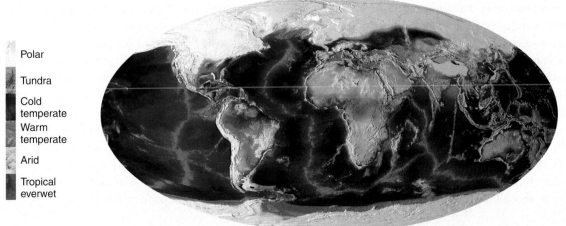

Polar

Tundra

Cold temperate

Warm temperate

Arid

Tropical everwet

of the expansive ice sheet had a higher albedo than bare ground, so it reflected more solar energy, contributing to global cooling by reducing the amount of solar energy entering the lithosphere, hydrosphere, and biosphere. The maximum ice advance of the glaciation was about 20,000 years ago, but retreat set in shortly thereafter. Once the ice began to melt, the albedo was lowered again, which contributed to more warming.

This and earlier glaciations have resculpted the face of North America's landscape. The moving ice sheet flattened and scraped the landscape and redefined the continent's major drainage systems. Great inland lakes were created behind tremendous ice dams. When these ice dams broke, the abrupt drainage of the lakes carved out new rivers. Glaciologists note that a series of these great floods have occurred during the Ice Age.

During glaciations, much more of the Earth's water was contained in ice, so the sea level was as much as 125 m (410 ft) lower than it is today, exposing large areas of the continental shelf on both sides of the Atlantic Basin. Rivers draining the continents deepened their valleys to meet the lowered sea level. The exposed continental shelf had a highly productive ecosystem and vegetated landscape, richly populated with animal life.

The weights of the continental ice sheets, covering vast areas with ice masses several kilometers thick, pushed down on the crust. This created depressions of hundreds of meters at some locations. When the ice melted, the crust began to rebound—and it is still rebounding today in some locations.

The Holocene Epoch Since the Wisconsin Glaciation ended, the Earth has been in a period of interglaciation called the Holocene Epoch. This period began with a rapid warming of ocean-surface temperatures. Continental climate zones then quickly shifted poleward, and plants recolonized the formerly glaciated areas.

There were three major climatic periods during the Holocene Epoch leading up to the last 2000 years. These periods are inferred from studies of fossil pollen and spores preserved in glacial bogs that show changes in vegetation cover over time. The earliest of the three is the Boreal stage, characterized by boreal forest vegetation in midlatitude regions. There followed a general warming until the Atlantic stage, with temperatures somewhat warmer than today, was reached about 8000 years ago (–8000 years). Next came a period of temperatures that were below average—the Subboreal stage. This stage spanned the age range –5000 to –2000 years.

We can describe the climate of the past 2000 years on a finer scale, thanks to historical records and detailed evidence. A secondary warm period occurred in the period CE 1000 to 1200 (–1000 to –800 years). This warm episode was followed by the Little Ice Age, CE 1450–1850 (–550 to –150 years), when valley glaciers made new advances and extended to lower elevations.

Triggering the Ice Age

About 50 million years ago, the Earth's climate was quite uniform and mild. Today, however, we are in an interglaciation of an ice age that began at least 2.5 million years ago, and possibly earlier. Most earth scientists believe the primary cause of this ice age was plate tectonics, which changed the configuration of the continents and oceans, shifting the circulation patterns of the oceans and atmosphere.

By the middle of the Cenozoic Era, North America, Eurasia, and Greenland had moved north toward the pole, creating a polar ocean largely surrounded by land. This reduced, and at times totally cut off, the flow of warm ocean currents into the polar ocean, leaving it ice-covered much of the time. The high albedo of the ice then lowered temperatures in high latitudes, setting the stage for formation of ice sheets on the encircling continents.

In the southern hemisphere, Antarctica had taken up a position over the South Pole, where it was ideally placed to develop an ice cap. Australia, migrating toward the equator, left a southern ocean free to circulate around the Antarctic continent, isolating it climatically from the other continents. By about 15 million years ago, both the East and West Antarctic Ice Sheets were growing. The collision of the Indian plate with the Eurasian plate pushed up the Himalaya Mountains and the Plateau of Tibet, changing atmospheric circulation patterns and adding to the world's snowy highland area. About 3 million years ago, the Isthmus of Panama emerged and the Bering Strait opened, providing the final configuration of land, ocean, and ice for the Ice Age to begin in earnest.

Cycles of Glaciation

During the present Ice Age, glaciers have come and gone many times. What mechanism is responsible for the many cycles of glaciation and interglaciation that the Earth has experienced during the present Ice Age? Although many causes have been proposed, the best evidence points to cyclic changes in the geometry of the Earth–Sun relationship. This theory, called the *astronomical hypothesis*, was first presented in the early 20th century. However, it was not widely accepted by the scientific community until the 1970s, when deep-ocean cores provided climate records that were accurate enough to confirm the theory.

Recall from Chapter 2 that the Earth's orbit around the Sun is an ellipse, not a circle. At perihelion, the Earth is closest to the Sun and receives the strongest insolation; insolation is weakest at aphelion, when the Earth–Sun distance is greatest. Astronomers have observed that the orbit itself slowly rotates on a 108,000-year cycle, so the time of year of the strongest and weakest global insolation changes by a very small amount each year. In addition, the orbit's shape varies on a cycle of 92,000 years, becoming more elliptical and then more circular. This changes the Earth–Sun distance and therefore the amount of solar energy the Earth receives through the annual cycle (**Figure 14.19**). The Earth's axis of rotation also experiences cyclic changes. The tilt angle of the axis varies from about 22 to 24 degrees on a 41,000-year cycle. The axis also wobbles on a 26,000-year cycle, moving in a slow circular motion much like a spinning top or toy gyroscope.

Changes in Earth–Sun geometry • Figure 14.19

Changes in the intensity of solar radiation with latitude and the season occur in response to variations in the Earth's orbit around the Sun and variations in the tilt and position of the Earth's axis of rotation.

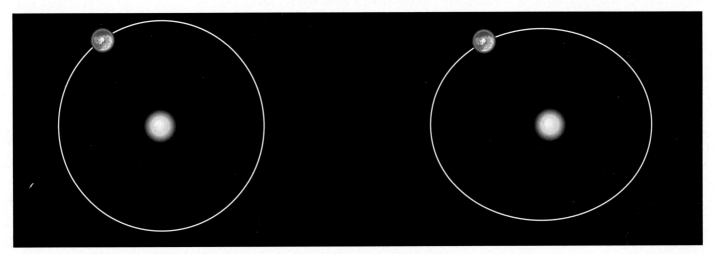

a. The shape of the Earth's orbit around the Sun varies from nearly circular to slightly elliptical, with a period of about 92,000 years. The orbit itself also slowly rotates, changing aphelion and perihelion on a cycle of 108,000 years (not shown).

b. The Earth's axis of rotation slowly revolves, tracing a circle over a period of about 26,000 years. As the axis of rotation moves along the circle, it also varies its angle slightly, from 22.1° to 24.5°, with a period of about 41,000 years.

24.5°
22.1°

Ask Yourself

When the tilt of the Earth is greatest (24.5°), what happens to the intensity of the Earth's seasons?
a. Summers are cooler and winters are warmer.
b. Summers are warmer and winters are cooler.
c. Summers are warmer and winters are warmer.
d. Summers are cooler and winters are cooler.

Global Climate and Glaciation **447**

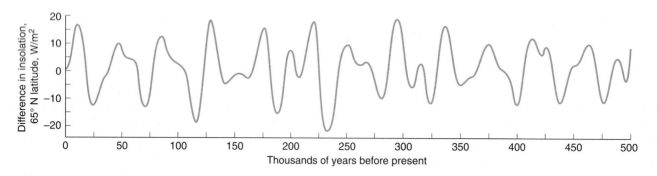

The Milankovitch curve • Figure 14.20

The vertical axis shows fluctuations in summer daily insolation at latitude 65° N for the past 500,000 years. Negative shifts prevent summer snowmelt and allow ice to build up over a period of years, while positive shifts encourage snowmelt and reduce ice buildup. These fluctuations are calculated from mathematical models of the change in Earth–Sun distance and the change in axial tilt with time. The zero value represents the present value.

These cycles in solar revolution and axial rotation mean that the annual insolation experienced at each latitude changes from year to year. Glacial onsets are thought to begin with cooler summers that allow snow to build up over many years in the northern hemisphere. This allows building ice sheets to further reflect solar insolation, creating a positive feedback loop. The cycles are plotted on a graph known as the Milankovitch curve, named for the engineer and mathematician who first calculated it in 1938 (**Figure 14.20**). The dominant cycle shown here, for summer daily insolation at 65° N latitude, indicates periods of 40,000 years. However, the combination of Milankovitch variations demonstrates major climate shifting cycles of 100,000 and 400,000 years. Dating methods from the deep-ocean cores and antarctic ice cores confirm the prevalence of these major cycles for the onset of climate change that creates glaciation and deglaciation.

Glaciation and Global Warming

Global temperatures have been slowly warming within the past century, as measured by an international community of meteorologists. Measurements have shown a noticeable decrease in polar sea ice since 1950 that corresponds with the temperature increase (**Figure 14.21**). *Video Explorations* discusses the implications of such changes for the planet. As the polar ice has melted, with its bright surface replaced by ocean water and land surfaces, the albedo of

Shrinking polar ice cap • Figure 14.21

Over the past 50 years, the extent of polar sea ice has noticeably decreased. Since 1970 alone, an area larger than Norway, Sweden, and Denmark combined has melted. This trend is predicted to accelerate as temperatures rise in the arctic and around the globe.

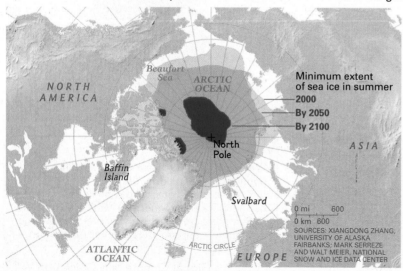

Put It Together

Review Figure 14.12 and complete this statement.
If polar sea ice shrinks as predicted by 2050, the zone of continuous permafrost will____.
a. increase in area
b. decrease in area
c. move farther south
d. remain the same

Video Explorations

Antarctic Ice

Antarctica is on the frontline of climate change. Surface temperatures have risen 2.5°C (4.5°F) in just 50 years, and ice shelves are disintegrating due to an excess amount of melting during the summer. Over the past 20,000 years, Antarctica has lost two-thirds of its mass. This video shows how geographers monitor the ice and the repercussions of melting on sea levels around the world.

arctic regions has significantly decreased, resulting in the absorption of more solar insolation. This in turn is creating a positive feedback loop that further increases warming. This warming feedback loop is accelerating deglaciation at both poles as well as in alpine environments.

Glacial melting is responsible for a series of events that affect humankind and the biosphere. Glacial lakes are forming in alpine regions, especially the Himalayas, creating huge freshwater lakes that are subject to flooding. Such flooding could affect the lives of hundreds of thousands living downslope in India and Nepal. Rising sea levels are enhanced by the glacial melt. Sea-level rise will affect coastal zones where the majority of the population lives. The influx of large volumes of fresh water from melting ice in Greenland and Antarctica is projected to disrupt long-standing ocean current systems that control regional and global weather patterns. The complex interactions among all these factors make it difficult to project the effects of glacial melting with any precision. International cooperation among scientists and policymakers will be critical as the science community learns more about the Earth's current climate change.

CONCEPT CHECK

1. **What** is the relationship between glaciation and sea-level rise?

2. **Why** does solar insolation on the Earth experience cyclical variations?

3. **How** does the Earth's surface albedo create a feedback loop during periods of glaciation and deglaciation?

THE PLANNER ✓

Summary

1 Types of Glaciers 426

- 70% of the world's fresh water is stored as ice in the **cryosphere**.

- **Glaciers** form in high latitudes and high mountain regions.

- **Alpine glaciers** develop in high mountains. **Ice sheets** are huge plates of ice and are present today in Greenland, shown in the photo, and Antarctica.

Greenland Ice Sheet • Figure 14.2

- The Antarctic Ice Sheet includes ice shelves—great plates of floating glacial ice.

---

2 Glacial Processes 429

- Glaciers form when snow accumulates in layers from year to year. The pressure of compacting snow transforms snow first into **firn** and then into glacial ice.

- As a result of the weight of overlying layers, glaciers are plastic in their lower layers and flow outward or downhill. Most glacial movement occurs by slippage on a thin water layer at the base of the glacier.

(Continued on next page.)

- As shown in the photograph, glaciers can erode bedrock by removing fractured blocks and using embedded rock fragments to chip or grind the bedrock. They leave depositional landforms, constructed of **glacial drift**, when the ice melts.

Glacial abrasion • Figure 14.6

3 Glacial Landforms 433

- Alpine glaciers erode and enlarge the heads of mountain valleys to form **cirques**, leaving peaks called **horns** and sharp ridges called **arêtes**.

- Through erosion, glaciers shape stream valleys into **glacial troughs**, which can become **fiords** if later submerged by rising sea level.

- **Moraines** of rubble and debris mark the sides (**lateral moraines**) and ends (**terminal moraines**) of glaciers, as shown in the diagram. When two tributary glaciers come together, they form a **medial moraine** down the middle of the trunk glacier.

Formation of a moraine • Figure 14.10

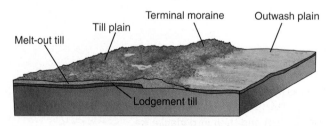

- Moving ice sheets leave behind till plains and **outwash plains**. Till plains are often marked with **eskers** and **drumlins**. Outwash plains contain **kettles** and **kames**.

- Plains shaped by glacial action can be fertile and productive. Glacial sand and gravel are important resources.

4 Periglacial Processes and Landforms 440

- **Periglacial** landscapes occur in regions at the edges of areas of glaciation, both in high latitudes and at high elevations. They are characterized by cold temperatures, low snowfall, and limited vegetation.

- **Permafrost** is ground that is perennially below freezing. Permafrost is continuous in the highest latitudes, as shown on the map. In lower latitudes, discontinuous permafrost includes areas of unfrozen ground called *taliks*.

Distribution of permafrost • Figure 14.12

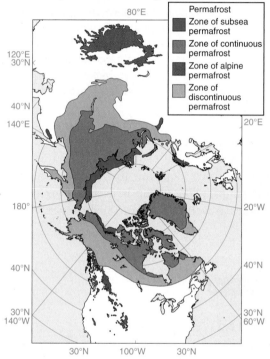

- Above the permafrost is an active layer that undergoes a seasonal cycle of freezing and thawing.

- Water that freezes in the ground is ground ice. Cycles of freezing and thawing of ground ice cause frost action including frost heaving and frost thrusting. In the process of gelifluction, saturated soil can flow slowly downhill on a bed of permafrost.

- Ice wedges are a form of ground ice found in permafrost terrains. Freezing pockets of water in lake beds push up hills called pingos.

- Patterned ground occurs when freeze–thaw cycles distribute rocks into patterns.

- When the permafrost thaws, the ground can collapse to form a landscape of shallow lakes and trenches called **thermokarst**. Increased permafrost thawing results from global warming as well as human development on permafrost terrain.

- An **ice age** includes alternating periods of **glaciation**, deglaciation, and interglaciation. During the past 2.5 million years or so, the Earth has been experiencing the Late-Cenozoic Ice Age.

- The most recent glaciation, the Wisconsin Glaciation, changed the course of North American rivers. Huge ice-dammed lakes filled and then drained abruptly. Sea level dropped to about 125 m (410 ft) below present.

- The Ice Age was most likely brought on by the motions of the continents, which changed patterns of oceanic and atmospheric circulation to produce ice covers at the poles.

- Individual periods of glaciation and interglaciation are most likely related to cyclic changes in Earth–Sun distance and the Earth's axial tilt, as shown in the diagram.

Changes in Earth–Sun geometry • Figure 14.19

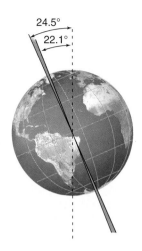

- The current warming of the Earth's climate is causing glacial ice to melt. Glacial melt contributes to sea-level rise and flooding. It may also change global ocean currents, which influence regional weather patterns.

Key Terms

- alpine glacier 426
- arête 435
- cirque 432
- col 435
- crevasse 432
- cryosphere 426
- drumlin 438
- equilibrium line 430
- esker 438
- fiord 435

- firn 429
- glacial drift 432
- glacial trough 435
- glaciation 433
- glacier 428
- hanging valley 435
- horn 435
- ice age 445
- ice cap 428

- ice field 428
- ice sheet 428
- kame 439
- kettle 438
- lateral moraine 435
- medial moraine 435
- moraine 433
- outwash plain 438
- periglacial 440

- permafrost 440
- stratified drift 432
- tarn 435
- terminal moraine 437
- thermokarst 444
- till 433
- till plain 438
- zone of ablation 430
- zone of accumulation 429

Critical and Creative Thinking Questions

1. Use Google Earth to find an alpine glacier. What are the latitude and elevation of the glacier? Explain how the elevation of the glacier would change if the glacier were located at a lower latitude.

2. Many glaciers are shrinking due to increased temperatures associated with climate change, but some glaciers are expanding. Formulate a hypothesis to explain these expanding glaciers that is consistent with observations of global climate change.

3. Why is the study of terminal moraines useful for understanding the dynamics of retreating glaciers but less useful for understanding advancing glaciers?

4. Many national parks have landscapes formed by alpine glaciers. Using the terminology of Figure 14.7, describe three such visually striking features in a specific national park.

5. Describe two techniques for building responsibly on permafrost and then predict the impact on the building when climate change melts the permafrost.

6. A friend tells you that global warming is part of a natural Earth cycle. Referring to Milankovitch theory, give a response that explains the difference between natural cycles of glaciation and current global climate change.

7. Guided by this map, describe an impact of retreating sea ice that will occur by 2050 and an additional impact that will occur by 2100.

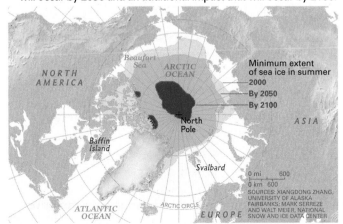

What is happening in this picture?

This photo, taken by an astronaut on the Space Shuttle, shows the Finger Lakes region of New York. The long, deep, parallel basins, which now hold lakes, have been called "inland fiords." The lakes are oriented roughly in a north–south direction.

Self-Test

(Check your answers in the Appendix.)

1. Of the following locations, which is not covered with an ice sheet?

   a. North Pole   c. Greenland

   b. South Pole   d. Antarctica

2. _____ are bodies of land ice that have broken free from glaciers that terminate in the ocean.

   a. Bergs   c. Sea ice

   b. Icebergs   d. Pack ice

3. Except with the coldest glaciers, _____ is the primary mechanism of glacial motion.

   a. plastic flow   c. abrasion

   b. ablation   d. basal sliding

4. Add the following labels to the accompanying illustration: zone of accumulation, equilibrium line, and zone of ablation.

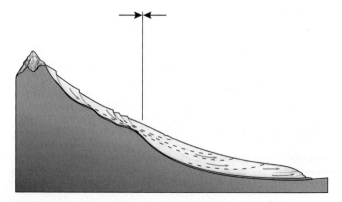

5. Glacial _____ produces grooved and polished bedrock surfaces that mark the path of advancing glacial ice.

   a. plucking

   b. abrasion

   c. deposition

   d. erosion

6. Where two cirque headwalls intersect from opposite sides, a jagged, knife-like ridge called a(n) _____ is formed.

   a. arête

   b. tarn

   c. horn

   d. col

7. A ridge or pile of rock debris and sediment left by glacial action that marks the terminus of a glacier is called a _____.

   a. medial moraine

   b. recessional moraine

   c. terminal moraine

   d. lateral moraine

8. The diagram shows some of the landforms produced by ice sheets. Label the following features: drumlins, eskers, kettles, and the outwash plain.

9. The _____ layer of permafrost terrains thaws and refreezes each year.

   a. continuous permafrost

   b. permafrost

   c. active

   d. discontinuous permafrost

10. Ice wedges _____.

   a. form when spring meltwater freezes in winter cracks that develop in permafrost

   b. are found only at high elevations

   c. are only rarely wider than 10 cm

   d. are inverted in thermokarst terrains

11. What type of periglacial landform is most prominent in the photograph?

   a. talik

   c. gelifluction zone

   b. pingo

   d. stone polygon

12. A succession of glaciations regularly interrupted by warmer interglacial periods constitutes a(n) _____.

   a. glacial period

   b. freezing epoch

   c. ice age

   d. interstadial period

13. What may have caused the Earth to enter into an ice age in the late Cenozoic Era?

   a. changes in oceanic and atmospheric circulation patterns

   b. plate tectonics

   c. formation of perennial ice cover in polar regions

   d. all of the above

14. The most likely explanation for the cyclical nature of glaciations and interglaciations during the Late-Cenozoic Ice Age involves _____.

   a. the changing distance between the Earth and the Sun

   b. the changing tilt of the Earth's axis of rotation

   c. both a and b

   d. none of the above

15. The elapsed time span of about 10,000 years since the Wisconsin Glaciation ended is called the _____.

   a. Holocene Epoch

   b. Miocene Epoch

   c. Paleocene Epoch

   d. Pleistocene Epoch

THE PLANNER ✓

Review your Chapter Planner on the chapter opener and check off your completed work.

Global Soils

The Sumerian culture developed between about 5000 and 3000 BCE, cradled between the Tigris and Euphrates Rivers that flow out across the deserts of modern-day Iraq and the Persian Gulf. The Sumerians are thought to have developed the first written language and built the first cities. They produced highly crafted pottery, jewelry, and ornate weapons of copper, silver, and gold. They also built a vast irrigation system to support their agricultural base, exploiting the nearby Tigris and Euphrates Rivers.

In about 2400 BCE Sumerian croplands began to deteriorate. Each year salt was added to the soil from evaporation of irrigation water. Over time, the salt buildup in the soil was high enough to cause serious declines in grain yields. In response, the Sumerian civilization withered in the south, shifting political control to the rising northern power of Babylon. In their application of water, the Sumerians unknowingly destroyed their own soils.

The great Babylonian civilization followed the same boom-to-bust pathway. Over time, its irrigation systems became clogged with silt and contaminated with salt, ultimately leading to the downfall of the Babylonians. Problems with irrigation systems have plagued desert populations for centuries. Today, Pakistan, Mexico, and Israel suffer similarly, highlighting the vital role of soil in agriculture.

CHAPTER OUTLINE

This ancient panel from Mesopotamia, decorated with shell, red limestone, and lapis lazuli, depicts a rich bounty of animals, fish, grain, and other goods brought in procession to a banquet.

CHAPTER PLANNER ✓

- ❑ Study the picture and read the opening story.
- ❑ Scan the Learning Objectives in each section:
  p. 456 ❑   p. 462 ❑   p. 467 ❑   p. 468 ❑
- ❑ Read the text and study all figures and visuals.
  Answer any questions.

Analyze key features

- ❑ Process Diagram p. 458 ❑   p. 466 ❑
- ❑ Video Explorations, p. 459
- ❑ What a Geographer Sees, p. 461
- ❑ Where Geographers Click, p. 469
- ❑ Geography InSight, p. 470
- ❑ Stop: Answer the Concept Checks before you go on.
  p. 462 ❑   p. 467 ❑   p. 468 ❑   p. 478 ❑

End of chapter

- ❑ Review the Summary and Key Terms.
- ❑ Answer the Critical and Creative Thinking Questions.
- ❑ Answer What is happening in this picture?
- ❑ Complete the Self-Test and check your answers.

Soil Formation

LEARNING OBJECTIVES

1. **Describe** the components of soil.
2. **Describe** the characteristics of the five primary soil horizons.
3. **Explain** the four processes of soil formation.
4. **Explain** the factors that influence soil development.

Soil is the uppermost layer of the lithosphere and serves as the biosphere's foundation for the vast majority of plants and animals.

> **soil** A natural terrestrial surface layer that contains minerals and organic matter and that supports, or is capable of supporting plants, animals, and microbial communities.

Components of Soil

Soil characteristics develop over a long period of time, through a combination of many processes acting together. Typical soil is 50% mineral and organic material and 50% open pore spaces filled equally with air and water. Physical processes break down rock fragments into smaller and smaller pieces. Chemical processes alter the mineral composition of the original **parent material**—rock or sediment—and produce new minerals. Microbes, invertebrates, and plant

> **parent material**
> The inorganic material base from which soil is formed.

roots mix and compost minerals and organic material into a new matrix of soil. Taken together, these physical, chemical, and biological processes are referred to as *weathering*.

Over time, weathering processes soften, disintegrate, and break bedrock apart, forming a layer of **regolith** (**Figure 15.1**). Other kinds of regolith are mineral particles transported by streams, glaciers, waves and water currents, or winds. For example, dunes formed of sand transported by wind are a type of regolith on which soil can form. In most cases, the upper layer of regolith weathers into soil, as shown in the figure.

Soil includes mineral matter from rock material and organic matter from living and dead plants, microorganisms, and invertebrate and vertebrate animals. Soil scientists use the term **humus** to describe finely divided, partially decomposed organic matter in soils. Some humus

A cross section through the land surface • Figure 15.1

In this cross section, vegetation and forest litter lie atop the soil. Below is regolith, produced by the breakup of the underlying bedrock.

Vegetation

Soil

Regolith

Bedrock

Forest litter

A column of soil normally shows a series of horizons, which are horizontal layers with different properties. The sequence of soil horizons shown here is typical of the horizons that commonly develop in moist, temperate climates.

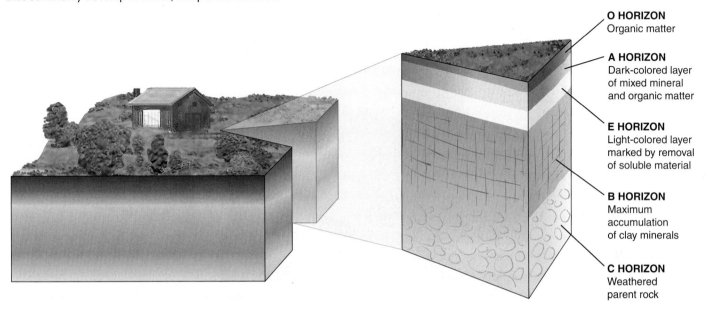

O HORIZON
Organic matter

A HORIZON
Dark-colored layer of mixed mineral and organic matter

E HORIZON
Light-colored layer marked by removal of soluble material

B HORIZON
Maximum accumulation of clay minerals

C HORIZON
Weathered parent rock

rests on the soil surface, and some is carried more deeply into the soil by percolating rainwater. Humus particles can make soil look brown or black as a result of high levels of organic carbon.

We also find air and water in soil. Water contains high levels of dissolved substances, such as nutrients and minerals. Air in soils typically has high levels of carbon dioxide or methane and low levels of oxygen.

Soil Horizons

Most soils have distinctive horizontal layers that are visually distinct because of their physical and chemical composition, organic content, or structure. We call these layers **soil horizons** and give them letter labels for easy reference: O, A, E, B, and C (**Figure 15.2**). Soil horizons develop over time through interactions between climate and subsequent leaching of water solutions. Living organisms, the land surface, topography, and parent materials are also involved in creation of soil horizons. At times, certain minerals and compounds can be washed or blown from the soil, while at other times they can accumulate in horizons. This removal or accumulation

> **soil horizon** A distinctive layer of soil, more or less horizontal, set apart from other layers by differences in physical or chemical composition.

is normally facilitated by water seeping through from the surface to deeper layers.

Horizons often have different textures and colors that reflect the way each layer was formed. The O horizon is the surface layer formed from *organic* plant and animal matter. Below the O horizon are four main mineral horizons. Soil scientists limit the term *soil* to the A, E, and B horizons, which plant roots can readily penetrate. The A horizon is rich in organic matter, washed downward from the organic horizons. Next is the E horizon. Clay particles and oxides of aluminum and iron are removed from the E horizon by downward-seeping water, leaving behind pure grains of sand or coarse silt.

The B horizon receives the clay particles, aluminum, and iron oxides, as well as organic matter washed down from the A and E horizons. This horizon is often dense and tough because its natural spaces are filled with clays and oxides.

Beneath the B horizon is the C horizon. It consists of the parent mineral matter of the soil. Below this regolith lies the R horizon, which consists of bedrock or sediments of much older age than the soil. A fresh vertical cut in the earth, called a **soil profile**, displays the layers in a cross section through the soil.

> **soil profile** A display of soil horizons on the face of a freshly cut vertical exposure through the soil.

PROCESS DIAGRAM

Enrichment and removal of upper soil • Figure 15.3

Wind and water loosen and carry sediment away from the uppermost soil layer and deposit it elsewhere.

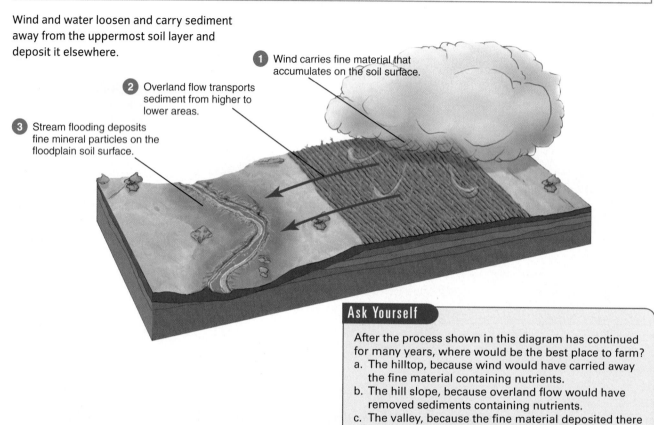

1 Wind carries fine material that accumulates on the soil surface.

2 Overland flow transports sediment from higher to lower areas.

3 Stream flooding deposits fine mineral particles on the floodplain soil surface.

Ask Yourself

After the process shown in this diagram has continued for many years, where would be the best place to farm?
a. The hilltop, because wind would have carried away the fine material containing nutrients.
b. The hill slope, because overland flow would have removed sediments containing nutrients.
c. The valley, because the fine material deposited there would contain nutrients.
d. Everywhere, because nutrients are continually formed in older soils.

Soil-Forming Processes

Soil formation is the result of interactions among the lithosphere, hydrosphere, biosphere, and atmosphere as the soil is enriched, removed, transformed, and translocated.

Enrichment and removal

In **soil enrichment**, matter—organic or inorganic—is added to the soil. In removal processes, material is either eroded from the top of the soil or carried out of the soil by seeping water and taken to the ground water below. In the upper soil, as covered in Chapters 10 through 13, wind and water can remove soil from one area and deposit it in another area through erosion, transport, and deposition (**Figure 15.3**).

Soils are also enriched by the addition of decaying plant and animal matter, rainwater, and atmospheric gases. Soils lose material when plants take up water and nutrients through evapotranspiration. Gases containing carbon, sulfur, and nitrogen are released by soil organisms and returned to the atmosphere.

Transformation

Soil materials are altered within horizons through processes of transformation. An important example of transformation is the decomposition of organic matter to produce humus, a mixture of long-lived organic residues that decomposes slowly. In warm, moist climates, organic matter can decompose completely to yield carbon dioxide and water, leaving virtually no organic matter in the soil. Minerals are transformed into smaller particles by weathering. Some minerals are transformed into different minerals through chemical alteration, and some material is broken down by microbial activity in the digestive tracts of invertebrates and mechanically by burrowing insects and animals.

Translocation

Materials are moved vertically through soil layers through processes of translocation. Much of the chemical movement occurs when seeping water suspends or dissolves soil materials, moving them to deeper horizons or into ground water. **Eluviation** describes this

eluviation The downward transport of dissolved or suspended materials to lower soil horizons or out of the soil completely.

illuviation The accumulation in a lower soil horizon of materials brought down from upper horizons.

downward movement of material. **Illuviation** refers to the process by which dissolved or suspended minerals recrystallize or accumulate in lower layers of the soil. *Leaching* refers to the downward loss of dissolved materials, which are carried out of the soil to ground water.

Translocation has different effects on soil layers, depending on climate conditions. In dry climates, calcium carbonate is leached and carried down into the B horizon, where it is deposited in crystalline form, a process called **calcification**. Calcification can create a cemented layer, known as a *hard pan*, that interferes with both eluviation and illuviation. This renders the soil less fertile by preventing the exchange of nutrients. In colder climates, a pan can also form from the accumulation of oxides of iron and aluminum by illuviation. This type of pan can block drainage and keep the soil saturated for long periods, creating chemical-reducing conditions in a process called **gleyzation**. The reducing conditions affect iron and manganese compounds, creating spots or streaks of blue and green colors in the soil.

In cold, moist climates under conifer forests, organic matter decomposes slowly and releases organic acids. In a process called **podzolization**, the acids translocate humus, aluminum, and iron compounds from the E horizon to the B horizon, leaving behind a distinctive white or gray E horizon of mineral matter.

In moist climates, a large amount of surplus soil water moves downward to the groundwater zone. This water movement leaches calcium carbonate from the entire soil in a process called **decalcification**. Decalcified soils usually have limited agricultural productivity unless plant nutrients are added.

In warm, moist climates, vegetation litter falling on the ground decomposes rapidly to CO_2 and water. The remaining components are rapidly taken up by plant roots or leached from the soil, so that almost no organic matter accumulates in the soil. To grow crops, indigenous peoples carry out slash-and-burn agriculture, clearing small patches of forest and burning the vegetation in place. The ash provides soil nutrients for crops that last only one or two seasons before requiring the subsistence farmers to slash and burn another farm plot.

In warm to hot climates with abundant rainfall, silica (SiO_2) becomes slightly soluble in water, and over long periods of time, it is leached out of the soil. The oxides of aluminum and iron remain, forming oxide-rich clay minerals. In some circumstances, the iron becomes enriched and takes on a form that hardens irreversibly with drying. This process, called **laterization**, can yield a soil horizon

of soft clay that can be cut into bricks and dried to form building materials and road foundations.

In desert climates, translocation can lead to **salinization** of the soil through the upward wicking of salt-laden ground water. In some low or irrigated situations, ground water lies close to the surface, producing a flat, poorly drained area. As water at or near the soil surface evaporates, ground water is drawn upward to replace it by capillary tension. This ground water is often rich in dissolved salts and minerals. When the salt-rich water evaporates, the mineral salts are deposited and accumulate. Large amounts of these salts are toxic to most plants. When salinization occurs in irrigated lands in a desert climate, the soil can be rendered unsuitable for further agriculture. This is exactly what happened to the ancient Sumerian civilization described in the chapter opener (see *Video Explorations*).

Video Explorations

Mesopotamia

Mesopotamia, often called the cradle of civilization, developed in what is now the desert of modern Iraq, between the Tigris and Euphrates Rivers. Archaeologists have identified Sumerian agriculture, the result of engineered irrigation systems, as the foundation for this early civilization.

Factors in Soil Formation

Various factors determine soil formation. Five natural factors—biological organisms, climate, time, topography, and parent material—represent the principal components of the soil-formation process. Human activity, including farming practices and the development of cities over **arable soils**, represents a sixth significant influence on soil development.

> **arable soil** Land suitable for vegetation, including forest or crops.

Biological organisms Living plants and animals, as well as their nonliving organic products, have an important effect on soil. Organisms living in the soil include many species—from ubiquitous microorganisms to tree roots and burrowing mammals (**Figure 15.4**). Invertebrates, especially earthworms and nematodes, continually rework the soil by burrowing and passing soil through their intestinal tracts.

Over time, earthworms ingest large volumes of decaying leaf matter, carry it down from the surface, and incorporate it into the mineral soil horizons. Many forms of insect larvae perform a similar function. Burrowing animals such as moles, gophers, rabbits, badgers, and prairie dogs make even larger tube-like openings in the soil.

Scientists recently discovered that beavers were one of the most important animals for soil formation in North America prior to human occupation. By creating beaver ponds across the landscape, these mammals radically altered the hydrology and ecology and influenced the soil formation process (see *What a Geographer Sees*).

Climate Moisture and temperature play an important role in determining the chemical development of soil and the formation of horizons. Biological activity is slowed at low temperatures. At or below the freezing point (0°C [32°F]), almost all biological activity stops, and chemical processes affecting minerals are inactive. The root growth

Soil organisms • Figure 15.4

Biological organisms represent the greatest influence on soil development. (Soil horizons are not drawn to scale.)

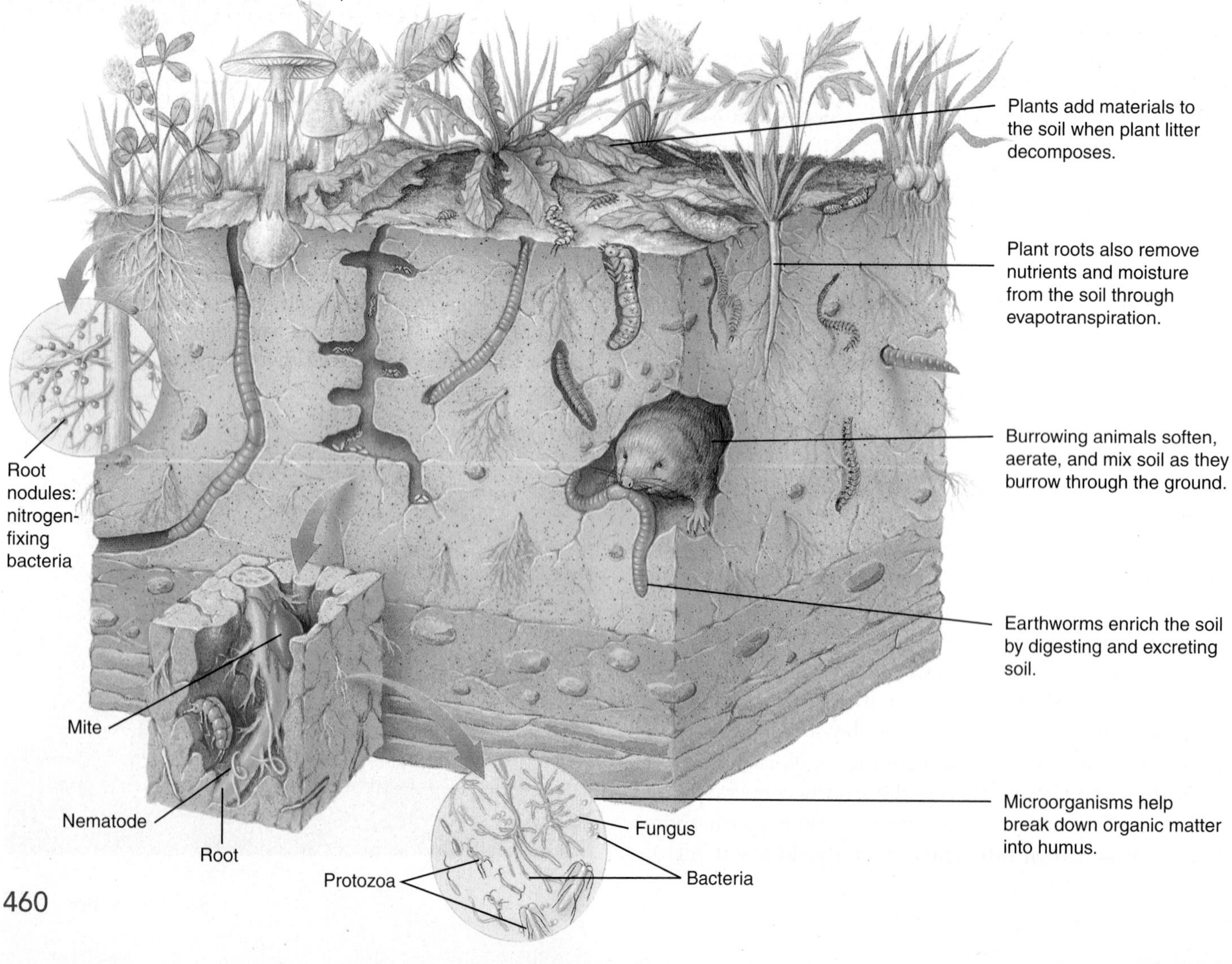

Plants add materials to the soil when plant litter decomposes.

Plant roots also remove nutrients and moisture from the soil through evapotranspiration.

Burrowing animals soften, aerate, and mix soil as they burrow through the ground.

Earthworms enrich the soil by digesting and excreting soil.

Microorganisms help break down organic matter into humus.

Root nodules: nitrogen-fixing bacteria

Mite

Nematode

Root

Protozoa

Fungus

Bacteria

WHAT A GEOGRAPHER SEES

Nature's Soil Engineers

A camper may come across a series of ponds that appear to have been created by debris dams and consider testing the local fishing, as here in the Grand Tetons National Park in Wyoming. Geographers would quickly recognize the engineer behind these dams as the American beaver (*Castor canadensis*). Prior to the human occupation of North America, and lacking any major predators, beavers proliferated and transformed the landscape with millions of small ponds and lakes that kept soils waterlogged and retained water-borne sediment. Paleobiologists analyzing pollen in soil cores from ancient ponds have uncovered a new chapter regarding the beavers' major role in the soil-formation process. Ragweed pollen in the soil cores tells the scientists the extent of the ragweed range over time by the abundance of the microscopic pollen in each of the soil layers. Because ragweed proliferates where former beaver ponds existed, the soil cores can be used to map the areas that were previously covered by ponds and wetlands.

Think Critically

What properties would you expect to find in soils that were formed from dried lakebeds?

of most plants and germination of their seeds require soil temperatures above 5°C (41°F). Plants adapted to the warm, wet, low-latitude climates may need temperatures of at least 24°C (75°F) for their seeds to germinate.

The temperature of the uppermost soil layer and the soil surface also strongly affects the rate at which organic matter is decomposed by microorganisms and invertebrates. In cold climates, decomposition is slow, and organic matter accumulates to form a thick O horizon. This material decomposes further to form humus, which is carried downward to enrich the A horizon. But in warm, moist climates of low latitudes, bacteria and invertebrates rapidly decompose plant material, preventing the formation of an O horizon layer, and the entire soil profile contains very little organic matter.

Time The term *climate* is associated with long time periods. Soil formation is based on processes that occur under a set of climatic conditions for moisture and temperature regimes that may continue for centuries. A soil scientist's rule of thumb is that it takes about 500 years to form 2.5 centimeters (1 inch) of topsoil.

Topography The configuration, or shape, of the ground surface also influences soil formation. Generally speaking, soil horizons are thick on flat or gentle slopes but thin on steep slopes. This is because the soil is more rapidly removed by erosion on the steeper slopes. In addition, slopes facing away from the Sun are sheltered from direct insolation and tend to have cooler, moister soils. Slopes facing toward the Sun are exposed to direct solar rays, raising soil temperatures and increasing evapotranspiration.

Parent material Soil chemistry is based primarily on the original source of parent material. For example, iron-rich bedrock creates soils rich in iron oxides, whereas limestone forms calcium-rich soils.

Paving paradise • Figure 15.5

NASA researchers used GIS software tools to analyze urbanization of productive agricultural land in California. The team found that more than 16% of the area of best crop soils is already under urban land use in this fast-growing region.

b. This map of urban development in California was generated using Earth observation satellite data. Peri-urban areas are areas of rapid growth in urban-development corridors.

a. The United Nations Food and Agriculture Organization generated this map showing the quality of agricultural soils (where 1 indicates best soils) for California.

Number of soil limiting factors

1 2 3 4 5 8

■ Peri-urban areas
☐ Urban areas

Human activity Human activity has influenced the physical and chemical nature of the soil since the dawn of agriculture about 10,000 years ago. Large areas of soils have been cultivated for centuries. As a result, the structure and composition of these agricultural soils have undergone great changes. These altered soils are often recognized as distinct soil classes that are just as important as natural soils. Research into the effects of the development of cities, where 50% of the world's population live, demonstrates that urban soils are being irreversibly altered by uprooting, compaction, and disruption of the ground water infiltration from impermeable surfaces. These findings are troubling because today's cities are located on some of the world's most productive agricultural soils (**Figure 15.5**).

CONCEPT CHECK   STOP

1. **What** is the effect of weathering on soil components?

2. **What** distinguishes the O horizon from the other horizons?

3. **How** does calcification occur?

4. **How** does soil temperature affect soil formation?

Soil Properties

LEARNING OBJECTIVES

1. **Explain** what can be learned from soil color.
2. **Describe** the three distinct sizes of soil grains.
3. **Define** the four basic soil structures.
4. **Explain** the factors that determine soil moisture.

oils can be classified by several different measurable physical properties, including color, texture, structure, and moisture. Soil scientists have used these characteristics for a variety of soil classification systems.

Soil Color

Color is the most obvious feature of a soil. Some color relationships are quite simple. For example, Midwest prairie soils are black or dark brown because they contain many humus particles (**Figure 15.6a**); the red or yellow soils of the Southeast are colored by the presence of iron oxides (**Figure 15.6b**).

In some areas, soil color is inherited from the mineral parent material, but, more often, the color is generated during soil formation. For example, dry climates often have soils with a white surface layer of mineral salts that have been transported upward by capillary action and evaporation of salt-rich ground water (**Figure 15.6c**). In the cold, moist climate of the boreal forest, a pale, ash-gray layer near the top of the soil is created when organic matter and colored minerals are washed downward, leaving behind only pure, light-colored mineral matter.

Soil Texture

The mineral matter of soil consists of individual mineral particles that vary widely in size. The term **soil texture** refers to the proportion of particles that fall into each of three size grades— sand, silt, and clay. Clays are the finest particles, and sand is the coarsest. Silt particles are intermediate in size. In describing soil texture, we do not include gravel and larger particles because they play a less important role in soil processes.

> **soil texture** A descriptive property of the mineral portion of soil based on the proportions of sand, silt, and clay.

Soil color • Figure 15.6

The color of soil provides vital clues to its type and condition.

a. Black soil
Dark soil colors normally indicate the abundance of organic matter in the form of humus. Here students are weeding rows of onions in a field near Rome, New York.

b. Red soil
The red-brown color of the soil near Cedar Mountain, Culpeper County, Virginia, is caused by iron oxides. The ancient soils of this region are highly productive with proper treatment.

c. White soil
Salts form a white surface deposit on a plateau in Western Argentina. Salts in the soil are carried upward and accumulate as water from infrequent rainstorms evaporates at the surface. Salt-tolerant plants provide a spotty vegetation cover.

Soil texture classification • Figure 15.7

Soils can be classified based on their proportion of sand, silt, and clay. Loams contain a proportion of sand, silt, and clay that is particularly effective for holding water. To read the diagram, follow the three grid lines at a point to the sides of the triangle and read the proportion. For example, the texture at point A is 60% sand, 15% clay, and 25% silt.

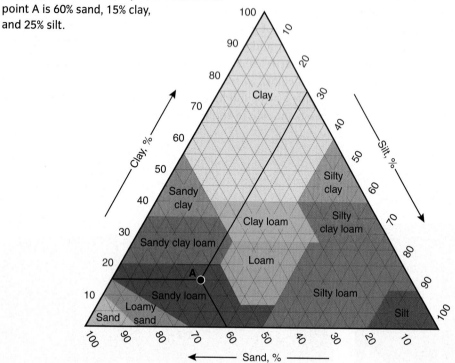

Ask Yourself

Soil with a texture classified as sandy clay loam could have which portions of sand, silt, and clay?
a. sand = 60%; clay = 30%; silt = 10%
b. sand = 30%; clay = 60%; silt = 10%
c. sand = 60%; clay = 10%; silt = 30%
d. sand = 10%; clay = 30%; silt = 60%

Soils can be characterized by their proportion of sand, silt, and clay (**Figure 15.7**). Soil texture determines the ability of the soil to hold water and support microbial and animal life. A **loam** is a soil that contains a proportion of each of the three grades that is most effective at storing moisture. This mixture is often added to gardens because it is particularly suitable for plants. Loams can be further classified as sandy, silty, or clay-rich, depending on which grade is dominant.

Soil Structure

Soil structure refers to the way in which soil grains are clumped together into larger masses, called **peds**, ranging from small grains to large blocks (**Figure 15.8**). Peds form when clays shrink as they dry out. Cracks form in the soil, which define the surfaces of the peds. Small peds, shaped roughly like spheres, give the soil a granular or crumb structure. Larger peds provide an angular, blocky structure.

Soils with a well-developed granular or blocky structure are easy to cultivate—a factor that is especially important for non-mechanized farming using animal-drawn

plows. Soils with a high clay content can lack peds. They are sticky and heavy when wet and are difficult to cultivate. When dry, they become too hard to be worked.

Soil Moisture

Soils are also a reservoir of moisture for plants. Soil moisture is a key factor in how the soils of a region support vegetation and crops. The interrelationships between soil moisture, climate, and vegetation help us understand how changes in climate may affect agriculture and native ecosystems.

The soil-water budget The soil receives water from rain and melting snow. Some of the water simply runs off, flowing into brooks, streams, and rivers, eventually reaching the sea. Some water sinks into the soil. Of this water, some is held in storage in the soil, while excess water, called *gravitational water*, flows down through the regolith to recharge ground water. Water that is held in storage is available for evaporation and plant use. When plant transpiration lifts soil water from roots to leaves, where it evaporates, it is called *evapotranspiration*.

The amount of water available at any given time is determined by the soil-water balance, which includes the gain, loss, and storage of soil water. This is similar to the glacier mass-balance process discussed in Chapter 14. Water held in storage is recharged during precipitation but depleted by evapotranspiration. Monitoring these values on a monthly basis provides a **soil-water budget** for the year. Based on regional climate, some landscapes have a surplus of soil water during the rainy season and a deficit in the dry season. Native vegetation exhibits a range of adaptations in order to survive during dry seasons, such as inhibiting respiration and sending tap roots down as far as 50 meters (164 feet) to access water.

Soil-water storage As gravitational water drains downward, some clings to the soil particles by cohesion. This water resists the pull of gravity because of the force of microscopic capillary tension. To understand this force, think about a droplet of condensation that has formed on the cold surface of a glass of ice water. The water droplet seems to be enclosed in a film of surface molecules, drawing the droplet together into a rounded shape. That surface is produced by capillary tension, which keeps the drop clinging to the side of the glass, defying the force of gravity. Similarly, tiny films of water stick to individual soil grains, particularly at the points of grain contact. The portion of water held in the soil by capillary tension is available for evaporation or absorption by plant rootlets, so it is sometimes called *available water*.

Soil also stores water in another way, holding it tightly by molecular adhesion. This *hygroscopic water* is located on the surfaces of individual soil particles. Hygroscopic water is largely unavailable to plants because it is held firmly by molecular attractions so strong that it cannot be broken by plants or through evaporation.

When a soil has been saturated by water and then allowed to drain under gravity until no more water moves downward, the soil is said to be holding its **field capacity** of water. Maintaining soil moisture at or near field capacity is important for agricultural productivity. In times of drought, when all available water has been removed from

Soil structure • Figure 15.8

Soils can be classified by the way particles clump together in peds.

a. Granular structure
Looking much like cookie crumbs, granular peds are rounded like little pellets the size of BBs. They are loosely packed and located in the surface layers along with organic material and roots.

b. Blocky structure
These cube-shaped peds form in the subsoil and are 5 to 50 mm (0.2 to 2 in.). These are typical of B horizons with high clay content.

c. Platy structure
Larger than granular peds, these thin, flat soil masses, looking much like a stack of pancakes, occur near the surface and are associated with soil compaction.

d. Columnar structure
Found in dry regions and sandy soils, these peds are vertical columns found in the lower soil horizons. Often several centimeters in length, they frequently occur with a salt cap from arid illuviation.

PROCESS DIAGRAM

Three forms of soil moisture • Figure 15.9
THE PLANNER ✓

Soil holds water in three forms: gravitational, capillary, and hygroscopic. Capillary water is the most important source of water for plants.

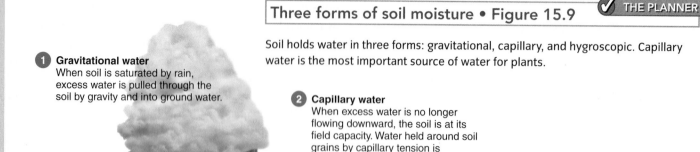

1 Gravitational water
When soil is saturated by rain, excess water is pulled through the soil by gravity and into ground water.

2 Capillary water
When excess water is no longer flowing downward, the soil is at its field capacity. Water held around soil grains by capillary tension is available to plants.

3 Hygroscopic water
After a period without rain, when there is no more available water, the soil is at its wilting point. Water held by adhesion is not available to plants.

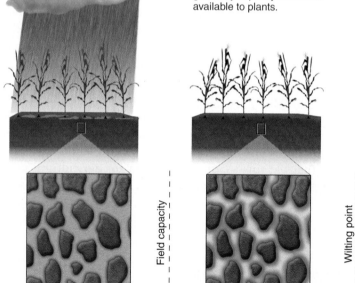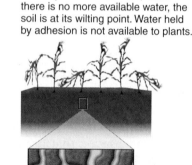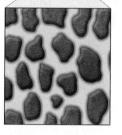

Field capacity

Wilting point

Decreasing soil moisture

Available water and soil texture • Figure 15.10

Finer-textured soils hold more water, increasing their field capacity. They also hold water more tightly, raising their wilting point. The soils with the most available water are loamy.

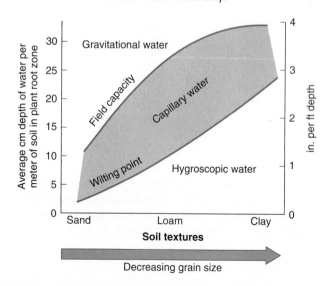

the soil, the soil is said to be at its **wilting point**. Although there is still hygroscopic water present in the soil, it is inaccessible to plants, and they begin to wilt. **Figure 15.9** shows the three forms of soil water in relation to field capacity and wilting point.

The availability and flow of water in the soil depends largely on its texture (**Figure 15.10**). Coarse-textured (sandy) soils store less water because they are more porous, allowing water to drain through to deeper layers. Because fine-textured (silty) soils have far smaller passages and spaces, they are less porous, allowing water to penetrate down more slowly, with some water held longer in the upper layers. As a result, field capacity increases as grain size decreases.

Finer-textured soils hold more water than coarser-textured soils because fine particles have a much larger surface area in a unit of volume than do coarse particles. Although this increases hygroscopic water, such water remains unavailable to plants.

466 CHAPTER 15   Global Soils

Capillary water has significantly more spaces to occupy in finer-textured soils. However, because fine particles also hold water more tightly, it is more difficult for plants to extract moisture from fine soils. As a result, the wilting point also increases as grain size decreases. This means that plants can wilt in fine-textured soils even though they hold more soil water than coarse-textured soils. Loamy soils are best for plant growth because they offer the most available water.

CONCEPT CHECK

1. **What** does a red color indicate about soil?
2. **What** soil texture is generally most productive for plants?
3. **What** soil conditions are associated with platy structure in soils?
4. **Why** do clay soils have a high wilting point?

Soil Chemistry

LEARNING OBJECTIVES

1. **Explain** the importance of soil pH to soil fertility.
2. **Describe** the role of colloids in retaining soil nutrients.

In addition to physical properties such a soil texture and structure, soil chemistry is another factor that determines the suitability of soil for supporting vegetation. The chemistry of a soil determines how well the soil can provide nutrients to plants.

Acidity and Alkalinity

One important factor in soil chemistry is pH, which is a measure of how acidic or alkaline the soil is. Soil pH is an important indicator of soil fertility (**Figure 15.11**). The pH of soil affects how minerals dissolve in water solutions and their availability for uptake by plant roots. Recall that

pH is a measure of the free hydrogen ions in water or soil. A pH of 7 indicates a neutral balance between positively charged hydrogen (H^+) and negatively charged hydroxide (OH^-). Acidic soils (with more H^+ ions and fewer OH^- ions) can be formed when an excess of organic decomposition occurs because of insufficient leaching or drainage. Limestone parent material, on the other hand, can be a factor in development of alkaline soils (with fewer H^+ ions and more OH^- ions). Most plants prefer a pH level between 6 and 8 for maximum growth.

Soils in cold, humid climates are often highly acidic, while in arid climates, soils are typically alkaline. Acidic soils can be modified by applying lime, a compound of calcium, carbon, and oxygen ($CaCO_3$), which removes acid ions and replaces them with the alkaline cation, calcium. Such soil modification strategies are sometimes used to combat the effects of acid rain on lakes and soils located downwind from industrial emissions sources.

The pH scale and soil fertility • Figure 15.11

Acidity and alkalinity ranges are used to define optimal soil pH for various agricultural and native plants. The optimal pH for most farm crops is 6.5.

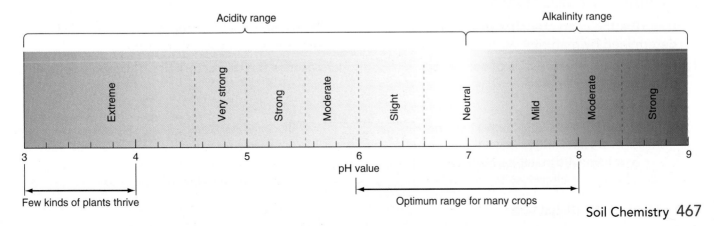

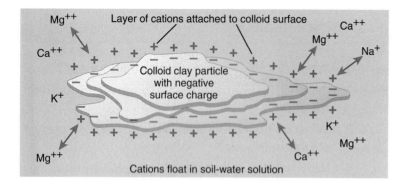

Cation-exchange capacity • Figure 15.12

Cation-exchange capacity is defined by the quantity of positively charged mineral ions (nutrients) that move between the negatively charged colloid particle and the soil-water solution. This capacity determines soil fertility.

Labels in figure: Mg++ Layer of cations attached to colloid surface Ca++ Mg++ Na+ Ca++ Colloid clay particle with negative surface charge K+ K+ Mg++ Mg++ Ca++ Cations float in soil-water solution

Colloids

soil colloid An extremely small mineral or organic particle that can remain suspended in water indefinitely.

Soil colloids are particles smaller than one ten-thousandth of a millimeter (0.0001 mm [0.000,004 in.]). Like other soil particles, some colloids are mineral, whereas others are organic. Mineral colloids are usually very fine clay particles. Under a microscope, mineral colloids display thin, plate-like bodies. When these particles are well mixed in water, they remain suspended indefinitely, turning the water murky. Organic colloids are tiny bits of organic matter that are resistant to decay.

Soil colloids are important because their surfaces attract soil nutrients dissolved in soil water as positively charged mineral ions (or cations) (**Figure 15.12**). Some cations are needed for plant growth, including calcium (Ca^{++}), magnesium (Mg^{++}), potassium (K^+), and sodium (Na^+). These are referred to as **base cations**, or simply **bases**. They need to be dissolved in a soil-water solution to be available for plant uptake. Colloids hold these ions but also exchange them with plants when they are in close contact with root membranes. The fertility of the soil-water solution for plants is based on the ability of the soil to hold and exchange cations, which is called the **cation-exchange capacity**. Without soil colloids, most vital nutrients would be leached out of the soil by percolating gravitational water and carried away in streams. The relative amount of bases held by a soil determines the base status of the soil. A high base status means that the soil holds an abundant supply of base cations necessary for plant growth.

CONCEPT CHECK

1. **What** factors influence soil pH?
2. **What** is the role of colloids in determining a soil's cation-exchange capacity?

The Global Distribution of Soils

LEARNING OBJECTIVES

1. **Describe** the characteristics of soil orders determined by maturity.
2. **Discuss** the characteristics of soil orders determined by climate.
3. **Describe** the characteristics of soil order determined by parent material.
4. **Discuss** the characteristics of organic soils.

S oils can be classified into 12 **soil orders**. These orders range from soils with very little soil development to highly developed soils (see *Where*

soil order A soil class that forms the highest category in soil classification.

Geographers Click). Some soil orders are related primarily to the maturity of the soil and some to parent material, and some are closely tied to a specific climate type (**Table 15.1**).

Where Geographers CLICK

Understanding Soils

http://soils.usda.gov

Go to this Web Site to explore an interactive map that analyzes soil data from around the world. The National Cooperative Soil Survey, part of the U.S. Department of Agriculture, has developed this collaborative effort of federal and state agencies, universities, and professional societies to deliver science-based soil information.

Global Soil Regions

Soil orders   Table 15.1

Soil order	Distribution	Characteristics
Soils characterized by maturity		
Entisols	Recently deposited materials or materials that don't form horizons	No distinct horizons
Inceptisols	Young soils	Weakly developed horizons
Alfisols	Humid climates with deciduous forest cover	• Subsurface accumulation of clay • High base status
Spodosols	Cold, boreal forest climate with coniferous forest cover	• Low in humus • Low base status
Ultisols	Warm, wet climates with some dry season	• Subsurface accumulation of clay • Low base status
Oxisols	Warm climates, wet year-round, with rainforest vegetation	• Subsurface accumulation of mineral oxides • Very low base status
Soils Characterized by Climate		
Mollisols	Semiarid to subhumid midlatitude grasslands	• Loose texture • Very high base status
Aridisols	Dry climates	• Low content of organic matter • Accumulation of salts
Gelisols	Cold, frozen climates	• Underlain by permafrost • Subject to churning by freeze–thaw
Soils characterized by parent material		
Vertisols	Clay parent material in climates with alternating wet and dry seasons	• Clay-rich • High base status • Subject to churning and cracking
Andisols	Soils developed on volcanic ash	• Dark, fertile soils with high amounts of carbon • Found near active volcanoes
Soils high in organic matter		
Histosols	Cool, poorly drained areas	• Very high content of organic matter

Soils vary greatly from continent to continent, from region to region, and even from field to field. This is because they are influenced by many factors and processes that in combination can vary widely from place to place.

How soils are distributed around the world helps to determine the quality of environments and arable soils around the globe for agricultural production. (See **Figure 15.13** on the next page).

Soil type, along with precipitation, is an important determinant of the ability of an environmental region to produce food for human consumption. Notice that large regions on the map, representing dunes of moving sand, areas of bare rock, high mountains, and surfaces of fresh lava, do not possess soil at all.

a. Histosols
After appropriate drainage and application of lime and fertilizers, Histosols are remarkably productive. This carrot crop is growing in a Histosol of a former glacial lakebed in Ontario, Canada.

b. Mollisols
With their loose texture and high base status, the Mollisols have favorable properties for growing cereals in large-scale mechanized farming, as shown here in Washington State.

c. Aridisols
Although these soils are low in organic matter and high in salts, Aridisols can be highly productive where there are water supplies from mountain streams or ground water. On this broad floodplain near the Colorado River in Arizona, canals are used for irrigating fields of cotton.

d. Vertisols
These nutrient-rich soils are formed from a particular type of clay mineral in climates with alternating wet and dry seasons. Despite their fertility, their high clay content makes them difficult to till, as they develop deep, wide cracks when dry, as shown in this close-up from a farm field in Texas.

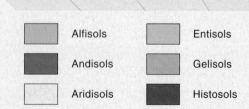

Alfisols		Entisols	
Andisols		Gelisols	
Aridisols		Histosols	

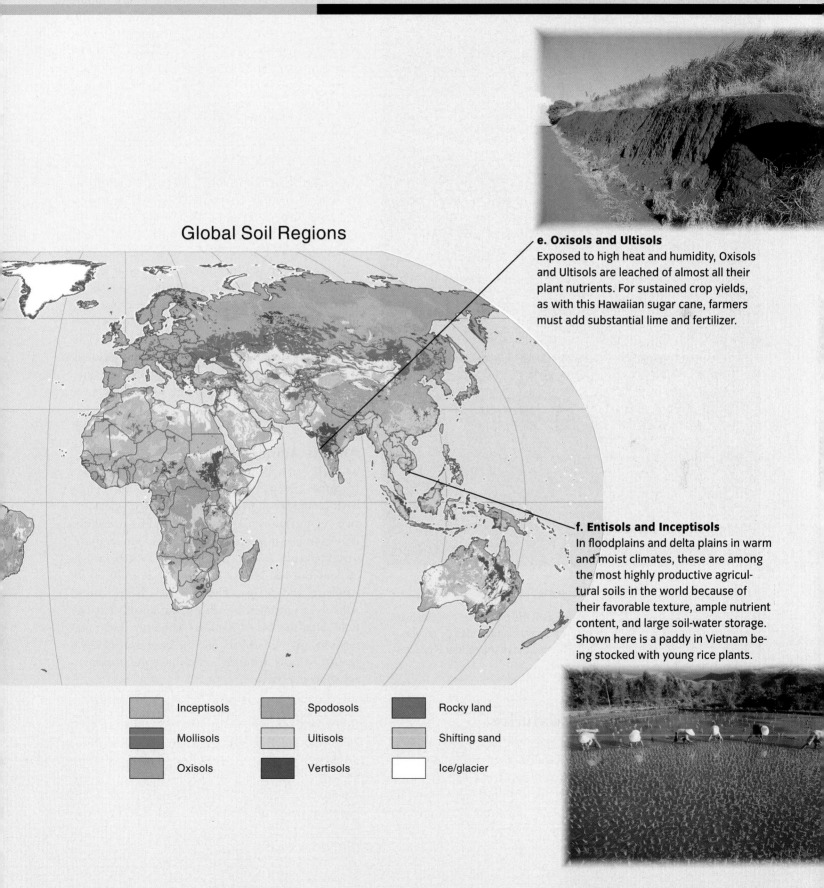

Global Soil Regions

e. Oxisols and Ultisols
Exposed to high heat and humidity, Oxisols and Ultisols are leached of almost all their plant nutrients. For sustained crop yields, as with this Hawaiian sugar cane, farmers must add substantial lime and fertilizer.

f. Entisols and Inceptisols
In floodplains and delta plains in warm and moist climates, these are among the most highly productive agricultural soils in the world because of their favorable texture, ample nutrient content, and large soil-water storage. Shown here is a paddy in Vietnam being stocked with young rice plants.

Inceptisols	Spodosols	Rocky land
Mollisols	Ultisols	Shifting sand
Oxisols	Vertisols	Ice/glacier

The Global Distribution of Soils **471**

Soil profile of Alfisols • Figure 15.14 _____

In this soil profile of an Alfisol from Michigan, the pale E horizon has lost some of its original nutrient bases, clay minerals, and oxides by eluviation. These materials have become concentrated in the redder B horizon.

Soils Characterized by Maturity

Where materials have been recently deposited, soils have poorly developed horizons or no horizons and are capable of further mineral alteration. This situation can also occur when parent materials are not appropriate for horizon formation. In contrast, soils that have experienced long-continued adjustment to prevailing soil temperature and water conditions have well-developed horizons or fully weathered materials.

Entisols **Entisols** are mineral soils without distinct horizons. They may be found in any climate and under any vegetation. They don't have distinct horizons for two reasons: Either the parent material, such as quartz dune sand, is not appropriate for horizon formation or not enough time has passed for horizons to form in recent deposits of alluvium or on actively eroding slopes. Entisols can be found anywhere from equatorial to arctic latitudes.

Inceptisols **Inceptisols** have weakly developed horizons, usually because the soil is young. These areas occur within some of the regions shown on the world map as Oxisols and Ultisols.

Along with Entisols of the subarctic and tropical deserts, arctic Inceptisols are some of the world's poorest soils. In contrast, in Southeast Asia, Inceptisols of floodplains support dense populations of rice farmers. Annual river floods cover low-lying plains and deposit layers of fine silt. This sediment is rich in primary minerals that yield nutrient bases as they weather chemically over time. The constant enrichment of the soil explains the high soil fertility of these alluvial soils.

Alfisols **Alfisols** are mature soils characterized by a clay-rich B horizon produced by illuviation (**Figure 15.14**). They typically form in regions with moderate moisture and humidity, where native vegetation is deciduous trees. The world distribution of Alfisols is extremely wide in latitude, ranging from as high as 60° N in North America and Eurasia to the equatorial zone in South America and Africa. The clay-rich B horizon of Alfisols is enriched by silicate minerals that can hold alkaline bases such as calcium and magnesium, giving Alfisols a high base status. As a result, Alfisols are fertile and have good agricultural productivity.

Spodosols Poleward of the Alfisols in North America and Eurasia lies a great belt of **Spodosols**, formed in the cold boreal forest climate beneath a needleleaf forest. They are distinguished by a spodic B horizon—a dense accumulation of iron and aluminum oxides and organic matter—and a gray or white albic E horizon (**Figure 15.15**).

Soil profile of a Spodosol • Figure 15.15

This Spodosol profile from France is marked by a light gray or white albic E horizon that has been bleached by organic acids percolating downward from a slowly decomposing O horizon. Below this is the reddish spodic horizon, a dense mixture of organic matter and compounds of aluminum and iron, all brought downward by eluviation from the overlying E horizon.

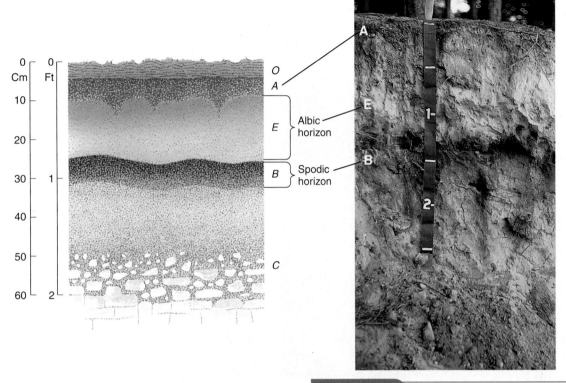

Put It Together

Review Figure 15.9 and answer this question.
There is no visible water leaching out of this soil profile. Which types of water are most likely still present?
a. gravitational and capillary
b. gravitational and hygroscopic
c. capillary and hygroscopic
d. granular and gravitational

Spodosols are closely associated with regions recently covered by the great ice sheets of the Late Cenozoic Ice Age. These soils are therefore very young. Typically, the parent material is coarse sand consisting largely of the mineral quartz. This mineral cannot weather to form clay minerals, so Spodosols are naturally agriculturally unproductive, although they do support forests of pine and spruce trees.

Spodosols are strongly acidic and are low in nutrient bases. They are also low in humus. Because they are acidic, lime application is essential for cultivation, as are heavy applications of fertilizers. With proper management, Spodosols can be highly productive, if the soil texture is favorable; for example, they give high yields of potatoes in Maine and New Brunswick.

Ultisols As a result of their exposure to warm temperatures and high rainfall, **Ultisols** are highly weathered and leached of nutrients. Ultisols are common in humid subtropical climates throughout Southeast Asia and the East Indies. Other important areas are in eastern Australia, Central America, South America, and the southeastern United States. Each of these regions has a dry season, even though it may be short, which gives Ultisols a period in which rainwater does not infiltrate and leach away nutrients.

Soil profile of an Ultisol • Figure 15.16 _____

In this soil profile of an Ultisol from North Carolina, the thin, dark top layer is all that remains of the A horizon after erosion; the pale layer is the E horizon, which has been leached of clays; and the reddish layer is the B horizon, where clays have accumulated through illuviation.

Ultisols are rich in aluminum and iron, with a layer of subsurface clay that accumulates through illuviation (**Figure 15.16**). The iron gives them their usual reddish color. Ultisols are acidic, with most of the minerals found in a shallow surface layer, where they are released by the decay of plant matter. These nutrients are quickly taken up and recycled by the shallow roots of trees and shrubs. As a result, base status is low. Without fertilizers, these soils can sustain crops on freshly cleared areas for only two or three years, at most, before the nutrients are exhausted. For sustained high crop yields, farmers must add substantial amounts of lime and fertilizers. Ultisols are also vulnerable to devastating soil erosion, particularly on steep hill slopes.

Oxisols **Oxisols** have developed in the moist climates of the equatorial, tropical, and subtropical zones on land surfaces that have been stable for long periods. We find these soils over vast areas of South America and Africa, in regions where the native vegetation is rainforest.

Because of their year-round hot, wet environment, Oxisols are the most weathered of all soil types, and they are heavily leached of nutrients. Soil minerals are weathered to an extreme degree and are dominated by oxides of aluminum and iron, giving them a red, yellow, and yellowish-brown color (**Figure 15.17**). As a result

Soil profile of an Oxisol • Figure 15.17 _____

Oxisols usually lack distinct horizons, except for dark surface layers. The intense red color in this soil profile from Hawaii is produced by iron oxides.

of leaching, these soils are not usually fertile, but they do contain a small store of nutrient bases very close to the soil surface. Extreme activity by insects and microbes rapidly recycles nutrients from the surface litter. The soil is quite easily broken up, so rainwater and plant roots can easily penetrate.

Soils Characterized by Climate Type

Some soil orders are more readily classified by the climate type in which they form than by their maturity.

Mollisols **Mollisols** are grassland soils that occupy vast areas of semiarid and subhumid climates in midlatitudes. They are unique in having a very thick dark brown to black surface A horizon (**Figure 15.18**) that can be several meters thick with a loose, crumby or granular structure, or a soft consistency when dry. Mollisols are dominated by calcium among the bases of the A and B horizons, giving them a very high base status.

In North America, Mollisols dominate the Great Plains region, the Columbia Plateau, and the northern Great Basin. In South America, a large area of Mollisols covers the Pampas of Argentina and Uruguay. In Eurasia, a great belt of Mollisols stretches from Romania eastward across the steppes of Russia, Siberia, and Mongolia. Because of their loose texture and very high base status, Mollisols are among the most naturally fertile soils in the world. They now produce most of the world's commercial grain crop.

Soil profile of a Mollisol • Figure 15.18

This soil profile of a Mollisol from Argentina demonstrates the dark brown to black layer lying within the A horizon. The B horizon contains high levels of calcium.

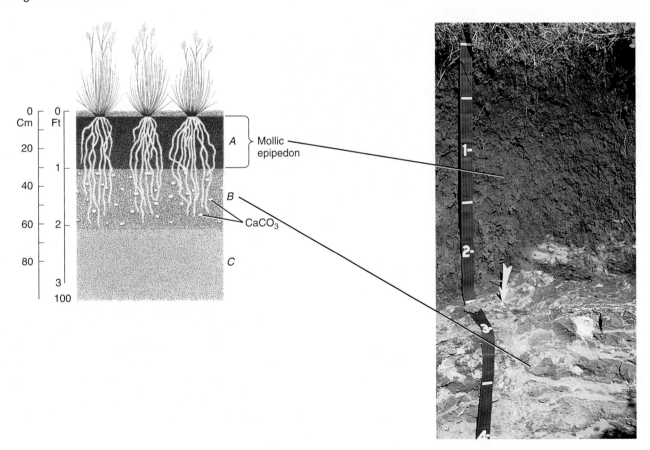

Soil profile of an Aridisol • Figure 15.19

The soil horizons of Aridisols are often weakly developed, but there may be important subsurface horizons of accumulated calcium carbonate or soluble salts. The white layer close to the surface in this Aridisol profile in the Nevada desert shows the accumulation of salts.

Aridisols **Aridisols** characterize desert regions and are dry for much or all of the year. Because the climate supports only sparse vegetation, the soils do not contain much humus, and they are pale, ranging from gray to red. They are low in organic matter and high in mineral salts. The salts, mostly containing sodium, make the soil very alkaline. In these dry climates, salt-rich ground water can be wicked up through the soil to accumulate in a concentrated layer that is toxic to plants (**Figure 15.19**).

Most Aridisols are used for nomadic grazing, just as they have been through the ages. This is because without much rainfall, it is difficult to cultivate crops without irrigation. However, major irrigation systems, such as in the Imperial Valley of the United States, the Nile Valley of Egypt, and the Indus Valley of Pakistan, make Aridisols fertile. Unfortunately, these irrigation systems have limited life spans because of salt buildup and waterlogging. Waterlogging can occur when a calcified hard pan develops from repeated flooding of the fields, which causes illuviation of calcium carbonate to form an impermeable layer. When too much water is applied in these conditions, it will not drain, and it drowns the plant roots.

Gelisols **Gelisols** are soils of cold regions that are underlain by perennially frozen ground, or permafrost. They usually consist of very recent parent material, left behind by glacial activity during the Ice Age, along with organic matter that decays very slowly at low temperatures. They are subjected to frequent freezing and thawing, which causes ice lenses and wedges to grow and melt in the soil. This freeze–thaw action churns the soil, mixing parent material and organic matter irregularly and inhibiting development of a soil profile.

Gelisols are mostly formed from minerals, ranging in size from silt to clay, that are broken down by frost action and glacial grinding. We often find layers of peat between mineral layers. The Gelisol lies on top of permafrost, and the annual summer thaw affects only a shallow surface layer. As a result, soil water can't easily drain away, and the soil is saturated with water over large areas. As the shallow surface layer repeatedly freezes and thaws, plant roots are disrupted and only small, tundra-adapted, shallow-rooted plants can establish themselves.

Soils Characterized by Parent Material

Some soils are more readily characterized by their parent material than by their climate or relative maturity.

Vertisols **Vertisols** have a unique set of properties that contrast sharply with Oxisols and Ultisols. They are black and have a high clay content (**Figure 15.20**). Vertisols typically form under grass and savanna vegetation in subtropical and tropical climates with a pronounced dry season.

Because Vertisols require a particular clay mineral as a parent material, regions of this soil are scattered and show no distinctive pattern on the world map. An important region of Vertisols is the Deccan Plateau of western India, where basalt, a dark variety of igneous rock, supplies the silicate minerals that are altered into the necessary clay minerals.

Vertisols have a high base status and are particularly rich in calcium and magnesium nutrients, with a moderate content of organic matter. The soil retains large amounts of water because of its fine texture, but much of this water is held tightly by the clay particles as hygroscopic water and is not available to plants. The clay minerals that are abundant in Vertisols shrink when they dry out, producing deep cracks in the soil surface. As the dry soil blocks are wetted and softened by rain, some fragments of surface soil drop into the cracks before they close, so that the soil is constantly being mixed.

Andisols **Andisols** are soils in which more than half of the parent mineral matter is volcanic ash, which is spewed high into the air from the craters of active volcanoes and comes to rest in layers over the surrounding landscape. The fine ash particles are glass-like shards. Andisols have a high proportion of carbon, formed from decaying plant matter, so the soil is very dark. They form over a wide range of latitudes and climates and are generally fertile soils. In moist climates, they support a dense natural vegetation cover.

Andisols are associated with individual volcanoes that are located mostly in the Ring of Fire—the chain of volcanic mountains and islands that surrounds the Pacific Ocean. Andisols are also found on the island of Hawaii, where volcanoes are presently active.

Soil profile of a Vertisol • Figure 15.20

The dark color and shiny surfaces of the clay particles are typical of Vertisols, as in this profile of a Vertisol in India.

Soil profile of a Histosol • Figure 15.21

This soil profile of a Histosol in Minnesota consists almost entirely of a deep layer of weathered peat.

Organic Soils

Most of the soils we have discussed so far have a significant mineral component. **Histosols** are a unique soil order with a very high content of organic matter in a thick, dark upper layer (**Figure 15.21**). Most Histosols go by common names, such as peats or mucks. They form in shallow lakes and ponds by accumulation of partially decayed plant matter. In time, the water is replaced by a layer of organic matter, or peat, and becomes a bog. Peat bogs are used extensively to cultivate cranberries. Sphagnum peat from bogs is dried and baled for sale as mulch. For centuries, Europeans have used dried peat from bogs of glacial origin as a low-grade fuel for cooking and heating.

Some Histosols are mucks—organic soils composed of fine black materials of sticky consistency. These are agriculturally valuable in midlatitudes, where they occur as beds of former lakes in glaciated regions. Histosols are also found in low latitudes, where poor drainage has favored thick accumulations of plant matter.

Put It Together

Review Figure 15.8 and answer this question.
Considering the soil peds at the bottom of the profile, what is the soil texture of this profile?
a. granular
b. blocky
c. platy
d. columnar

CONCEPT CHECK   STOP

1. **Why** are Ultisols and Oxisols less fertile than Alfisols?

2. **What** explains the accumulation of a salt layer in Aridisols?

3. **Why** are Vertisols difficult to cultivate despite being rich in plant nutrients?

4. **How** are Histosols unique among soil orders?

Summary

1 Soil Formation 456

• The **soil** layer is a complex mixture of solid, liquid, and gaseous components. Most soils, which lie atop **regolith**, as shown in the diagram, possess distinctive horizontal layers called **soil horizons**. Horizons can be seen in a fresh vertical cut through the surface, called a **soil profile**.

A cross section through the land surface • Figure 15.1

• Horizons develop through **enrichment**, removal, translocation, and transformation.

• **Eluviation** describes the downward movement of material through the soil. The accumulation of minerals in lower layers of the soil is called **illuviation**. Translocation can lead to depletion or buildup of mineral salts, through processes such as **calcification**, **gleyzation**, **podzolization**, **lateriza-tion**, and **salinization**.

• Climate, **parent material**, the configuration of the land, time, and the presence of plants and animals all influence soil development.

• Human activities over thousands of years, including cultivation of **arable soils**, have altered agricultural soils to form distinctive soil classes.

2 Soil Properties 462

• Soil color reflects the mineral or organic content of the soil.

• **Soil texture** refers to the proportions of sand, silt, and clay in the soil, as shown in the diagram. **Loam** has fairly balanced proportions of each type and is ideal for plant growth.

Soil texture classification • Figure 15.7

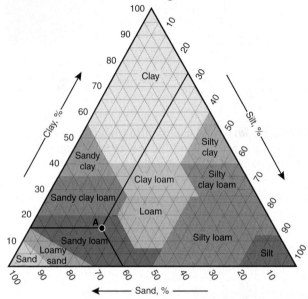

• Soil structure is determined by the way particles clump together into **peds**. Soil structure can be granular, platy, blocky, or columnar.

• Evapotranspiration removes soil water, and precipitation recharges it. The **field capacity** of soil is its maximum storage capacity, and its **wilting point** is the point at which there is no moisture available to plants.

3 Soil Chemistry 467

- Soils show a wide range of pH values. As seen in the accompanying diagram, highly acidic soils are not good for plant growth. Moderately alkaline soils are the most fertile.

The pH scale and soil fertility • Figure 15.11

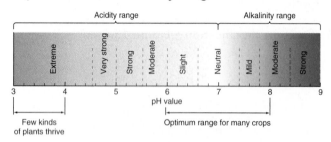

- **Soil colloids**, the finest soil particles, help retain nutrient ions that are used by plants. Soils with high **base** status have abundant available nutrients.

4 The Global Distribution of Soils 468

- Global soils are classified into 12 **soil orders**.

- **Entisols** are composed of fresh parent material and have no horizons. The horizons of **Inceptisols** are only weakly developed.

- **Alfisols**, **Spodosols**, **Oxisols**, and **Ultisols** are more mature soils that show well-developed profiles and greater weathering.

- Some soils are characterized by climate. As shown in the accompanying diagram, **Mollisols** are soils of midlatitude grasslands. **Aridisols** are soils of arid regions, marked by horizons of carbonate minerals or salts, and **Gelisols** are soils of permafrost regions that are churned by the freezing and thawing of water in the soil.

Soil profile of a Mollisol • Figure 15.18

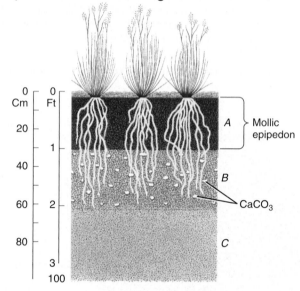

- Some soils are best characterized by their parent material. **Vertisols** are formed from a distinctive clay mineral. **Andisols** are formed from young volcanic ash deposits.

- **Histosols** have a thick upper layer formed almost entirely of organic matter.

Key Terms

Critical and Creative Thinking Questions

1. Successful agriculture requires both favorable climate and soils. Describe a location that potentially has good soil for growing crops but a poor climate. Next, describe a location with the opposite problem: good climate but poor soils.

2. Considering the processes of soil formation, discuss three specific ways that soil formation could differ in the future due to climate change. Assume that the world will be warmer and that there will be more variability in precipitation (that is, more frequent droughts and floods).

3. Soil properties are important for determining the type of vegetation an area can support, but vegetation also plays an important role in the process of soil formation. Describe a situation in which a sudden loss of vegetation could affect soil development and how subsequent vegetation would be influenced.

4. When growing plants in a home garden, why is it important to test the pH of the soil? What step can be taken if the soil pH is too low?

5. Examine the world soils map (see Figure 15.13) and identify three soil types that are found near your location. Develop a short list of characteristics that would help you tell them apart.

6. What is the type of soil in the soil profile shown here? Predict how this soil would perform in an agricultural setting.

What is happening in this picture?

This bog in Connemara, County Galway, Ireland, has been trenched, revealing its soil profile. The soil blocks, seen to the sides of the trench, are drying for use as fuel.

Self-Test

(Check your answers in the Appendix.)

1. Soil scientists use the term _____ to describe finely divided, partially decomposed organic matter in soils.

   a. colloid

   b. peat

   c. organic residue

   d. humus

2. The term _____ describes all forms of mineral matter that are suitable for transformation into soil.

   a. regolith

   b. weathered rock

   c. parent material

   d. bedrock

3. The diagram shows a sequence of horizons that could be found in forest soil. Label the following horizons: A, B, C, E, and O.

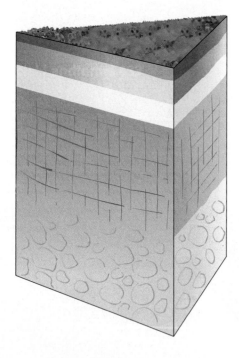

4. Two processes of _____ that operate simultaneously are eluviation and illuviation.

   a. relocation

   b. soil enrichment

   c. leaching

   d. translocation

5. In _____ climates, organic matter accumulates in soils, while it is relatively scarce in _____ climates.

   a. cold; warm

   b. dry; wet

   c. warm; cold

   d. wet; dry

6. A _____ is a soil that contains the proportion of each of the three grades that is most effective at storing moisture.

   a. humus

   b. ped

   c. loam

   d. regolith

7. Label the charge of this clay particle. Give examples of plant nutrients that will be attracted to the particle's charged surface.

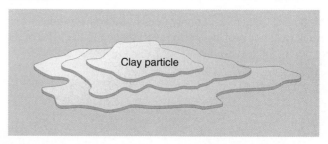

Clay particle

8. High soil acidity is typical of _____ climates.

   a. cold, humid

   b. cold, dry

   c. warm, humid

   d. warm, dry

9. Soils with _____ lack soil-structure peds.

   a. low clay content

   b. high clay content

   c. high sand content

   d. low sand content

10. _____, soils of desert climates, are dry for long periods of time.

    a. Spodosols

    b. Mollisols

    c. Vertisols

    d. Aridisols

11. Tundra soils fall largely into the order of _____, soils underlain by permanently frozen ground.

    a. Spodosols

    b. Andisols

    c. Gelisols

    d. Histosols

12. Which soil, shown in this photo, is usually reddish to yellowish and is closely related to the Oxisols in appearance and environment of origin?

13. _____ have a particular clay mineral as a parent material.

    a. Vertisols

    b. Spodosols

    c. Entisols

    d. Aridisols

14. _____ are soils in which more than half of the parent mineral matter is volcanic ash. This soil type has a high carbon content, so its color is very dark.

    a. Oxisols

    b. Andisols

    c. Alfisols

    d. Entisols

15. The soil profile shown in this diagram has a unique property—a dense layer of iron and aluminum oxides and organic matter. Identify the soil and describe where it may have formed.

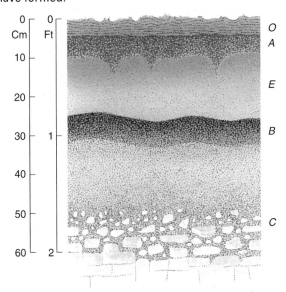

THE PLANNER ✓

Review your Chapter Planner on the chapter opener and check off your completed work.

Biogeographic Processes

Lying west of Ecuador in South America, the Galápagos Islands are named for the giant tortoises that inhabit them. The tortoise shells reminded the 16th-century Spanish explorers who discovered the islands of a kind of saddle they called a *galápago*—and the name stuck for both the animals and the region.

The remarkable birds and animals of these islands inspired the young naturalist Charles Darwin, who visited the islands in 1835. Impressed by "the amount of creative force displayed on these small, barren and rocky islands," he proposed that the life-forms could have evolved from common ancestors through natural selection.

As many as 250,000 giant tortoises inhabited the islands when they were discovered. Today only about 15,000 are left. This is mainly because 18th- and 19th-century pirates used the animals—who can survive for several months without food and water—as a source of fresh meat on their whaling ships. Goats introduced to the island by humans have also nearly destroyed the tortoises' food supply.

The tortoises are part of the Earth's large, diverse, and constantly changing biological realm, the biosphere, which includes millions of species of organisms, from the ocean's abyssal depths to the land's highest peaks.

CHAPTER OUTLINE

A Galápagos tortoise ambles over rocky terrain.

CHAPTER PLANNER ✔

- ☐ Study the picture and read the opening story.
- ☐ Scan the Learning Objectives in each section:
  p. 486 ☐ p. 495 ☐ p. 502 ☐ p. 504 ☐
- ☐ Read the text and study all figures and visuals. Answer any questions.

Analyze key features
- ☐ Process Diagram p. 487 ☐ p. 503 ☐ p. 506 ☐
- ☐ What a Geographer Sees, p. 501
- ☐ Video Explorations, p. 507
- ☐ Where Geographers Click, p. 509
- ☐ Geography InSight, p. 510
- ☐ Stop: Answer the Concept Checks before you go on.
  p. 495 ☐ p. 501 ☐ p. 504 ☐ p. 512 ☐

End of chapter
- ☐ Review the Summary and Key Terms.
- ☐ Answer the Critical and Creative Thinking Questions.
- ☐ Answer What is happening in this picture?
- ☐ Complete the Self-Test and check your answers.

Energy and Matter Flow in Ecosystems

LEARNING OBJECTIVES

1. **Outline** how energy travels through a food web.
2. **Explain** the processes of photosynthesis and respiration.
3. **Define** net primary production.
4. **Describe** the carbon and nitrogen cycles.

---

biogeography The study of the distribution patterns of organisms over space and time and of the processes that produced these patterns.

Biogeography seeks to explain the distribution of plants and animals—the **biota**—over the Earth. Biogeographers identify and describe the processes that influence plant and animal distribution patterns throughout the biosphere. They study how the environment influences the distribution patterns of organisms and ecological communities and how their spatial distribution patterns have changed over time. Biogeography represents a sub-discipline of both geography and ecology. It has become increasingly important due to the rise of species extinction rates and the proliferation of harmful invasive species throughout the planet. This chapter is the first of two that look at biogeography; this one introduces biological and ecological processes, and the next chapter addresses the global diversity of life on Earth.

Before we turn to those processes, we begin by looking at how organisms live and interact as **ecosystems** and how energy and matter flow through and drive the processes of ecosystems (**Figure 16.1**). An ecosystem can be as small as a drop of water on a plant leaf or as large as a forest. All ecosystems are parts of the biosphere.

ecosystem A collection of organisms and the environment with which they interact.

Ecosystems • Figure 16.1

An ecosystem includes a collection of organisms and the environment they interact with.

b. Freshwater marsh ecosystem
The marsh or swamp ecosystem supports a wide variety of life-forms, both plant and animal. Here, a group of white ibises forages for food in the shallow waters of Okefenokee Swamp, Georgia.

a. Tundra ecosystem
The caribou, a large grazing mammal, is one of the important primary consumers in the tundra ecosystem, shown here on Alaska's North Slope.

c. Savanna ecosystem
The savanna ecosystem, with its abundance of grazing mammals and predators, has a rich and complex network of interrelationships. Here African elephants travel single-file across the savanna plains of the Savuti region, Botswana.

A food chain • Figure 16.2

The food chain of an ecosystem describes how energy flows from one trophic level to the next. This example shows some of the species found in a salt marsh ecosystem.

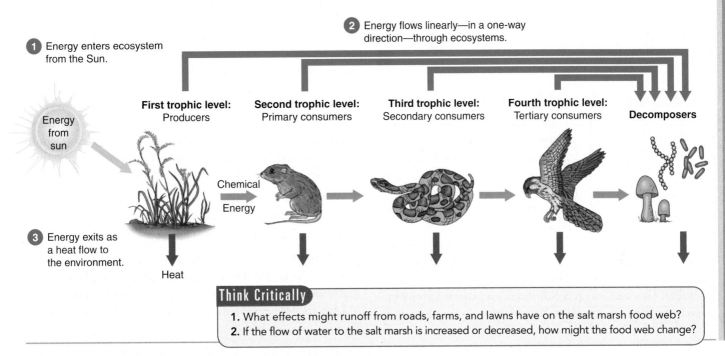

1 Energy enters ecosystem from the Sun.

2 Energy flows linearly—in a one-way direction—through ecosystems.

3 Energy exits as a heat flow to the environment.

Energy from sun

First trophic level: Producers   **Second trophic level:** Primary consumers   **Third trophic level:** Secondary consumers   **Fourth trophic level:** Tertiary consumers   **Decomposers**

Chemical Energy

Heat

Think Critically
1. What effects might runoff from roads, farms, and lawns have on the salt marsh food web?
2. If the flow of water to the salt marsh is increased or decreased, how might the food web change?

Energy Flows and the Food Chain

All ecosystems take up matter and energy as their component plants and animals grow, reproduce, sustain life, and are recycled after death. Beginning with the Sun's energy captured by plants, matter and energy are constantly cycled within the ecosystem. Matter and energy are exchanged within an ecosystem and are exported and imported across ecosystem boundaries. Ecosystems maintain dynamic balances with the various processes and activities within them. Many of these balances are robust and self-regulating, but some are sensitive and can be readily altered or destroyed.

Solar radiation floods the Earth's surface and drives the flow of energy throughout Earth's ecosystems. The **food chain** describes how energy in an ecosystem flows from one **trophic level** to the next (**Figure 16.2**). Solar energy is absorbed by the plants, as primary producers, and stored in the chemical products of photosynthesis. As these photosynthetic organisms are eaten and digested by consumers higher in the food chain, chemical

food chain The fundamental levels through which energy flows through an ecosystem as the organisms at each level consume energy from the bodies of organisms in the level below.

trophic level The position an organism occupies within a food web.

energy is transferred. This chemical energy, a derivative of the Sun's energy, powers new biochemical reactions in the consumer organisms, which again store energy in new chemical forms and lose energy to the surrounding environment. A more complete picture of energy flow through an ecosystem is a **food web**, which shows all the energy flow links among species in an ecosystem.

Energy that flows through the ecosystem is either captured or lost with each interaction of species. For example, when cows graze, they capture the Sun's energy stored in the grass. They store more energy than they use while grazing, and they then use that energy to produce milk and sustain their offspring. In a steady-state, or stable, ecosystem, energy is lost as it passes from one trophic level to consumers at the next higher level. Up to 90% of the energy is lost in each transfer across trophic levels. Much of the energy consumed at each trophic level in the food web is lost through the **respiration** of organisms at that level. You can think of this energy as fuel burned to keep the organisms operating. Energy expended in respiration cannot be stored for use by other organisms higher in the food chain. This means that, generally, both the numbers of organisms and their total amount of living tissue decrease greatly as we move up the food chain. Most ecosystems can be viewed as containing one level of producers and four fundamental levels of consumers.

The number of individuals of any species present in an ecosystem depends on the resources available to support them. If these resources provide a steady supply of energy, the population size will normally stay steady, or maintain equilibrium. But resources can vary with time—for example, in an annual cycle. In such cases, the population of a species depending on these resources may fluctuate in a corresponding cycle called a *dynamic equilibrium*.

Photosynthesis and Respiration

photosynthesis
The production of carbohydrates from water and carbon dioxide, using light energy and releasing oxygen.

At the base of the food chain, plants and algae absorb and store solar energy in the chemical bonds of carbohydrates that they manufacture. This process is called **photosynthesis** (**Figure 16.3**). Oxygen gas molecules (O_2) are a byproduct of photosynthesis. Carbohydrate is a class of organic compounds, including sugars, which are made from the elements carbon (C), hydrogen (H), and oxygen (O). Sugar molecules are composed of short chains of carbon bonded to one another.

Hydrogen atoms and hydroxyl (OH) molecules are also attached to the carbon atoms. We can symbolize a single carbon atom with its attached hydrogen atom and hydroxyl molecule as –CHOH–. The leading and trailing dashes indicate that the unit is a portion of a longer chain of connected carbon atoms.

Photosynthesis of carbohydrate requires a series of complex biochemical reactions using water (H_2O) and carbon dioxide (CO_2), as well as visible light energy. Photosynthesis takes place in chloroplasts, tiny energy factories in plant cells that have layers of chlorophyll, enzymes, and other molecules in close contact. This process requires chlorophyll, a complex organic molecule that absorbs light energy for use by the plant cell. A simplified chemical reaction for photosynthesis can be written as:

$$H_2O + CO_2 + \text{Light energy} \rightarrow \text{–CHOH–} + O_2$$
Water + carbon dioxide + light energy
produces carbohydrate + oxygen

Respiration, the counterpart of photosynthesis, occurs in both plants and animals. In this process, carbohydrates are broken down in the organism's cells and combine with oxygen to yield carbon dioxide and water. The overall reaction is:

$$\text{–CHOH–} + O_2 \rightarrow CO_2 + H_2O + \text{Chemical energy}$$
Carbohydrate + oxygen produces
carbon dioxide + water + chemical energy

As with photosynthesis, the actual reactions involved in respiration are not this simple. The chemical energy released is stored in several types of energy-carrying molecules in living cells and is used later to synthesize all the biological molecules used to sustain life.

Although much of the chemical energy released in respiration is taken up in the new chemical bonds of these molecules, a certain portion raises the temperature of the new molecules. As the molecules cool, this internal energy is lost to the environment as heat flow. In a similar way, each following biochemical reaction releases internal energy and adds to the heat flow. Eventually, all the chemical energy gained by respiration is lost to the surroundings in these transformations.

Because respiration consumes carbohydrate, we have to take respiration into account when talking about the amount of new carbohydrate placed in storage. Gross photosynthesis is the total amount of carbohydrate produced by photosynthesis. Net photosynthesis is the amount of a plant's carbohydrate remaining after respiration has used a portion of its carbohydrate budget to grow and maintain the life of the plant.

The rate of net photosynthesis depends on various factors, including the intensity of light energy available,

Photosynthesis • Figure 16.3

Leaves take in CO_2 from the air and H_2O from their roots; they use solar energy absorbed by chlorophyll to combine them and form carbohydrate. In the process, O_2 is released.

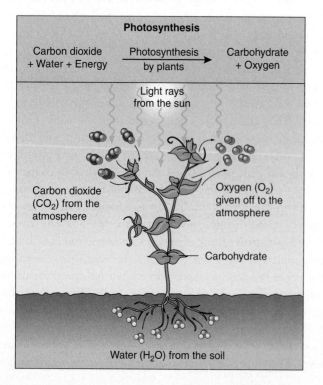

Photosynthesis

| Carbon dioxide + Water + Energy | Photosynthesis by plants → | Carbohydrate + Oxygen |

Light rays from the sun

Carbon dioxide (CO_2) from the atmosphere

Oxygen (O_2) given off to the atmosphere

Carbohydrate

Water (H_2O) from the soil

Global distribution of biomass • Figure 16.4

The Earth's green biomass is the photosynthetic machinery of the planet, producing 50 to 60 billion metric tonnes (55 to 66 billion tons) of carbon each year. On land, equatorial and wet low-latitude regions are most productive (dark green on the map). Land is least productive where limited by temperature (polar regions) or moisture (deserts). In the oceans, coastal waters and upwelling zones are most productive (red on the map).

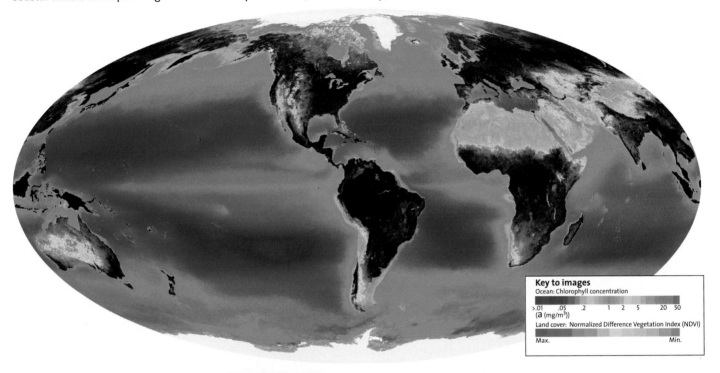

Key to images
Ocean: Chlorophyll concentration

>.01 .05 .2 1 2 5 20 50
(a (mg/m^3))

Land cover: Normalized Difference Vegetation Index (NDVI)

Max. Min.

Ask Yourself

What accounts for the band of low productivity on the western edge of South America?
a. Air coming off the Pacific Ocean is very moist, and the intense rains suppress plant growth.
b. Air coming off the Pacific Ocean is very cold and suppresses plant growth.
c. The Andes Mountains span this area, and here temperatures are cool and precipitation is low.
d. The Andes Mountains span this area, and here temperatures are warm and precipitation is high.

up to a maximum limit. Most green plants need only about 10 to 30% of full summer sunlight for maximum net photosynthesis. Once the intensity of light is high enough for maximum net photosynthesis, the duration of daylight becomes an important factor in determining the rate at which the products of photosynthesis build up in plant tissues. The rate of photosynthesis increases as air temperature increases, up to a limit. Moisture availability, nutrients, and levels of carbon dioxide also influence the rate of photosynthesis. Increased levels of atmospheric carbon dioxide, as in urban areas, have been shown to result in plant growth rates three times higher than normal.

biomass The dry weight of living organic matter in an ecosystem within a specified surface area.

Net Primary Production

Plant ecologists measure the accumulated net production by photosynthesis in terms of the **biomass**, which is the dry weight of organic matter. This quantity could, of course, be stated for a single plant or animal, but a more useful measurement is the biomass per unit of surface area within the ecosystem—that is, kilograms of biomass per square meter or (metric) tons of biomass per hectare (1 hectare = 10^4 m^2) (**Figure 16.4**).

Of all ecosystems, forests have the greatest biomass because of the large amount of wood that the trees accumulate through time. The biomass of grasslands and croplands is much smaller in comparison. The biomass of

Net primary production of ecosystems ranges widely across lands and oceans. On land, freshwater swamps and marshes are most productive, on average. In marine environments, coral reefs and algal beds head the list.

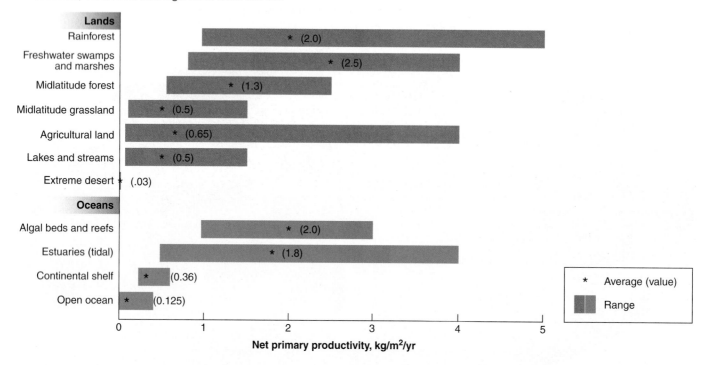

freshwater bodies and the open oceans is about one-hundredth that of the grasslands and croplands. The highest values are in two dissimilar environments: forests and wetlands (swamps and marshes). Agricultural land compares favorably with grassland, but the range is very large in agricultural land, reflecting many factors, such as availability of soil water, soil fertility, and use of fertilizers and machinery.

In marine environments, open oceans aren't generally very productive because they lack nutrients. Continental shelf areas, which receive nutrients from rivers draining the land, are more productive, supporting much of the world's fishing industry. Estuaries and coral reefs are highly productive, serving as nurseries for many aquatic species. Upwelling zones, where nutrient-rich bottom waters rise to the surface, are also highly productive.

The amount of biomass per unit area indicates the amount of photosynthetic activity, but this number can be misleading. In some ecosystems, biomass is broken down very quickly by consumers and decomposers. To know how productive the ecosystem is, we need to calculate the annual yield of useful energy produced by the ecosystem, or the **net primary production** (**Figure 16.5**). Net primary production represents a potential source of renewable energy derived from the Sun that can be exploited by humans

for their energy needs. Humans exploit stored plant energy in a number of ways, such as burning wood and distilling grains or plant wastes into ethanol for fuel.

Biogeochemical Cycles

We have seen how energy from the Sun flows through ecosystems, passing from one level of the food chain to the next. Ultimately, respiration consumes all of this energy and releases it to the environment as a flow of heat. This flow is part of the Earth's global energy budget, which maintains a dynamic balance within the Sun–Earth atmosphere system. Matter also moves through ecosystems, but because gravity keeps surface material earthbound, matter is not lost by the Earth system. As molecules are formed and reformed by chemical and biochemical reactions within an ecosystem, the atoms that compose them remain the same and are not changed or lost. In this way, matter is conserved, and atoms and molecules are used and reused, or cycled continuously, within ecosystems.

Atoms and molecules move through ecosystems under the influence of both physical and biological processes. We call the pathways that a particular type of matter takes through the Earth's ecosystem a **biogeochemical cycle** (or a *material cycle* or *nutrient cycle*). Ecologists have studied and documented biogeochemical cycles for many elements, including the primary life-supporting elements of carbon, oxygen, nitrogen, sulfur, and phosphorus.

The carbon cycle Of all the biogeochemical cycles, the **carbon cycle** is the most important. All life is composed of carbon compounds of one form or another. Human activities since the Industrial Revolution have modified the carbon cycle in significant ways.

> **carbon cycle** The biogeochemical cycle in which carbon moves through the biosphere, hydrosphere, atmosphere, and lithosphere.

Carbon moves through the cycle as a gas, as a liquid, and as a solid (**Figure 16.6**). In the gaseous portion of the cycle, carbon moves largely as carbon dioxide (CO_2), which is a free gas in the atmosphere and a dissolved gas in fresh and salt water. Atmospheric carbon dioxide makes up less than 2% of all the carbon, excluding carbonate rocks and sediments. The atmospheric pool of carbon is supplied by plant and animal respiration in the oceans and on land, by outgassing volcanoes, and by fossil fuel combustion in industry.

In the sedimentary portion of its cycle, we find carbon in carbohydrate molecules in organic matter, as hydrocarbon compounds in rock (petroleum and coal), and as mineral carbonate compounds such as calcium carbonate ($CaCO_3$). A lot of carbon is incorporated into shells of marine organisms, large and small. When they die, their shells settle to the ocean floor, where they dissolve or accumulate as layers of sediment. This provides an enormous carbon storage pool, but it is not available to organisms until it is later released by rock weathering. Organic compounds synthesized by phytoplankton also settle to the ocean floor and eventually are transformed into the hydrocarbon compounds that make up petroleum and natural gas.

The carbon cycle • Figure 16.6

Ecosystems cycle carbon through photosynthesis, respiration, decomposition, and combustion. The movement of carbon between the atmosphere, ocean, and living organisms is known as the carbon cycle.

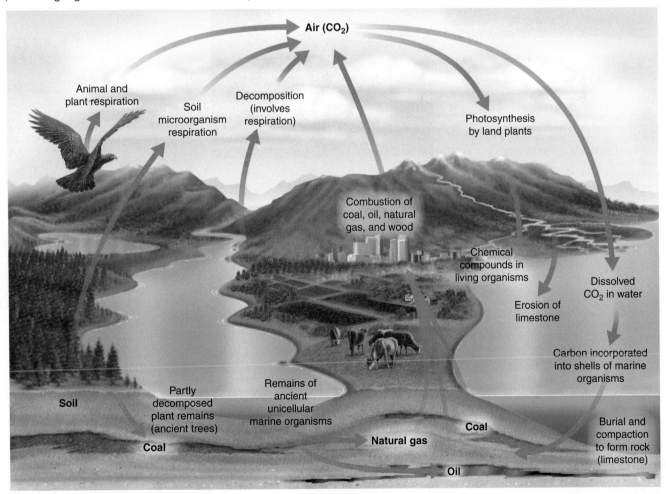

The landscape of Baltimore, Maryland, is mapped for a 100-year period of urban expansion using GIS and satellite data. These images portray the rate of urban expansion from 1900 to 1992, as humans removed forest biomass containing carbon to locate industry and homes. The yellow areas define where urban development has removed the natural vegetation for human habitation. What happened in Baltimore has been repeated throughout the world, causing a major shift in the carbon budget from the forest biomass to the atmosphere.

1900

1966

1992

Today, human activities are significantly affecting the carbon cycle. As a result of the burning of fossil fuels, CO_2 is being released into the atmosphere at a rate far beyond that of any natural process. Another important human impact lies in changing the Earth's land cover. For example, over the past several millennia, humans have been active in clearing forests and building urban areas (**Figure 16.7**).

What happens to the extra carbon liberated by fossil fuel burning in these cities? More than half goes into the oceans and biosphere, and the rest goes into the atmosphere. A small amount is absorbed by the oceans. But the rest must flow into the biosphere, prompting scientists to seek out and determine exactly what effect this carbon has on ecosystems (**Figure 16.8**).

Carbon budget in the boreal forest ecosystem • Figure 16.8

Since the early 1990s, a team of international scientists, led by NASA, has been conducting the Boreal Ecosystem-Atmosphere Study (BOREAS) to find out how the boreal forests of northern Canada interact with the atmosphere, how much CO_2 they can store, and how climate change will affect them.

a. The map depicts the two long-term research sites (red boxes) located in Canada that are used to characterize the full range of boreal forests.

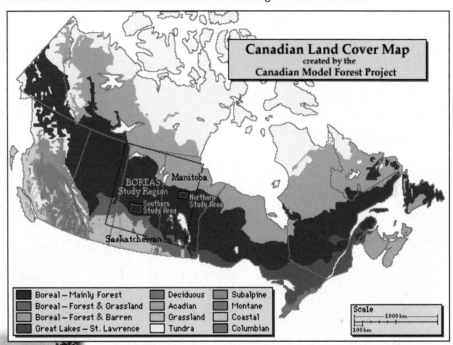

b. Using instruments mounted on 30-meter (98-foot) towers, scientists continuously monitor how carbon dioxide, methane, nitrous oxide, and carbon monoxide flow through the canopy of the boreal forests.

c. Stations located on the ground in the black spruce and jack pine forest measure key gases to provide a three-dimensional model of forest photosynthesis and respiration.

At the present time, nitrogen fixation far exceeds natural levels of denitrification due to human activity. Nitrogen is fixed naturally by oceanic blue-green algae, soil bacteria, and wild legumes. Human-caused nitrogen fixation is due to activities such as fertilizer manufacture, fuel combustion, and growth of legumes as crops. Bacteria in soil and sediments returns N_2 to the atmosphere through denitrification. Nitrates from fertilization run off into freshwater and marine aquatic ecosystems, creating pollution problems.

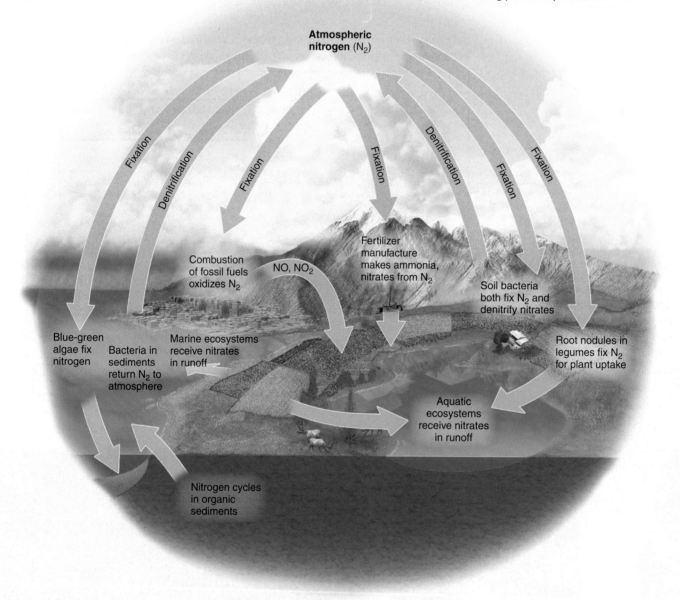

The nitrogen cycle The **nitrogen cycle** is another important biogeochemical cycle. Nitrogen makes up 78% of the atmosphere by volume, so the atmosphere is a vast storage pool in this cycle. Nitrogen, as N_2 in the atmosphere cannot be assimilated directly by plants or animals, but certain microorganisms supply nitrogen to plants through **nitrogen fixation**. Nitrogen-fixing bacteria live primarily in the soil and are associated with the roots of certain plants, including legumes such as beans and peas. Once these bacteria have fixed the nitrogen into ammonia (NH_3), it can be assimilated by plants. Animals then assimilate the nitrogen when they eat the plants. Nitrogen is returned to the soil in the waste of the animals. Other soil bacteria convert nitrogen from usable forms back to N_2, in a process called **denitrification**, completing the organic portion of the nitrogen cycle.

Human activities such as agriculture and industry add more nitrogen to the system, disrupting the nitrogen cycle (**Figure 16.9**). At present rates, nitrogen fixation from human activity nearly equals all natural biological fixation, and usable nitrogen is accumulating, resulting in a number of negative effects. Nitrous oxides in the atmosphere

(NO and NO_2), formed by fuel combustion, contribute to climate change. Industrially emitted nitrogen may fall back to the surface as acid rain, harming plants, corroding buildings, and acidifying soils, lakes, and streams. Reactive nitrogen is also transferred from the land into rivers, where it reduces biodiversity, and then to the seas and oceans, where it can create zones devoid of life. These problems will increase in years to come due to the fact that industrial fixation of nitrogen in fertilizer manufacture is currently doubling about every six years. The global impact of such large amounts of nitrogen reaching rivers, lakes, and oceans remains uncertain.

CONCEPT CHECK	STOP

1. **What** happens to the number of organisms at each level of the food chain?

2. **What** factors control the levels of net photosynthesis and gross photosynthesis?

3. **What** ecosystems have the highest net primary production?

4. **How** are the carbon and nitrogen cycles affected by human activities?

Ecological Biogeography

LEARNING OBJECTIVES

1. **Explain** the physical factors that influence plant and animal distribution.

2. **Describe** how species interact through competition, predation, and symbiosis.

We have seen how energy and matter move through ecosystems. But if we want to fully understand ecosystems, we also need to look at **ecological biogeography**, which examines the distribution and abundance patterns of plants and animals from the viewpoint of their physiological needs. That is, we must examine how the individual organisms of an ecosystem interact with their environment. From fungi digesting organic matter on a forest floor to ospreys fishing in a coastal estuary, each organism has a range of environmental conditions that establishes the boundaries for its survival. Species' behavioral ranges dictate the adaptations available for the individual to meet its energy requirements. Each organism has a particular **habitat**, a specific kind of environment to which it has adapted (**Figure 16.10**).

> **habitat** A subdivision of the environment according to the needs and preferences of organisms or groups of organisms.

Habitats • Figure 16.10

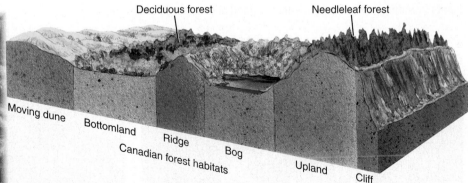

We can distinguish six distinct habitats across the Canadian boreal forest near the Mackenzie River: moving dune, bottomland, ridge, bog, upland, and cliff. These different habitats each demonstrate a characteristic composition of plants and animals that thrive in these conditions. Deciduous forest, shown in this photo about to shed leaves in the fall, occupies moist bottomland habitats, while spruce and fir occupy the uplands. Swamps and bogs show low grasses and sedges, which are still green.

We use the term **ecological niche** to describe the functional role played by an organism as well as the physical space it inhabits. If the habitat can be compared to an individual's address, then the niche would be its profession, including how and where it obtains its energy and how it influences other species and the surrounding environment. When describing the ecological niche, we talk about an organism's tolerances and responses to changes in moisture, temperature, soil chemistry, illumination, and other factors. Although many different species may occupy the same habitat, only a few of these species will ever share the same ecological niche, for, as we will see shortly, evolution tends to separate those that do. As we move from habitat to habitat, we find that each is the home of a group of organisms that occupy different but interrelated ecological niches. We now turn to the environmental factors that help determine where organisms, as individuals and species, are found.

Climatic Factors Affecting Ecosystems

Taken separately or together, moisture, temperature, light, and wind can limit the distribution of plant and animal species. Biogeographers recognize that there is a critical level of climatic stress beyond which a species cannot thrive. This means that we can use geographic boundaries to define the spatial limits of the potential distribution of a species.

Moisture Plants and animals have adapted to the abundance or scarcity of water in a variety of ways (**Figure 16.11**). For example, some desert plants produce a widespread, but shallow, root system, so they can absorb water from short desert downpours that saturate only the uppermost soil layer. Others adapt to a desert environment by greatly reducing their leaf area or by bearing no leaves at all in order to reduce transpiration. Still others locate close to dry stream beds and send deep taproots down to access ground water.

Certain climates have one season in which water is unavailable to plants because of lack of precipitation or because the soil water is frozen. This season alternates with one in which there is abundant water. Plants adapted to these climates are *deciduous*: They drop their leaves at the close of the favorable season and are dormant during the dry or cold season. When water is again available, they leaf out and grow rapidly.

Animals have evolved a variety of mechanisms to conserve water and survive in harsh environments. Many invertebrates stay dormant during the dry period. When rain falls, they emerge to take advantage of the new and short-lived vegetation that often results. Many species of birds nest only when the rains occur, the time of most abundant food for their offspring. The tiny brine shrimp of the Great Basin can wait many years in dormancy until normally dry lakebeds fill with water, an event that occurs perhaps three or four times a century. The shrimp then emerge and complete their life cycles before the lake evaporates. Mammals are by nature poorly adapted to desert environments, but many survive through a variety of mechanisms that enable them to avoid water loss. Many desert mammals do not sweat through skin glands; they remain in underground burrows where humidity is higher and temperatures are cooler, and they move about only at night.

Organisms adapted to water scarcity • Figure 16.11

Plants in water-scarce environments are good at obtaining and storing water.

a. The stems of prickly pear cactus are thickened by a spongy tissue that stores water. The plant also conserves water by having spines instead of leaves.

b. Plants that are adapted to drought conditions, such as this Californian live oak, have a thick layer of waxy material on leaves and stems, helping them to seal water vapor inside.

Temperature adaptations • Figure 16.12

Animals that live in extreme temperatures have adapted by regulating their body temperature.

b. Warm-blooded animals maintain tissues at a constant temperature by internal metabolism. A heavy coat and a thick layer of body fat insulate this Alaskan brown bear, allowing the mammal to maintain a constant body temperature.

a. Some vertebrates enter a state called hibernation, in which their metabolic processes virtually stop and their body temperatures closely parallel those of the surroundings. These little brown bats can survive for almost half the year in this state.

c. This namaqua chameleon in southern Africa changes its skin color to regulate its body temperature, turning black in the morning to absorb solar rays and light gray during the day to reflect them.

Temperature The temperature of the air and soil directly influences the rates of physiological processes in plant and animal tissues. Temperature can also act indirectly on plants and animals. Higher air temperatures reduce the relative humidity of the air, enhancing transpiration from plant leaves as well as increasing direct evaporation of soil water. In general, each plant species has an optimum temperature associated with each of its functions, such as photosynthesis, flowering, fruiting, or seed germination. There are also limiting lower and upper temperatures for these individual functions and for the survival of the plant itself. Animals also have a temperature range to which they have adapted for survival (**Figure 16.12**).

In general, the colder the climate, the smaller the number of species capable of surviving in it. Polar ecosystems operate with a much lower annual energy input than those in the tropics, and therefore there is less energy flowing through a simpler food web. That is why we find only a few plants and animals in the severely cold arctic and alpine environments of high latitudes and high altitudes.

This also helps explain why an equatorial rainforest with a much higher energy budget and a more complex food web contains such a diverse array of plants and animals.

Sunlight Light also helps determine local plant distribution patterns. Some plants are adapted to bright sunlight, whereas others require shade. The amount of light available to a plant will depend in large part on the plant's position. Tree crowns in the upper layer of a forest receive maximum light but correspondingly reduce the amount available to lower layers. In extreme cases, forest trees so effectively cut off light that the forest floor is almost free of shrubs and smaller plants. In certain deciduous forests of the midlatitudes, the period of early spring, before the trees are in leaf, is one of high light intensity at ground level, permitting the smaller plants to go through a rapid growth cycle. These plants largely disappear as the leaf canopy is completed. Other low plants in the same habitat require shade and do not appear until the leaf canopy is well developed.

Adaptation to wind • Figure 16.13

This tree above a broad plain in Tierra del Fuego, Argentina, has been shaped by long exposure to wind in a constant direction (from the right side in this photo).

The light available for plant growth varies by latitude and season. As we saw earlier, the number of daylight hours in summer increases rapidly with higher latitude and reaches its maximum poleward of the Arctic and Antarctic Circles, where the Sun may be above the horizon for 24 hours. Although frost in these regions shortens the growing season for plants, the rate of plant growth in the short frost-free summer is greatly accelerated by the prolonged daylight.

In the midlatitudes, where many species are deciduous, the annual rhythm of increasing and decreasing periods of daylight determines the timing of budding, flowering, fruiting, and leaf shedding. Even on overcast days, there is usually enough light for most plants to carry out photosynthesis at their maximum rates.

Light also influences animal behavior. The day–night cycle controls the activity patterns of many animals. Birds, for example, are generally active during the day, whereas small foraging mammals, such as weasels, skunks, and chipmunks, are more active at night. In the midlatitudes, as autumn days grow shorter and shorter, squirrels and other rodents hoard food for the coming winter season. Later, increasing hours of daylight in the spring trigger such activities as mating and reproduction.

Wind Wind is also an important environmental factor in the structure of vegetation in highly exposed positions. Wind causes excessive drying, damaging the exposed side of the plant and killing its leaves and shoots. Trees of high-mountain summits can be severely distorted by wind, with trunks bent to near-horizontal, facing away from the prevailing wind direction (**Figure 16.13**). Winds also have a beneficial role in the environment, carrying pollen and seeds to facilitate the reproduction and spread of plants.

Other Factors Affecting Ecosystems

Although climate is the most important factor determining the global distribution of plant and animal species, other factors also play a role, including geomorphic factors, soil conditions, and disturbances such as wildfire.

Geomorphic factors Geomorphic, or landform, factors such as the steepness and orientation of a slope and differences in elevation can affect ecosystems. In a much broader sense, geomorphic factors include the sculpting of the landscape by erosion, transportation, and deposition by streams, waves, wind, and ice.

Slope steepness affects the rate at which precipitation drains from a surface, which indirectly influences plants and animals. On steep slopes, surface runoff is rapid, but on gentle slopes, more precipitation can penetrate the soil and be held there. Steep slopes often have thin soil because they are more easily eroded, while soil on gentler slopes is thicker. The orientation of a slope controls plants' exposure to sunlight and prevailing winds. Slopes facing the Sun have a warmer, drier environment than slopes that face away from the Sun. In the midlatitudes, these contrasts may be strong enough to support distinctly different ecosystems or biotic communities on north-facing and south-facing slopes.

On peaks and ridge crests, rapid drainage dries the soil, which is also more exposed to sunlight and drying winds. By contrast, the valley floors are wetter because water converges there. In humid climates, the water table in valley floors may lie close to or at the ground surface, producing marshes, swamps, ponds, and bogs.

Soil conditions Soil conditions can vary widely from one small area to the next, influencing the local distribution of plants and animals. For example, sandy soils store less water than soils with abundant silt and clay, and they are often home to plants and animals adapted to dry conditions. If there is a high amount of organic matter in the soil, then the soil will be rich in nutrients and will harbor more plant species. The relationship can work in the opposite direction, too: Biota can change soil conditions, as when a prairie grassland builds a rich, fertile soil beneath it, or when beavers colonize and reengineer stream ecosystems.

Disturbances Fires, floods, volcanic eruptions, storm waves, high winds, and other infrequent catastrophic events can damage or destroy ecosystems and modify habitats. Although such disturbance can greatly alter the nature of an ecosystem, it is often part of a natural cycle of regeneration that gives short-lived or specialized species an opportunity to grow and reproduce.

In many cases, fire is beneficial to an ecosystem, clearing out undergrowth and promoting the germination of new plants (**Figure 16.14**). Among tree species, pines are typically well adapted to germinating under such conditions. In fact, the jack pine of eastern North America and the lodgepole pine of the intermountain West have cones that remain tightly closed until the heat of a fire opens them, allowing the seeds to be released. Fires also preserve grasslands. Grasses are fire resistant because they have extensive root systems below ground and leaf buds located at or just below the surface. But woody plants that might otherwise invade grassland areas are not so resistant and are usually killed by grass fires. Fires are crucial for nutrient cycling in fire-adapted ecosystems, such as the chaparral ecosystem in the southwestern United States.

In many regions, active fire suppression has reduced the frequency of burning to well below natural levels. This might sound like a good thing, but it causes dead wood to build up as fire fuel on the forest floor. When a fire occurs under these conditions, it is especially destructive—burning hotter and more rapidly, and consuming trees and shrubs that would otherwise have survived and flourished.

Flooding is another important disturbance. It displaces animal communities and also deprives plant roots of oxygen. Where flooding brings a swift current, mechanical damage rips limbs from trees and scours out roots.

Effects of fire on a forest environment • Figure 16.14

Fire cleans out the understory and consumes dead and decaying organic matter while leaving most of the overstory trees untouched. Without a cover of shrubs, more sunlight reaches the soil, which is also enriched by ash from the fire.

Interactions Among Species

Species do not interact only with their physical surroundings. They also interact with each other. Such interactions may benefit at least one of the species, may be negative to one or both species, or may have no effect on either species (Figure 16.15).

Competition happens when two species need a common resource that is in short supply. Both populations suffer from lower growth rates than they would have if only one species were present. Sometimes one species will win the competition and crowd out its competitor. At other times, the two species may remain in competition indefinitely, in dynamic equilibrium. Competition is a constant and shifting condition that can become severe when critical resources are limited. If a genetic strain within one of the populations emerges and can use a substitute resource, its survival rate will be higher than that of the remaining strain, which still competes. The original strain may become extinct. In this way, evolutionary mechanisms tend to reduce competition among species.

Predation occurs when one species feeds on another. There are obvious benefits for the predator species, which obtains energy for survival, but, of course, the interaction has a negative outcome for individuals of the prey species. Although predation is negative for individuals in the prey species, it sometimes plays an important role in keeping populations in balance with available resources (see *What a Geographer Sees*). Predation also selectively removes the weaker individuals, improving the genetic composition of the attacked species. Plants often use the behavior of their predators (called herbivores) to their advantage for seed dispersal, such as when birds carry seeds long distances in their intestines.

Interactions among species • Figure 16.15

Ecosystem dynamics involve a variety of interrelationships between species.

b. Predation
When one species feeds on another species, it is called predation. This giant anteater enjoys a lunch of Brazilian termites.

a. Competition
When two species both rely on a limited resource, they are in a competitive relationship. Here a pride of lions and a herd of elephants share a water hole in Chobe National Park, Botswana.

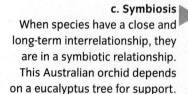

c. Symbiosis ▶
When species have a close and long-term interrelationship, they are in a symbiotic relationship. This Australian orchid depends on a eucalyptus tree for support.

WHAT A GEOGRAPHER SEES

Positive Effects of Predation

This buck is a member of today's Kaibab deer herd, which is a population of mule deer. They are named for their large ears, which resemble a mule's, and are common throughout western North America. Animal lovers may feel sad that such deer are routinely killed by wolves and other predators. But a geographer knows that predation is an important part of what keeps this ecosystem in equilibrium.

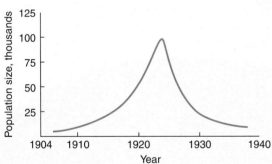

This graph plots the population size of the deer herd in the Kaibab National Forest, Arizona. In 1907, the government began controlling predatory wolves, coyotes, and mountain lions. As a result, the herd grew from about 4000 to nearly 100,000 by 1924. However, confined to an area of 283,000 hectares (700,000 acres), the huge deer population proved too much for the land, and overgrazing led to a population crash. In one year, half the animals starved to death; by the late 1930s, the population had declined to a stable level near 10,000. Predation had maintained the deer population at levels that were in harmony with the supportive ability of the environment.

---

symbiosis A close, coexisting interaction between species that is beneficial to at least one of the species and usually does not harm the other.

A close and long-term interrelationship between two species is called **symbiosis**. One species may benefit from the relationship, or both may benefit. If the relationship reaches a point where one or both species cannot survive alone, it is called **mutualism**. The relationship between the nitrogen-fixing bacterium and legumes mentioned earlier in the chapter is a classic example of mutualism because the bacteria need the plant to survive and the plant needs the bacteria to fix nitrogen for use by the plant. **Parasitism** is a type of symbiotic relationship in which one species gains nutrition from another, typically when the parasite organism invades or attaches to the body of the host in some way. In most cases, parasitism does not cause great harm to the host species, but in some cases, it can weaken or kill the host.

CONCEPT CHECK

1. **How** do plants and animals adapt to water scarcity?

2. **How** can predation have positive effects for the prey species?

Ecological Biogeography **501**

Ecological Succession

LEARNING OBJECTIVES

1. **Describe** ecological succession.
2. **Explain** how succession can be interrupted.

Plant and animal communities change constantly through time. Walk through the country, and you will see patches of vegetation in many stages of development—from open, cultivated fields through grassy shrublands to forests. Clear lakes gradually fill with sediment and become bogs. We call these changes—in which biotic communities succeed one another on the way to a more mature assemblage within a dynamic life cycle—**ecological succession**.

> **ecological succession** A sequence of distinctive plant and animal communities that first colonize and then replace each preceding stage, leading to a stable mature forested community, often after clearance by burning, clear cutting, or other agents.

Primary and Secondary Succession

Biogeographers use the concept of ecological succession to identify and analyze the state of equilibrium in an ecosystem. It provides a model for comparison when assessing ecological conditions in an area. In general, succession forms increasingly complex communities of organisms, based on the physical conditions of the area. The relatively stable and most complex community, which is the end point of succession, is called the *climax stage*. Climax stages are not really terminal stages, as no ecosystems are permanent; rather, they represent long-term, thriving communities. If succession begins on a newly constructed deposit of mineral sediment, it is called **primary succession**. If, on the other hand, succession occurs on a previously vegetated area that has been recently disturbed—perhaps by fire, flood, windstorm, or humans—it is referred to as **secondary succession**.

Primary succession could happen on a sand dune, a sand beach, the surface of a new lava flow or freshly fallen layer of volcanic ash, or the deposits of silt on the inside of a river bend that is gradually shifting, for example (**Figure 16.16**). Such sites are often little more than deposits of coarse mineral fragments.

Primary succession • Figure 16.16

Sand-dune colonization is a good example of primary succession. Animal species change as succession proceeds, from sand spiders and grasshoppers on the open dunes to sowbugs and earthworms in the dune forest.

a. In the earliest stages of succession on coastal dunes, beach grass colonizes the barren habitat in Cape Cod, Massachusetts. It propagates by underground stems that creep beneath the surface of the sand and send up shoots and leaves.

b. Once the dunes are stabilized, low tough shrubs take over, paving the way for drought-resistant tree species, such as pines and hollies. This coastal dune forest is found on Dauphin Island, Alabama.

Secondary succession • Figure 16.17

Abandoned farmland is a good example of secondary succession. This example is characteristic of vegetation in the southeastern United States.

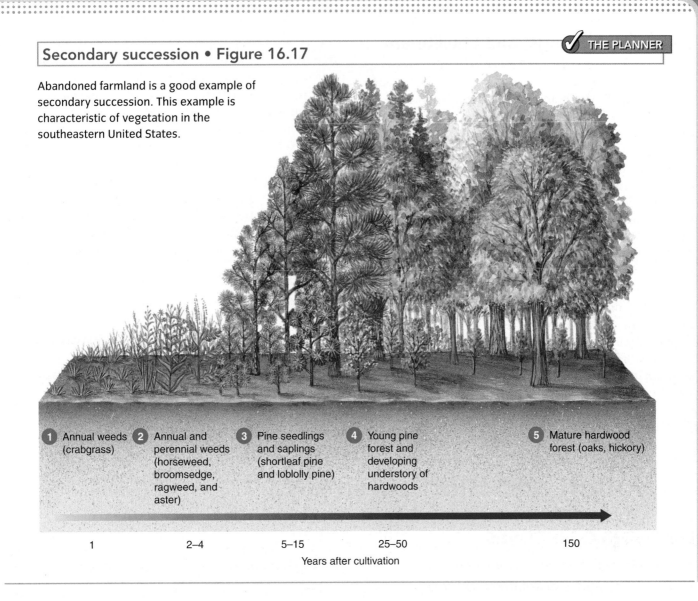

1. Annual weeds (crabgrass)
2. Annual and perennial weeds (horseweed, broomsedge, ragweed, and aster)
3. Pine seedlings and saplings (shortleaf pine and loblolly pine)
4. Young pine forest and developing understory of hardwoods
5. Mature hardwood forest (oaks, hickory)

| 1 | 2–4 | 5–15 | 25–50 | 150 |

Years after cultivation

In other cases—such as with floodplain silt deposits—the surface layer is composed of redeposited soil, with substantial amounts of organic matter and nutrients.

Succession begins with the **pioneer stage**. This stage includes a few plant and animal pioneers that are unusually well adapted to colonizing otherwise inhospitable conditions or cleared areas. As pioneer plants grow, their roots penetrate the soil. When the plants decay, their roots add organic matter directly to the soil, while their fallen leaves and stems add an organic layer to the ground surface. Large numbers of bacteria and invertebrates begin to live in the soil. Herbivores feed on the small plants, and birds forage the newly vegetated area for seeds and grubs.

As pioneer species transform prevailing undesirable conditions, they make them favorable for other species that invade the area and then displace the pioneers. The new arrivals may be larger plants with foliage that covers the ground more extensively. If this happens, the climate near the ground will have less extreme air and soil temperatures, higher humidity, and less intense insolation. These changes allow still other species to invade and thrive. After many years, when the succession has finally run its course, a mature, or climax, community of plant and animal species will exist in a more-or-less stable composition.

Secondary succession can occur after a disturbance alters an existing community or when farmland is abandoned (**Figure 16.17**). The first stages of succession can depend on the last use of the land—such as whether it was cultivated or grazed—or even on the weather in the year or two following abandonment. A multiyear drought can stunt the progress of succession, resulting in modified ecosystems. Usually, if left undisturbed, farmland quickly grows tall weeds with occasional shrubs. As shown in Figure 16.17, shrubs are joined by trees for an eventual climax forest community.

Succession, Change, and Equilibrium

Successional changes have so far been described as changes caused by the dynamic actions of the plants and animals that make up the community. One set of inhabitants paves the way for the next. As long as nearby populations of species provide a ready supply of colonizers, the changes lead progressively from bare soil or fallow field to climax forest.

But in many cases, succession does not follow this conceptual model. Environmental disturbances—such as wind, fire, flood, or clearing for agriculture—constantly interrupt succession, temporarily or even permanently. For example, winds and waves can disturb succession on seaside dunes, or a mature forest may be destroyed by insects or fire. In addition, inhospitable habitat conditions such as site exposure, unusual bedrock, or impeded drainage can hold back or divert the course of succession so successfully that the climax is never reached.

Introducing a new *invasive species* can also greatly alter existing ecosystems and successional pathways. The parasitic chestnut blight fungus was introduced from Asia to New York City in 1904. From there, it spread across the eastern states, decimating populations of the American chestnut tree within a period of about 40 years. This tree species, which may have accounted for as many as one-fourth of the mature trees in eastern forests, is now found only as small blighted stems sprouting from old root systems.

While succession is a reasonable model to explain many of the changes that we see in ecosystems with time, we must also take into account other effects. External forces can reverse or redirect succession temporarily or permanently. Most natural landscapes are a mosaic of distinctive biotic communities with different biological potentials and different histories. These developing ecosystems constantly interact in a dynamic equilibrium.

CONCEPT CHECK

1. **What** is the difference between primary and secondary succession?
2. **How** does environmental disturbance affect succession?

Evolution and Biodiversity

LEARNING OBJECTIVES

1. **Explain** evolution and natural selection.
2. **Describe** species distribution patterns.
3. **Explain** the causes and results of extinction.

An astonishing number of organisms exist on Earth, each adapted to the ecosystem in which it carries out its life cycle (**Figure 16.18**). The variety of life on Earth is known as **biodiversity**. How has life gained this astonishing diversity?

> **biodiversity** The variety of biological life supported by the Earth's biosphere or within a region.

Evolution

> **evolution** The creation of diversity of life-forms through the process of natural selection.

Through the process of **evolution**, organisms continue to respond to a changing environment to create living diversity. The famed biologist Sir Charles Darwin knew from careful field observations and specimen collections that all life possesses *variation*—the differences that arise between parent and offspring. He proposed that the environment acts on variation in organisms, in much the same fashion as a plant or an animal breeder, selecting individuals with qualities that are best suited to their environment. These individuals are more likely to go on and propagate. Darwin termed this survival and reproduction of the fittest **natural selection**. The theory of evolution based on natural selection describes the mechanism for this evolutionary process.

Darwin observed that, when acted upon by natural selection through time, variation could bring about the formation of new species whose individuals' ability to survive and successfully reproduce differed greatly from their ancestors' ability. He noted that the

> **natural selection** The selection of organisms by their survival in the environment in a process similar to selection of plants or animals by breeders.

Diversity of life-forms on Earth • Figure 16.18

About 40,000 species of microorganisms, 350,000 species of plants, and 2.2 million species of animals, including some 800,000 insect species, have been described and identified. The Census of Marine Life released in 2010 described 250,000 aquatic species. This is probably only a fraction of the actual number of species existing on the Earth.

a. Microorganism
Aquatic microorganisms include diatoms—a class of algae with more than 70,000 known species. They contain silica skeletons and are extremely abundant in fresh and ocean water.

b. Fungus
This stinkhorn fungus was photographed on the floor of a Costa Rican rainforest. Fungi are organisms without chlorophyll that gain their nutrition from dead and decaying organic matter.

d. Insect
These blue-nose caterpillars are placid herbivores. They are well protected from predation by defensive spines that sting and detachable burrs that work their way into the skin of a would-be predator.

c. Reptile
An Australian thorny devil lizard makes its way across an arid landscape. Inhabiting the sand plains of South Australia, this small reptile lives on ants. It drinks dew that condenses on its body and is carried to its mouth in fine grooves between its spines.

Darwin's finches • Figure 16.19

Charles Darwin studied the variety of finches during his 1835 visit to the Galápagos Islands. He hypothesized that natural selection in beak shape allowed finches to develop separate species that could successfully compete for limited resources by foraging with beaks adapted to specific food resources.

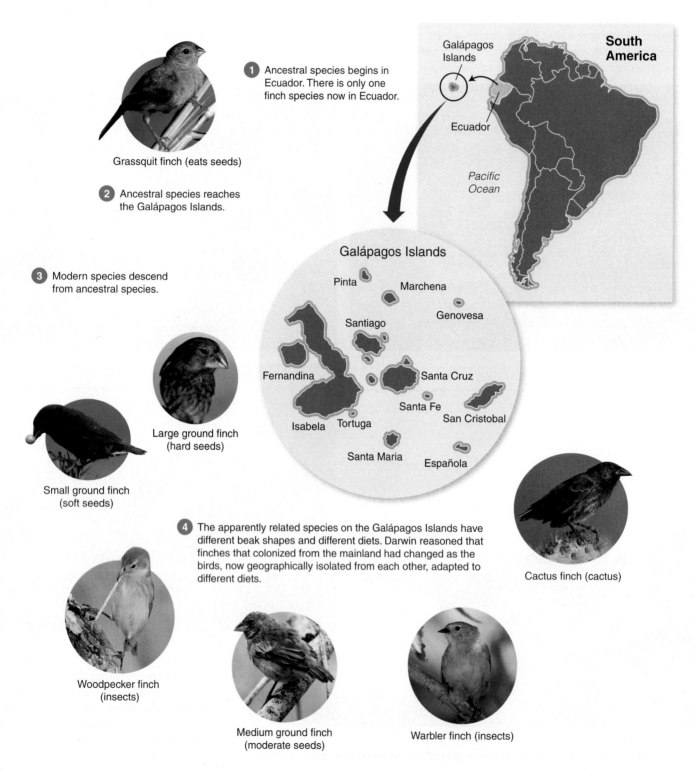

1 Ancestral species begins in Ecuador. There is only one finch species now in Ecuador.

Galápagos Islands

South America

Ecuador

Pacific Ocean

Grassquit finch (eats seeds)

2 Ancestral species reaches the Galápagos Islands.

3 Modern species descend from ancestral species.

Galápagos Islands

Pinta

Marchena

Genovesa

Santiago

Fernandina

Santa Cruz

Santa Fe

San Cristobal

Isabela   Tortuga

Santa Maria

Española

Large ground finch (hard seeds)

Small ground finch (soft seeds)

4 The apparently related species on the Galápagos Islands have different beak shapes and different diets. Darwin reasoned that finches that colonized from the mainland had changed as the birds, now geographically isolated from each other, adapted to different diets.

Cactus finch (cactus)

Woodpecker finch (insects)

Medium ground finch (moderate seeds)

Warbler finch (insects)

variation in the beaks of finches that had become isolated on the Galápagos Islands enabled the finches to successfully exploit a greater variety of food sources. After thousands of years, the finches had evolved into distinct species that specialized in specific food types (**Figure 16.19**). Although Darwin was not aware of DNA and the process by which mutations occur at the cellular level, he correctly identified how nature selected those mutations to provide advantages for the survival of a species.

A reproductive cell's genetic material (DNA) can mutate when chemical bonds in the DNA are broken and reassembled. Most mutations either have no effect or are harmful, but a small proportion of mutations have a positive effect on an individual's genetic makeup. If that positive effect makes the individual organism more likely to survive and reproduce, then the altered gene is likely to survive as well and be passed on to offspring.

> **species** A collection of individual organisms that are capable of interbreeding to produce fertile offspring.

Mutations change the nature of species through time. We can define a **species** as a collection of individuals capable of interbreeding to produce fertile offspring. A **genus** is a collection of closely related species that share a similar evolutionary history. Today, developments in genetic research allow scientists to trace the evolutionary paths of today's species back through the millennia to their prehistoric ancestry (see *Video Explorations*).

Dispersal

Most organisms can move from their home territory to new sites in a process called **dispersal**. Often this movement is confined to one life stage, as in the dispersal of plants as seeds, or coral reef polyps. Even for animals, there is often a developmental stage in which the animals are more likely to move from one site to the next. Normally, dispersal doesn't change the geographic range of a species. Seeds fall near their sources, and animals seek out nearby habitats to which they are adjusted. When land is cleared or new land is formed, new colonists move into the new environment, as we saw in our discussion of succession. Species can also slowly extend their range from year to year.

Species may encounter barriers that limit their dispersal, such as the presence of an ocean, an ice sheet, or a desert. It is beyond the physiological limits of a species to cross such a barrier. There may be ecological barriers as well—for example, a zone filled with predators or a region occupied by strongly and successfully competing species.

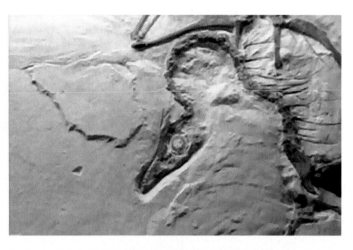
There are also corridors that help dispersal. For example, Central America forms a present-day land bridge, or biological corridor, connecting North and South America that has been in place for about 3.5 million years. Other corridors existed in the recent past. The Bering Strait region between Alaska and easternmost Siberia was dry land during the early Cenozoic Era (about 60 million years ago) and during the glaciations of the most recent ice age, when sea level dropped by more than 100 m (325 ft). Many plant and animal species of Asia are known to have crossed this bridge and then spread southward into the Americas. One notable migrant species of the last continental glaciation was the aboriginal human, and evidence suggests that these skilled hunters caused the extinction of many of the large animals, including wooly mammoths and ground sloths, in the Americas about 10,000 years ago.

Over time, evolution and dispersal have distributed many species across the Earth, creating a number of spatial distribution patterns. Often, members of the same lineage have similar adaptations to the environment, and so they are found in similar climatic regions. Closely related species tend to be nearby or to occupy similar regions.

Evolution of species and plate tectonics • Figure 16.20

These birds share a common ancestry in the ratite group, which scientists theorize originated on the southern supercontinent Gondwana before it split apart. As Gondwana split into South America, Africa, Australia, and New Zealand, ratite bird populations were isolated, allowing separate but related species to evolve into their present-day distribution.

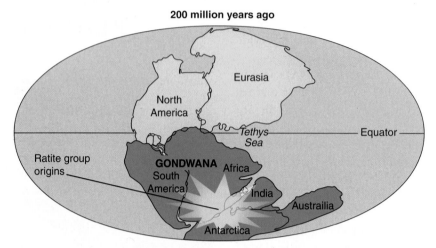

200 million years ago

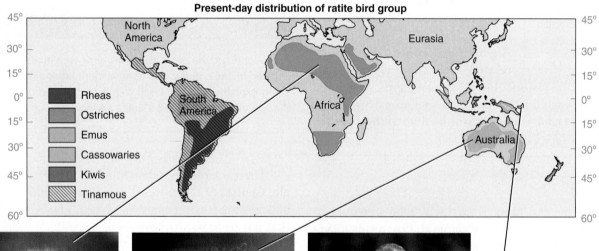

Present-day distribution of ratite bird group

- Rheas
- Ostriches
- Emus
- Cassowaries
- Kiwis
- Tinamous

a. Ostrich
The ostrich is restricted to Africa and the Middle East, although it was formerly found in Asia.

b. Emu
The emu inhabits most of Australia and is commonly encountered in the wild.

c. Cassowary
The cassowary hails from New Guinea and northeastern Australia. It is a rainforest dweller.

Ask Yourself

Which of the following patterns would also be well explained by the split of the supercontinent Gondwana?
a. Australia and Eurasia both have palm trees.
b. Identical fossilized leaves are found on the east coast of South America and the west coast of Africa.
c. Alaska and Antarctica both host tundra ecosystems.
d. India and North America both have arid regions.

Sometimes families of organisms are found in widely separated regions as a result of the breaking up of continents long ago. When parts of a population become geographically isolated on separate continents, evolution takes a different course in each location (**Figure 16.20**). Eventually the subpopulations diverge genetically and lose the ability to interbreed as a single species.

Extinction and Threats to Biodiversity

Over geologic time, all species are doomed to extinction, either by mutating into new species or by dying out. When conditions change more quickly than populations can evolve new adaptations, population size falls. When that happens, the population is more vulnerable to chance perturbations, such as a fire, a rare climatic event, or an outbreak of disease. Ultimately, populations are wiped out. Some extinctions occur very rapidly. The fossil record provides convincing data showing that the Earth has experienced at least five major episodes of species extinction. For example, strong evidence suggests that a meteorite struck the Earth about 65 million years ago, wiping out the dinosaurs and many other groups of terrestrial and marine organisms.

Over recent centuries, human activity has ushered in a wave of extinctions unlike any seen before. We are now

Degree of endangerment of species
• Figure 16.21

The number of plant and animal species is decreasing at a rate not equaled since about 65 million years ago, during the fifth mass species extinction. Amphibians are especially sensitive to subtle changes in the environment and may be considered indicator species that point to further trends in extinction rates. The Red List of the International Union for Conservation of Nature and Natural Resources documents those species currently identified as critical, endangered, or vulnerable.

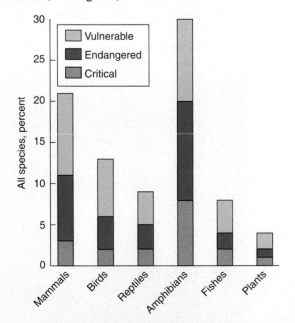

Where Geographers CLICK

The IUCN Red List

The rate of animal and plant extinction and natural habitat destruction is increasing every day. The International Union for the Conservation of Nature and Natural Resources (IUCN) maintains a "Red List" to keep track of the plant and animal species that are under threat. The IUCN's Web site describes the habitats that are at greatest risk and consequently which species are in danger of extinction.

www.iucnredlist.org

experiencing the sixth mass species extinction, with as many as 10% of all species expected to be extinct by 2100 (**Figure 16.21**). Two out of every five species on the planet that have been assessed by scientists face extinction, according to the International Union for Conservation of Nature and Natural Resources (see *Where Geographer's Click*). In the past 40 years, several hundred land animal species have disappeared. Aquatic species have also been severely affected, with 40 species or subspecies of freshwater fish lost in North America alone in the past few decades. In the plant kingdom, botanists estimate that more than 600 species have become extinct in the past four centuries. These documented extinctions may be only the tip of the iceberg. Many hundreds of thousands of species have not yet been discovered and may become extinct without ever being described by science.

How has human activity caused extinctions? The human population has reached 7 billion, over three times the population just 100 years ago. We are monopolizing the world's resources—and altering habitats along the way. We have doubled the natural rate of nitrogen fixation, used more than half of the Earth's supply of surface water, and transformed more than 40% of the land surface. Over our history, we have dispersed new organisms to regions where they outcompete or prey on existing organisms. Areas that were formerly wilderness have been subjected to waves of invading species, ranging from rats to weeds, brought first by prehistoric humans and later by explorers and conquerors. Hunting, burning, and habitat alteration by humans have exterminated many species. Today, some of the richest areas of biodiversity on the planet are becoming isolated habitat patches under threat as human development encroaches on them. (See **Figure 16.22** on the next page.)

Functioning ecological systems require a resilient web of interacting organisms. Conservation International has identified regions of the world where biodiversity is critically endangered.

Projectec status of biodiversity, 1998–2018
- Critical and endangered
- Threatened
- Relatively stable/intact

a. Bering Sea
Nutrient-rich currents flowing into the Bering Sea create one of the world's most diverse marine environments, with polar bears, seals, sea lions, walruses, whales, enormous populations of sea birds, and more than 400 species of fish, crustaceans, and mollusks. Warming temperatures are disrupting the dynamic equilibrium of the arctic ecosystems affecting all species, especially keystone predators such as the polar bear.

b. Southeastern U.S. streams and rivers
Eastern deciduous forests are threatened by air pollution, increasing drought episodes linked to climate change, and growing population pressures on the region.

c. Amazon River and flooded forests
One of the most biodiverse regions on the planet, the Amazon Basin is experiencing the pressures of economic development. Such development is displacing species, like this rare black jaguar, which has limited options for living elsewhere.

d. Eastern Himalayan broadleaf and conifer forests
Snaking across the lowlands and foothills of the Himalaya, this ecosystem supports a remarkable diversity of plants and animals, including endangered mammals such as the clouded leopard shown here, the Himalayan black bear, the golden langur, and many bird species. Rapid resource exploitation and increasing temperatures threaten the biota.

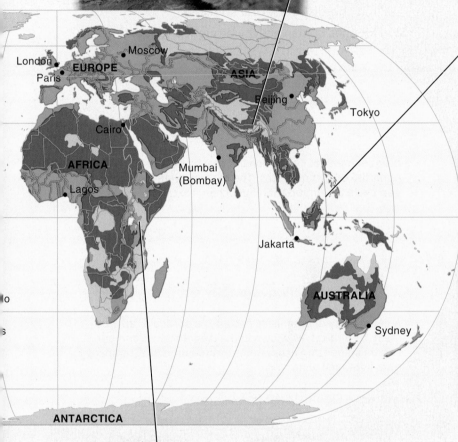

e. Sulu-Sulawesi Seas
One of the world's most biologically diverse marine ecosystems is under great pressure from growing populations of fishermen and pollution runoff from the land. Coral reefs here, considered some of the world's most pristine, are susceptible to warming and rising waters.

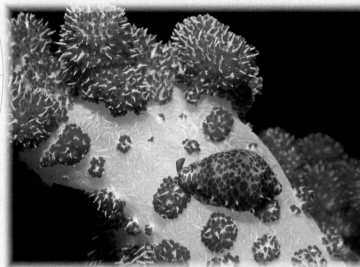

f. Rift Valley lakes
Drainage of lakes for a growing population and pollution from nearby agriculture are causing drops in the bird and mammal populations. The fate of pelicans and flamingos is tied to the protection of the lakes.

Put It Together

Review Figure 16.4 and complete this statement.
The highlighted areas threatened by biodiversity loss _____.

a. occur in a range of biomass densities
b. occur only in regions of low biomass density
c. occur only in regions of the highest biomass density
d. occur only in regions of moderate biomass density

Fed by fossil fuels and by the international demand for growth, today's population can be expected to outstrip most conservation and protection gains of the past half century through the sheer scale of increased consumption and pollution. Many scientists have termed the present geologic age the Anthropocene Period, in recognition of the human footprint on our planet.

Why is biodiversity important? Nature has provided an incredibly rich array of organisms that interact with each other in a seamless web of organic life. When we cause the extinction of a species, we break a link in that web. Ultimately, that web will fray, with unknown consequences for both the human species and continued life on the Earth.

CONCEPT CHECK STOP

1. **What** might cause two species with the same ancestor to lose the ability to interbreed?
2. **What** geographic or ecological factors might limit the dispersal of a species?
3. **How** has human activity brought about extinctions?

Summary

THE PLANNER ✓

1 Energy and Matter Flow in Ecosystems 486

- **Biogeography** is the study of how plants and animals are distributed around the Earth.

- The **food web** of an **ecosystem** details how food energy flows from one **trophic level** to the next.

- **Photosynthesis**, as shown in the diagram, is the production of carbohydrates from water, carbon dioxide, and light energy by primary producers. **Respiration** is the opposite process, in which carbohydrates are broken down into carbon dioxide and water to yield chemical energy for use by organisms.

Photosynthesis • Figure 16.3

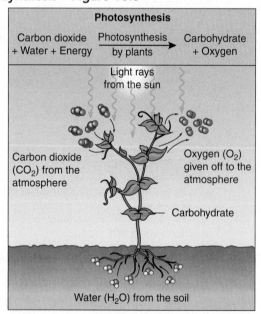

- The productivity of an ecosystem is measured in terms of **biomass**, the dry weight of organic matter produced by an ecosystem.

- **Biogeochemical cycles** consist of active pools and storage pools linked by flow paths. The **carbon cycle** and the **nitrogen cycle** are important biogeochemical cycles.

2 Ecological Biogeography 495

- A community of organisms occupies a particular environment, or **habitat**, as shown in the diagram. Environmental factors influencing ecosystems include moisture, temperature, light, and wind.

Habitats • Figure 16.10

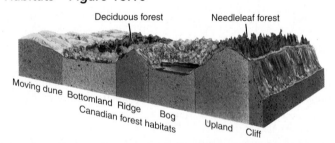

- Geomorphic (landform) factors affect both the moisture and temperature environment of the habitat. Soil conditions can also limit the distribution of organisms or affect community composition.

- Species interactions include **competition**, **predation**, **symbiosis**, and **parasitism**.

3 Ecological Succession 502

• As shown in the diagram, **ecological succession** comes about as ecosystems change in predictable ways through time.

Secondary succession • Figure 16.17

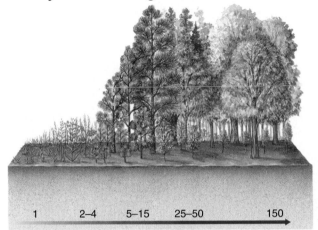

Years after cultivation

• **Primary succession** occurs on new surfaces, and **secondary succession** occurs on disturbed habitats.

• Succession may be interrupted by natural disturbances and limited by local environmental conditions.

4 Evolution and Biodiversity 504

• Life has attained its astonishing **biodiversity** through **evolution**. **Natural selection** acts on variation to produce populations that are progressively better adjusted to their environment.

• **Species** differentiate themselves over time through gene mutation.

• Species maintain or expand their ranges by **dispersal**.

• Extinction occurs when populations become very small and thus are vulnerable to chance occurrences of fire, disease, or climate anomaly. Rare but extreme events can cause mass extinctions.

• Humans disperse predators, parasites, and competitors widely, disrupting long-established species and threatening their survival, as shown in the diagram. Hunting, burning, and habitat alteration have exterminated many species.

Degree of endangerment of species • Figure 16.21

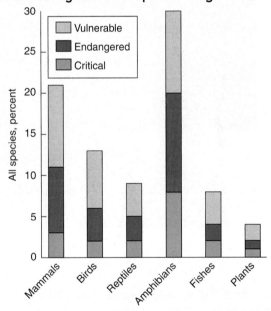

Key Terms

Critical and Creative Thinking Questions

1. Humans can eat at several different trophic levels on the food chain. What is an environmental advantage of consuming primary producers as opposed to consuming organisms from higher trophic levels?

2. Look at the diagram. What are three processes that will be accelerated in the global carbon cycle as atmospheric CO_2 levels increase due to fossil fuel combustion? What is one environmental impact of these accelerated processes?

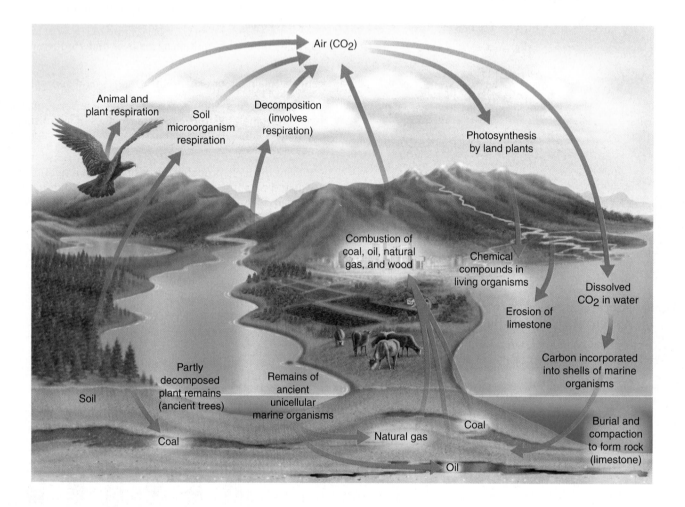

3. Select three distinctive habitats for plants and animals that occur near where you live and with which you are familiar. If you were to visit these habitats, select one characteristic of each that could be used to distinguish it from the others.

4. For your locality, explain which climatic factor—moisture, temperature, light, or wind—is most important in limiting the distribution of plant species.

5. By Darwin's time, scientists already understood that species evolve. What exactly is the breakthrough with Darwin's theory of evolution? What important piece of evidence was discovered in the 20th century that added support for Darwin's theory of evolution?

6. Imagine yourself as a biogeographer discovering a new group of islands. Select a global location for your island group, including climate and proximity to nearby continents or land masses. What types of organisms would you expect to find within your island group, and why?

What is happening in this picture?

The north-facing side of this slope in the Chisos Mountains of Big Bend National Park, on the right of the photo, supports a much different biotic community from the south-facing slope, on the left.

Think Critically

1. What is different about the biotic communities on the two slopes?
2. What has caused this extreme difference?

Self-Test

(Check your answers in the Appendix.)

1. Within an ecosystem, energy transformations occur through a series of _____ levels, commonly referred to as the food web.

   a. photosynthesis

   b. respiration

   c. trophic

   d. mutualism

2. During the nitrogen cycle, the process in which certain soil bacteria convert nitrogen from forms usable by plants back to N_2 is called _____.

   a. nitrification

   b. nitrogen fixation

   c. denitrification

   d. nitrogen consumption

3. _____ is a subdivision of the environment according to the needs and preferences of organisms or groups of organisms.

a. Pioneer stage

b. Genus

c. Ecological niche

d. Habitat

4. The process of _____ removes carbon from the atmosphere and incorporates it into plants.

a. photosynthesis

b. respiration

c. denitrification

d. nitrogen fixation

5. The process of _____ removes nitrogen from the atmosphere and incorporates it into soil microorganisms and eventually plants.

a. photosynthesis

b. respiration

c. denitrification

d. nitrogen fixation

6. As climate change progresses, what habitat most likely will migrate to the northern study area shown on the map?

a. Boreal—Mainly Forest

b. Boreal—Forest & Grassland

c. Boreal—Forest & Barren

d. Tundra

Canadian Land Cover Map
created by the
Canadian Model Forest Project

BOREAS Study Region

Manitoba

Northern Study Area

Southern Study Area

Saskatchewan

Scale
1000 km
100 km

Legend:
- Boreal – Mainly Forest
- Boreal – Forest & Grassland
- Boreal – Forest & Barren
- Great Lakes – St. Lawrence
- Deciduous
- Acadian
- Grassland
- Tundra
- Subalpine
- Montane
- Coastal
- Columbian

7. _____ is a concept that describes how a species obtains its energy and influences other species in its own environment.

   a. Habitat

   b. Ecological niche

   c. Community

   d. Ecosystem

8. An orchid living on the branch of another plant is an example of _____.

   a. competition

   b. predation

   c. symbiosis

   d. biodiversity

9. Which of the following is not a climatic factor that limits the distribution of plant and animal species?

   a. soil

   b. temperature

   c. moisture

   d. light

10. The graph shows the changes in the population size of the Kaibab deer herd in Arizona between about 1904 and 1940. What would best explain the rapid population growth between 1910 and 1920?

    a. decreased competition

    b. decreased predation

    c. increased symbiosis

    d. increased predation

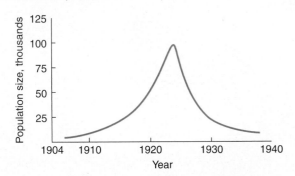

11. Succession that occurs on a site that has been burned in a forest fire is _____ succession.

    a. primary

    b. secondary

    c. tertiary

    d. old-field

12. The pioneer species shown in this photograph are part of _____ succession.

    a. primary

    b. secondary

    c. tertiary

    d. old-field

13. Darwin used the variation in beaks of finches on the Galápagos Islands as evidence of _____.

    a. dispersal

    b. natural selection

    c. mutualism

    d. habitat

14. The dispersal of organisms is assisted by _____ and restricted by _____.

    a. corridors; land bridges

    b. ice sheets; barriers

    c. corridors; barriers

    d. land bridges; corridors

15. Extinction is _____.

    a. the extermination of a species from one area that may still exist in other areas

    b. always the result of human activities

    c. the fate of all species over geologic time

    d. a process that has become relatively infrequent during the past 100 years

THE PLANNER ✓

Review your Chapter Planner on the chapter opener and check off your completed work.

Global Biogeography

On the island of Borneo in Southeast Asia, a majestic rainforest boasts an astonishing range of biodiversity, including more than 15,000 known species of plants. Rhinoceroses and elephants roam the ground, while the world's largest collection of gliding animals, including flying snakes, take to the air, and orangutans make their homes in the trees.

As Borneo's forest is destroyed to make way for plantations of oil palm trees, orangutans are being driven to the point of extinction. Geographers estimate that 8000 square kilometers (about 2 million acres) of lowland forests have been cleared each year over the past two decades. The government has cracked down on illegal logging and exports, but the drive to meet the world's demand for palm oil for cooking, cosmetics, and biofuels puts pressure on the forests. Indonesia and Malaysia currently provide 86% of the world's palm oil supply. Indonesia now has about 60,000 km^2 (about 15 million acres) of land under cultivation, a figure that could double by 2020. If the palm oil rush cannot be slowed, time may soon run out for the forests of Borneo and the orangutan.

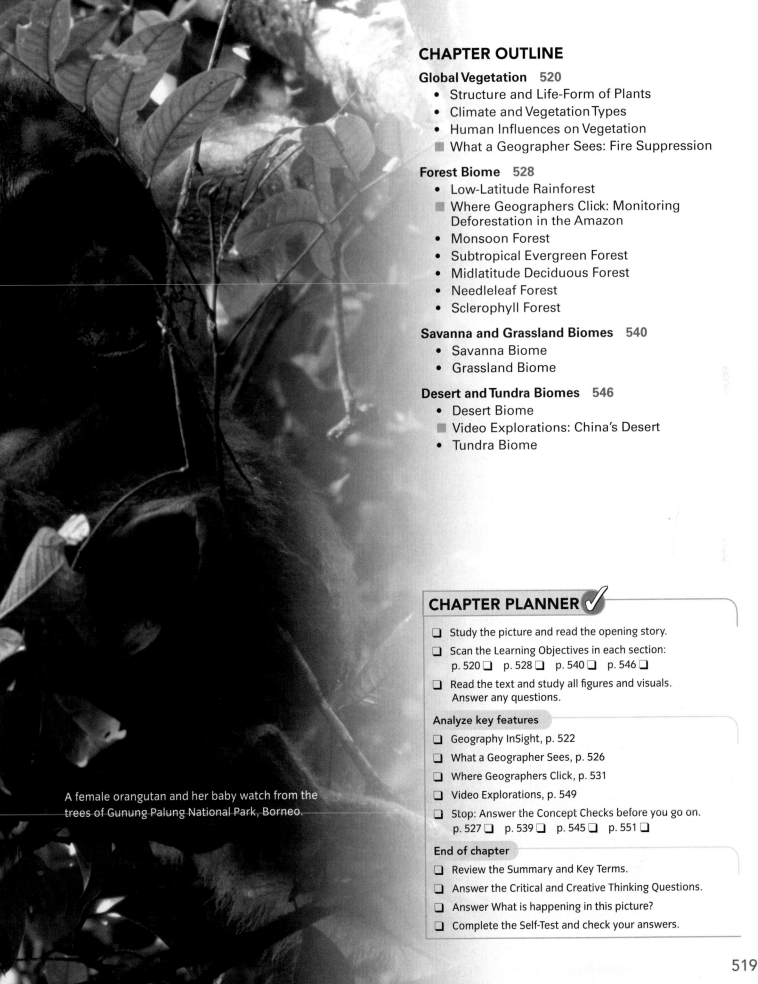

CHAPTER OUTLINE

A female orangutan and her baby watch from the trees of Gunung Palung National Park, Borneo.

CHAPTER PLANNER ✓

- ❑ Study the picture and read the opening story.
- ❑ Scan the Learning Objectives in each section:
  p. 520 ❑   p. 528 ❑   p. 540 ❑   p. 546 ❑
- ❑ Read the text and study all figures and visuals. Answer any questions.

Analyze key features

- ❑ Geography InSight, p. 522
- ❑ What a Geographer Sees, p. 526
- ❑ Where Geographers Click, p. 531
- ❑ Video Explorations, p. 549
- ❑ Stop: Answer the Concept Checks before you go on.
  p. 527 ❑   p. 539 ❑   p. 545 ❑   p. 551 ❑

End of chapter

- ❑ Review the Summary and Key Terms.
- ❑ Answer the Critical and Creative Thinking Questions.
- ❑ Answer What is happening in this picture?
- ❑ Complete the Self-Test and check your answers.

Global Vegetation

LEARNING OBJECTIVES

1. **Describe** the structure of different plant life-forms.

2. **Describe** the climatic factors that influence the five principal biomes.

3. **Outline** major human influences on vegetation.

The land plants that spread widely over the continents dominate **terrestrial ecosystems**. These ecosystems, which include the soils they utilize, are directly influenced by climate and are therefore closely woven into the fabric of physical geography. We divide terrestrial ecosystems into **biomes**. Although a biome includes both plant and animal life, green plants dominate the description of a biome simply because of their enormous biomass. Plant geographers concentrate on the characteristic life-forms of the green plants within a biome—principally trees, shrubs, lianas, and herbs—but also other life-forms in certain biomes. Soil's critical role in terrestrial ecosystems has been overlooked until recently, as geographers have helped uncover both the amount of living biomass that exists in a cubic meter of soil and the important role soils play in the carbon and nitrogen cycles.

terrestrial ecosystem An ecosystem of land plants and animals found on the land surface of the continents.

biome The largest recognizable subdivision of terrestrial ecosystems.

Structure and Life-Form of Plants

Plants come in many types, shapes, and sizes. Botanists recognize and classify plants by species. In contrast, geographers are less concerned with individual species than they are with larger classes of **plant life-forms**. Most life-form names are in common use, and you're probably familiar with them already, but we will quickly review them now.

plant life-form The characteristic physical structure, size, and shape of a plant or of an assemblage of plants.

Figure 17.1 illustrates various plant life-forms. Trees and shrubs are erect, woody plants. They are **perennial**, meaning that their woody tissues endure from year to year. Most have life spans of many years. A tree is a large plant with a single upright main trunk, often with few branches in the lower part but branching in the upper part to form a crown. A shrub has several stems branching from a base near the soil surface, creating a mass of foliage close to ground level.

Layers of a beech–maple–hemlock forest • Figure 17.1

Tree crowns form the uppermost layer, shrubs an intermediate layer, and herbs a lower layer. Mosses and lichen grow very close to the ground. In this schematic diagram, the vertical dimensions of the lower layers are greatly exaggerated.

Lianas are also woody plants, but they take the form of vines supported on trees and shrubs. Lianas include tall, heavy vines in the wet equatorial and tropical rainforests and also some woody vines of midlatitude forests. English ivy, poison ivy and poison oak, and Virginia creeper are familiar North American examples of lianas.

Herbs are a major class of plant life-forms. They lack woody stems and are usually small, tender plants. They occur in a wide range of shapes and leaf types. Some are **annuals**, living only for a single season, while others are perennials, living for multiple seasons. Some herbs are broad leaved, and others, such as grasses, are narrow leaved. Herbs share few characteristics with each other, except that they usually form a lower layer than shrubs and trees. **Lichens** also grow close to the ground (**Figure 17.2**). They are plant forms in which algae and fungi live together, forming a single plant structure. Lichens dominate the vegetation in some alpine and arctic environments.

> **forest** An assemblage of trees growing close together, their crowns forming a layer of foliage that largely shades the ground.

Forest is a vegetation structure in which trees grow close together. The crowns of forest trees often touch, so that their foliage largely shades the ground. Many forests in moist climates show at least three layers of life-forms—the tree, shrub, and herb layers. There is sometimes a fourth, lowermost, layer of mosses and related very small plants. In **woodland**, tree crowns are separated by open areas that usually have a low herb or shrub layer.

Climate and Vegetation Types

The structure and appearance of plants tend to conform within their native habitats. Plants have limited tolerance for different conditions of soil water, temperature, and soil nutrients. As a result, each vegetation type is associated with a characteristic geographic and climatic region. There are five principal biomes: forest, savanna, grassland, desert, and tundra. Biogeographers subdivide the biomes further into smaller vegetation units, called **formation classes**, based on the life-form of the plants. For example, at least four and perhaps as many as six kinds of forests can be distinguished within the forest biome. At least three kinds of grasslands are easily recognizable. Deserts, too, span a wide range in terms of the abundance and life-form of plants.

Precipitation and temperature patterns are key factors in determining the type of vegetation characteristic of each biome. (See **Figure 17.3** on the

Lichens • Figure 17.2 _____

Lichens are plant forms that combine algae and fungi in a single symbiotic organism. They are abundant in some high-latitude habitats, and they also range to the tropics. Pictured here are lichen growing on rocks and tree trunks in Ribeiro Frio on the island of Madeira.

next page.) In both low- and midlatitude environments, strong precipitation gradients produce vegetation types grading from forest to desert. At high latitudes, decreasing temperatures control the transition from forest to tundra. In low latitudes, savannas and grasslands are found in regions with a distinct dry season. In the midlatitudes, west coasts are marked with a strong summer dry period, supporting hard-leaved evergreen trees and shrubs in coastal regions and, farther poleward, encouraging the growth of lush conifer forests.

Geography InSight

How climate influences terrestrial biomes
• **Figure 17.3**

Temperature and precipitation are the two most important factors in determining the pattern of natural vegetation types.

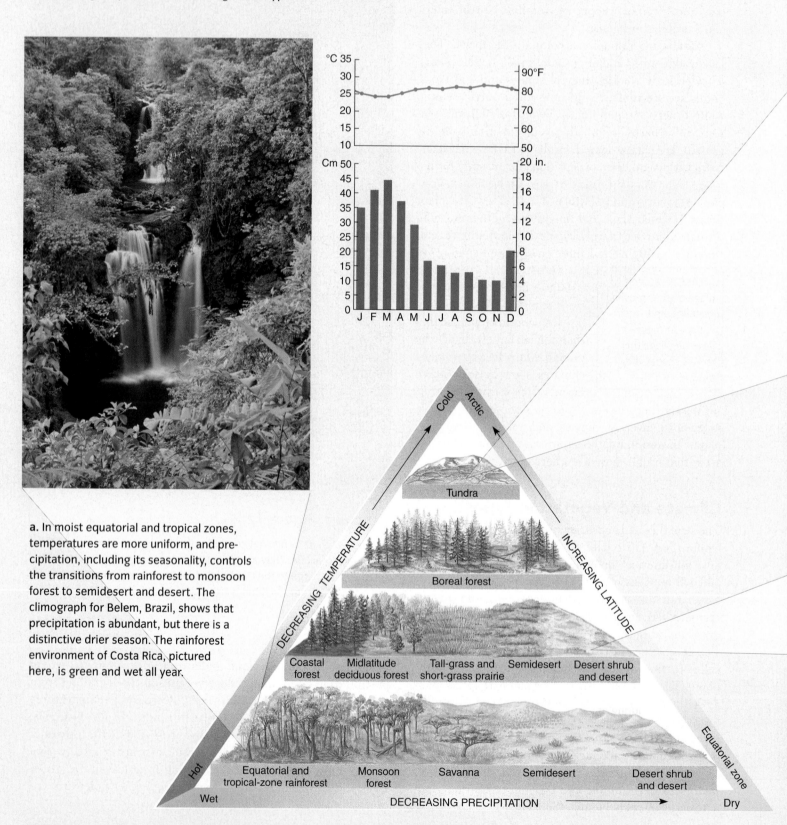

a. In moist equatorial and tropical zones, temperatures are more uniform, and precipitation, including its seasonality, controls the transitions from rainforest to monsoon forest to semidesert and desert. The climograph for Belem, Brazil, shows that precipitation is abundant, but there is a distinctive drier season. The rainforest environment of Costa Rica, pictured here, is green and wet all year.

522 CHAPTER 17   Global Biogeography

b. At higher latitudes, temperature is more important than precipitation in determining the boundary between tundra and boreal forest. Alaskan tundra in fall colors is pictured here with the climograph for Fort Yukon, Alaska.

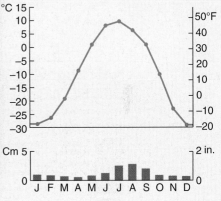

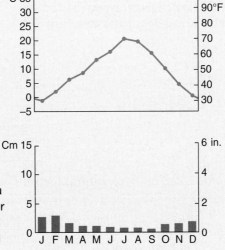

c. In hot, dry biomes such as the Arizona desert pictured here, low precipitation or precipitation during only a short period of the year limits the types of vegetation that can thrive. The climograph of Sedona, Arizona, shows very low precipitation and a very dry summer.

Ask Yourself

Which climograph has the largest annual range of temperature?
a. tundra
b. rainforest
c. desert
d. All have approximately the same annual variability.

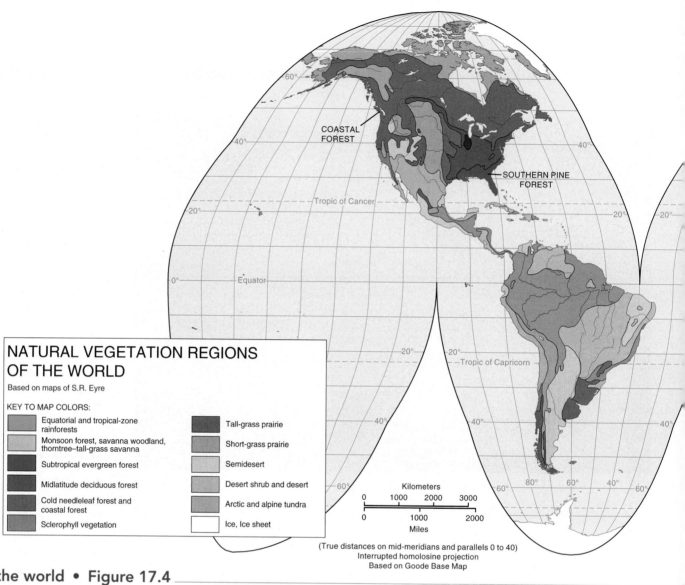

NATURAL VEGETATION REGIONS
OF THE WORLD

Based on maps of S.R. Eyre

KEY TO MAP COLORS:

Equatorial and tropical-zone
rainforests

Monsoon forest, savanna woodland,
thorntree–tall-grass savanna

Subtropical evergreen forest

Midlatitude deciduous forest

Cold needleleaf forest and
coastal forest

Sclerophyll vegetation

Tall-grass prairie

Short-grass prairie

Semidesert

Desert shrub and desert

Arctic and alpine tundra

Ice, Ice sheet

COASTAL FOREST

SOUTHERN PINE FOREST

Tropic of Cancer

Equator

Tropic of Capricorn

Kilometers
0   1000   2000   3000
0   1000   2000
Miles

(True distances on mid-meridians and parallels 0 to 40)
Interrupted homolosine projection
Based on Goode Base Map

Biomes of the world • Figure 17.4

This generalized world map simplifies the very complex patterns of natural vegetation to show large uniform regions in which a given formation class might be expected to occur. When studying any map of natural features, keep in mind that boundaries are almost always approximate and gradational.

The formation classes described in the remainder of this chapter are major, widespread types that are clearly associated with specific climate types (**Figure 17.4**).

Human Influences on Vegetation

Natural vegetation is plant cover that develops with little or no human interference. It is subject

natural vegetation
Stable, mature plant cover, largely free from the influences of human activities.

to natural forces, storms, or fires that can modify or even destroy it. In contrast, there is also human-influenced vegetation. Over the past few thousand years, human societies have modified and cultivated much of the land area of our planet. We have changed the natural vegetation of many regions, in some cases drastically (**Figure 17.5**).

Other areas, such as national parks and national forests, appear to be untouched but

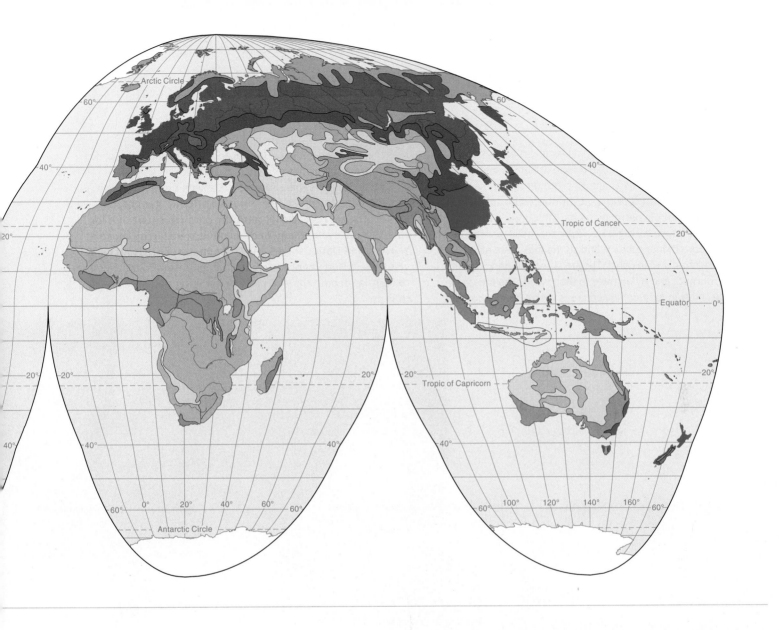

Human influences on vegetation • Figure 17.5

Much of the midlatitude land surface is totally under human control, through intensive agriculture, grazing, or urbanization. Natural vegetation can still be seen over vast areas of the wet equatorial climate, although the rainforests there are being rapidly cleared. Much of the subarctic zones is also in a natural state.

Put It Together

Review Figure 17.4 and answer this question.
Which biome has experienced the greatest extent of human influence?
a. tundra
b. desert
c. midlatitude deciduous forest
d. rainforest

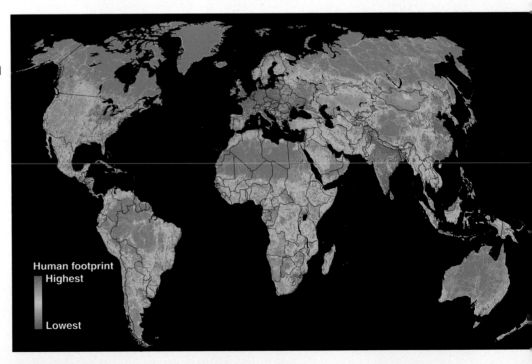

Human footprint
Highest

Lowest

WHAT A GEOGRAPHER SEES

Fire Suppression

Yellowstone National Park is a priceless forest ecosystem, little disturbed by natural catastrophe or human interference for at least the past two centuries. But through August and September 1988, 45 forest fires—mostly started by lightning strikes—burned out of control in the park (Figure a). The number of fires was not unusually high. However, they followed the driest summer of more than a century, and so the fires were able to spread more rapidly than usual. They ravaged the park, leaving scorched, burned, and dying trees in their wake. The fires were most destructive in long-unburned areas where dead wood and branches had accumulated for decades as a result of the park's fire-suppression program.

After 10 years, new forests were well on their way to replacing the burned ones (Figure b). Today, geographers and ecologists realize that periodic burning is part of a forest's natural cycle. Many park managers, including those at Yellowstone National Park, have stopped suppressing most wildfires. When vegetation burns, the nutrient-rich ashes remain, enriching the soil and encouraging new tree species to grow.

a.

b.

Think Critically

1. What would be the advantages and disadvantages of deliberately starting fires in forests under supervised conditions?
2. What are the implications of such prescribed burns for both the forest ecosystem and the neighboring human communities?

are actually dominated by human activity in a subtle manner. For example, most national parks and national forests have been protected from fire for many decades. Although fire protection would seem to be a beneficial human influence on the biome, the reality is more complex (see *What a Geographer Sees*).

Humans have also moved plant species from their original habitats to foreign lands and environments. Species dispersal, whether deliberate or accidental, can significantly disturb native ecosystems. Sometimes these exotic plants thrive like weeds, forcing out natural species and becoming major nuisances. Other human activities, such as clear-cutting, slash-and-burn agriculture, overgrazing, and wood gathering, have had profound effects on the plant species and the productivity of the land (**Figure 17.6**).

The true dimensions of human policies and economic development in the tropics and around the globe are being mapped and monitored by teams of scientists using satellite data and GIS to provide citizens and decision makers with objective information to chart the future. As international corporations heavily invest in resource exploitation, local citizens look to national and international governance to stem the onslaught of deforestation. At a minimum, through the use of satellite monitoring technology, the world can watch the continued decline of remnant habitats for some of nature's most intriguing and threatened species (**Figure 17.7**).

Overgrazing • Figure 17.6

Overgrazing strips vegetation from the land, reducing evapotranspiration, increasing temperatures, and leaving the soil without cover. The result is land degradation that may require decades to reverse.

SCIENCE TOOLS

Deforestation in Borneo • Figure 17.7

Both tropical lowland and highland forests of Borneo, containing a rich biodiversity of plants and animals, have decreased rapidly since 1950. Scientists from the United Nations (UN) used a combination of historic maps and aerial photography to create a base map of Borneo's original terrestrial ecosystems (shown in green). Next, they analyzed Landsat data to create a series of GIS maps delineating the extent of deforestation for a 60-year period. Field investigations showed that the burned, logged, and cleared areas, previously habitat for species such as orangutans and elephants, were converted to agricultural land, developed areas, or palm oil plantations (shown in beige). The UN team made projections to 2020, based on national policies and agricultural land deeds.

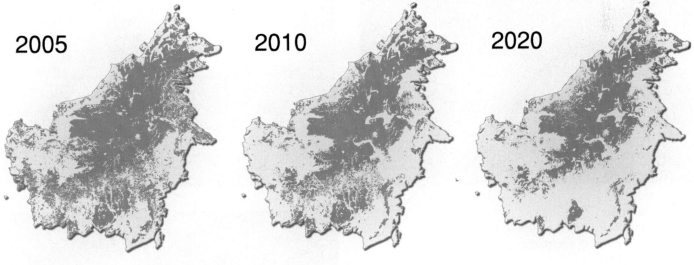

2005   2010   2020

CONCEPT CHECK   **STOP**

1. **What** are the characteristic vegetation patterns of woodland?

2. **What** is the primary climatic factor that determines the pattern of biomes in higher latitudes?

3. **How** might human-induced plant dispersal affect an ecosystem?

Forest Biome

LEARNING OBJECTIVES

1. **Define** forest biome.
2. **Distinguish** between the six types of forest biomes.
3. **Explain** how the forest types are related to climate.

Within the **forest biome**, we recognize six major formations: low-latitude rainforest, monsoon forest, subtropical evergreen forest, midlatitude deciduous forest, needleleaf forest, and sclerophyll forest. Ecologists sometimes recognize three principal types of forest as separate biomes, based on their widespread nature and occurrence

> **forest biome**
> A biome that includes all regions of forest over the lands of the Earth.

in different latitude belts: low-latitude rainforest, midlatitude deciduous and evergreen forest, and boreal forest.

Since humans took up agriculture, the Earth's forests have been diminishing through the process of **deforestation** (**Figure 17.8**). At present, about 13 million hectares (32 million acres) of forest are lost each year, largely to agriculture. More than half

Deforestation • Figure 17.8

Forest clearing, if carried out unsustainably, reduces global biomass and biodiversity while contributing to global warming.

a. Clear-cutting
Clear-cutting of large tracts of timber, without sustainable replanting, contributes to erosion and loss of habitat.

b. Wood gathering
Where fuel is in short supply, firewood is stripped from living trees, reducing the vegetation cover. When trees are gone, dung is burned, further impoverishing the soil.

c. Land use conversion
Large areas of equatorial rainforest in the Amazon basin are now being converted to grazing land and agriculture. The first step in this process is cutting the forest for timber and burning the debris to release nutrients to soil. The nutrients are soon exhausted, leaving the land impoverished and unproductive without fertilization.

Plants and animals of the low-latitude rainforest • Figure 17.9

The prodigious rainfall and warm temperatures of rainforests provide the most biologically diverse habitats on the planet.

a. Rainforest interior
Crowns form a continuous canopy shading lower layers. The lower two-thirds of the trees are characteristically smooth-barked and unbranched. Many rainforest species, especially in low or wet areas, have buttress roots extending out from the base of the tree. Here a researcher makes observations in a rainforest near Kalimantan, Borneo.

b. Rainforest dweller
Many animal species of the rainforest are adapted to tree life. Here a red leaf monkey in the jungles of Borneo looks upon its tropical world. Toucans, parrots, and fruit-eating bats also exploit the resources of the rainforest canopy.

of this area is in South America, Africa, and equatorial Asia. Much of the deforestation can be linked to economic development policies that promote cultivation of cash crops such as soybeans and palm oil. Loss of habitat in these species-rich areas takes a toll on the Earth's biodiversity. Slashing and burning of forest also releases carbon dioxide, and loss of evapotranspiration from trees leads to decreased rainfall and higher temperatures.

Low-Latitude Rainforest

Low-latitude rainforest, found in the equatorial and tropical latitude zones, consists of tall, closely spaced trees (**Figure 17.9**). Equatorial and tropical rainforests are not jungles of impenetrable plant thickets. Rather, the floor of the low-latitude rainforest is usually so densely shaded by a canopy of tree crowns that plant foliage is sparse close to the ground.

This world map of low-latitude rainforest shows equatorial and tropical rainforest types. A large area of rainforest lies astride the equator and extends poleward through the tropical zone (lat. 10° to 25° N and S) along monsoon and trade-wind coasts. Within the low-latitude rainforest are many highland regions where temperatures are cooler and rainfall is increased by the orographic effect.

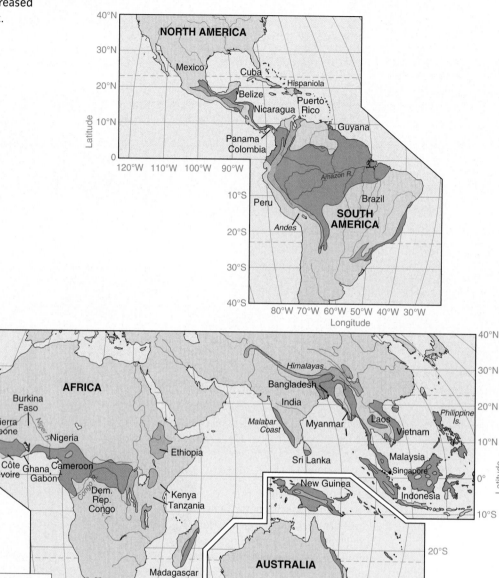

Figure 17.10 shows the world distribution of low-latitude rainforests. These rainforests develop in a climate that is continuously warm and frost-free and that has abundant precipitation in all months of the year (or, at most, has only one or two dry months). These conditions occur in the wet equatorial climate and the monsoon and trade-wind coastal climate. Plants grow in these rainforests continuously throughout the year.

A particularly important characteristic of the low-latitude rainforest is the large number of species of plants and animals that coexist. For example, equatorial rainforests contain as many as 3000 different tree species within a few square kilometers. The trees define a number of distinct rainforest layers (**Figure 17.11**).

Rainforest layers • Figure 17.11

This diagram shows the typical structure of equatorial rainforests. Crowns of the trees of low-latitude rainforests tend to form two or three layers. The highest layer consists of scattered "emergent" crowns that protrude from the closed canopy below, often rising to 40 m (130 ft). Some emergent species develop wide buttress roots, which provide physical support. Below the layer of emergents is a second, continuous layer, which is 15 to 30 m (about 50 to 100 ft) high. A third, lower layer consists of small, slender trees 5 to 15 m (about 15 to 50 ft) high, with narrow crowns.

Put It Together

Review Figure 17.1 and answer this question. Which of the following is a common feature of midlatitude forest and rainforest structures?
a. emergent crowns   c. herb layer
b. layers   d. moss layer

Biologists have not yet identified many of the plants and animals that live in low-latitude rainforests. Smithsonian scientists have determined that a single tree may contain more than 1000 species of invertebrates, many of which have never been identified. These areas, and the species within them, are now under threat. Slowly but surely, the rainforest is being conquered by logging, clear-cutting, and conversion to grazing and farming (see *Where Geographers Click*).

Where Geographers CLICK

Monitoring Deforestation in the Amazon

http://earthobservatory.nasa.gov/Features/WorldOfChange/deforestation.php

Rondônia, Brazil 2000

Rondônia in western Brazil is one of the most deforested parts of the Amazon. Thirty years ago, the state contained 208,000 km² of forest (about 51.4 million acres). By 2003, an estimated 67,764 km² (16.7 million acres) of rainforest had been cleared. Since then, however, the rate of deforestation has diminished significantly.

NASA's Moderate Resolution Imaging Spectroradiometer (MODIS) has been monitoring the changing landscape over the past decade. NASA's Web site shows how deforestation has progressed over the years. The first clearings appear in a fish-bone pattern, arrayed along the edges of roads. Over time, the fish bones collapse into a mixture of forest remnants, cleared areas, and settlements. Small farmers slowly migrate into more and more remote areas of forest and clear patches for crops. When crop yields fall, farmers convert the degraded land to cattle pasture and clear more forest for crops.

Rondônia, Brazil 2009

Forest Biome 531

Monsoon Forest

Monsoon forest is typically open (**Figure 17.12**). It grades into woodland, with open areas occupied by shrubs and grasses. Monsoon forest of the tropical latitude zone differs from tropical rainforest in that it is **deciduous**; that is, most of the trees of the monsoon forest shed their leaves because of stress during the long dry season that occurs at the time of low Sun and cool temperatures.

Figure 17.13 is a world map of the monsoon forest. This forest develops in the wet-dry tropical climate, in which a long rainy season alternates with a dry, rather cool season. The typical regions of monsoon forest are in Myanmar, Thailand, and Cambodia. Large areas of monsoon forest also occur in south-central Africa and in Central and South America, bordering the equatorial and tropical rainforests. Meteorologists have been monitoring shifts in the monsoon cycles that move the rain system over northeastern Australia, causing record floods in recent years and reducing precipitation over traditional forest of southeastern Asia.

Monsoon forest • Figure 17.12

Trees of the monsoon forest shed their leaves in the dry season. The forest cover is sparser and the trees are shorter than in rainforest. Tree trunks are massive, often with thick, rough bark. Branching starts at a comparatively low level and produces large, round crowns. Shown here is the monsoon forest of the Ranthambore Tiger Reserve, Rajasthan, India.

Global Locator

Rajasthan, India

NATIONAL GEOGRAPHIC

Global distribution of monsoon forest • Figure 17.13

For thousands of years, monsoon forests have adapted to the tropical wet-dry climate, with its seasonal deluges of rain followed by months of little or no precipitation. Recent studies have raised concerns that climate change may alter monsoon weather patterns and affect monsoon forest ecosystems.

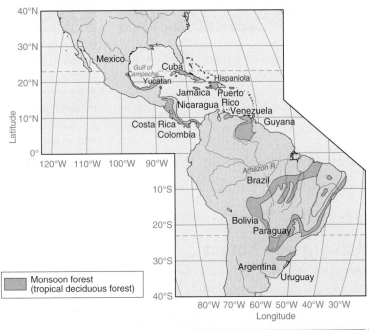

Monsoon forest
(tropical deciduous forest)

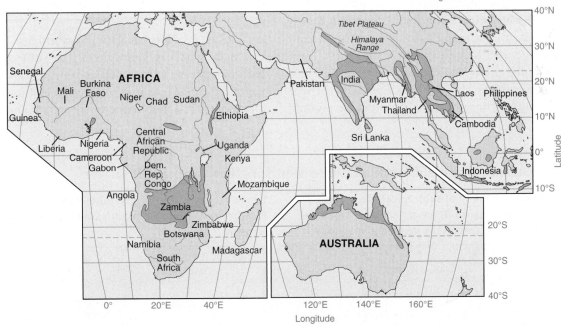

Ask Yourself

Which continent has the highest-latitude monsoon forest?
a. Africa
b. South America
c. Australia
d. Asia

a. Broadleaf
Here broadleaved species dominate over a lower layer of smaller plants. This example is from New South Wales, Australia, and includes many species of eucalyptus.

b. Needleleaf
The subtropical needleleaf evergreen forest inhabits dry, sandy soils, and it experiences occasional droughts that lead to fires. Pines are well adapted to these conditions and form a stable vegetation cover in many areas. Where fire is less frequent, a layer of broadleaf shrubs and small trees often occurs beneath the pines. Shown here is a longleaf pine stand near Aiken, South Carolina.

Plants of the subtropical evergreen forest • Figure 17.14

This forest of mild and moist climates includes both broadleaf and needleleaf types.

Subtropical Evergreen Forest

Subtropical evergreen forest is generally found in regions of moist subtropical climate, where winters are mild and there is ample rainfall throughout the year (**Figure 17.14**). This forest occurs in two forms: broadleaf and needleleaf. The **subtropical broadleaf evergreen forest** doesn't have as many tree species as the low-latitude rainforests, which are also broadleaf evergreen types. Trees are also not as tall as in the low-latitude rainforests. Their leaves tend to be smaller and more leathery, and the leaf canopy is less dense. The subtropical broadleaf evergreen forest often has a well-developed lower layer of vegetation. Depending on the location, this layer may include tree ferns, small palms, bamboos, shrubs, and herbaceous plants. There are also many lianas and *epiphytes*—plants that perch on tree branches and grow without roots in the soil.

Figure 17.15 is a map of the subtropical evergreen forests of the northern hemisphere. The **subtropical needleleaf evergreen forest** occurs only in the southeastern United States. Here it is referred to as the southern pine forest, since it is dominated by species of pine.

Global distribution of the subtropical evergreen forest • Figure 17.15

This northern hemisphere map of subtropical evergreen forests shows the needleleaf southern pine forest. Broadleaf evergreen forest includes trees of the laurel and magnolia families, so it is also termed *laurel forest*. Because the laurel forest region is intensely cultivated, little natural laurel forest remains.

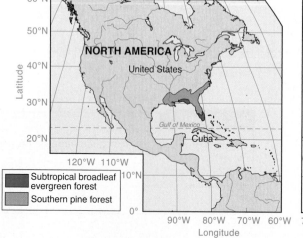

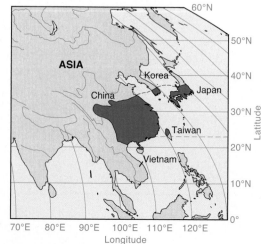

Midlatitude Deciduous Forest

Midlatitude deciduous forest is the native forest type of eastern North America and Western Europe (**Figure 17.16**). It is dominated by tall, broadleaf trees that provide a continuous and dense canopy in summer but shed their leaves completely in the winter. Lower layers of small trees and shrubs are weakly developed. In the spring, a lush layer of lowermost herbs quickly develops. This layer of herbs soon fades after the trees reach full foliage and shade the ground. The deciduous forest includes a great variety of animal life, stratified according to canopy layers.

Plants and animals of the deciduous forest • Figure 17.16

Four distinct seasons are notably displayed in the annual spring-through-winter vegetative growth cycles of these forest ecosystems.

a. Trees of the forest
Common trees of the deciduous forest of eastern North America, southeastern Europe, and eastern Asia are oak, beech, birch, hickory, walnut, maple, elm, and ash. Where the deciduous forests have been cleared by lumbering, pines readily develop as second-growth forest. Shown here is a forest of maple, oak, and hickory in fall colors in the Monongahela National Forest, West Virginia.

b. Fox squirrel
Scampering freely from the trees to the forest floor, the fox squirrel feeds primarily on nuts and seeds of the deciduous forest. In the canopy, squirrels are joined by many species of birds and insects.

c. Woodchuck
Many small mammals burrow in forest soils for shelter or food, including the woodchuck, or groundhog, shown here. They are joined by ground squirrels, mice, and shrews.

d. White-tailed deer
Among the large herbivores that graze in the deciduous forest are the white-tailed deer of North America and the red deer and roe deer of Eurasia.

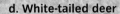

Northern hemisphere distribution of midlatitude deciduous forest • Figure 17.17

In Western Europe, the midlatitude deciduous forest is associated with the marine west-coast climate. Here the dominant trees are mostly oak and ash, with beech in cooler and moister areas. Regions of deciduous forest are also found in a belt across Asia and near the southern tip of South America.

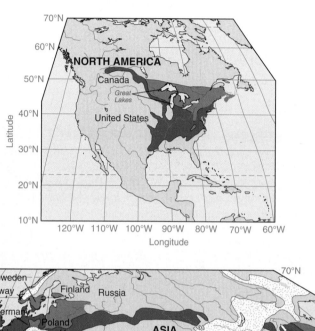

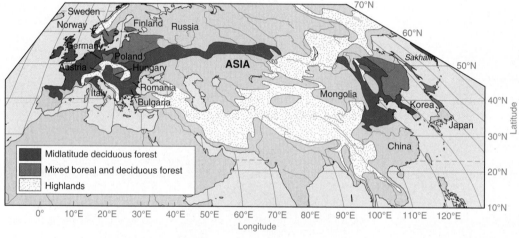

Figure 17.17 is a map of midlatitude deciduous forests, which are found almost entirely in the northern hemisphere. Throughout much of its range, this forest type is associated with the moist continental climate. This climate receives adequate precipitation in all months, normally with a summer maximum. This forest type has a strong annual temperature cycle, with a cold winter season and a warm summer.

Needleleaf Forest

Needleleaf forest is composed largely of straight-trunked, cone-shaped trees with relatively short branches and small, narrow, needle-like leaves (**Figure 17.18**). These trees are conifers. Most are evergreen, retaining their needles for several years before shedding them. Where the needleleaf forest is dense, it provides continuous and deep shade to the ground. Lower layers of vegetation are sparse or absent, except for possibly a thick carpet of mosses. Species are few; in fact, large tracts of needleleaf forest consist almost entirely of only one or two species.

Boreal forest is the cold-climate needleleaf forest of high latitudes. It occurs in two great continental belts, one in North America and one in Eurasia (**Figure 17.19**). These belts closely correspond to the region of boreal forest climate. The boreal forest of North America, Europe, and western Siberia is composed of evergreen conifers such as spruce and fir.

Plants of the needleleaf forest • Figure 17.18

In this forest at the shore of Caribou Lake in the Mackenzie River Valley, Northwest Territories, Canada, reds and yellows mark maples and deciduous birches in their fall colors. The needleleaf trees are white and black spruce.

Global distribution of needleleaf forest • Figure 17.19

This northern hemisphere map shows cold-climate needleleaf forests, including coastal forest.

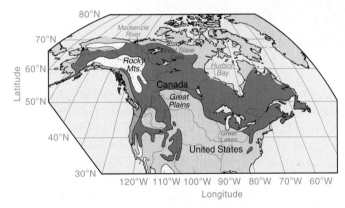

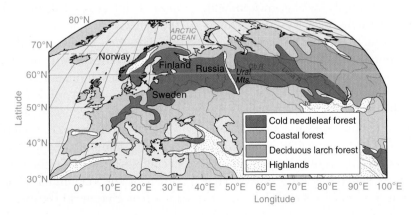

Plants of the coastal forest • Figure 17.20

Along the western coast of North America, from central California to Alaska, in a band of heavy orographic precipitation, mild temperatures, and high humidity, is a forest dominated by many unique species of needleleaf trees. Shown here is the coast redwood, *Sequoia sempervirens*, which is found at the southern end of this range. It is generally considered the tallest of tree species, attaining a height of 115 m (377 ft) and diameter of 7 m (23 ft) at the base. These redwoods are at Muir Woods National Monument, not far from San Francisco.

Mammals of the boreal needleleaf forest in North America include deer, moose, black bear, marten, mink, wolf, wolverine, and fisher. Common birds include jays, ravens, chickadees, nuthatches, and a number of warblers. The caribou, lemming, and snowshoe hare inhabit both needleleaf forest and the adjacent tundra biome. The boreal forest often experiences large fluctuations in animal species populations, a result of its highly variable environment.

Coastal forest is a distinctive needleleaf evergreen forest of the Pacific Northwest coastal belt, ranging in latitude from northern California to southern Alaska (**Figure 17.20**).

Sclerophyll Forest

The native vegetation of the Mediterranean climate is adapted for survival through the long summer drought. Shrubs and trees that can survive such drought are equipped with small, hard, or thick leaves that resist water loss through transpiration. These plants are called sclerophylls.

Sclerophyll forest is made up of trees with small, hard, leathery leaves. The trees are often low branched and gnarled, with thick bark. The formation class includes sclerophyll woodland, an open forest in which only 25 to 60% of the ground is covered by trees. Also included are extensive areas of scrub, a plant formation type consisting of shrubs covering somewhat less than half of the ground area. The trees and shrubs are evergreen, retaining their thickened leaves despite the severe annual drought.

The map of sclerophyll vegetation shown in **Figure 17.21** includes forest, woodland, and scrub types. Over the centuries, human activity has reduced the sclerophyll forest to woodland or destroyed it entirely. Today, large areas of this former forest consist of dense scrub. In the California coastal range, the sclerophyll forest or woodland is typically dominated by live oak and white oak. Grassland occupies the open ground between the scattered oaks. Much of the remaining vegetation is sclerophyll scrub, known as **chaparral** (**Figure 17.22**).

Global distribution of sclerophyll forest • Figure 17.21

Sclerophyll forest is closely associated with the Mediterranean climate and is narrowly limited to west coasts between 30° and 40° to 45° N and S latitude. In the Mediterranean lands, the sclerophyll forest forms a narrow coastal belt ringing the Mediterranean Sea. Here, the Mediterranean forest consists of trees such as cork oak, live oak, Aleppo pine, stone pine, and olive. The other northern hemisphere region of sclerophyll vegetation is the California coast ranges. Important areas of sclerophyll forest, woodland, and scrub are found in southeast, south-central, and southwest Australia and Chile, as well as the Cape region of South Africa.

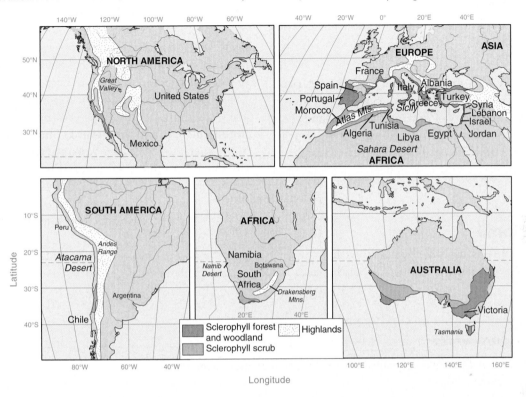

Chaparral • Figure 17.22

The tall, tough, wiry shrubs in the middle of this photo are typical of California coastal chaparral. They also cover the mountain slopes seen in the background. The pool in the foreground, normally dry, has been filled by a winter rainstorm.

| CONCEPT CHECK | STOP |

1. **What** is the global distribution of the forest biome?
2. **Which** forest formation classes shed their leaves in an annual cycle?
3. **Which** climate is associated with sclerophyll forest?

Savanna and Grassland Biomes

LEARNING OBJECTIVES

1. **Describe** the climate, plants, and animals of the savanna biome.

2. **Describe** the climate, plants, and animals of the grassland biome.

The savanna and grassland biomes of African safaris and Argentine gauchos support an important variety of grazing mammals and their predators, as well as the world's great migrating herds. These lands are easily converted to agriculture when irrigated by water drawn from deep wells, adding to their rapid conversion.

Savanna Biome

The **savanna biome** is usually associated with the tropical wet-dry climate of Africa and South

> **savanna biome** A biome that consists of a combination of trees and grassland in various proportions.

America. Its vegetation ranges from woodland to grassland. In savanna woodland, the trees are spaced rather widely apart because there is too little soil moisture during the dry season to support full tree cover (**Figure 17.23**). The open spacing allows for the development of a dense lower layer, which usually consists of grasses. The woodland has an open, park-like appearance. Savanna woodland usually lies in a broad belt adjacent to equatorial rainforest.

In the tropical savanna woodland of Africa, the trees are of medium height. Tree crowns are flattened or umbrella shaped, and tree trunks have thick, rough bark. Some species of trees are **xerophytic** forms—adapted to the dry environment with small leaves and thorns. Others are broadleaved deciduous species that shed their leaves in the dry season. In this respect, savanna woodland resembles monsoon forest.

Fires occur frequently in the savanna woodland during the dry season, but the tree species

Savanna • Figure 17.23

Savannas include several different types, depending on the amount of tree cover and the nature of plants underneath.

a. Savanna woodland
Where the trees are more closely spaced, we have savanna woodland. This example is from the central Kalahari Desert, Botswana. The rich green of the landscape identifies the time of year as just after the rainy season.

b. Thorntree–tall-grass savanna
The long dry season of the wet-dry climate restricts the vegetation to grasses with an open canopy of drought-resistant trees, such as the acacias shown in this photo of the Serengeti National Park, Tanzania. Zebras are one of more than a dozen species of hoofed grazers that inhabit the savanna.

Global distribution of savanna • Figure 17.24

Savanna biome vegetation is described as rain-green. That's because the thorntree–tall-grass savanna is closely identified with the semiarid subtype of the dry tropical and subtropical climates. In the semiarid climate, soil-water storage is only enough for plants during the brief rainy season. After rains begin, the trees and grasses quickly green. Vegetation of the monsoon forest is also rain-green.

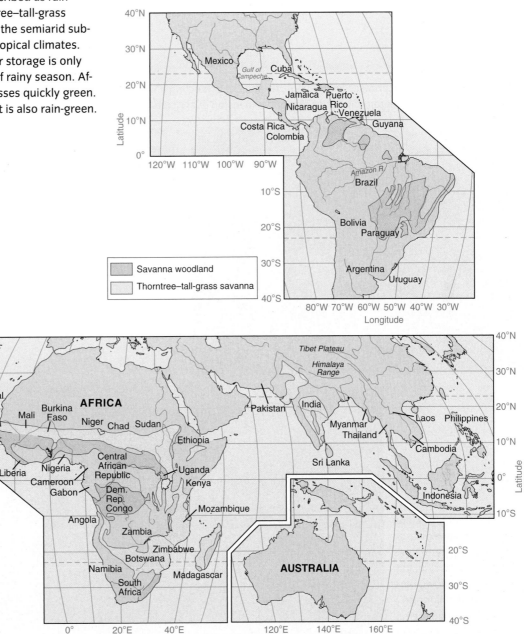

are particularly resistant to fire. Many geographers think that periodic burning of the savanna grasses keeps forest from invading the grassland. Fire doesn't kill the underground parts of grass plants, but it limits tree growth to fire-resistant species. This means that many rainforest tree species that might otherwise grow in the wet-dry climate regime are suppressed by fires. Browsing animals also kill many young trees, helping maintain grassland at the expense of forest.

The regions of savanna woodland are shown in **Figure 17.24**. In Africa, the savanna woodland grades into a belt of **thorntree–tall-grass savanna**, a formation class transitional to the desert biome. The trees are largely of thorny species. They are more widely scattered, and the open grassland is more extensive than in the savanna woodland. A characteristic tree is the flat-topped acacia. Elephant grass is a common species. It can grow to a height of 5 m (16 ft) and can form an impenetrable thicket.

The African savanna is widely known for the diversity of its large grazing mammals (**Figure 17.25**). With these grazers come a large variety of predators—lions, leopards, cheetahs, hyenas, and jackals. Elephants are the largest animals of the savanna and adjacent woodland regions.

Animals of the African savanna • Figure 17.25

A tourism economy is the primary foundation available for protecting a wild African landscape constantly threatened by poaching and expansion of agriculture.

a. Wildebeest
These strange-looking animals are actually antelopes. More than a dozen antelope species graze on the savanna. Each one has a particular preference for eating either the grass blade, sheath, or stem. Grazing stimulates the grasses to continue to grow, and so the ecosystem is more productive when grazed than when left alone.

b. Lions
The abundance of large grazing herbivores brings predators, including the lion. Shown here is a pride of lions on a walk through the tall grass of Masai Mara National Reserve, Kenya.

c. Spotted hyena
Another savanna predator is the hyena, which does not attack its prey directly like the lion but runs it to exhaustion. Small packs of 6 to 12 animals do the hunting, employing different strategies for different antelope prey. They fear only the big cats, such as lions. Shown here is a hyena in the Ngorongoro Conservation Area, Tanzania, carrying a dead wildebeest calf in its vise-grip jaws.

Grassland Biome

The **grassland biome** includes two major formation classes that we will discuss here: tall-grass prairie and steppe (**Figure 17.26**). **Tall-grass prairie** consists largely of tall grasses. There

> **grassland biome**
> A biome consisting largely, or entirely, of herbs, which may include grasses, grass-like plants, and forbs.

are also some broadleaved herbs, called *forbs*. We don't see trees or shrubs in the prairie, but they do occur in narrow patches of forest in stream valleys.

Plants of the grasslands • Figure 17.26

Low precipitation and frequent fires maintain a shrub- and tree-free horizon for most of the grasslands. Botanists continue to discover medically important shrubs and herbs in these ecosystems.

a. Tall-grass prairie
In addition to grasses, tall-grass prairie vegetation includes many forbs, such as the wildflowers shown in this photo. The grasses are deeply rooted and form a thick and continuous turf.

b. Steppe
Buffalo grass and blue grama grass are typical of the American steppe, seen here at the Pawnee National Grassland, near Fort Collins, Colorado. There may also be some scattered shrubs and low trees near watercourses. Bare soil is often exposed between the low grasses and weeds.

Steppe, or short-grass prairie, consists of sparse clumps of short grasses. Steppe grades into semidesert in dry environments and into prairie where rainfall is higher.

Steppe grassland is concentrated largely in the midlatitude areas of North America and Eurasia. **Figure 17.27** shows the distribution of grassland around the world.

Global distribution of grassland • Figure 17.27

This world map shows the grassland biome in subtropical and midlatitude zones. Prairie grasslands are associated with the drier areas of moist continental climate. The tall-grass prairies once lay in a belt from the Texas Gulf coast to southern Saskatchewan and extended eastward into Illinois. Now they have been converted almost entirely to agricultural land. Steppe grasslands correspond well with the semiarid subtype of the dry continental climate.

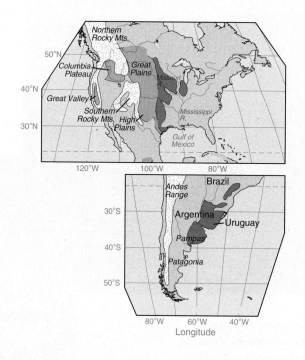

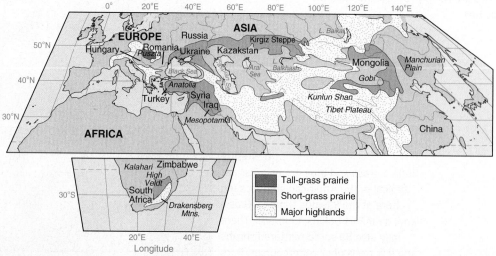

Grassland animals • Figure 17.28

The animals of the grassland are distinctive, including many grazing mammals.

a. American bison
Prairie grasslands are the home of many types of grazing animals, including the American bison, also known as the buffalo. Once extremely widespread throughout the Great Plains, it was hunted to near extinction in the 19th century. These animals are part of a managed herd in Kentucky. Other prairie grazers include the elk and the pronghorn antelope.

c. Prairie dog
Many animals, such as the prairie dog, burrow into the prairie soil for shelter. These highly social animals live in colonies, or *dogtowns*, of hundreds of animals. They feed on grasses, forbs, and insects.

b. Jackrabbit
The jackrabbit is a common grazer of the prairies and steppes. It has developed a leaping habit that allows it to see above the grass as it moves. The pronghorn and jumping mouse have also developed this habit.

The grassland ecosystem supports some unique adaptations to life (**Figure 17.28**). Animals have learned to jump or leap, to get an unimpeded view of their surroundings. We see jackrabbits and jumping mice, and the pronghorn combines the leap with great speed, which allows it to avoid predators and fire. Many animals burrow because the soil provides the only shelter in the exposed grasslands. Examples are burrowing rodents, including prairie dogs, gophers, and field mice. Rabbits exploit old burrows, using them for nesting or shelter. Invertebrates also seek shelter in the soil, and many are adapted to living in the burrows of rodents, where extremes of moisture and temperature are substantially moderated.

CONCEPT CHECK STOP

1. **What** is the role of wildfires in maintaining the savanna biome?

2. **How** have animals adapted to the grassland biome?

Desert and Tundra Biomes

LEARNING OBJECTIVES

1. **Describe** the climate, plants, and animals of the desert biome.

2. **Describe** the climate, plants, and animals of the tundra biome.

Global distribution of the desert biome • Figure 17.29 _____

This world map of the desert biome includes desert and semidesert formation classes.

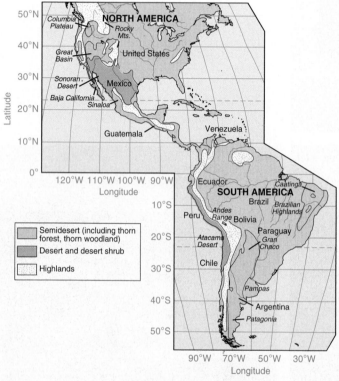

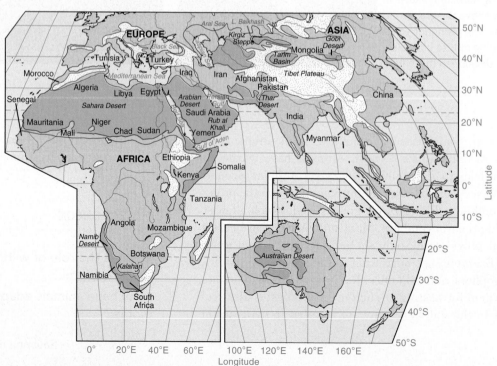

he desert is a highly evolved ecosystem that contains a multitude of plants and animals. Insects, reptiles, mammals, and birds can occasionally be spotted at night when the Sun ceases its unfiltered radiation of the sparse vegetation. Rare and fantastic plants may flower after many decades, when conditions allow for sufficient rains to trigger long-dormant seeds to germinate. These desert blooms may last only a few days or weeks for lucky witnesses.

Desert Biome

The **desert biome** includes several formation classes that are transitional from grassland and savanna biomes into vegetation of the arid desert. Here we recognize two basic formation classes: semidesert and dry desert.

> **desert biome** A biome of dry climates consisting of thinly dispersed shrubs, grasses, and perennial herbs but lacking in trees.

Semidesert is found in a wide latitude range—from the tropical zone to the midlatitude zone (**Figure 17.29**). The arid subtypes of all three dry climates can be thought of as semidesert. In recent times, overgrazing and trampling by livestock have helped semidesert shrub vegetation expand widely into areas of the western United States that used to be steppe grasslands.

Thorntree semidesert of the tropical zone is made up of xerophytic trees and shrubs (**Figure 17.30**). These plants are adapted to a climate with a very long, hot dry season and only a very brief, but intense, rainy season. We find these conditions in the semiarid and arid subtypes of the dry tropical and dry subtropical climates. The thorny trees and shrubs are known locally as thorn forest, thornbush, or thornwoods. In some places, you can also see cactus plants.

Desert vegetation • Figure 17.30

Semidesert landscape is covered with low grasses and sagebrush, creating ecosystems with adaptations for extreme dryness and short, intense rain episodes.

a. Sagebrush semidesert
Sparse grasses and shrubs, largely sagebrush, provide vegetation cover near Monument Valley, Arizona. You can also see mesas and buttes.

Butte

Mesa

b. Thorntree semidesert
The thorntree semidesert formation is found in tropical climates with very long dry seasons and short, but intense, rainy seasons. It consists of sparse vegetation cover of grasses and thorny shrubs that are dormant for much of the year. This photo shows a steenbok (a small African antelope) in Etosha National Park, Namibia.

Dry desert is a formation class of plants that are widely dispersed over only a very small proportion of the ground. It consists of small, hard-leaved, or spiny shrubs; succulent plants (such as cactus); and hard grasses. Many species of small annual plants appear only after rare and heavy downpours. In fact, many of the areas designated as desert vegetation on maps have no plant cover at all because the surface consists of shifting dune sands or sterile salt flats.

Desert plants around the world look very different from each other. In the Mojave and Sonoran deserts of the southwestern United States, plants are often large, giving the appearance of a woodland. There, you can find the tree-like saguaro cactus, prickly pear cactus, ocotillo, creosote bush, and smoke tree. Desert animals, like the plants, are typically adapted to the dry conditions of the desert (**Figure 17.31**).

In recent decades, the desert biome has been expanding into arid, semiarid, and dry subhumid areas (see *Video Explorations*). Climate change, such as drought, and human activities, such as overgrazing, and conversion of natural areas to agricultural use, are leading causes of **desertification**. Desertification brings loss of topsoil, increased soil salinity, damaged vegetation, regional climate change, and a decline in biodiversity.

Desert animals • Figure 17.31

Small desert mammals live mostly out of sight in underground burrows, surviving on a low water budget. Many are noctural, avoiding the hot part of the day. Shown in the background is a scene from the Kalahari Desert, featuring the endangered quiver tree, a species of Aloe, on a landscape of fractured red rock.

a. Kangaroo rat
This small, nocturnal desert dweller is well adapted to the desert environment. Rarely drinking water, it has a metabolism that is very efficient at retaining water and excreting salty wastes. Not a true rat, it is closely related to the pocket gopher. It gets its name from its powerful hind feet and its jumping habit.

b. Meerkat
The meerkat, a denizen of the Kalahari Desert, is related to the mongoose. Meerkats are very social animals, living in large colonies in underground burrows. They are immune to many of the poisons and stingers of desert reptiles, such as scorpions and snakes. This photo shows a young meerkat snacking on a lizard.

Video Explorations
China's Desert

  WILEY PLUS · NATIONAL GEOGRAPHIC · Video

Over the past century, much of the farmland in northern China has been devoured by sand. Today, 17% of China's land mass is desert. Years of deforestation are to blame, and as the country continues its fast-paced development, the problem may well worsen.

This video shows that desertification is taking a toll on China's population. Beijing is often blasted by sand and dust storms, damaging people's health and reducing visibility. The government is working to reforest areas—building walls of trees to try to hold back the encroaching sand. But some experts argue that these steps will not be enough in the face of development pressures in China that are unparalleled anywhere else in the world.

Tundra Biome

Arctic tundra is a formation class of the **tundra biome**, with a tundra climate. In this climate, plants grow during the brief summer of long days and short (or absent) nights. At this time of year, air temperatures rise above freezing, and a shallow surface layer of ground ice thaws. The permafrost beneath, however, remains frozen, keeping the meltwater at the surface. These conditions create a marshy environment for a short time over wide areas. Because plant remains decay very slowly in the cold meltwater, layers of organic matter build up in the marshy ground. The tundra is not usually home to large plants because frost action in the soil fractures and breaks large roots, keeping tundra plants small (**Figure 17.32**). In winter, wind-driven snow and extreme cold also injure plant parts that project above the snow. Figure 17.4 shows the distribution of arctic and alpine tundra. You can also refer to Figure 7.31, which shows the tundra climate type.

> **tundra biome** A biome of the cold regions of arctic tundra and alpine tundra, consisting of grasses, grass-like plants, flowering herbs, dwarf shrubs, mosses, and lichens.

Vegetation of the arctic tundra • Figure 17.32

Found in areas of extreme winter cold with little or no true summer, arctic tundra consists of low perennial grasses, sedges, herbs, and dwarf shrubs, as well as lichens and mosses. This photo shows tundra in fall colors in Lapland, Finland. In the background, a sparse stand of trees grows in a sheltered spot.

Tundra animals demonstrate a variety of strategies to survive in a hostile and often frozen landscape. Seasonally changing insulating layers and fine-tuned migration behavior enable these animals to survive.

a. Caribou

Barren ground caribou roam the tundra, constantly grazing the lichens and plants of the tundra and boreal zone. They migrate long distances between calving and feeding grounds.

b. Musk oxen

These wooly tundra-grazers are more closely related to goats than to cattle. They feed on grasses, sedges, and other ground plants, scratching their way through the snow to find them in winter. Hunted close to extinction, musk oxen are now protected. Originally restricted to Alaska, Canada, and Greenland, they have been introduced into northern Europe.

c. Sandpiper

This small migratory bird, a dunlin sandpiper, travels long distances to return to the tundra to nest and fledge its young. It probes the tundra with its sensitive beak, searching for insects. The boggy tundra presents an ideal summer environment for many other migratory birds, such as waterfowl and plovers.

Tundra vegetation is also found at high elevations, above the limit of tree growth and below the vegetation-free zone of bare rock and perpetual snow. This **alpine tundra** resembles arctic tundra in many physical respects.

The tundra does not have many diverse species but it does have a large number of individual animals (**Figure 17.33**). The food web of the tundra ecosystem is simple and direct. The important producer is reindeer moss, the lichen *Cladonia rangifera*. Caribou, reindeer, lemmings, ptarmigan (arctic grouse), and snowshoe hares all graze on this lichen. Foxes, wolves, and lynxes prey on those animals, although they may all feed directly on plants as well. During the summer, the abundant insects help support the migratory waterfowl populations.

Biomes provide an organizational framework for our understanding of the world's major terrestrial ecosystems. These ecosystems have been formed over the millennia by the prevailing climates and the interacting components of soil, plants, and animals in patterns of dynamic equilibrium. When the equilibrium is perturbed, whether by nature or human activities, a biome makes adjustments with species shifts, and sometimes losses, as the biome boundaries move, grow, or shrink in size. Because climate plays such a pivotal role in setting the physical conditions for these biomes to thrive, we can expect that as the planet warms and the weather patterns change, these biomes will experience a cascade of adjustments, much as they have over geologic time, but more rapidly. The growth of human populations and increasing land use activities, however, will limit the adjustments available for these biomes. Scientists will need to focus on understanding and managing the effects of change in biomes and natural cycles on socioeconomic policy, with the aim of maintaining the ecological goods and services provided by these major ecosystems for our survival as a species.

1. **What** is the difference between semidesert and dry desert vegetation?

2. **What** are two formation classes of the tundra biome?

THE PLANNER ✓

Summary

1 Global Vegetation 520

- The largest unit of **terrestrial ecosystems** is the **biome**.

- **Plant life-form** refers to the physical structure, size, and shape of a plant. Life-forms include trees, shrubs, **lianas**, **herbs**, and **lichens**. **Forests** are dense groups of trees; in grasslands, trees are separated by open spaces.

- The five principal biomes are **forest biome**, **grassland biome**, **savanna biome**, **desert biome**, and **tundra biome**. Precipitation and temperature patterns are key factors in determining the type of vegetation characteristic of each biome, as shown in the diagram.

How climate influences terrestrial biomes • Figure 17.3

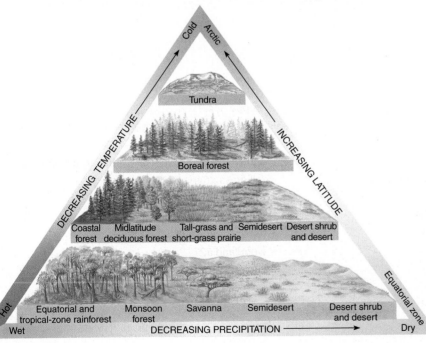

- Climate changes gradually with latitude; biome changes are therefore typically gradual.

- **Natural vegetation** is plant cover that develops with little or no human interference. However, over the past few thousand years, human societies have modified and cultivated much of the land area of our planet, drastically changing the natural vegetation of many regions.

2 Forest Biome 528

- The **low-latitude rainforest** exhibits a dense canopy and an open floor with a very large number of species.

- **Monsoon forest** is largely **deciduous**, with most species shedding their leaves after the wet season.

- **Subtropical evergreen forest** occurs in **broadleaf** and **needleleaf** forms in the moist subtropical climate.

- As shown on the map, **midlatitude deciduous forest** is associated with the moist continental climate.

Northern hemisphere distribution of midlatitude deciduous forest • Figure 17.17

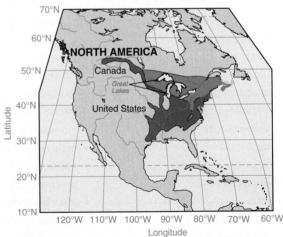

- **Needleleaf forest** consists largely of evergreen conifers.

- **Sclerophyll forest**, comprising trees with small, hard, leathery leaves, is found in the Mediterranean climate region.

3 Savanna and Grassland Biomes 540

- The savanna biome consists of widely spaced trees with an understory, often of grasses. Dry-season fire is frequent, limiting the number of trees and encouraging the growth of grasses.

- The grassland biome of midlatitude regions, as shown on the map, includes **tall-grass prairie** in moister environments and short-grass prairie, or **steppe**, in semiarid areas.

Global distribution of grassland • Figure 17.27

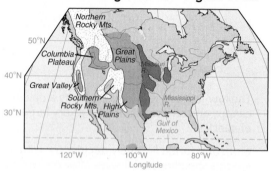

4 Desert and Tundra Biomes 546

- Vegetation of the desert biome, as shown in the photo, ranges from thorny shrubs and small trees to **dry desert** vegetation.

Desert vegetation • Figure 17.30

- Deserts expand through the process of **desertification**, as a result of drought, overgrazing, or poor agricultural practices.

- Tundra biome vegetation is limited largely to low herbs that are adapted to the severe drying cold.

Key Terms

- alpine tundra 550
- annual 521
- arctic tundra 549
- biome 520
- chaparral 538
- deciduous 532
- deforestation 528
- desert biome 547
- desertification 548
- dry desert 548
- forest 521
- forest biome 528
- formation class 521

- grassland biome 543
- herb 521
- liana 521
- lichen 521
- low-latitude rain forest 529
- midlatitude deciduous forest 535
- monsoon forest 532
- natural vegetation 524
- needleleaf forest 536
- perennial 520
- plant life-form 520
- savanna biome 540
- sclerophyll forest 538

- semidesert 547
- steppe 544
- subtropical broadleaf evergreen forest 534
- subtropical evergreen forest 534
- subtropical needleleaf evergreen forest 534
- tall-grass prairie 543
- terrestrial ecosystem 520
- thorntree–tall-grass savanna 541
- tundra biome 549
- woodland 521
- xerophytic 540

Critical and Creative Thinking Questions

1. For each of the five major biomes, predict how the areal extent and latitudinal location of each would be modified by climate change.

2. Which major biome are you located in? List three features of the biome that you have observed locally. What characteristics of other biomes do you observe either locally or within a reasonable distance? What does this tell you about the globally defined biome maps?

3. Monsoon forest and midlatitude deciduous forest are both deciduous—but for different reasons. Compare the characteristics of these two formation classes and their climates and describe how these characteristics favor deciduous trees in each case.

4. Subtropical broadleaf evergreen forest and tall-grass prairie are two vegetation formation classes that have been greatly altered by human activities. How was this done and why?

5. What are the features of arctic and alpine tundra? How does the cold tundra climate influence the vegetation cover?

6. As shown on the map, the savanna woodland formation class is closely associated with areas such as central Africa that experience dramatic crop failures that lead to famines. What characteristics of this formation class make agriculture problematic?

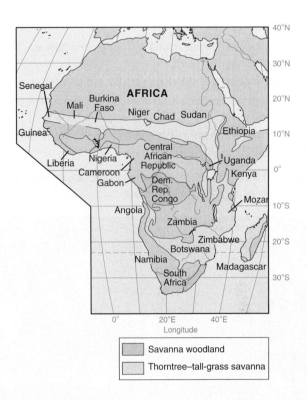

Savanna woodland

Thorntree–tall-grass savanna

What is happening in this picture?

This recently installed section of a border wall between the United States and Mexico sits in the Cabeza Prieta National Wildlife Refuge, which was created in 1939. The U.S. government is planning to build a border wall 2000 km (1240 mi) long across four desert ecosystems, multiple mountain ecosystems, and the Rio Grande riparian ecosystem.

Think Critically

1. What biome is shown here?
2. What effect will the border wall have on the plants and animals in this refuge?
3. What adaptations might occur over time if this fence separates migratory species?

Self-Test

(Check your answers in the Appendix.)

1. Natural vegetation _____.
   a. is plant cover that develops with little or no human interference
   b. is plant cover that develops with no human interference
   c. no longer exists because humans have affected every part of the globe's surface
   d. is an ideal state from which human modifications can be judged

2. Name the plant form in the photo, in which algae and fungi live together as a single plant structure.

3. The _____ biome develops in regions with moderate shortages of soil water.
   a. forest
   b. savanna
   c. desert
   d. grassland

4. Which biome is represented by a sparse and open cover of trees with grasses and herbs underneath?
   a. forest
   b. desert
   c. grassland
   d. savanna

5. _____ are plants that are well adapted to drought conditions.
   a. Forbs
   b. Xerophytes
   c. Deciduous trees
   d. Evergreens

6. Of the following climates, which has a strong wet-dry alternation and many xerophytic plants?
   a. Mediterranean
   b. tundra
   c. wet equatorial
   d. moist subtropical

7. _____ is a disturbance that is part of a forest's natural cycle.
   a. Overgrazing
   b. Clear-cutting
   c. Forest fire
   d. Slash-and-burn agriculture

8. Biogeographers split the five principal biomes into smaller vegetation units, called _____ classes, based on the life-form of plants.
   a. formation
   b. climate
   c. climax
   d. life-form

9. The diagram shows the layer structure of which type of forest?

10. A forest formation class with broadleaf and needleleaf forms is _____ forest.

a. low-latitude

b. monsoon

c. midlatitude deciduous

d. subtropical evergreen

11. Needleleaf forests are noted for _____.

a. generally having few tree species

b. the low level of shade they provide

c. evergreens that hold on to their needles for several years

d. poor-quality pulp wood

12. The _____ is dominated by low trees with thick, leathery leaves that are well adapted to the long summer drought of the Mediterranean climate.

a. deciduous forest

b. coastal forest

c. low-latitude rainforest

d. sclerophyll forest

13. Many plants of the _____ biome grow, bloom, and set seed during a short summer thaw following harsh cold winters.

a. desert

b. tundra

c. grassland

d. savanna

14. Which major formation class found in the grassland biome is shown in this photo?

15. A problem with maps in the display of natural features is that _____.

a. they create a large amount of distortion

b. it is impossible to display a natural feature on a map

c. many boundaries are approximate and gradational

d. maps are two-dimensional

THE PLANNER ✓

Review your Chapter Planner on the chapter opener and check off your completed work.

Answers to *Ask Yourself* and *Put It Together* Questions

Chapter 1
Put It Together, Figure 1.14: b
Ask Yourself, Figure 1.15: (a) 3:00 PM, (b) 2:00 PM, and (c) noon
Ask Yourself, Figure 1.16: Kiribati
Put It Together, Figure 1.19: b

Chapter 2
Ask Yourself, Figure 2.2: d
Put It Together, Figure 2.7: b
Put It Together, Figure 2.10: c
Ask Yourself, What a Geographer Sees: a

Chapter 3
Ask Yourself, Figure 3.2: 84°F
Put It Together, Figure 3.9: d
Ask Yourself, Figure 3.10: a
Put It Together , Figure 3.23: e

Chapter 4
Ask Yourself, Figure 4.1: a
Put It Together, Figure 4.6: At 40° N, temperature is 12° C and absolute humidity is 8 g/kg. From Figure 4.5, the saturation humidity is 10 g/kg at 12° C, so the relative humidity is 8/10 = 80%.
Put It Together, Figure 4.13: c
Ask Yourself, Figure 4.18: c

Chapter 5
Ask Yourself, Figure 5.5: c
Put It Together, Figure 5.13: d
Ask Yourself, Figure 5.15: b
Put It Together, Figure 5.22: a

Chapter 6
Ask Yourself, Figure 6.4: d
Put It Together, Figure 6.5: a
Ask Yourself, Figure 6.10: b
Put It Together, Figure 6.23: b

Chapter 7
Ask Yourself, Figure 7.2: d
Put It Together, Figure 7.17: b
Put It Together, Figure 7.35: c

Chapter 8
Ask Yourself, What a Geographer Sees: Approximately 4.5
Ask Yourself, Figure 8.5: b
Put It Together, Figure 8.9: b
Put It Together, Figure 8.12: smaller

Chapter 9
Put It Together, Figure 9.11: d
Ask Yourself, Figure 9.17: d
Ask Yourself, Figure 9.29: b
Put It Together, Figure 9.30: b

Chapter 10
Ask Yourself, Figure 10.2: b
Put It Together, Figure 10.7: c
Put It Together, Figure 10.13: a, but elements of b are also evident
Ask Yourself, Figure 10.14: translational

Chapter 11
Ask Yourself, Figure 11.1: a
Ask Yourself, Figure 11.18: d
Put It Together, Figure 11.21: a
Put It Together, Figure 11.26: aquiclude

Chapter 12
Ask Yourself, Figure 12.5: c
Put It Together, Figure 12.7: b
Ask Yourself, Figure 12.12: d
Put It Together, Figure 12.16: d

Chapter 13
Ask Yourself, Figure 13.1: b
Put It Together, Figure 13.8: b
Put It Together, Figure 13.23: c

Chapter 14
Ask Yourself, Figure 14.7: c
Put It Together, Figure 14.11: a
Ask Yourself, Figure 14.19: b
Put It Together, Figure 14.21: b

Chapter 15
Ask Yourself, Figure 15.3: c
Ask Yourself, Figure 15.7: a
Put It Together, Figure 15.15: c
Put It Together, Figure 15.21: b

Chapter 16
Ask Yourself, Figure 16.4: c
Ask Yourself, Figure 16.20: b
Put It Together, Figure 16.22: a

Chapter 17
Ask Yourself, Figure 17.3: a
Put It Together, Figure 17.5: c
Put It Together, Figure 17.11: b
Ask Yourself, Figure 17.13: d

Answers to Self-Tests

Chapter 1

1. c; **2.** b; **3.** d; **4.** c; **5.** latitude 50° N, longitude 60° W; **6.** the great circle route because it follows the shortest line possible between points on the globe; **7.** d; **8.** d; **9.** 9:00 AM; **10.** b; **11.** c; **12.** c; **13.** b; **14.** a; **15.** d

Chapter 2

1. (a) visible light, 0.4–0.7 μm; (b) gamma rays, below 0.0015 μm; (c) X-rays, 0.0015–0.01 μm; and (d) ultraviolet light, 0.01–0.35 μm; **2.** a; **3.** c; **4.** c; **5.** a; **6.** c; **7.** c; **8.** Northern hemisphere winter, see Figure 2.9 for label. At the Tropic of Capricorn, the Sun is directly overhead; **9.** b; **10.** a; **11.** See Figure 2.12; **12.** c; **13.** c; **14.** a; **15.** c

Chapter 3

1. a; **2.** d; **3.** See Figure 3.6 **4.** b; **5.** See Figure 3.13; **6.** b; **7.** a; **8.** b; **9.** a; **10.** d; **11.** See Figure 3.11; **12.** c; **13.** c; **14.** d; **15.** c

Chapter 4

1. a; **2.** c; **3.** See Figure 4.1; **4.** d; **5.** c; **6.** 50%; **7.** d; **8.** Cirrus clouds that form at high levels in the sky; **9.** c; **10.** c; **11.** b; **12.** a; **13.** c; **14.** c; **15.** d

Chapter 5

1. a; **2.** c; **3.** See Figure 5.5; **4.** b; **5.** a; **6.** b; **7.** See Figure 5.9; **8.** d; **9.** c; **10.** c; **11.** b; **12.** See Figure 5.16; **13.** a; **14.** c; **15.** d

Chapter 6

1. b; **2.** See Figure 6.1; **3.** c; **4.** d; **5.** Occluded front; **6.** a; **7.** a; **8.** c; **9.** See Figure 6.5; **10.** c; **11.** b; **12.** a; **13.** d; **14.** b; **15.** a

Chapter 7

1. a; **2.** c; **3.** c; **4.** e; **5.** d; **6.** d; **7.** c; **8.** a; **9.** b; **10.** c; **11.** d; **12.** See Figure 7.28 (boreal forest climate); **13.** a. 3; b. 1; c. 2; **14.** c; **15.** d

Chapter 8

1. d; **2.** b; **3.** See Figure 8.4; **4.** d; **5.** b; **6.** b; **7.** d; **8.** a; **9.** b; **10.** a; **11.** Batholiths are created as plutons when magma cools below the surface. They are often resistant to weathering and become exposed when surrounding rocks weather away; **12.** Fossil fuels are formed from organic deposits on land or ocean bottoms. The burning of fossil fuels has raised levels of carbon dioxide in the atmosphere; **13.** c; **14.** c; **15.** c

Chapter 9

1. The middle segment represent a converging boundary, and the bottom segment represents a diverging boundary. Two separating oceanic plates would be a diverging boundary, and the middle figure is a converging plate boundary; **2.** d; **3.** c; **4.** a; **5.** a; **6.** d; **7.** d; **8.** reverse fault **9.** graben (top) and horst (bottom); **10.** c; **11.** d; **12.** b; **13.** a; **14.** c; **15.** d

Chapter 10

1. c; **2.** d; **3.** b; **4.** d; **5.** c; **6.** a; **7.** a; **8.** b; **9.** c; **10.** d; **11.** a; **12.** a; **13.** c; **14.** b; **15.** c

Chapter 11

1. See Figure 11.4; **2.** b; **3.** c; **4.** b; **5.** a; **6.** c; **7.** See Figure 11.14; **8.** d; **9.** a; **10.** a; **11.** d; **12.** b; **13.** c; **14.** a; **15.** b

Chapter 12

1. d; **2.** b; **3.** a; **4.** b; **5.** c; **6.** d; **7.** See Figure 12.5; **8.** c; **9.** d; **10.** c; **11.** c; **12.** See Figure 12.14; **13.** b; **14.** a; **15.** See Figure 12.16

Chapter 13

1. b; **2.** c; **3.** b; **4.** d; **5.** a; **6.** d; **7.** c; **8.** b; **9.** c; **10.** See Figure 13.18; **11.** b; **12.** c; **13.** b; **14.** See Figure 13.28; **15.** b

Chapter 14

1. a; **2.** b; **3.** d; **4.** See Figure 14.3; **5.** b; **6.** a; **7.** c; **8.** See Figure 14.9; **9.** c; **10.** a; **11.** b; **12.** c; **13.** d; **14.** c; **15.** a

Chapter 15

1. d; **2.** c; **3.** See Figure 15.2; **4.** d; **5.** a; **6.** c; **7.** See Figure 15.12; **8.** a; **9.** b; **10.** d; **11.** c; **12.** Ultisols; **13.** a; **14.** b; **15.** Spodosol, boreal forest region

Chapter 16

1. c; **2.** c; **3.** d; **4.** a; **5.** d; **6.** b; **7.** b; **8.** c; **9.** a; **10.** b; **11.** b; **12.** a; **13.** b; **14.** c; **15.** c

Chapter 17

1. a; **2.** lichen; **3.** d; **4.** d; **5.** b; **6.** a; **7.** c; **8.** a; **9.** rainforest; **10.** d; **11.** c; **12.** d; **13.** b; **14.** tall-grass prairie; **15.** c

This glossary contains definitions of terms shown in the text in both black and green boldface.

abrasion The erosion of bedrock of a stream channel by impact of particles carried in a stream and by rolling of larger rock fragments over the stream bed; abrasion is also an activity of glacial ice, waves, and wind.

absorption The process in which electromagnetic energy is absorbed when radiation strikes the molecules or particles of a gas, liquid, or solid, raising its energy content.

acid rain Rainwater having an abnormally low pH as a result of air pollution by sulfur oxides and nitrogen oxides.

adiabatic process A process in which the temperature of a parcel of air changes in response to a change in atmospheric pressure.

advection A process whereby a mass of air moves to a new location, bringing its properties, such as temperature or moisture content, along with it.

advection fog Fog produced by condensation within a moist air layer moving over a cold land or water surface.

aerosols Tiny particles present in the atmosphere that are so small and light that the slightest air movements keep them aloft.

aggradation The raising of stream channel altitude by continued deposition of bed load.

air mass An extensive body of air in which temperature and moisture characteristics are fairly uniform over a large area.

albedo The proportion of solar radiation reflected upward from a surface.

Alfisols A soil order consisting of soils of humid and subhumid climates, with high base status and subsurface accumulation of clay.

alluvial fan A gently sloping, conical accumulation of alluvium deposited by a stream emerging from a rock gorge or rock notch.

alluvial terrace A bench-like landform carved by a stream during degradation.

alluvium Sediment laid by a stream that is found in a stream channel or in low parts of a stream valley subject to flooding.

alpine glacier A glacier formed at high elevation, typically flowing down steep mountain slopes and filling valleys below.

alpine tundra A plant formation class within the tundra biome, found at high altitudes above the limit of tree growth and below the vegetation-free zone of rock and snow.

Andisols A soil order that includes soils formed on volcanic ash; often enriched by organic matter, yielding a dark soil color.

anemometer A weather instrument used to indicate wind speed.

angle of repose The maximum slope angle that allows rock, sand, clay, and debris to stay in place without movement; depends on slope material and moisture content.

annuals Plants that live only a single growing season, passing the unfavorable season as a seed or spore.

anticlines The arch in folds of the Earth's crust.

anticyclone A center of high atmospheric pressure where upper troposphere air spirals down to the surface and diverges outward.

aphelion A point on the Earth's elliptical orbit at which the Earth is farthest from the Sun.

aquiclude A relatively impermeable rock or sediment layer that impedes or prevents the movement of ground water.

aquifer A layer of rock or sediment that contains abundant freely flowing ground water.

arable soil Land suitable for vegetation, including forest or crops.

arctic tundra A plant formation class within the tundra biome, consisting of low, mostly herbaceous plants, but with some very small stunted trees, associated with the tundra climate.

arête A sharp, knife-like divide or crest formed between two cirques by alpine glaciation.

arid climates A subtype of the dry climates; extremely dry and supports little or no vegetation cover.

Aridisols A soil order of dry climates with accumulations of carbonates or soluble salts.

artesian well A drilled well in which water rises under hydraulic pressure above the level of the surrounding water table and reaches the surface.

asthenosphere A soft layer of the upper mantle, beneath the rigid lithosphere.

atmosphere The envelope of gases surrounding the Earth, held by gravity.

atmospheric pressure The pressure exerted by the atmosphere because of the force of gravity acting on the overlying column of air.

axis An imaginary straight line through the center of the Earth around which the Earth rotates.

backswamp An area of low, swampy ground on the floodplain of an alluvial river between the natural levee and the bluffs.

badlands An area of deep gullies carved into clay-rich sediment in arid or semiarid regions where sparse or absent vegetation cannot protect the surface from erosion.

barchan dune A type of sand dune with a crescent-shaped base outline, a sharp crest, and a steep lee slip face, with crescent points (horns) pointing downwind.

barometer An instrument that measures atmospheric pressure.

barrier island A long narrow island, built largely of beach sand and dune sand, parallel with the mainland and separated from it by a lagoon.

base cation (base) A positively charged mineral ion (calcium, magnesium, potassium and sodium) dissolved in soil water and needed for plant growth.

base flow That portion of the discharge of a stream contributed by ground water seepage.

base level The lower limit to how far a stream can erode its bed.

bases Positively-charged plant nutrient ions; base cations.

batholith A large, deep-seated body of intrusive igneous rock, usually with an area of surface exposure greater than 100 km^2 (40 mi^2).

beach A thick, wedge-shaped deposit of sand, gravel, or larger stones in the zone of breaking waves.

beach drift The transport of sand on a beach parallel with the shoreline, produced by a succession of landward and seaward water movements at times when swash approaches obliquely.

bedrock The solid rock layer under soil and regolith, which is relatively unchanged by weathering.

biodiversity The variety of biological life supported by the Earth's biosphere or within a region.

biogeochemical cycle The total system of pathways by which a particular type of matter (a given element, compound, or ion, for example) moves through the Earth's ecosystem or biosphere; also called a material cycle or nutrient cycle.

biogeography The study of the distribution patterns of organisms over space and time and of the processes that produced these patterns.

biomass The dry weight of living organic matter in an ecosystem within a specified surface area.

biome The largest recognizable subdivision of terrestrial ecosystems.

biosphere All living organisms of the Earth and the environments with which they interact.

biota Plants and animals, referred to collectively; a list of plants and animals found at a location or in a region.

boreal forest climate The cold climate of the subarctic zone in the northern hemisphere, with long, severe winters and generally low annual precipitation.

braided channel A type of stream channel with many individual threads of flow separated by shallow bars of sand or gravel.

braided stream A stream with a shallow channel carrying multiple threads of flow that subdivide and rejoin repeatedly and continually shift in position.

calcification The accumulation of calcium carbonate in a soil, usually occurring in the B or C horizons.

caldera A large, steep-sided circular depression resulting from the explosion and subsidence of a stratovolcano.

carbon cycle The biogeochemical cycle in which carbon moves through the biosphere, hydrosphere, atmosphere, and lithosphere.

carbonation The process whereby carbonic acid weathers rocks by slowly dissolving minerals.

cation-exchange capacity The ability of soil to hold and exchange cations (positively charged mineral ions); determines fertility of soilwater solution for plants.

chaparral The sclerophyll scrub and dwarf forest plant formation class found throughout the coastal mountain ranges and hills of central and Southern California.

chemical weathering The decomposition or decay of minerals in rocks by chemical reactions with water, oxygen, and acids.

Chinook wind A local wind occurring at certain times to the lee of the Rocky Mountains; a very dry wind with a high capacity to evaporate snow.

cinder cone A small, conical volcano formed when magma is ejected under high pressure from a narrow vent.

circle of illumination A great circle that divides the globe at all times into a sunlit hemisphere and a shadowed hemisphere.

cirque A bowl-shaped depression carved in rock by glacial processes at the upper end of an alpine glacier.

climate The annual cycle of prevailing weather conditions at a given place, based on statistics taken over an extended period.

climatology The science of analyzing weather over the short term or the long term as it varies over time and around the globe.

climograph A graph on which two or more climate variables are plotted for each month of the year.

coastline (coast) The zone in which coastal processes operate or have a strong influence.

col A natural pass or low notch in an arête between opposed cirques.

cold front A moving weather front along which a cold air mass moves underneath a warm air mass, causing the warm air mass to lift rapidly.

colluvium A deposit of sediment or rock particles accumulating from overland flow at the base of a slope and originating from higher slopes where sheet erosion is in progress. (See also *alluvium*.)

competition A form of interaction among plant or animal species in which both draw resources from the same pool.

condensation nucleus A tiny bit of solid matter, or dust particle, in the atmosphere on which water vapor condenses to form a water droplet.

condense The process of change of matter in the gaseous state (e.g., water vapor) to the liquid state (e.g., water) or solid state (e.g., ice).

conduction The process whereby internal energy flows from warm objects to colder ones when placed in contact; neighboring particles exchange energy until all particles have same level of internal energy.

cone of depression The conical configuration of the lowered water table around a well from which water is being rapidly withdrawn.

convection The flow of internal energy that occurs when matter moves from one place to another; at the global scale, convection helps drive the circulation of the Earth's atmosphere and oceans.

convective precipitation Precipitation that is induced when warm, moist air is heated at the ground surface, rises, cools, and condenses to form water droplets, raindrops, and eventually rainfall.

convergent boundary A boundary where two lithospheric plates come together.

coordinated universal time (UTC) The primary time standard by which the world regulates clocks and time; based on atomic time and synchronized with the Earth's rotation.

coral bleaching When corals turn white because of a change in the local environment that expels the algae that live inside the coral structure; leads to coral death.

coral reef A rock-like accumulation of carbonates secreted by corals and algae in shallow water along a marine shoreline.

core The spherical central mass of the Earth, composed largely of iron; consists of an outer liquid zone and an inner solid zone.

Coriolis effect An effect of the Earth's rotation that acts like a force to deflect a moving object on the Earth's surface to the right in the northern hemisphere and to the left in the southern hemisphere.

corrosion The erosion of bedrock of a stream channel (or other rock surface) by chemical reactions between solutions in stream water and mineral surfaces.

counterradiation Longwave atmospheric radiation moving downward toward the Earth's surface.

crevasse A deep crack in the surface ice of a glacier, formed where the glacier moved over a ridge or cliff.

crust The outermost solid layer of the Earth, composed largely of silicate materials.

cryosphere The portion of the hydrosphere in which water is stored as ice.

cut bank The outside bank of the channel wall of a meander, where erosion is continually occurring.

cyclone A center of low atmospheric pressure where surface air converges into a spiral and is uplifted to the upper troposphere.

cyclonic storm An intense weather disturbance within a moving cyclone that generates strong winds, cloudiness, and precipitation.

daylight saving time The practice of adjusting time to achieve longer evening daylight; clocks are set ahead by an hour (sometimes two) for part of the year.

debris flow A stream-like flow of muddy water heavily charged with sediment of a wide range of size grades, including boulders, generated by sporadic torrential rains upon steep mountain watersheds.

decalcification The removal of calcium carbonate from a soil horizon as carbonic acid reacts with carbonate mineral matter.

deciduous Referring to a tree or shrub that sheds its leaves seasonally.

declination The latitude at which the Sun is directly overhead; varies from −23 1/2° (23 1/2° S lat.) to +23 1/2° (23 1/2° N lat.).

deflation The lifting and transport in turbulent suspension by wind of loose particles of soil or regolith from dry ground surfaces.

deforestation The process of forest clearing that, when carried out unsustainably, reduces global biodiversity and contributes to global warming.

delta A sediment deposit built by a stream entering a body of standing water.

dentrification The chemical reduction of nitrates to nitrites, ammonia, and free nitrogen in the soil under anaerobic conditions.

deposition The change of state from a gas (water vapor) to a solid (ice). (See also *sublimation*.)

desalination The process that separates fresh water from sea water.

desert biome A biome of dry climates consisting of thinly dispersed shrubs, grasses, and perennial herbs but lacking in trees.

desert pavement A surface layer of closely fitted pebbles or coarse sand from which finer particles have been removed.

desertification The process through which non-desert areas are converted to desert due to drought or poor land-use practices, such as overgrazing and overpopulation.

dew-point temperature The temperature at which air with a given humidity will reach saturation when cooled without changing its pressure.

discharge The volume of flow moving through a given cross section of a stream in a given unit of time; commonly given in cubic meters (feet) per second.

dispersal The capacity of a species to move from a location of birth or origin to new sites.

distributary A branching stream channel that crosses a delta to discharge into open water.

downcutting The process through which water flowing through a channel erodes segments of the stream back to the average gradient of the stream.

drainage basin (watershed) The total land surface occupied by a drainage system, bounded by a drainage divide or watershed.

drainage divide The imaginary line following a crest of high land such that overland flow on opposite sides of the line enters different streams.

drainage pattern The plan of a network of interconnected stream channels.

drainage system A branched network of stream channels and adjacent land slopes that converge to a single channel at the outlet.

drawdown In describing a water well, the difference in height between base of cone of depression and original water table surface.

drumlin A hill of glacial till, oval or elliptical in basal outline and with smoothly rounded summit, formed by plastering of till beneath moving, debris-laden glacial ice.

dry adiabatic lapse rate The rate at which rising air cools or descending air warms when no condensation is occurring; 10°C per 1000 m (or 5.5°F per 1000 ft).

dry desert A plant formation class in the desert biome consisting of widely dispersed xerophytic plants that may be small, hard-leaved or spiny shrubs, succulent plants (cacti), or hard grasses.

dry midlatitude climate The dry climate of the midlatitude zone, with a strong annual temperature cycle and cold winters.

dry subtropical climate The dry climate of the subtropical zone, transitional between the dry tropical climate and the dry midlatitude climate.

dry tropical climate The climate of the tropical zone, with high temperatures and low rainfall.

earthflow The moderately rapid downhill flow of water-saturated soil, regolith, or weak shale.

earthquake A trembling or shaking of the ground produced by movements along a fault.

easterly wave A weak, slowly moving trough of low pressure within the belt of tropical easterlies; causes a weather disturbance with rain showers.

ebb tide The oceanward flow of tidal current in a bay or tidal stream.

ecological biogeography A subfield of biogeography that examines how relationships between organisms and environment determine when and where organisms are found.

ecological niche The functional role of an organism within its community; its "profession"—how and where it obtains energy, how it influences other species and the environment around it.

ecological succession A sequence of distinctive plant and animal communities that first colonize and then replace each preceding stage, leading to a stable mature forested community, often after clearance by burning, clear cutting, or other agents.

ecosystem A collection of organisms and the environment with which they interact.

El Niño–Southern Oscillation (ENSO) The episodic cessation of the typical upwelling of cold deep water off the coast of Peru; literally, "The Christ Child," for its occurrence in the Christmas season once every few years.

electromagnetic radiation A wave form of energy radiated by any substance possessing internal energy; it travels through space at the speed of light.

electromagnetic spectrum The total wavelength range of electromagnetic energy.

eluviation The downward transport of dissolved or suspended materials to lower soil horizons or out of the soil completely.

endogenic processes The internal Earth processes that create landforms, such as tectonics and volcanism.

Enhanced Fujita Scale A measure of tornado intensity, based on damage to structures and surrounding vegetation.

Entisols A soil order consisting of mineral soils lacking soil horizons that would persist after normal plowing.

entrenched meanders A winding, sinuous valley produced by degradation of a stream with trenching into the bedrock by downcutting.

environmental temperature lapse rate The average rate at which air temperature drops with increasing altitude.

eolian landforms Landforms eroded or deposited by the action of wind.

epicenter The location on the Earth's surface directly above where a fault slipped to produce an earthquake.

equal area projection A type of map projection on which any given area of the Earth's surface is shown to the correct relative areal extent, regardless of position on the globe.

equator The parallel that lies midway between the Earth's poles.

equatorial trough The zone of rising warm air and low pressure centered more or less over the equator and situated between the two belts of trade winds.

equilibrium line The line between the zone of accumulation and the zone of ablation on a glacier where the rate of snow accumulation balances the rate of evaporation and melting.

equinox The instant in time when the Sun is directly overhead at the equator and the circle of illumination passes through both poles.

erg A large expanse of active sand dunes in the Sahara Desert of North Africa.

erosion The wearing away of the Earth's surface materials by wind and water.

esker A narrow, often sinuous embankment of coarse gravel and boulders deposited in the bed of a meltwater stream enclosed in a tunnel within stagnant ice of an ice sheet.

eutrophication The excessive growth of algae and other related organisms in a stream or lake as a result of the input of large amounts of nutrient ions, especially phosphate and nitrate.

evaporate To change state from liquid to vapor.

evapotranspiration The combined water loss to the atmosphere by evaporation from soil and transpiration from plants.

evolution The creation of diversity of life-forms through the process of natural selection.

exfoliation The process of removal of overlying rock load from bedrock by processes of denudation, accompanied by expansion and often leading to the development of sheeting structure.

exogenic processes Landform-making processes that are active at the Earth's surface, such as erosion by water, waves and currents, glacial ice, and wind.

extrusive igneous rock Rock produced by the solidification of lava or ejected fragments of igneous rock.

eye The center of a tropical cyclone in which clear skies and calm winds prevail; a cloud-free vortex produced by the intense spiraling of a tropical storm.

fault A sharp break in rock associated with a slippage of the crustal block on one side of a tectonic plate with respect to the block on the other.

fault-line scarp An erosion scarp developed upon an inactive fault line.

fault scarp A cliff-like surface feature produced by faulting and exposing the fault plane; commonly associated with a normal fault.

field capacity The amount of soil moisture held in soil when it has been saturated by water and allowed to drain under gravity until no more water moves downward.

fiord A narrow, deep ocean inlet that partially fills a glacial trough.

fiord coast A deeply embayed, rugged coast formed by partial submergence of glacial troughs.

firn Granular old snow forming a surface layer in the zone of accumulation of a glacier.

flash flood A flood in which heavy rainfall causes a stream or river to rise very rapidly.

flood basalts Large-scale outpourings of basalt lava to produce thick accumulations of basalt over large areas.

flood stage The designated stream-surface level for a particular point on a stream, above which overbank flooding occurs.

flood tide The incoming, high tide of coastal regions.

floodplain A broad belt of low, flat ground bordering a river channel that floods regularly.

fluvial landforms Landforms shaped by running water.

focus The point within the Earth where a fault first slips to cause an earthquake.

fog A cloud layer at or very close to the Earth's surface.

folds Corrugations of strata caused by crustal compression.

food chain The fundamental levels through which energy flows through an ecosystem as the organisms at each level consume energy from the bodies of organisms in the level below.

food web A diagram of organization of an ecosystem, showing how food energy flows among the various species of the ecosystem.

forest An assemblage of trees growing close together, their crowns forming a layer of foliage that largely shades the ground.

forest biome A biome that includes all regions of forest over the lands of the Earth.

formation classes Subdivisions within a biome based on the size, shape, and structure of the plants that dominate the vegetation.

front The surface or boundary of contact between two different air masses.

frost cracking The physical weathering process that occurs when water penetrates joints or pores in the bedrock and freezes under certain conditions to break the rock by expansion.

Gelisols A soil order of cold regions, including soils underlain by permafrost with organic and mineral materials churned by frost action.

genus A group of closely related species that share a similar evolutionary history.

geographic grid The complete network of intersecting parallels and meridians on the surface of the globe, used to fix the locations of surface points.

geographic information system (GIS) A combination of software, data, and operational organization that provides the capacity to capture and communicate spatial relationships among geographic features, values, and objects in digital databases.

geography The study of the Earth's features, places and phenomena, and how they change over time.

geologic timescale The time frame of geologic events that divides the 4.5 billion years since the Earth formed into eons, eras, and periods.

geomorphic cycle A theory which suggests that landscapes go through stages of development and that the rejuvenation of landscapes arises from tectonic uplift of the land.

geomorphology The study of the Earth's surface processes and landforms, including their history and origin.

geostrophic wind Wind at high levels above the Earth's surface moving parallel to the isobars, at a right angle to the pressure gradient.

glacial drift A general term for all varieties and forms of rock debris deposited by ice sheets.

glacial trough A deep, steep-sided valley shaped by the action of alpine glaciers.

glaciation A single episode or time period in which ice sheets and alpine glaciers formed, spread, and disappeared.

glacier Any large natural accumulation of land ice affected by present or past motion.

gleyzation A condition in which iron and manganese compounds are reduced in the soil, creating streaks of blue; results from a hard pan of calcification that occurs in cold climates.

global climate model A computer-generated model of the physical, chemical, and biological conditions in the Earth's atmosphere, oceans, land, and ice; used to investigate the Earth's past, current and future weather patterns.

Global Positioning System (GPS) The system of satellites and ground instruments that locates the global position of an observer.

graben A crustal block dropped down between two normal faults.

graded stream A stream with a gradient adjusted to carry its average sediment load.

grassland biome A biome consisting largely, or entirely, of herbs, which may include grasses, grass-like plants, and forbs.

great circles The circles formed by passing a plane through the exact center of a perfect sphere; the largest circles that can be drawn on the surface of a sphere.

greenhouse effect Absorption of outgoing longwave radiation by components of the atmosphere and reradiation back to the surface, which raises surface temperatures.

greenhouse gases The collection of gases in the atmosphere that absorb the Earth's emitted longwave radiation, raising temperatures in the Earth's lower atmosphere.

groin A wall or embankment built out into the water at right angles to the shoreline.

ground water The subsurface water in the saturated zone that can move as part of the hydrologic cycle.

gully A deep, V-shaped trench carved by overland flow and/or headward growth of a small stream during advanced stages of accelerated soil erosion.

gyres The large circular ocean current systems centered on the oceanic subtropical high-pressure cells.

habitat A subdivision of the environment according to the needs and preferences of organisms or groups of organisms.

Hadley cell A low-latitude atmospheric circulation cell with rising air over the equatorial trough and sinking air over the subtropical high-pressure belts.

hanging valley A stream valley that has been truncated by marine erosion so as to appear in cross section in a marine cliff, or truncated by glacial erosion so as to appear in cross section in the upper wall of a glacial trough.

heat Flow of internal energy from one substance to another as a result of their differences in temperature.

heat index The measure of apparent temperature based on actual air temperature and relative humidity, designed to account for inability to remove heat through perspiration.

herb A tender plant, lacking woody stems, usually small or low; may be annual or perennial.

heterosphere A region of the atmosphere above about 100 km in which gas molecules tend to become increasingly sorted into layers by molecular weight and electric charge.

highland climate A climate type found in mountains and high plateaus at any latitude, characterized by cooler temperatures due to altitude and greater moisture from orographic precipitation, becoming wetter with elevation.

high-pressure system A weather system created when atmospheric pressure at the surface of the planet is greater than its surrounding environment; air spirals downward and outward (divergence) to form an anticyclone; often associated with cooler, drier air and clear skies.

Histosols A soil order found in cool, poorly drained areas, containing a very high content of organic matter.

homosphere The lower portion of the atmosphere, below about 100 km altitude, in which atmospheric gases are uniformly mixed.

horn A steep-sided peak formed by glacial erosion from three sides.

horst A crustal block pushed up between two normal faults.

hotspot A center of volcanic activity thought to be located over a rising mantle plume.

human geography The part of systematic geography that deals with social, economic, and behavioral processes that differentiate places.

humidity The amount of water vapor in the air.

humus Dark brown to black organic matter on or in the soil, consisting of fragmented plant tissues partly digested by organisms.

hurricane A tropical cyclone of the western North Atlantic and Caribbean Sea.

hydraulic action Stream erosion by the force of impact of flowing water upon the bed and banks of the stream channel.

hydrologic cycle The total plan of movement, exchange, and storage of the Earth's free water in gaseous state, liquid state, and solid state.

hydrolysis The chemical union of water molecules with minerals to form different, more stable mineral compounds; a type of chemical weathering.

hydrosphere The total water realm of the Earth's surface, including the oceans, surface waters of the lands, ground water, and water held in the atmosphere.

hypothesis A logical explanation for a process or phenomenon that allows prediction and testing by experiment.

hypoxia When microorganisms use up oxygen in water to the point at which other organisms, including desirable types of fish, cannot survive.

ice age A span of geological time, usually on the order of one to three million years, or longer, in which glaciations alternate with interglaciations in rhythm with cyclic global climate changes.

ice cap The large, continuous mass of ice at the very tops of mountain ranges characterized by a distinctive dome shape.

ice field Interconnected valley glaciers with protruding rock ridges or summits called nunataks.

ice sheet A large, thick plate of glacial ice that moves outward in all directions.

ice sheet climate The severely cold climate found on the Greenland and Antarctic ice sheets.

igneous rock Rock formed from the cooling of magma.

illuviation The accumulation in a lower soil horizon of materials brought down from upper horizons.

Inceptisols A soil order consisting of soils having weakly developed soil horizons and containing weatherable minerals.

infiltration Absorption and downward movement of precipitation into the soil and regolith.

initial landforms Landforms produced directly by internal Earth processes of volcanism and tectonic activity. Examples: volcano, fault scarp.

insolation The flow rate of incoming solar energy, as measured at the top of the atmosphere.

international date line The imaginary line on the surface of the Earth at roughly the 180th meridian, passing through the middle of the Pacific Ocean, that designates the place where each calendar day begins.

interrupted projection A map projection subdivided into a number of sectors, each of which is centered on a different central meridian.

intertropical convergence zone (ITCZ) A zone of convergence of air masses along the equatorial trough.

intrusive igneous rock An igneous rock body produced by solidification of magma beneath the surface, surrounded by preexisting rock.

isobar A line on a map drawn through all points having the same atmospheric pressure.

isostasy The principle describing the flotation of the lithosphere, which is less dense, on the asthenosphere, which is more dense.

isotherm A line on a map drawn through all points with equal temperatures.

jet stream A high-speed airflow in a narrow band within the upper-air westerlies and along certain other global latitude zones at high altitudes.

jet stream disturbances Broad, wave-like undulations of the jet stream.

jetty A breakwater structure built to keep navigation channels open, but which can disrupt littoral drift.

joint A fracture within bedrock, usually occurring in parallel and intersecting sets of planes.

kame A steep-sided mound of stratified sand and gravel deposited at, or near, the terminus of a glacier.

karst topography A landscape or topography dominated by surface features of limestone solution and underlain by a limestone cavern system.

kettle A depression formed as outwash sand and gravel build up around a block of stagnant ice.

La Niña The cooling of the ocean surface off the western coast of South America, occurring periodically every 4 to 12 years.

lagoon A shallow body of open water lying between a barrier island or a barrier reef and the mainland.

lahar A rapid downslope or downvalley movement of a tongue-like mass of water-saturated tephra (volcanic ash) originating high up on a steep-sided volcanic cone; a variety of mudflow.

lake A body of standing water without an appreciable gradient or current.

land breeze A local wind blowing from land to water during the night.

landforms The specific shapes of the Earth's surface, including those freshly created and those shaped by weathering, erosion, and mass wasting.

landslide A rapid sliding of large masses of bedrock on a steep mountain slope or from a high cliff.

latent heat A flow of heat taken up or released when a substance changes from one state (solid, liquid, or gas) to another.

lateral moraine Parallel ridges of debris deposited along the sides of an alpine glacier.

laterization The process in which oxide-rich clay hardens, yielding a soil horizon of hard clay that can be cut into bricks and dried to form building materials and road foundations.

latitude An angular distance for a point north or south of the equator, as measured from the Earth's center.

lava Magma emerging on the Earth's solid surface, exposed to air or water.

liana A woody vine supported on the trunk or branches of a tree.

lichens Plant forms in which algae and fungi live together, forming a single plant structure that grows close to the ground.

lifting condensation level The level at which a rising air mass cools to the dew point and condensation begins.

lithosphere The strong, brittle, outermost rock layer of the Earth, lying above the asthenosphere.

lithospheric plate A segment of lithosphere moving as a unit, sliding on top of the asthenosphere in contact with adjacent lithospheric plates along plate boundaries.

littoral drift The transport of sediment parallel with the shoreline by the combined action of beach drift and longshore current transport.

loam The soil-texture class in which no one of the three size grades (sand, silt, clay) dominates over the other two.

loess A surface deposit of wind-transported silt.

longitude The angular distance for a point east or west of the prime meridian (Greenwich), as measured from the Earth's center.

longitudinal dunes A class of sand dunes in which the dune ridges are oriented parallel with the prevailing wind.

longshore drift Littoral drift caused by the action of a longshore current, resulting in transport of sediment for long distances along a coast.

longwave radiation Electromagnetic energy in the range of 3 to 5 μm.

low-latitude rainforest The evergreen broadleaf forest of the wet equatorial and tropical climate zones.

low-pressure system A weather system created when air moves toward low pressure in an inward, spiraling motion (convergence), forming a cyclone; often associated with cloudy or rainy weather.

magma Rock in a mobile, high-temperature molten state.

magnitude The amount of energy released by an earthquake, which can be measured by the amplitude of the seismic waves produced.

mantle The rock layer of the Earth beneath the crust and surrounding the core, composed of mafic igneous rock containing silicate minerals.

map A graphic, scaled representative view of the Earth, or any portion of the Earth, as viewed from above, depicting various features of interest.

map projection A system of parallels and meridians representing the Earth's curved surface drawn on a flat surface.

map scale The relationship between distance on a map and distance on the ground, given as a fraction or a ratio.

marine terrace A former shore platform elevated to become a step-like coastal landform.

marine west-coast climate The cool, moist climate of west coasts in the midlatitude zone, usually with abundant precipitation and a distinct winter precipitation maximum.

mass wasting The abrupt or incremental downhill movement of soil, regolith, and bedrock under the influence of gravity.

meander scar A cut into the bank of a bluff or valley wall produced by a meandering stream that has been cut off.

meanders The sinuous bends of a graded stream flowing in the alluvial deposit of a floodplain.

medial moraine A long, narrow deposit of fragments on the surface of a glacier; created by the merging of lateral moraines when two glaciers join into a single stream of ice flow.

Mediterranean climate The climate type of the midlatitudes, characterized by the alternation of a very dry summer and a mild, rainy winter.

meridian A north–south line on the Earth's surface that connects the poles.

mesosphere An atmospheric layer of upwardly diminishing temperature, situated above the stratopause and below the mesopause.

metamorphic rock Rock altered in physical or chemical composition by heat, pressure, or other endogenic processes taking place at a substantial depth below the surface.

microclimates Local atmospheric zones where the climate differs from surrounding areas.

midlatitude cyclone A traveling cyclone of the midlatitudes that involves repeated interaction of cold and warm air masses along sharply defined fronts.

midlatitude deciduous forest A plant formation class within the forest biome dominated by tall, broadleaf deciduous trees, found mostly in the moist continental climate and marine west-coast climate.

midoceanic ridge The central belt of submarine mountains where two plates separate.

millibar A unit of atmospheric pressure; one-thousandth of a bar. Bar is a force of one million dynes per square centimeter.

mineral A naturally occurring inorganic substance, usually having a definite chemical composition and a characteristic atomic structure.

Moho The boundary that separates the Earth's crust from the mantle, short for Mohorovičić discontinuity.

moist adiabatic lapse rate The rate at which rising air is cooled by expansion when condensation is occurring; ranges from 4 to 9°C per 1000 m (2.2 to 4.9°F per 1000 ft).

moist continental climate The moist climate of the midlatitudes with strongly defined winter and summer seasons and adequate precipitation throughout the year.

moist subtropical climate The moist climate of the subtropical zone, characterized by abundant rainfall and a strong seasonal temperature cycle.

Mollisols A soil order found in semiarid to subhumid midlatitude grasslands, characterized by a loose texture and high base status.

monsoon A seasonal wind in southern Asia formed by southwesterly flow across the region, which draws the ITCZ well to the north, changing the pressure pattern with the seasons (wet monsoons in summer and dry monsoons in winter).

monsoon and trade-wind coastal climate The moist climate of low latitudes that shows a strong rainfall peak in the high-Sun season and a short period of reduced rainfall in the low-Sun season.

monsoon forest A formation class within the forest biome consisting in part of deciduous trees adapted to a long dry season in the wet–dry tropical climate.

moraine The accumulation of rock debris carried by an alpine glacier of an ice sheet and deposited to become a depositional landform.

mountain breeze The daytime movement of air up the gradient of valleys and mountain slopes; alternating with nocturnal valley winds.

mudflow A flowing mixture of water and soil or regolith that flows rapidly downhill.

mutualism The interaction between two different species, in which both species benefit.

natural levee The belt of higher ground paralleling a meandering river on both sides of a stream channel and built up by deposition of sediment during flooding.

natural selection The selection of organisms by their survival in the environment in a process similar to selection of plants or animals by breeders.

natural vegetation Stable, mature plant cover, largely free from the influences of human activities.

needleleaf forest A plant formation class within the forest biome, consisting largely of needleleaf trees.

net primary production The rate at which carbohydrate is accumulated in the tissues of plants within a given ecosystem; units are kilograms of dry organic matter per year per square meter of surface area.

nickpoint A portion of a stream where gradient changes abruptly, such as a waterfall or rapids.

nitrogen cycle The biogeochemical cycle in which nitrogen moves through the biosphere by the processes of nitrogen fixation and denitrification.

nitrogen fixation The process through which microorganisms supply nitrogen to plants via nitrogen-fixing bacteria living in the soil; these bacteria fix nitrogen into ammonia, which is then assimilated by plants.

normal fault A variety of fault in which the fault plane inclines (dips) toward the downthrown block and a major component of the motion is vertical.

occluded front A weather front along which a fast-moving cold front overtakes a warm front, forcing the warm air mass aloft.

ocean current A persistent, dominantly horizontal flow of water.

ocean tide The periodic rise and fall of the ocean level induced by gravitational attraction between the Earth and Moon in combination with Earth rotation.

oceanic trench A narrow, deep depression in the seafloor representing the line of subduction of an oceanic lithospheric plate beneath the margin of a continental lithospheric plate.

orogeny A major episode of tectonic activity resulting in strata being deformed by folding and faulting.

orographic precipitation Precipitation that is induced when moist air is forced to rise over a mountain barrier.

outwash plain A flat, gently sloping plain built up of sand and gravel by the aggradation of meltwater streams in front of the margin of an ice sheet.

overland flow The motion of a surface layer of water over a sloping ground surface at times when the infiltration rate is exceeded by the precipitation rate; a form of runoff.

overthrust fault A fault characterized by the overriding of one crustal block (or thrust sheet) over another along a gently inclined fault plane; associated with crustal compression.

ox-bow lake A crescent-shaped lake representing the abandoned channel left by the cutoff of a meander.

oxidation The chemical union of free oxygen with metallic elements in minerals, which makes minerals unstable and causing rock to degrade in strength and crumble.

Oxisols A soil order consisting of very old, highly weathered soils of low latitudes, with an oxic horizon and low base status.

ozone layer A layer of the stratosphere containing ozone (O_3), which absorbs ultraviolet radiation from the Sun.

Pangea A hypothetical parent continent, enduring until near the close of the Mesozoic Era, consisting of the continental shields of Laurasia and Gondwana joined into a single unit.

parabolic dunes Isolated low sand dunes of parabolic outline, with points directed into the prevailing wind.

parallel An east–west circle on the Earth's surface, lying in a plane parallel to the equator.

parasitism A form of negative interaction between species in which a small species (parasite) feeds on a larger one (host) without necessarily killing it.

parent material The inorganic material base from which soil is formed.

pascal (Pa) A metric unit of pressure, defined as a force of one newton per square meter ($1 N/m^2$); symbol, Pa; 100 Pa = 1 mb, 10^5 Pa = 1 bar.

ped An individual natural soil aggregate.

percolation The slow passage of liquid through a filtering system, such as the movement of water down through soil.

perennials Plants that live for more than one growing season.

periglacial In an environment located in cold climates or near the margins of alpine glaciers or large ice sheets.

perihelion The point on the Earth's elliptical orbit at which the Earth is nearest to the Sun.

permafrost Soil, regolith, and bedrock at a temperature that remains below 0°C (32°F) on a perennial basis.

photosynthesis The production of carbohydrates from water and carbon dioxide, using light energy and releasing oxygen.

physical geography The study of the features of the Earth's living and nonliving systems, such as its landscapes and the natural processes that formed them, as well as its planetary origins.

physical weathering The breakup of massive rock (bedrock) by physical forces at or near the Earth's surface.

pioneer stage The first stage of an ecological succession.

plane of the ecliptic The imaginary plane in which the Earth's orbit lies.

plant life-form The characteristic physical structure, size, and shape of a plant or of an assemblage of plants.

plate tectonics The theory of tectonic activity, which deals with lithospheric plates and their motions.

playa A flat land surface underlain by fine sediment or evaporite minerals deposited from shallow lake waters in a dry climate in the floor of a closed topographic depression.

pluton Any body of intrusive igneous rock that has solidified below the surface, enclosed in preexisting rock.

podzolization A process in which soils are depleted of bases, becoming acidic; occurring in cold, moist climates under conifer forests where organic matter decomposes slowly and releases organic acids.

point bar Sediment accumulation on the inside of a meandering stream bend, where flow is slower.

polar front The boundary between cold polar air masses and warm subtropical air masses.

polar-front jet stream A jet stream found along the polar front, where cold polar air and warm tropical air are in contact.

pole One of the two points on the Earth's surface where the axis emerges.

pore A small space between grains in a rock surface that makes the rock vulnerable to weathering.

precipitation Particles of liquid water or ice that fall from the atmosphere and reach the ground.

predation A form of negative interaction among animal species in which one species (predator) kills and consumes the other (prey).

pressure gradient Change of atmospheric pressure measured along a line at right angles to the isobars.

primary succession Ecological succession that begins on a newly constructed ground base, such as a sand dune or stream deposit.

prime meridian The reference meridian of zero longitude; universally accepted as the Greenwich meridian.

progradation The shoreward building of a beach, bar, or sandspit by addition of coarse sediment carried by littoral drift or brought from deeper water offshore.

pyroclastic material Collective name for ejected particles from an erupting volcano.

radiation fog Fog produced by radiation cooling of air at or just above the surface.

radiometric dating A method of determining the geologic age of a rock or mineral by measuring the proportions of certain of its elements in their different isotopic forms.

regolith The layer of rock and mineral particles that lies above bedrock.

rejuvenated stream A graded stream that has begun a new cycle of downcutting due to tectonic uplift, creating new, steeper stream segments.

relative humidity The ratio of water vapor present in the air to the maximum quantity possible for saturated air at the same temperature.

remote sensing The use of technology (on the ground or on ships, aircraft, satellites and other spacecraft) to record observations from a distance.

respiration The oxidation of organic compounds by organisms that power bodily functions.

retrogradation The cutting back (retreat) of a shoreline, beach, marine cliff, or marine scarp by wave action.

reverse fault A type of fault in which one fault block rides up over the other on a steep fault plane.

revolution The motion of a planet in its orbit around the Sun, or of a planetary body around a planet.

ria coast A deeply embayed coast formed by partial submergence of a land mass previously shaped by fluvial denudation.

rifting The pulling apart of the Earth's crust and lithosphere due to movement of tectonic plates away from each other.

rill A small channel in the soil or regolith caused by erosion from heavy rain.

rock A solid aggregate of two or more minerals; divided into three major classes: igneous, sedimentary and metamorphic.

rockfall The most visible form of mass wasting, in which rocks fall down steep slopes or cliff sides, bringing with them soil or regolith.

Rossby waves Horizontal undulations in the flow path of the upper-westerlies; also known as jet stream disturbances.

rotational slide A type of landslide in which the mass rotates as it slides downslope, resulting in displacement of soil from the top of the rupture.

runoff The flow of water leaving an area through surface, subsurface, or ground-water flow.

Saffir-Simpson scale A scale of tropical cyclone intensity, based on wind speeds and central atmospheric pressure of the storm, ranging from 1 (minimal damage) to 5 (intense).

salinization The precipitation of soluble salts within the soil.

saltation Leaping, impacting, and rebounding of sand grains transported over a sand or pebble surface by wind.

salt-crystal growth A form of weathering in which rock is disintegrated by the expansive pressure of growing salt crystals during dry weather periods when evaporation is rapid.

sand dune A hill or ridge of loose, well-sorted sand shaped by wind and usually capable of downwind motion.

Santa Ana winds Easterly winds, often hot and dry, that blow from the interior desert region of Southern California and pass over the coastal mountain ranges to reach the Pacific Ocean.

saturated zone The zone beneath the land surface in which all pores of the bedrock or regolith are filled with ground water.

savanna biome A biome that consists of a combination of trees and grassland in various proportions.

scarification A general term for artificial excavations and other land disturbances produced for extracting or processing mineral resources.

scattering The process by which particles and molecules deflect incoming solar radiation in different directions on collision; atmospheric scattering can redirect solar radiation back to space.

scientific method The formal process that a scientist uses to solve a problem, which involves first observing and formulating a hypothesis and then testing and evaluating results.

sclerophyll forest A plant formation class of the forest biome, consisting of low sclerophyll trees and often including sclerophyll woodland or scrub, associated with regions of Mediterranean climate.

sea breeze A local wind blowing from sea to land during the day.

sea cliff A rock cliff shaped and maintained by the weathering and erosion of breaking waves at its base.

secondary succession Ecological succession beginning on a previously vegetated area that has been recently disturbed by such agents as fire, flood, windstorm, or humans.

sedimentary rock Rock formed from the accumulation of sediment.

semiarid climate Dry climate subtype of the dry climates exhibiting a short wet season supporting the growth of grasses and annual plants.

semidesert A plant formation class of the desert biome; characterized by low grasses, sagebrush and ecosystems with adaptations for extreme dryness and short, intense rain episodes.

sensible heat A flow of heat that results in a temperature change of an object or its surroundings; it is measured by a thermometer.

sequential landforms Landforms produced by external Earth processes in the total activity of denudation; examples: gorge, alluvial fan, floodplain.

sheet flow Overland flow taking the form of a continuous thin film of water over a smooth surface of soil, regolith, or rock.

shield volcano A low, often large, dome-like accumulation of basalt lava flows emerging from long, radial fissures on flanks.

shoreline The shifting line of contact between water and land.

shortwave radiation Electromagnetic energy in the range of 0.2 to 3 μm.

sinkhole A surface depression in limestone, caused by the collapse of underground caverns formed by the solution of limestone in ground water.

small circles Circles on the surface of the Earth formed by a plane that does not pass through the exact center of the Earth.

soil A natural terrestrial surface layer that contains minerals and organic matter and that supports, or is capable of supporting plants, animals, and microbial communities.

soil colloid An extremely small mineral or organic particle that can remain suspended in water indefinitely.

soil creep The extremely slow downhill movement of soil and regolith.

soil enrichment Additions of materials to the soil body.

soil horizon A distinctive layer of soil, more or less horizontal, set apart from other layers by differences in physical or chemical composition.

soil order A soil class that forms the highest category in soil classification.

soil profile A display of soil horizons on the face of a freshly cut vertical exposure through the soil.

soil texture A descriptive property of the mineral portion of soil based on the proportions of sand, silt, and clay.

soil-water belt The region of the soil in which infiltrating rain water is stored temporarily and available to plants.

soil water budget The accounting system evaluating the daily, monthly, or yearly amounts of precipitation, evapotranspiration, soil-water storage, water deficit, and water surplus.

solar constant The average flow rate of solar energy received by the Earth, measured just beyond the outer limits of the Earth's atmosphere.

solifluction A type of earthflow, found in arctic permafrost regions caused by soil that is saturated with water and then deformed into terraces.

solstice The instant in time when the Earth's axis of rotation is fully tilted (23 1/2°) either toward or away from the Sun.

species A collection of individual organisms that are capable of interbreeding to produce fertile offspring.

specific heat A physical property of a material that describes the amount of heat energy in joules required to raise the temperature of one gram of the substance by 1° C.

specific humidity The mass of water vapor contained in a parcel of air, expressed as grams of water vapor per kilogram of air (g/kg).

spit A long, finger-like embankment of sand extending from the shore to the open water of a bay, formed by littoral drift.

Spodosols A soil order consisting of soils with a spodic horizon, an albic horizon, with low base status, and lacking in carbonate materials.

spreading boundary A boundary where two lithospheric plates move apart.

stalactite An icicle-shaped formation created when water containing dissolved minerals drips through the ceiling of an underground cavern.

stalagmite An upward-pointing formation created when water containing dissolved minerals hits the floor of an underground cavern.

star dune A large, isolated sand dune with radial ridges culminating in a peaked summit; largely found in the deserts of North Africa and the Arabian Peninsula.

stationary front A transition zone between two nearly stationary air masses of different properties.

steppe Semiarid grassland occurring largely in dry continental interiors.

storm surge A rapid rise of coastal water level accompanying the onshore arrival of a tropical cyclone.

strata Layers of sediment or sedimentary rock in which individual beds are separated from one another along bedding planes.

stratified drift Glacial drift made up of sorted and layered clay, silt, sand, or gravel deposited from meltwater in stream channels, or in marginal lakes close to the ice front.

stratosphere The layer of atmosphere directly above the troposphere, where temperature slowly increases with height.

stratovolcano A volcano constructed of multiple layers of lava and volcanic ash.

stream A long, narrow body of flowing water moving along a channel to lower levels under the force of gravity.

stream capacity The maximum stream load of solid matter that can be carried by a stream for a given discharge.

stream gradient The rate of descent to lower elevations along the length of a stream channel.

stream load The solid matter carried by a stream in dissolved form, in suspension, and as bed load.

strike-slip faults A variety of fault on which the motion is dominantly horizontal along a near-vertical fault plane.

subduction Descent of the edge of a lithospheric plate under an adjoining plate and into the asthenosphere.

sublimation The process of change of ice (solid state) to water vapor (gaseous state). (See also *deposition*.)

subsidence Sinking of the landscape caused by collapse of empty pore spaces in subsurface sediment when ground water is removed.

subsolar point The point on the Earth's surface at which solar rays are perpendicular to the surface.

subtropical broadleaf evergreen forest A formation class of the forest biome composed of broadleaf evergreen trees; occurs primarily in the regions of the moist subtropical climate.

subtropical evergreen forest A subdivision of the forest biome composed of both broadleaf and needleleaf evergreen trees.

subtropical high-pressure belt An area of persistent high atmospheric pressure centered at about 30° N and 30° S latitude.

subtropical jet stream The jet stream that occurs at the tropopause, just above subtropical high-pressure cells in the northern and southern hemispheres.

subtropical needleleaf evergreen forest A formation class of the forest biome composed of needleleaf evergreen trees occurring in the moist subtropical climate of the southeastern United States; also referred to as the southern pine forest.

symbiosis A close, coexisting interaction between species that is beneficial to at least one of the species and usually does not harm the other.

synclines The U-shaped base of folds in the Earth's crust.

tall-grass prairie A formation class of the grassland biome that consists of tall grasses with broad-leaved herbs.

talus An accumulation of rock fragments falling from a rock face or cliff, collecting at the bottom in a cone-shaped pile.

tarn A small lake occupying a rock basin in a cirque of glacial trough.

temperature A measure of the level of internal energy of a substance (the internal motion of atoms and molecules that make up all matter).

temperature gradient The rate of temperature change along a selected line or direction.

temperature inversion A state of the atmosphere in which air temperature increases with elevation.

terminal moraine The irregular jumbled heaps of debris, sand, and gravel that accumulates at the front edge of a glacier or ice sheet.

terrestrial ecosystem An ecosystem of land plants and animals found on the land surface of the continents.

theory A hypothesis that has been tested and is strongly supported by experimentation, observation, and scientific evidence.

thermal action Cracking of rocks by unequal stresses between mineral grains that develop as rocks heat and cool.

thermokarst In arctic environments, an uneven terrain produced by thawing of the upper layer of permafrost, with settling of sediment and related water erosion; often occurs when the natural surface cover is disturbed by fire or human activity.

thermosphere An atmospheric layer of upwardly increasing temperature, lying above the mesopause.

thorntree–tall-grass savanna A plant formation class, transitional between the savanna biome and the grassland biome, consisting of widely scattered trees in an open grassland.

thunderstorm An intense local storm associated with a tall, dense cumulonimbus cloud in which there are very strong updrafts of air.

till A heterogeneous mixture of rock fragments ranging in size from clay to boulders, deposited beneath moving glacial ice or directly from the melting in place of stagnant glacial ice.

till plain The undulating, plain-like land surface underlain by glacial till.

tornado A small, very intense wind vortex with extremely low air pressure in the center, formed below a dense cumulonimbus cloud in proximity to a cold front.

trade winds Surface winds in low latitudes that move air toward the equator, blowing from the northeast in the northern hemisphere (northeast trade winds) or the southeast in the southern hemisphere (southeast trade winds).

transform boundary A boundary where two lithospheric plates slide past each other.

translational slide A type of landslide in which the mass moves out and down the slope with little or no rotation.

transpiration The process by which plants lose water to the atmosphere by evaporation through leaf pores.

transverse dunes A field of wave-like sand dunes with crests running at right angles to the direction of the prevailing wind.

trophic level The position an organism occupies within a food web.

tropical cyclone An intense traveling cyclone of tropical and subtropical latitudes, accompanied by high winds and heavy rainfall.

tropopause The boundary between the troposphere and the stratosphere.

troposphere The lowest layer of the atmosphere, where human activity and most weather takes place.

true-shape projection A type of map projection that shows the true shape of areas, even though their size may be distorted (e.g., the Mercator projection).

tsunami A train of sea waves triggered by an earthquake or another seafloor disturbance.

tundra biome A biome of the cold regions of arctic tundra and alpine tundra, consisting of grasses, grass-like plants, flowering herbs, dwarf shrubs, mosses, and lichens.

tundra climate The cold climate of the arctic zone, which has eight or more months of frozen ground.

typhoon A tropical cyclone of the western North Pacific and coastal waters of Southeast Asia.

Ultisols A soil order that is weathered and leached of nutrients, common in humid subtropical climates throughout Southeast Asia and the East Indies.

uniformitarianism The idea that the same geologic processes we can observe today have operated since the beginning of the Earth's history.

urban heat island An area at the center of a city that has a higher temperature than surrounding regions.

valley breeze Air movement at night down the gradient of valleys and the enclosing mountainsides; alternating with daytime mountain winds.

Vertisols A soil order consisting of soils of the subtropical zone and the tropical zone with high clay content, developing deep, wide cracks when dry.

volcano A conical or dome-shaped structure built by accumulation of lava flows and volcanic ash.

warm front A moving weather front along which a warm air mass slides over a cold air mass, producing stratiform clouds and precipitation.

water table The upper limit of the body of ground water, marking the boundary between the saturated and unsaturated zones.

weather The physical state of the atmosphere at one location and at one time, as characterized by temperature, atmospheric moisture, precipitation, and winds.

weather system A recurring pattern of atmospheric circulation that leads to distinctive weather events; often associated with the motion of air masses.

weathering All the processes that physically disintegrate or chemically decompose a rock at or near the Earth's surface.

westerly winds Winds produced as air spirals outward and poleward from subtropical high pressure cells, producing southwesterly winds in the northern hemisphere and northwesterly winds in the southern hemisphere.

wet-dry tropical climate The climate of the tropical zone characterized by a very wet season alternating with a very dry season.

wet equatorial climate The moist climate of the equatorial zone, with abundant precipitation and uniformly warm temperatures throughout the year.

wilting point The quantity of stored soil water, below which the foliage of plants not adapted to drought will wilt.

wind Air motion, moving predominantly horizontally over the Earth's surface.

wind chill index Index used to express how cold the air feels to the skin when the wind is blowing.

wind shear A change in wind speed or direction with height.

wind vane A weather instrument used to indicate wind direction.

woodland A plant formation class, transitional between forest biome and savanna biome, consisting of widely spaced trees with canopy coverage between 25 and 60 percent.

xerophytic Plant forms that have adapted to dry environments with small leaves and thorns.

zone of ablation The lower portion of a glacier, where increased evaporation and melting exceed snowfall.

zone of accumulation The upper portion of a glacier, where snowfall exceeds evaporation and melting, snow accumulates, and new glacial ice is formed.

Chapter 1

Chapter Opener: Brenda Kean/Alamy; figure 1.4: MODIS Rapid Response Team, NASA-GSFC/Courtesy NASA; figure 1.5b: Marc Moritsch/NG Image Collection; figure 1.5c: Courtesy NASA; figure 1.9: Dennis di Cicco/Corbis; figure 1.9 inset: NG Maps; figure 1.11: Frank Zullo/Photo Researchers, Inc.; figure 1.17: Charles Colbeck. Longmans, Green; New York; London; Bombay. 1905; figure 1.18: NG Maps; page 25: Courtesy NASA; figure 1.23: US Geological Survey (USGS)Courtesy NASA/Goddard Space Flight Center; figure 1.24: Jenny Mottar/Courtesy NASA; page 26: NG Image Collection; figure 1.25a: USGS; figure 1.25b: Geodatabase Mexico figure and text courtesy of Esri. Data layers courtesy of Sistemas de Informacion Geografica, S.A. de C.V. and NASA. Reproduced by permission. All rights reserved; page 29: Reto Stîckli and Robert Simmon/Courtesy NASA; page 31: Courtesy NASA; page 32: Courtesy NASA; page 33: Courtesy NASA.

Chapter 2

Chapter Opener: Hyoung Chang/The Denver Post/Zuma Press; figure 2.11a: Kike Calvo/V&W/The Image Works; figure 2.11b: Richard Nowitz/NG Image Collection; figure 2.11c: George Steinmetz/NG Image Collection; figure 2.11d: Tim Laman/NG Image Collection; figure 2.12: SPL/Photo Researchers, Inc.; page 49: Science Central; figure 2.14a: John Dunn/Arctic Light/NG Image Collection; figure 2.14b: Jeremy Woodhoue/Masterfile; page 56: UNEP Assesment Report; page 57: Jesse Allen, Earth Observatory, using data obtained courtesy of the MODIS Rapid Response team/Courtesy NASA; figure 2.19a: Courtesy NOAA; figure 2.19b: Courtesy NOAA; figure 2.19c: Courtesy NASA; page 61: Courtesy Argon ST, Inc.

Chapter 3

Chapter Opener: Alberto Garcia/©Corbis; page 65 inset: NG Maps; figure 3.8a: Melissa Farlow/NG Image Collection; figure 3.8b: George F. Mobley/NG Image Collection; figure 3.9a: Frank & Helen Schreider/NG Image Collection; figure 3.9b: Richard Nowitz/NG Image Collection; figure 3.9c: Michael Poliza/NG Image Collection; figure 3.9d: Gerd Ludwig/INSTITUTE; figure 3.10: Colin Monteath/Minden Pictures/NG Image Collection; page 79: Courtesy NASA; figure 3.17a: MODIS/Courtesy NASA; figure 3.17b: Goddard Space Flight Center Scientific Visualization Studio/Courtesy NASA; figure 3.18a: George F. Mobley/NG Image Collection; page 85: EPA/Berit Roald/Landov LLC; page 87: National Geographic; figure 3.23a: National Snow and Ice Data Center; figure 3.23b: imagebroker/Alamy; figure 3.23c: Courtesy NOAA; figure 3.23d: Mitsuaki Iwago/Minden Pictures/NG Image Collection; figure 3.23e: Chris Newbert/Minden Pictures/NG Image Collection; page 93 (center): Field, William Osgood. 1941. Muir Glacier: From the Glacier Photograph Collection. Boulder, Colorado USA: National Snow and Ice Data Center/World Data Center for Glaciology; page 93 (bottom): Molnia, Bruce F. 2004. Muir Glacier: From the Glacier Photograph Collection. Boulder, Colorado USA: National Snow and Ice Data Center/World Data Center for Glacioulogy.

Chapter 4

Chapter Opener: Daniel Berehulak/Getty Images; figure 4.2a: Bill Curtsinger/NG Image Collection; figure 4.2c: Dr. Maurice G. Hornocker/NG Image Collection; figure 4.2b: George Steinmetz/NG Image Collection; figure 4.3: SPL/Photo Researchers; page 103 (left): Dawn Kish/NG Image Collection; page 103 (right): Raymond Gehman/NG Image Collection; figure 4.9: Raymond Gehman/NG Image Collection; page 111: National Geographic; figure 4.10a: Norbert Rosing/NG Image Collection; figure 4.10b: Gordon Wiltsie/NG Image Collection; figure 4.10c: Jason Edwards/NG Image Collection; figure 4.10d: Gordon Wiltsie/NG Image Collection; figure 4.11a: Norbert Rosing/NG Image Collection; figure 4.11b: Thomas Marent/Minden Pictures/NG Image Collection; figure 4.11c: Alaska Stock Images/NG Image Collection; figure 4.12b: Tom Murphy/NG Image Collection; figure 4.12c: James P. Blair/NG Image Collection; figure 4.12d: Medford Taylor/NG Image Collection; figure 4.12e: Medford Taylor/NG Image Collection; figure 4.12f: Richard Olsenius/NG Image Collection; figure 4.15a: Courtesy NOAA; figure 4.15b: Courtesy NOAA; figure 4.20a: Gerd Ludwig/INSTITUTE; figure 4.20b: Al Petteway/NG Image Collection; page 121: ISCCP/NASA Images; page 123 (center left): Gordon Wiltsie/NG Image Collection; page 123 (bottom right): Al Petteway/NG Image Collection; page 125: USGS; page 127: Norbert Rosing/NG Image Collection.

Chapter 5

Chapter Opener: Greg Dale/NG Image Collection; figure 5.2: Nick Caloyianis/NG Image Collection; figure 5.4: Courtesy Taylor Instrument Company and Wards Natural Science Establishment, Rochester, New York; figure 5.13b: Steve Raymer/NG Image Collection; page 142: Courtesy NASA; figure 5.14: Ralph Lauer/Zuma Press/NewsCom; figure 5.21: NASA/JPL; figure 5.24: NG Maps; figure 5.26a: NG Maps; figure 5.26b: NG Maps; figure 5.26c: NG Maps; figure 5.26d: NG Maps; page 154: National Geographic; page 156: NG Maps; page 157 (top): NG Maps; page 157 (bottom): Courtesy Otis B. Brown, Robert Evans, and M. Carle, University of Miami, Rosenstiel School of Marine and Atmospheric Science, Florida, and NOAA/Satellite Data Services Division.

Chapter 6

Chapter Opener: Tyrone Turner/NG Image Collection; figure 6.1e: Jason Edwards/NG Image Collection; figure 6.1b: David Doubilet/NG Image Collection; figure 6.1a: Melissa Farlow/NG Image Collection; figure 6.1d: GOES/NASA; figure 6.1c: John Dunn/NG Image Collection; page 167 (bottom left): Mark Downey/Masterfile; page 167 (bottom right): The Image Bank/Getty Images; figure 6.7: Robert F. Bukaty/AP/Wide World Photos; figure 6.8: Courtesy NASA/MODIS Rapid Response Team at Goddard Space Flight Center; page 173: National Geographic; figure 6.11: NG Maps; figure 6.13: NG Image Collection; figure 6.14a: Ty Harrington/FEMA News; figure 6.14b: Andrea Booher/FEMA; figure 6.14c: FEMA News Photo; figure 6.14d: Mark Wolfe/FEMA; figure 6.14e: FEMA News Photo; figure 6.15: Vincent J. Musi/NG Image Collection; figure 6.17: Courtesy NOAA;

figure 6.19: Image courtesy of the Image Science & Analysis Laboratory, NASA Johnson Space Center; page 184: Courtesy NOAA; figure 6.22b: Peter Carsten/NG Image Collection; figure 6.22c: Priit Vesilind/NG Image Collection; figure 6.22e: Win Henderson/FEMA; figure 6.22f: Greg Henshal/FEMA; figure 6.22a: AFP Photo/Yasser Al-Zayyat/Getty Images; figure 6.22d: John McCombe/Getty Images; page 189: Alen Kadr/Shutterstock.

Chapter 7

Chapter Opener: Paul Nicklan/NG Image Collection; figure 7.1a: John Dunn/NG Image Collection; figure 7.1b: Jim Richardson/NG Image Collection; figure 7.1c: George F. Mobley/NG Image Collection; page 236: Ashley Cooper/Alamy Images; figure 7.3: Barbara Summey, NASA/GSFC Visualization Analysis Laboratory; figure 7.3d: Thomas Marent/Minden Pictures/NG Image Collection; figure 7.3a: John Dunn/NG Image Collection; figure 7.3c: Rich Reid/NG Image Collection; figure 7.3e: Peter Carsten/NG Image Collection; figure 7.3b: James P. Blair/NG Image Collection; figure 7.6a: Jason Edwards/NG Image Collection; figure 7.6b: NOAA images; figure 7.6c: Yamanaka, Cindy/NG Image Collection; figure 7.7f: Kari Niemelainen/Alamy; figure 7.7e: James P. Blair/NG Image Collection; figure 7.7b: Tim Fitzharris/Minden Pictures, Inc.; figure 7.7d: Stephen Alvarez/NG Image Collection; figure 7.7e: Will & Deni McIntyre/Photo Researchers, Inc.; figure 7.7a: J. Baylor Roberts/NG Image Collection; figure 7.9a: Justin Guariglia/NG Image Collection; figure 7.9b: Maciej Wojtkowiak/Alamy Images; figure 7.11b: William Albert Allard/NG Image Collection; figure 7.12: Raymond Gehman/NG Image Collection; figure 7.13: Sajeev Krishnan/Alamy Limited; figure 7.15: AP/Wide World Photos; figure 7.17: Roger Wood/Corbis; page 214 (left): Remi Benali/The Image Bank/Getty Images, Inc.; page 214 (right): Earth Observatory/NASA Images; figure 7.18: Karsten Wrobel/Alamy Images; figure 7.19: Richard Cummins/Superstock; figure 7.21: Raymond Gehman/NG Image Collection; figure 7.23: Sisse Brimberg/NG Image Collection; figure 7.24: David Alan Harvey/NG Image Collection; figure 7.25: Andre Jenny/The Image Works; figure 7.27: Brand X/Superstock; figure 7.29: Gerd Ludwig/INSTITUTE; figure 7.30: Doug Cheeseman/Alamy Images; figure 7.32: Hinrich Baesemann/Landov LLC; unnumbered figure 7.32: National Geographic; figure 7.33: Maria Stenzel/NG Image Collection; figure 7.35: NASA Images; figure 7.37 (2): Earth Observatory/NASA Images; figure 7.37 (4): GeoMatrix© Earth Visualization System, GeoFusion, Inc.; figure 7.37 (5): NASA Earth Observatory, based on IPCC Fourth Assessment Report (2007).

Chapter 8

Chapter Opener: Martin Gray/NG Image Collection; figure 8.1: NG Maps; page 242 (left): Copyright Bruce Watson, Rensselaer Polytechnic Institute; page 242 (right): DeAgostini/Getty Images, Inc.; figure 8.1a: David Hutt/Alamy; figure 8.1b: courtesy Didier Descouens; figure 8.1d: Gordon Wiltsie/NG Image Collection; figure 8.1e: Tui De Roy/Minden Pictures, Inc.; figure 8.1c: iStockphoto; figure 8.6: National Geographic; figure 8.8: Michael Nichols/NG Image Collection; figure 8.9b: Jerry Monkman/Nature Picture Library; figure 8.9c: Corbis Images; figure 8.10a: J.D. Griggs/USGS; figure 8.10b: Douglas Peebles Pho-

tography/Alamy; figure 8.11a: Courtesy Brian J. Skinner; figure 8.11d: Courtesy Brian J. Skinner; figure 8.11f: Courtesy Brian J. Skinner; figure 8.13b: Reni Burri/Magnum Photos, Inc.; figure 8.14a: Bill Hatcher/NG Image Collection; figure 8.14b: Walter Meayers Edwards/NG Image Collection; figure 8.14c: Norbert Rosing/NG Image Collection; figure 8.14d: Melissa Farlow/NG Image Collection; figure 8.15a: Bill Hatcher/NG Image Collection; figure 8.15b: Gordon Wiltsie/NG Image Collection; figure 8.15c: Walter Meayers Edwards/NG Image Collection; page 253: NG Maps; figure 8.16a: Norbert Rosing/NG Image Collection; figure 8.16b: Eric Condliffe, University of Leeds Electron Optics Image Laboratory; figure 8.17a: Melissa Farlow/NG Image Collection; figure 8.18a: Ralph Lee Hopkins/NG Image Collection; figure 8.18b: Raymond Gehman/NG Image Collection; figure 8.18c: Robert Sisson/NG Image Collection; page 258: Courtesy JPL/NASA Images; page 259: Raymond Gehman/NG Image Collection; figure 8.21 (center): NG Maps; figure 8.21a: Carr Clifton/Minden Pictures/NG Image Collection; figure 8.21c: Wolfgang Kaehler/SuperStock; figure 8.21f: Jon Arnold Images/Masterfile; figure 8.21d: Momatiuk, Yva & Eastcott, John/Minden Pictures/NG Image Collection; figure 8.21e: Georg Gerster/Photo Researchers; figure 8.21g: Dinodia/Age Fotostock America, Inc.; figure 8.21b: Courtesy NASA Johnson Space Center Collection; figure 8.22: Copyright © 1995, David T. Sandwell. Used by permission; page 263: National Geographic; figure 8.25: Christopher Scotese/NG Maps; page 269: Courtesy Brian J. Skinner; page 269 (inset): Walter Meayers Edwards/NG Image Collection.

Chapter 9

Chapter Opener: The Asahi Shimbun/Getty Images Inc.; figure 9.5: NG Maps; figure 9.5a: Emory Kristof/NG Image Collection; figure 9.5b: Alison Wright/NG Image Collection; figure 9.5c: NASA/Goddard Space Flight Center Scientific Visualization Studio; figure 9.5e: Courtesy NASA Images; figure 95d: Paul Zahl/NG Image Collection; figure 9.8a: Science Source/Photo Researchers; figure 9.12: A. N. Strahler; page 284 (right): Jacques Descloitres, MODIS Land Rapid Response Team; page 284 (left): Gordon Wiltsie/NG Image Collection; page 285: NG Image Collection; figure 9.17b: Alison Wright/NG Image Collection; figure 9.18: Courtesy NASA Images; figure 9.19: Peter Essick/Aurora Photos/Alamy; figure 9.20: Toppozada et al, 2000, California Geological Survey Map Sheet 49; figure 9.22a: NG Maps; figure 9.22b: Christine Osborne Pictures/Alamy; figure 9.022c: AFP/Getty Images, Inc.; figure 9.23: Nasa Jet Propulsion Laboratory; figure 9.25: Emory Kristof/NG Image Collection; page 297 (bottom): Image courtesy of AVO/U.S. Geological Survey; page 297 (top): Neal, C. A./Alaska Volcano Observatory/U.S. Geological Survey; figure 9.29a: Chris Johns/NG Image Collection; figure 9.29b: Kevin West/Liaison Agency, Inc./Getty Images; figure 9.30.3: John Richardson/NG Image Collection; figure 9.30.4: Paul Chesley/NG Image Collection; figure 9.30.2: Randy Wells/Stone/Getty Images, Inc.; figure 9.31: © Frans Lanting/www.lanting.com; figure 9.32d: Nasa Jet Propulsion Laboratory; figure 9.33: A. N. Strahler; page 304 (left): Science Source/Photo Researchers, Inc.; page 304 (right): Peter Essick/Aurora Photos/Alamy; page 305 (top): Chris Johns/NG Image Collection; page 305 (bottom): Thomas R. Fletcher/Alamy; page 306: NASA.

Chapter 10

Chapter Opener: J. Baylor Roberts/NG Image Collection; figure 10.1: William E. Ferguson; page 311 (top): NG Image Collection; page 311 (bottom): Peter Essick/Aurora Photo Inc.; figure 10.3: Gordon Wiltsie/NG Image Collection; figure 10.4b: Bruce Dale/NG Image Collection; figure 10.5: Judyth Platt; Ecoscene/©Corbis; figure 10.6: Marc Moritsch/NG Image Collection; figure 10.7a: Darlyne A. Murawski/NG Image Collection; figure 10.7b: Todd Gipstein/NG Image Collection; figure 10.10: NASA Images; figure 10.11c: ©AP/Wide World Photos; figure 10.11d: NG Image Collection; figure 10.11b: USGS; figure 10.13: Dr. Marli Miller/Visuals Unlimited/Getty Images Inc.; page 322: USGS; figure 10.16: Steve Raymer/NG Image Collection; figure 10.17: Arno Balzarini/epa/Corbis Images; figure 10.18: Courtesy Los Angeles County Department of Public Works; figure 10.19: Melissa Farlow/NG Image Collection; page 328: Photo by Mark Reid, U.S. Geological Survey; page 328 (bottom): William E. Ferguson; page 329 (top Left): Bruce Dale/NG Image Collection; page 329 (bottom right): George F. Mobley/NG Image Collection.

Chapter 11

Chapter Opener: Franco Debernardi//Getty Images; figure 11.1: Medford Taylor/NG Image Collection; page 336 (left): Gordon Wiltsie/NG Image Collection; page 336 (right): Thomas Mangelsen/Minden Pictures/NG Image Collection; figure 11.6b: Gordon Wiltsie/NG Image Collection; figure 11.8(1): Stephen Alvarez/NG Image Collection; figure 11.8(2): Bruno Barbey/Magnum Photos, Inc.; figure 11.9: George F. Mobley/NG Image Collection; figure 11.11: USGS; figure 11.12a: westphalia/iStockphoto; page 342: Courtesy of the EPA; figure 11.13: Tom McHugh/Photo Researchers; figure 11.21a and b: NASA Image Courtesy MODIS Rapid Response Team, Goddard Space Flight Center; figure 11.21c: Reuters/Landov LLC; figure 11.23: Jodi Cobb/NG Image Collection; figure 11.24b: James P. Blair/NG Image Collection; figure 11.24a: Mike Theiss/NG Image Collection; figure 11.25: Volkmar K. Wentzel/NG Image Collection; figure 11.27b: Ed Kashi/Corbis; figure 11.27: NG Maps; figure 11.27a: James L. Amos/NG Image Collection; figure 11.27d: Gilbert M. Grosvenor/NG Image Collection; figure 11.27c: J. Baylor Roberts/NG Image Collection; figure 11.28: Stephen Sharnoff/NG Image Collection; page 357: National Geographic; page 358: Stephen Sharnoff/NG Image Collection; page 360 (top): Jaap Hart/iStockphoto; page 360 (bottom): Stephen Alvarez/NG Image Collection.

Chapter 12

Chapter Opener: Annie Griffiths/NG Image Collection; figure 12.1a: Steve Winter/NG Image Collection; figure 12.1b: Jim Wark/Peter Arnold Images/Photolibrary; figure 12.1d: Alaska Stock Images/NG Image Collection; figure 12.1c: Digital Vision/SuperStock; page 366 (top): Ken Gilham/Alamy; figure 12.2a,b: Offical U.S. Navy Photographs; figure 12.3a: Sheldan Collins/©Corbis; figure 12.3b: Steve McCurry/Magnum Photos, Inc.; figure 12.3c: NG Maps; figure 12.4: Jeffrey Lepore/Photo Researchers, Inc.; page 370: NG Image Collection; figure 12.6a: Oxford Scientific/Photolibrary/Getty Images, Inc; figure 12.6b: Nikreates/Alamy; figure 12.9: Richard Nowitz/NG Image Collection; figure 12.10b: Joel Sartore/NG Image Collection; figure

12.10c: George H.H. Huey/Alamy; figure 12.11b: F. Kenneth Hare; figure 12.14b: Tim Fitzharris/Minden Pictures/NG Image Collection; figure 12.15: Frans Lanting/NG Image Collection; figure 12.16b: Raymond Gehman/NG Image Collection; page 382 (top): Christian Kober/Robert Harding/Getty Images; figure 12.17: © John S. Shelton/University of Washington Libraries, Special Collections; figure 12.18b: Walter Meayers Edwards/NG Image Collection; figure 12.19: NASA Image created by Robert Simmon and Jesse Allen, using SRTM elevation data provided courtesy of the University of Maryland's Global Land Cover Facility; page 387: NASA/Science Source/Photo Researchers.

Chapter 13

Chapter Opener: Eurasia/Robert Harding World Imagery/Getty Images, Inc.; figure 13.2a: Mark A. Wilson/Department of Geology/The College of Wooster; figure 13.2b: Mike P Shepherd/Alamy; figure 13.4: ©AP/Wide World Photos; figure 13.5: Lou Linwei/Alamy; figure 13.7: USGS/JPL; figure 13.8b: Annie Griffiths Belt/NG Image Collection; figure 13.8a: Gerry Ellis/Minden Pictures/NG Image Collection; figure 13.8c: George Steinmetz/NG Image Collection; figure 13.8e: Joe Scherschel/NG Image Collection; figure 13.8d: Annie Griffiths/NG Image Collection; figure 13.9: Phil Schermeister/©Corbis; page 402: Patrick McFeeley/NG Image Collection; figure 13.12ab: Digital Globe/HO/AFP/Getty Images; figure 13.13: The Asahi Shimbun/Getty Images; figure 13.14a: Carl & Ann Purcell/©Corbis; figure 13.15: Mark Karrass/Corbis; figure 13.17: Michael Medford/NG Image Collection; figure 13.19b: University of Washington Libraries, Special Collections Division; figure 13.20a: E.J. Baumeister Jr./Alamy; figure 13.20b: AHowden - USA Stock Photography/Alamy; figure 13.21: James P. Blair/NG Image Collection; figure 13.23: David Alan Harvey/NG Image Collection; figure 13.24c: age fotostock/Superstock; figure 13.25: Brian J. Skerry/NG Image Collection; figure 13.26a: Peter McBride/Getty Images, Inc; figure 13.26b: Mauricio Handler/NG Image Collection; figure 13.27a: ©Gary O'Brien/Newscom; figure 13.27b: Rick Wilking/©Corbis; figure 13.28: Jim Wark/AirPhoto-; page 418 (bottom): NG Image Collection; page 420: Jim Wark/AirPhoto-; page 421: ©Gary O'Brien/Newscom; page 422: James P. Blair/NG Image Collection.

Chapter 14

Chapter Opener: Kennan Ward/©Corbis; figure 14.1c: Gavin Hellier/Getty Images; figure 14.1a: Melissa Farlow/NG Image Collection; figure 14.1b: Ron Niebrugge/Alamy; figure 14.1d: Bryan and Cherry Alexander/Photo Researchers, Inc.; page 428: Reuters/ESA/Landov LLC; figure 14.2: Digital Vision/Getty Images, Inc; figure 14.4a: Austin Post/Matt Nolan; figure 14.4b: Matt Nolan; figure 14.4c: Dr. Matt Nolan; figure 14.5: Courtesy Stephen C. Porter; figure 14.6: Michael Sewell/Peter Arnold, Inc.; figure 14.8: Alan Majchrowicz/Age fotostock/Photolibrary; figure 14.11a: Grambo Photography/All Canda Photos/Photolibrary; figure 14.11b: Arthur N. Strahler; figure 14.12d: USGS; figure 14.12c: Galen Rowell/©Corbis; figure 14.13: Blickwinkel/Alamy; page 442: Courtesy Troy L. Pewe; figure 14.14: Paolo Koch/Photo Researchers, Inc.; figure 14.15a: Ralph Lee Hopkins/Alamy Limited; figure 14.16: Courtesy NASA; figure

14.17: imagebroker.net/SuperStock; figure 14.18: NG Maps; figure 14.21: National Snow and Ice Data Center; page 449 (top): National Geographic; page 449 (bottom): Digital Vision/Getty Images, Inc; page 450: Michael Sewell/Peter Arnold, Inc.; page 451: National Snow and Ice Data Center; page 452: Courtesy NASA; page 453: Paolo Koch/Photo Researchers, Inc.

Chapter 15

Chapter Opener: ©The Trustees of The British Museum/Art Resource, NY /Art Resource; page 459: NG Image Collection; page 461: Pete Oxford/Minden Pictures/NG Image Collection; page 461 (inset): Yva Momatiuk & John Eastcott/Minden Pictures/NG Image Collection; figure 15.5 a and b: Image courtesy Marc Imhoff, NASA GSFC, and Flashback Imaging Corporation, Ontario, Canada; figure 15.6a: B. Anthony Stewart/NG Image Collection; figure 15.6b: Sam Abell/National Geographic; figure 15.6c: ©Eduardo Pucheta Photo/Alamy; figure 15.13: Courtesy U.S. Department of Agriculture; figure 15.13f: Robert Harding Images/Masterfile; figure 15.13e: Alan H. Strahler; figure 15.16d: Dinodia Photos/Alamy; figure 15.13b: Anthony Boccaccio/Getty Images, Inc; figure 15.13c: Ron Chapple Stock/Alamy; figure 15.13a: Ron Erwin/AllCanadaPhotos/Getty; figure 15.14: Estate of Henry D. Foth; figure 15.15b: Estate of Henry D. Foth; figure 15.16: Estate of Henry D. Foth; figure 15.17: Estate of Henry D. Foth; figure 15.18b: Estate of Henry D. Foth; figure 15.19: Courtesy Soil Conservation Service; figure 15.20: Estate of Henry D. Foth; figure 15.021: Estate of Henry D. Foth; page 481 (bottom): Farrell Grehan/Photo Researchers, Inc.; page 481 (top right): Estate of Henry D. Foth; page 483: Estate of Henry D. Foth.

Chapter 16

Chapter Opener: Tim Laman/NG Image Collection; figure 16.1a: Joel Sartore/NG Image Collection; figure 16.1b: Raymond Gehman/NG Image Collection; figure 16.1c: Beverly Joubert/NG Image Collection; figure 16.4: NG Maps; figure 16.7: USGS; figure 16.8a: Newcomer, J., D. Landis, S. Conrad, S. Curd, K. Huemmrich, D. Knapp, A. Morrell, J.; figure 16.10: Raymond Gehman/NG Image Collection; figure 16.11a: Annie Griffiths Belt/NG Image Collection; figure 16.11b: Dennis Flaherty/Photo Researchers, Inc; figure 16.12a: Jeff Lepore/Photo Researchers, Inc.; figure 16.12b: Roy Toft/NG Image Collection; figure 16.12c: Michael and Patricia Fogden/Corbis; figure 16.13: Sam Abell/NG Image Collection; figure 16.14: Kent Dannen/Photo Researchers, Inc.; figure 16.15a: Berverly Joubert/NG Image Collection; figure 16.15b: All Canada Photos/Alamy; figure 16.15c: Des and Jen Bartlett/NG Image Collection; page 501: Mira/Alamy; figure 16.16a: Raymond Forbes/Masterfile; figure 16.16b: Raymond Gehman/NG Image Collection; figure 16.18a: Bill Curtsinger/NG Image Collection; figure 16.18b: Roy Toft/NG Image Collection; figure 16.18c: Jason Edwards/NG Image Collection; figure 16.18d: Darlyne A. Murawski/NG Image Collection; figure 16.19b: Mark Moffett/Minden Pictures/NG Image Collection; figure 16.19c: J. Dunning/Vireo; figure 16.19g: Tierbild Okapia/Photo Researchers, Inc.; figure 16.19h: Eric Hosking/Photo Researchers, Inc.; figure 16.19d: Tim Laman/NG Image Collection; figure 16.19e: Hiroya Minakuchi/Minden Pictures/NG Image Collection; figure 16.19f: Tui De Roy/Minden

Pictures/NG Image Collection; page 507: NG Image Collection; figure 16.20a: Bates Littlehales/NG Image Collection; figure 16.20b: Joel Sartore/NG Image Collection; figure 16.20c: Tim Laman/NG Image Collection; page 509: Albert Gea/RTR/NewsCom; figure 16.22a: Dan Guravich/Photo Researchers, Inc.; figure 16.22b: Carr Clifton/Minden Pictures /NG Image Collection; figure 16.22c: Frans Lanting/NG Image Collection; figure 16.22f: Tim Fitzharris/Minden Pictures/NG Image Collection; figure 16.22d: DLILLC/Corbis; page 515: Gregory G. Dimijian/Photo Researchers; figure 16.22e: Hiroya Minakuchi/Minden Pictures/NG Image Collection; page 516: Newcomer, J., D. Landis, S. Conrad, S. Curd, K. Huemmrich, D. Knapp, A. Morrell, J. Nickeson, A. Papagno, D. Rinker, R. Strub, T. Twine, F. Hall, and P. Sellers, eds. 2000. Collected Data of The Boreal Ecosystem-Atmosphere Study. NASA. CD-ROM; page 517: Raymond Forbes/Masterfile.

Chapter 17

Chapter Opener: Tim Laman/NG Image Collection; figure 17.2: Cyril Ruoso/JH Editorial/Minden Pictures/NG Image Collection; figure 17.3a: Michael Melford/NG Image Collection; figure 17.3b: Joel Sartore/NG Image Collection; figure 17.3c: Richard Nowitz/NG Image Collection; figure 17.5: NG Maps; page 526 (right): Raymond Gehman/Corbis Images; page 526 (left): M.P. Kahl/Photo Researchers; figure 17.6: Balance/Photoshot/NewsCom; figure 17.7: Hugo Ahlenius, UNEP/GRID-Arendal/; figure 17.8a: Digital Vision/Getty Images; figure 17.8b: Michael Nichols/NG Image Collection; figure 17.8c: Eric Baccega/Age Fotostock America, Inc. ; figure 17.9a: Tim Laman/NG Image Collection; figure 17.9b: Hiroya Minakuchi/Minden/NG Image Collection; page 531: Robert Simmon and Reto Stîckli./NASA; figure 17.12 inset: NG Maps; figure 17.12: Aditya Dicky Singh/Alamy; figure 17.14a: George Gall/NG Image Collection; figure 17.14b: Raymond Gehman/NG Image Collection; figure 17.16a: Raymond Gehman/NG Image Collection; figure 17.16b: Joel Sartore/NG Image Collection; figure 17.16c: Melissa Farlow/NG Image Collection; figure 17.16d: Raymond Gehman/NG Image Collection; figure 17.18: Raymond Gehman/NG Image Collection; figure 17.20: Raymond Gehman/NG Image Collection; figure 17.22: Eco Images/Getty Images, Inc.; figure 17.23a: Peter Johnson/Corbis; figure 17.23b: Annie Griffiths Belt/NG Image Collection; figure 17.25a: Norbert Rosing/NG Image Collection; figure 17.25b: Michael Nichols/NG Image Collection; figure 17.25c: DanitaDelimont.com/NewsCom; figure 17.26a: Joel Sartore/NG Image Collection; figure 17.26b: James Steinberg/Photo Researchers, Inc.; figure 17.28a: Raymond Gehman/NG Image Collection; figure 17.28b: Joel Sartore/NG Image Collection; figure 17.28c: Bates Littlehales/NG Image Collection; figure 17.30a: Rich Reid/NG Image Collection; figure 17.30b: Marcello Calandrini/Corbis; figure 17.31a: Bartlett, Des and Jen/NG Image Collection; figure 17.31: Focus_on_Nature/Vetta/Getty Images, Inc.; figure 17.31b: Mattias Klum/NG Image Collection; page 549 (top): NG Image Collection; figure 17.32: Chad Ehlers/Alamy; figure 17.33a: Michael S. Quinton/NG Image Collection; figure 17.33b: Joel Sartore/NG Image Collection; figure 17.33c: Joel Sartore/NG Image Collection; page 552: Rich Reid/NG Image Collection; page 553: ©Jeff Foott; page 554: Cyril Ruoso/JH Editorial/Minden Pictures/NG Image Collection; page 555: Joel Sartore/NG Image Collection.

Line Art Credits

Chapter 1

Figure 1.2: From *Visualizing Geology, 2e* by Barbara W. Murck, Brian J. Skinner, and Dana Mackenzie. Copyright © 2010 John Wiley & Sons, Inc. Reprinted with permission of John Wiley & Sons, Inc. Figure 1.7: From *Discovering Physical Geography, 2e* by Alan F. Arbogast. Copyright © 2011 John Wiley & Sons, Inc. Reprinted with permission of John Wiley & Sons, Inc. Figure 1.8: From *Introducing Physical Geography, 5e* by Alan Strahler. Copyright © 2011 John Wiley & Sons, Inc. Reprinted with permission of John Wiley & Sons, Inc. Figure 1.12: From *Introducing Physical Geography, 5e* by Alan Strahler. Copyright © 2011 John Wiley & Sons, Inc. Reprinted with permission of John Wiley & Sons, Inc. Figure 1.14: U.S. Navy Oceanographic Office. Figure 1.15: From *Introducing Physical Geography, 5e* by Alan Strahler. Copyright © 2011 John Wiley & Sons, Inc. Reprinted with permission of John Wiley & Sons, Inc. Figure 1.18: Adapted from *Visualizing Human Geography* by Allison L. Greiner. Copyright © 2011 John Wiley & Sons, Inc. Reprinted with permission of John Wiley & Sons, Inc. Figure 1.20b: From *Introducing Physical Geography, 5e* by Alan Strahler. Copyright © 2011 John Wiley & Sons, Inc. Reprinted with permission of John Wiley & Sons, Inc. Figure 1.21: Copyright © by the University of Chicago. Used by permission of the Committee on Geographical Studies, University of Chicago. Figure 1.22: From *Introducing Physical Geography, 5e* by Alan Strahler. Copyright © 2011 John Wiley & Sons, Inc. Reprinted with permission of John Wiley & Sons, Inc.

Chapter 2

Figure 2.1: Adapted from *Discovering Physical Geography, 2e* by Alan F. Arbogast. Copyright © 2011 John Wiley & Sons, Inc. Reprinted with permission of John Wiley & Sons, Inc. Figure 2.2: After W.D. Sellers, *Physical Climatology,* University of Chicago Press. Used by permission. Figure 2.3: The Life Layer, A. N. Strahler, *Journal of Geography,* copyright © National Council for Geographic Education, reprinted by permission of Taylor & Francis Ltd, http://www.informaworld.com on behalf of the National Council for Geographic Education. Figure 2.5: From *Introducing Physical Geography, 5e* by Alan Strahler. Copyright © 2011 John Wiley & Sons, Inc. Figure 2.9: From *Introducing Physical Geography, 5e* by Alan Strahler. Copyright © 2011 John Wiley & Sons, Inc. Reprinted with permission of John Wiley & Sons, Inc. Figure 2.10: From *Introducing Physical Geography, 5e* by Alan Strahler. Copyright © 2011 John Wiley & Sons, Inc. Reprinted with permission of John Wiley & Sons, Inc. Figure 2.13: From *Introducing Physical Geography, 5e* by Alan Strahler. Copyright © 2011 John Wiley & Sons, Inc. Reprinted with permission of John Wiley & Sons, Inc. Figure 2.15: Copyright © A. N. Strahler. Used by permission.

Chapter 3

Figure 3.1: Adapted from *Discovering Physical Geography, 2e* by Alan F. Arbogast. Copyright © 2011 John Wiley & Sons, Inc. Reprinted with permission of John Wiley & Sons, Inc. Figure 3.2: Courtesy NOAA. Figure 3.3: Adapted from *Visualizing Weather and Climate* by Bruce T. Anderson and Alan Strahler. Copyright © 2008 John Wiley & Sons, Inc. Reprinted with permission of John Wiley & Sons, Inc. Figure 3.5: From *Introducing Physical Geogra-*

phy, 5e by Alan Strahler. Copyright © 2011 John Wiley & Sons, Inc. Reprinted with permission of John Wiley & Sons, Inc. Figure 3.9, graphs: Data courtesy of David H. Miller. Figure 3.15: Data compiled by John E. Oliver. Figure 3.16: Data compiled by John E. Oliver. Figure 3.19: James Hanson/NASA Goddard Institute for Space Studies. Figure 3.20: Courtesy NASA. Figure 3.21: After James Hansen, Makiko Sato, Reto Ruedy, Andrew Lacis, and Valdar Oinas. "Global warming in the twenty-first century: An alternative scenario." *Proceedings of the National Academy of Sciences of the United States of America.* Copyright 2000 National Academy of Sciences, U.S.A. Figure 3.22: Climate Change 2007: Synthesis Report. Contribution of Working Groups I, II and III to the Fourth Assessment Report of the Intergovernmental Panel on Climate Change, Figure SPM.5. IPCC, Geneva, Switzerland.

Chapter 4

Figure 4.6: From *Introducing Physical Geography, 5e* by Alan Strahler. Copyright © 2011 John Wiley & Sons, Inc. Reprinted with permission of John Wiley & Sons, Inc. Data of J. von Hann, R. Süring, and J. Szava-Kovats as shown in Haurwitz and Austin, *Climatology.* Figure 4.8: Copyright © A. N. Strahler. Used by permission. Figure 4.10: Copyright © A. N. Strahler. Used by permission. Figure 4.13: Adapted from *Visualizing Weather and Climate* by Bruce T. Anderson and Alan Strahler. Copyright © 2008 John Wiley & Sons, Inc. Reprinted with permission of John Wiley & Sons, Inc. Figure 4.14: Adapted from *Visualizing Weather and Climate* by Bruce T. Anderson and Alan Strahler. Copyright © 2008 John Wiley & Sons, Inc. Reprinted with permission of John Wiley & Sons, Inc. Figure 4.16: Adapted from *Visualizing Weather and Climate* by Bruce T. Anderson and Alan Strahler. Copyright © 2008 John Wiley & Sons, Inc. Reprinted with permission of John Wiley & Sons, Inc. Figure 4.17: Adapted from *Discovering Physical Geography, 2e* by Alan F. Arbogast. Copyright © 2011 John Wiley & Sons, Inc. Reprinted with permission of John Wiley & Sons, Inc. Figure 4.18: Adapted from *Visualizing Weather and Climate* by Bruce T. Anderson and Alan Strahler. Copyright © 2008 John Wiley & Sons, Inc. Reprinted with permission of John Wiley & Sons, Inc. Figure 4.19: National Atmospheric Deposition Program (NRSp-3)/National Trends Network, Illinois State Water Survey.

Chapter 5

Figure 5.3: Adapted from *Discovering Physical Geography, 2e* by Alan F. Arbogast. Copyright © 2011 John Wiley & Sons, Inc. Reprinted with permission of John Wiley & Sons, Inc. Figure 5.5: Adapted from *Discovering Physical Geography, 2e* by Alan F. Arbogast. Copyright © 2011 John Wiley & Sons, Inc. Reprinted with permission of John Wiley & Sons, Inc. Figure 5.8: From *Introducing Physical Geography, 5e* by Alan Strahler. Copyright © 2011 John Wiley & Sons, Inc. Reprinted with permission of John Wiley & Sons, Inc. Figure 5.10: Adapted from *Visualizing Weather and Climate* by Bruce T. Anderson and Alan Strahler. Copyright © 2008 John Wiley & Sons, Inc. Reprinted with permission of John Wiley & Sons, Inc. Figure 5.12: Adapted from *Visualizing Weather and Climate* by Bruce T. Anderson and Alan Strahler. Copyright © 2008 John Wiley & Sons, Inc. Reprinted with permission of John Wiley & Sons, Inc. Figure 5.14: Adapted from *Visualizing Weather and Climate* by Bruce T. Anderson and Alan Strahler. Copyright © 2008 John Wiley & Sons, Inc. Reprinted with permission of John Wiley & Sons, Inc. Figure 5.15: Copyright © A. N. Strahler. Used by permission. Figure 5.16: Adapted from *Visualizing Weather and Climate* by Bruce T. Anderson and Alan Strahler. Copyright © 2008 John Wiley & Sons, Inc. Reprinted with permission of John Wiley & Sons, Inc. Figure 5.18: Data compiled by John E. Oliver. Figure 5.22: Copyright © A. N. Strahler. Used by permission. Figure 5.26: From *Introducing Physical Geography, 5e* by Alan Strahler. Copyright © 2011 John Wiley & Sons, Inc. Reprinted with permission of John Wiley & Sons, Inc.; Adapted from NOAA, National Weather Service.

Chapter 6

Figure 6.2: Data from U.S. Department of Commerce. Figure 6.3: Adapted from *Discovering Physical Geography, 2e* by Alan F. Arbogast. Copyright © 2011 John Wiley & Sons, Inc. Reprinted with permission of John Wiley & Sons, Inc. Figure 6.4: Adapted from *Discovering Physical Geography, 2e* by Alan F. Arbogast. Copyright © 2011 John Wiley & Sons, Inc. Reprinted with permission of John Wiley & Sons, Inc. Figure 6.5: From *Introducing Physical Geography, 5e* by Alan Strahler. Copyright © 2011 John Wiley & Sons, Inc. Reprinted with permission of John Wiley & Sons, Inc. Figure 6.8b: Adapted from NOAA, National Weather Service. Figure 6.9: Data from H. Riehl, *Tropical Meteorology*, McGraw-Hill. Figure 6.14: From *Visualizing Weather and Climate* by Bruce T. Anderson and Alan Strahler. Copyright © 2008 John Wiley & Sons, Inc. Reprinted with permission of John Wiley & Sons, Inc. Figure 6.16: From *Visualizing Weather and Climate* by Bruce T. Anderson and Alan Strahler. Copyright © 2008 John Wiley & Sons, Inc. Reprinted with permission of John Wiley & Sons, Inc. Figure 6.18: Adapted from *Visualizing Weather and Climate* by Bruce T. Anderson and Alan Strahler. Copyright © 2008 John Wiley & Sons, Inc. Reprinted with permission of John Wiley & Sons, Inc. Figure 6.20: Adapted from *Visualizing Weather and Climate* by Bruce T. Anderson and Alan Strahler. Copyright © 2008 John Wiley & Sons, Inc. Reprinted with permission of John Wiley & Sons, Inc. Figure 6.21: Adapted from *Visualizing Weather and Climate* by Bruce T. Anderson and Alan Strahler. Copyright © 2008 John Wiley & Sons, Inc. Reprinted with permission of John Wiley & Sons, Inc.

Chapter 7

Figure 7.3a: Simplified and modified from Plate 3, "World Climatology," Volume 1, *The Times Atlas*, Editor John Bartholomew, The Times Publishing Company, Ltd., London, 1958. Figure 7.7: Compiled from station data by A. N. Strahler. Figure 7.34, Top graph: Global temperatures: After Hansen et al., 2001, Proc. National Academy of Sciences. Used by permission. Top graph, CO_2 concentration: Courtesy Scripps CO_2 Program. Bottom graph: Adapted from EPA. Figure 7.36: Adapted from http://maps.grida.no/go/graphic/trends-in-natural-disasters, Centre for Research on the Epidemiology of Disasters (CRED) by Emmanuelle Bournay, UNEP/GRID-Arendal.

Chapter 8

Figure 8.3: Drawn by A. N. Strahler. What a Geographer Sees, p. 242: Diagram adapted from *Visualizing Geology, 2e* by Barbara

W. Murck, Brian J. Skinner, and Dana Mackenzie. Copyright © 2010 John Wiley & Sons, Inc. Reprinted with permission of John Wiley & Sons, Inc. Figure 8.5: Adapted from *Visualizing Geology, 2e* by Barbara W. Murck, Brian J. Skinner, and Dana Mackenzie. Copyright © 2010 John Wiley & Sons, Inc. Reprinted with permission of John Wiley & Sons, Inc. Figure 8.9: Adapted from a drawing by A. N. Strahler. Figure 8.11: Adapted from *Visualizing Geology, 2e* by Barbara W. Murck, Brian J. Skinner, and Dana Mackenzie. Copyright © 2010 John Wiley & Sons, Inc. Reprinted with permission of John Wiley & Sons, Inc. Figure 8.12: Adapted from *Visualizing Geology, 2e* by Barbara W. Murck, Brian J. Skinner, and Dana Mackenzie. Copyright © 2010 John Wiley & Sons, Inc. Reprinted with permission of John Wiley & Sons, Inc. Figure 8.13a: Drawn by A. N. Strahler. Figure 8.17b: Copyright © A. N. Strahler. Used by permission. Figure 8.19: From *Visualizing Geology* by Barbara W. Murck, Brian J. Skinner, and Dana Mackenzie. Copyright © 2007 John Wiley & Sons, Inc. Reprinted with permission of John Wiley & Sons, Inc. Figure 8.20a: Based in part on data of R. F. Murphy, P. M. Hurley, and others. Copyright © A. N. Strahler. Used by permission. Figure 8.22: Bottom: Copyright © A. N. Strahler. Used by permission. Figure 8.23: Based on A. Wegener,. 1915, *Die Entsechtung der Kontinente und Ozeane*, F. Vieweg, Braunschweig. Figure 8.24: From *Visualizing Geology, 2e* by Barbara W. Murck, Brian J. Skinner, and Dana Mackenzie. Copyright © 2010 John Wiley & Sons, Inc. Reprinted with permission of John Wiley & Sons, Inc.

Chapter 9

Figure 9.1: From *Physical Geology* by Charles Fletcher. Copyright © 2011 John Wiley & Sons, Inc. Reprinted with permission of John Wiley & Sons, Inc. Figure 9.2: Copyright © A. N. Strahler. Used by permission. Figure 9.3: Copyright © A. N. Strahler. Used by permission. Figure 9.4: Copyright © A. N. Strahler. Used by permission. Figure 9.6b: Copyright © A. N. Strahler. Used by permission. Figure 9.8b: Drawn by Erwin Raisz. Copyright © A. N. Strahler. Used by permission. Figure 9.9: Copyright © A. N. Strahler. Used by permission. Figure 9.10: Copyright © A. N. Strahler. Used by permission. Figure 9.11 b–d: Copyright © A. N. Strahler. Used by permission. Figure 9.13: A. N. Strahler. Figure 9.14: Copyright © A. N. Strahler. Used by permission. Figure 9.15: From *Physical Geology* by Charles Fletcher. Copyright © 2011 John Wiley & Sons, Inc. Reprinted with permission of John Wiley & Sons, Inc. Figure 9.16: Adapted from *Discovering Physical Geography, 2e* by Alan F. Arbogast. Copyright © 2011 John Wiley & Sons, Inc. Reprinted with permission of John Wiley & Sons, Inc. Figure 9.18: From *Physical Geology* by Charles Fletcher. Copyright © 2011 John Wiley & Sons, Inc. Reprinted with permission of John Wiley & Sons, Inc. Figure 9.21: *Physical Geology* by Charles Fletcher. Copyright © 2011 John Wiley & Sons, Inc. Reprinted with permission of John Wiley & Sons, Inc. Adapted from USGS, http://eqhazmaps.usgs.gov. Figure 9.24a: Compiled by A. N. Strahler from data from NOAA. Copyright © A. N. Strahler. Used by permission. Figure 9.24b: Copyright © A. N. Strahler. Used by permission. Figure 9.26, Top: Copyright © A. N. Strahler. Used by permission. Figure 9.26, Bottom: Illustration by Frank Ippolito. Figure 9.27: Data from NOAA. Figure 9.28: Illustration by Frank Ippolito. Figure 9.30: Drawn by Erwin Raisz. Copyright © A. N. Strahler. Used by permission. Figure 9.31: Drawn by A. N. Strahler.

Chapter 10

What a Geographer Sees, p. 311: Adapted from *Visualizing Geology*, *2e* by Barbara W. Murck, Brian J. Skinner, and Dana Mackenzie. Copyright © 2010 John Wiley & Sons, Inc. Reprinted with permission of John Wiley & Sons, Inc. Figure 10.2: Drawn by A. N. Strahler. Figure 10.8: Drawn by A. N. Strahler. Figure 10.9: From *Discovering Physical Geography*, *2e* by Alan F. Arbogast. Copyright © 2011 John Wiley & Sons, Inc. Reprinted with permission of John Wiley & Sons, Inc. Figure 10.12a: After C. F. S. Sharpe. Figure 10.14: Adapted from USGS. Figure 10.15: Adapted from July 2004, "Landslide Types and Processes," *U.S. Geological Survey Fact Sheet 2004-3072*, Version 1.0.

Chapter 11

Figure 11.2: Drawn by A. N. Strahler. Figure 11.3: Drawn by A. N. Strahler. Figure 11.5: From *Visualizing Geology*, *2e* by Barbara W. Murck, Brian J. Skinner, and Dana Mackenzie. Copyright © 2010 John Wiley & Sons, Inc. Reprinted with permission of John Wiley & Sons, Inc. Figure 11.7: From *Physical Geology* by Charles Fletcher. Copyright © 2011 John Wiley & Sons, Inc. Reprinted with permission of John Wiley & Sons, Inc. Figure 11.8: Drawn by Erwin Raisz. Copyright © A. N. Strahler. Used by permission. Figure 11.10: A. H. Strahler. Figure 11.12: Copyright © A. N. Strahler. Used by permission. Figure 11.14: Data from USGS and Mark A. Melton. Figure 11.15: From *Introducing Physical Geography*, *4e* by Alan Strahler. Copyright © 2006 John Wiley & Sons, Inc. Reprinted with permission of John Wiley & Sons, Inc. Figure 11.16b: Copyright © A. N. Strahler. Used by permission. Figure 11.17: From *Physical Geology* by Charles Fletcher. Copyright © 2011 John Wiley & Sons, Inc. Reprinted with permission of John Wiley & Sons, Inc. Figure 11.19: Data from USGS. Figure 11.20: Adapted from *Physical Geology* by Charles Fletcher. Copyright © 2011 John Wiley & Sons, Inc. Reprinted with permission of John Wiley & Sons, Inc. Figure 11.22: After Hoyt and Langbein, *Floods* © 1955 Princeton University Press. Used by permission. Figure 11.26: Redrawn from *A Geologist's View of Cape Cod*, © A.N. Strahler, 1966. Used by permission of Doubleday, a division of Bantam Doubleday Dell Publishing Group, Inc.

Chapter 12

Figure 12.1: A. N. Strahler; Figure 12.8: Drawn by Erwin Raisz. Copyright © A. N. Strahler. Used by permission. Figure 12.10: Drawn by Erwin Raisz. Copyright © A. N. Strahler. Used by permission. Figure 12.11: Drawn by A. N. Strahler. Figure 12.12: Drawn by A. N. Strahler. Figure 12.14a: Drawn by A. N. Strahler. Figure 12.16a: Drawn by A. N. Strahler. Figure 12.18a: Copyright © A. N. Strahler. Used by permission.

Chapter 13

Figure 13.1: Adapted from *Discovering Physical Geography*, *2e* by Alan F. Arbogast. Copyright © 2011 John Wiley & Sons, Inc. Reprinted with permission of John Wiley & Sons, Inc. Figure 13.6: Adapted from K. Pye and L. Tsoar, *Aeolian Sand and Sand Dunes*, Figure 7.1, Chapman and Hall, 1990. Reprinted with kind permission of Springer Science and Business Media. Figure 13.7:

Adapted from "Seeing under the Sahara: Spaceborne Imaging Radar," by Charles Elachi. *Engineering & Science*, September 1983. 47 (1). p. 5. Figure 13.8: Adapted from John T. Hack, *The Geographical Review*, vol. 31, fig. 19, page 260 by permission of the American Geographical Society. Figure 13.10: Data from *Map of Pleistocene Eolian Deposits of the United States*, Geological Society of America. What a Geographer Sees, p. 401: Adapted from *Visualizing Earth Science* by Zeeya Merali and Brian J. Skinner. Copyright © 2009 John Wiley & Sons, Inc. Reprinted with permission of John Wiley & Sons, Inc. Figure 13.11: After W. M. Davis. Figure 13.14: H. J. de Blij and Peter O. Muller, *Physical Geography and the Global Environment*, copyright © 1996 John Wiley & Sons, Inc. Reprinted by permission of John Wiley & Sons, Inc. Figure 13.15: Copyright © A. N. Strahler. Used by permission. Figure 13.16: From *Visualizing Earth Science* by Zeeya Merali and Brian J. Skinner. Copyright © 2009 John Wiley & Sons, Inc. Reprinted with permission of John Wiley & Sons, Inc. Figure 13.18: Drawn by E. Raisz. Figure 13.19a: Drawn by A. N. Strahler. Figure 13.22: Drawn by A. N. Strahler. Figure 13.23: Drawn by A. N. Strahler. Figure 13.24a.: Drawn by A. N. Strahler. Figure 13.28: Adapted from *Discovering Physical Geography*, *2e* by Alan F. Arbogast. Copyright © 2011 John Wiley & Sons, Inc. Reprinted with permission of John Wiley & Sons, Inc.

Chapter 14

Figure 14.1: From *Visualizing Geology*, *2e* by Barbara W. Murck, Brian J. Skinner, and Dana Mackenzie. Copyright © 2010 John Wiley & Sons, Inc. Reprinted with permission of John Wiley & Sons, Inc. Figure 14.3: After Bloom, Arthur L., *Geomorphology*, Third Edition, 1998, reissued 2004 with corrections, Waveland Press, Long Grove, Illinois, Fig. 16.2 p. 356. Figure 14.9: Drawn by A. N. Strahler. Figure 14.12: Adapted from Troy L. Pewe, *Geotimes*, Volume 29, no. 2, p 11. Copyright 1984 by the American Geological Institute. What a Geographer Sees, p. 442: Adapted by permission from A. H. Lachenbruch in Rhodes W. Fairbridge, Ed., *The Encyclopedia of Geomorphology*, New York, Reinhold Publishing Corp. Figure 14.15b: After C. F. S. Sharpe. Figure 14.21: Based on calculations by A. D. Vernekar, 1968. Copyright © A. N. Strahler. Used by permission.

Chapter 15

Figure 15.2: From *Visualizing Geology*, *2e* by Barbara W. Murck, Brian J. Skinner, and Dana Mackenzie. Copyright © 2010 John Wiley & Sons, Inc. Reprinted with permission of John Wiley & Sons, Inc. Figure 15.7: After http://soils.usda.gov/education/resources/lessons/texture/. Figure 15.8: From *Discovering Physical Geography*, *2e* by Alan F. Arbogast. Copyright © 2011 John Wiley & Sons, Inc. Reprinted with permission of John Wiley & Sons, Inc. Figure 15.9: Adapted from *Discovering Physical Geography*, *2e* by Alan F. Arbogast. Copyright © 2011 John Wiley & Sons, Inc. Reprinted with permission of John Wiley & Sons, Inc. Figure 15.10: From *Introducing Physical Geography*, *5e* by Alan Strahler. Copyright © 2011 John Wiley & Sons, Inc. Reprinted with permission of John Wiley & Sons, Inc. Figure 15.11: After http://www.omafra.gov.on.ca/english/environment/soil/chemical.htm.

Chapter 16

Figure 16.2: Adapted from *Visualizing Environmental Science, 3e* by Linda R. Berg, Mary Catherine Hager, and David M. Hassenzahl. Copyright © 2011 John Wiley & Sons, Inc. Reprinted with permission of John Wiley & Sons, Inc. Figure 16.3: From *Visualizing Earth Science* by Zeeya Merali and Brian J. Skinner. Copyright © 2009 John Wiley & Sons, Inc. Reprinted with permission of John Wiley & Sons, Inc. Figure 16.5: Compiled by the National Science Board, National Science Foundation. Figure 16.6: Values are from Schlesinger, W. H. *Biogeochemistry; An Analysis of Global Change, 2e.* Academic Press, San Diego (1997) and based on several sources. Figure 16.10: After P. Dansereau. What A Geographer Sees, p. 501: After D. I. Rasmussen, *Ecological monographs,* vol. 11, 1941, p. 23. Copyright 2012. Reproduced with permission of Ecological Society of America. Figure 16.17: From *Visualizing Environmental Science, 3e* by Linda R. Berg, Mary Catherine Hager, and David M. Hassenzahl. Copyright © 2011 John Wiley & Sons, Inc. Reprinted with permission of John Wiley & Sons, Inc. Figure 16.19: Adapted from *Visualizing Geology* by Barbara W. Murck, Brian J. Skinner, and Dana Mackenzie. Copyright © 2007 John Wiley & Sons, Inc. Reprinted with permission of John Wiley & Sons, Inc. Figure 16.20a: From *Visualizing Geology* by Barbara W. Murck, Brian J. Skinner, and Dana Mackenzie. Copyright © 2007 John Wiley & Sons, Inc. Reprinted with permission of John Wiley & Sons, Inc. Figure 16.20b: From J. H. Brown and M. V. Lomolino/ Biogeography, 2e/1998/Sinuaer, Sunderland, Massachusetts. Used by permission. Figure 16.21: Data from IUCN.

Chapter 17

Figure 17.1: After P. Dansereau. Figure 17.3: Based on Holdridge, L. Life Zone Ecology. Tropical Science Center, San Jose, Costa Rica (1967). Figure 17.4: Based on maps of S. R. Eyre. Figure 17.11: Reproduced by kind permission of the Oxford Forest Information Service, University of Oxford. Figures 17.13, 17.15, 17.17, 17.19, 17.21, 17.24, 17.27, 17.29: Based on maps of S. R. Eyre/196/ Figure 8.

Note: "f" after a page number indicates the entry is found in a figure; "t" indicates the entry is found in a table.

A

A horizons, 457, 457f
Abandoned farmland, 503f
Ablation zone, 430
Abrasion
  glacial, 432, 433f
  stream erosion and, 368, 368f
  wind erosion and, 393
Absorption, atmospheric, 50
Accumulation zone, 429–430
Acid rain, 118–119, 119f, 120f
Acidity, soil, 467, 467f
Active layer, of permafrost, 441
Adiabatic processes, 105–107
  cloud formation and, 107f
  cooling and heating, 105f
  dry adiabatic lapse rate, 106
  moist adiabatic lapse rate, 106–107
Advection, 68
Advection fog, 110
Aerosols, 49
African savanna, 542, 542f
Aggradation, 380
Aiken, South Carolina, 534f
Air masses, 162–164
  global precipitation and, 196f
  movement of, 164, 164f
  source regions of, 162, 162f–163f
Air pollution, 55–57
Air temperature. *See* Temperature
Alabama, Dauphin Island, 502f
Alaska
  brown bear, 497f
  Chitina River, 380f
  Denali National Park, 371f
  Fort Yukon, 523f
  McCall Glacier, 431f
  North Slope of, 486f
Alaska Volcano Observatory, 297
Albedo, 51, 51f, 56
Alfisols, 472, 472f
Alkalinity, soil, 467, 467f
Alluvial fans, 383, 383f
Alluvial terraces, 376, 376f
Alluvium, 369
Alpine chains, 258–259
Alpine debris avalanches, 324, 324f
Alpine glaciers, 426–428, 435
Alpine tundra, 550
Altitude, atmospheric pressure and, 131, 131f
Amazon River, threatened biodiversity and, 510f
American beaver, 461
American bison, 545f
American Dust Bowl, 395
Andesite, 250f

Andisols, 477
Anemometers, 132, 132f
Angle, sunlight, 41, 41f
Angle of repose, 318, 318f, 396
Annual cycles of air temperature, 71, 71f
Annual insolation by latitude, 44–47, 45f
Annuals, plants, 521
Annular pattern, drainage, 344f
Antarctic (AA) mass, air mass, 163, 163f
Antarctic ice, 449
Antarctica, 228f, 439f
Anticlinal valleys, 280, 280f
Anticlines, 279
Anticyclones, 136, 136f, 167
Anvil clouds, 183f
Aphelion, 42f
Appalachian Mountains, 280
Aquicludes, 335
Aquifers, 335, 340
Arable soils, 460
Archaeopteryx, 507
Arctic (A) mass, air mass, 163f
Arctic latitude, net radiation and air
    temperature by, 75f
Arctic thawing, 88f
Arctic tundra, 549, 549f
Aridisols, 470f, 476, 476f
Arêtes, 435
Argentina, Tierra del Fuego, 498f
Argentine Pampas, Russia, 401
Arid climates, 199, 380–383, 381f
Arizona
  Canyon de Chelly, 313f
  desert climate and, 523f
  Grand Canyon, 381f
  Kaibab National Forest, 501
  Monument Valley, 547f
  Yuma, 73f, 215f
Armero, Colombia, 323f
Artesian wells, 335, 335f
Artificial beach nourishment, 417
Ash, volcanic, 296
Asian atmosphere, 56–57
Asian monsoons, 139, 139f
Asthenosphere, 244
Astronomical hypothesis, 446
Aswan, Egypt, 75f
Atmosphere
  absorption and, 50
  aerosols and, 49
  component gases of, 48f
  composition of, 5, 5f, 47–49
  global energy balance, 50f
  layers of, 47–48
  permanent gases and, 48
  reflection and, 51–52, 51f
  scattering and, 51
  stratosphere, 47
  troposphere, 47
  variable gases and, 48–49

Atmospheric and oceanic circulation
  atmospheric pressure, 130–131
  global wind and pressure patterns,
      137–149
  oceanic circulation, 150–154
  wind speed and direction, 132–136
Atmospheric lifting, 116–118
  convective lifting, 117–118, 117f, 118f
  convergent lifting, 117f
  frontal lifting, 117f
  orographic lifting, 116–117, 117f
Atmospheric moisture and precipitation
  acid rain, 118–119, 119f, 120f
  adiabatic processes and, 105–107
  cloud cover, precipitation, and global
      warming, 120–121
  clouds and fog, 108–111
  humidity and, 102–105
  hydrologic cycle and, 100–101, 101f
  hydrosphere and, 98–100, 99f, 100f
  precipitation, 111–118
  three states of water, 98, 98f
Atmospheric pressure
  altitude and, 131, 131f
  measuring, 130–131, 130f, 131f
Atolls, 416
Australian thorny devil lizard, 505f
Autumnal equinox, 44
Available water, 465
Avon River, England, 371f
Axis, Earth, 11

B

B horizons, 457, 457f
Backswamps, 379, 379f
Backwash, of wave, 402, 403f
Badlands, 382
Baltimore, Maryland, 492f
Bam, Iran, 292, 292f
Barchan dunes, 398f
Barometers, 130, 130f
Barrier islands, 412–413, 413f
Barrier reefs, 415–416
Basal sliding, 432
Basalt, 250f
Base cations, 468
Base flow, discharge related to, 346, 347f
Base level, streams, 374
Bases, 468
Batholiths, 251, 251f
Bay of Fundy, Nova Scotia, Canada, 409f
Baymouth bars, 412
Beach drift, 405–406
Beach nourishment, artificial, 417
Beaches, 411, 411f
Beaver, 461
  role in soil formation, 460–461, 461f
Bedrock, 310, 312f
Beech-maple-hemlock forest, 520f
Beijing, China, 395, 395f

Forest (continued)
needleleaf forests, 536–538, 537f, 538f
sclerophyll forests, 538, 539f
subtropical evergreen forests, 534, 534f
Formation, precipitation, 114, 114f
Formation classes, 521
Fort Vermilion, Canada, climograph for, 225f
Fort Yukon, Alaska, 523f
Fossil fuels, 254–255, 255f
Fox squirrel, 535f
Freezing rain, 115
Fresh water
desalination and, 353
distribution of, 332–333, 332f
ground water, 335–342
hydrologic cycle and, 333–334, 333f, 334f
infiltration and, 334
lakes, 351–353
runoff and, 334
surface water and fluvial systems, 342–351
surface water pollution, 356–357, 356f
water access and supply, 353–356, 354f–355f
Freshwater marsh ecosystem, 486f
Frictional force, 134–136, 135f
Fringing reefs, 415
Fronts, weather, 165–166
cold fronts, 165, 165f
frontal lifting, 117f
occluded fronts, 166
stationary fronts, 166
warm fronts, 166, 166f
Frost action, physical weathering and, 312, 312f
Frost cracking, 312, 312f
Fungus, 505f

G

Gabbro, 250f
Galápagos Islands, 505f, 506
Galápagos tortoise, 485, 485f
Gamma radiation, 38, 38f
Gelifluction, 441–442, 441f
Gelisols, 476
Genus, 507
Geobrowsers, 29–30
Geocoding, 28
Geographic grid, 11–12, 12f
Geographic information systems (GIS), 27–28, 28f
Geography
essential themes of, 4–5
major fields and subfields, 4f, 5
methods and tools for, 6–7, 6f
science of, 4–5
scientific method and, 6–7, 6f, 8f–9f
testable hypothesis and, 6
theories and, 7
Geologic time, 240–242, 240f–241f

Geomorphic cycle, 376–378, 377f
Geomorphic factors, ecosystems and, 498–499
Geomorphology, 240
Georgia
Chattahoochee River, 347f
Okefenokee Swamp, 486f
Geostrophic wind, 134, 134f
Germany
Hamburg, 75f
Rugen Island, 254f
Geysers, 336
Gilf Kebir, Egypt, 393f
Glaciers
abrasion and, 432, 433f
alpine glaciers, 426–428
climate change and, 431f
crevasse and, 432, 432f
equilibrium line and, 430
erosion and deposition and, 432–433
fiords and, 435
formation of, 429–430
glacial drift, 432
glacial landforms, 433–439, 434f
glacial processes, 429–433
glacial trough, 435
glaciation cycles, 446–448, 447f
glaciation history, 445–446, 445f
global climate and glaciation, 445–449
global warming and glaciation, 448–449, 448f
human development of, 439
ice sheets, 428–429, 429f
lateral moraine and, 435
mass balance and, 430, 430f
medial moraine and, 435, 435f
moraine and, 433
movement of, 431–432
shrinking polar ice caps, 448, 448f
types of, 426–429, 426f–427f
zone of ablation and, 430
zone of accumulation and, 429–430
Gleyzation, 459
Global air temperature patterns, 80–82, 81f
Global biogeography
desert biome, 546–548
forest biome, 528–539
global vegetation, 520–527
grassland biome, 543–545
savanna biome, 540–542
tundra biome, 549
Global climate modeling, 232–233, 232f
Global energy balance, 50f
Global energy budget, 54, 54f
Global energy system, 53–59
air pollution and, 56–57
Earth's energy output, 52–54
greenhouse effect and, 52–54, 53t
human impacts and, 55–59

net radiation and global energy budget, 54, 54f
ozone layer and, 55, 57–59, 58f
rising greenhouse gases and, 54–56, 55f
surface albedo and, 56
Global location, 11–14
equator, 12
geographic grid, 11–12, 12f
latitude and longitude, 12–14, 13f, 14f
meridians, 11
parallels, 12
Global Positioning System (GPS), 29
Global precipitation, 112f–113f
Global time, 14–19
coordinated universal time (UTC), 16f, 18
daylight saving time, 19
international date line, 19, 19f
solar time, 14–15, 15f
standard time, 16–17, 16f
time zones of the world, 17f
U.S. time zones, 18f
Global warming
causes of, 86, 86f
climate change and, 90
clouds and precipitation and, 120–121
computer projections and, 87f
consequences of, 87, 88f–89f
glaciation and, 448–449, 448f
international response to, 90
mean annual surface temperature and, 85f
sea-level rise and, 418
temperature trends by latitude, 86f
Goode projection, 24–25, 24f
Google Earth, 30
Graben, 284, 284f
Gradation, stream, 370–372, 371f
Graded stream, 370–372, 371f
Gradient, temperature, 80
Grand Canyon, Arizona, 381f
Grand Tetons National Park, Wyoming, 461
Granite, 250f, 315, 315f
Granular disintegration, 312f
Granular structure, soil, 465f
Grassland biome, 543–545, 543f–545f
Gravitational force, 406–408, 407f
Gravitational water, 464
Great circles, 11
Great Tohoku earthquake, Japan, 288, 405
Greenhouse effect, 52–54, 53f
Greenhouse gases, 52, 53t, 54–56, 55f
Greenland, Upernavik, climograph for, 227f
Greenland Ice Cap, 228f
Greenland Ice Sheet, 429, 429f
Greenwich meridian, 13
Greenwich time, 17f
Groins, 417–418, 419f
Ground ice, 441–443
Ground water, 335–342
aquicludes and, 335
aquifer and, 335

Moist subtropical climate, 203f, 216–217, 216f, 217f
Moisture
    ecosystems and, 496, 496f
    soil, 464–467, 466f
Mojave Desert, California, 393f
Mollisols, 470f, 475, 475f
Monsoon and trade-wind coastal climate, 209, 209f
Monsoon forest, 532–533, 532f, 533f
Monsoons, 139–140, 139f
Monterey, California, climograph for, 219f
Montreal Protocol Treaty of 1987, 59
Monument Valley, Arizona, 547f
Moon, and ocean tides, 406
Moraines, 433, 435, 435f, 436, 436f
Mountain breezes, 148, 148f
Mt. Pelée volcano eruption, 300
Mt. Pinatubo, Philippines, 85
Mt. Shasta, California, 301f
Mt. St. Helens, Washington State, 300f
Mucks, 478
Mud flats, 413
Mudflows, 319f, 323
Muir Woods National Monument, California, 538f
Muskingum River, Ohio, 350f
Mutation, in evolution, 507
Mutualism, 501

N

Namaqua chameleon, 497f
Namibia
    Etosha National Park, 547f
    Walvis Bay, climograph for, 214f
National Bridges National Monument, Utah, 375f
National Oceanic and Atmospheric Administration (NOAA), 183, 201f
National Weather Service, 200f
Natural disaster frequency, related to climate change, 231, 231f
Natural levees, 379, 379f
Natural selection, 504–507
Natural vegetation, 524
Needleleaf forests, 536–538, 537f, 538f
Nepal, Katmandu, 367f
Net primary production, 489–490, 490f
Net radiation, global energy budget and, 54, 54f
Nevado del Ruiz, 323f
New Delhi, India, climograph for, 205f
New Madrid earthquake, Missouri, 291, 291f
New Orleans, Louisiana, 180, 180f
New York
    Rome, 463f
    Sodus, 438f
New Zealand
    Fiordland National Park, 415f

loess and, 400f
Rakaia River gorge, 376f
Ngorongoro Conservation Area, Tanzania, 542f
Niagara Falls, 374f
Nickpoints, 372–373, 373f
Nimbus clouds, 109f
Nitrogen cycle, 494–495, 494f
Nitrogen fixation, 494
Nitrogen gas, 48
Nitrous oxide, 53t
NOAA Advanced Hydrologic Prediction Service, 349
Nohoch Ch'en, Belize, 338f
Nor'easters, 171f
Normal faults, 282f
North Carolina, Rodanthe, 417f
Northridge earthquake, 291
Nunataks, 428

O

O horizons, 457, 457f
Occluded fronts, 166
Ocean aerosols, cloud formation and, 108f
Ocean basin relief features, 262–263, 262f
Ocean currents, 150–151, 150f, 196f
Ocean temperatures, cyclone formation and, 174, 174f
Ocean tides
    coastal landforms and, 406–407, 406f
    ebb tides, 408
    flood tides, 408
    tidal range, 408, 408f
Ocean trenches, 279
Oceanic circulation, 150–154
    cycles in, 152–154, 153f
    energy transfer and, 151–152, 152f
    ocean currents, 150–151, 150f
Oceanic crust, formation of, 275
Oceanic trenches, 278f, 279
Ohio, Muskingum River, 350f
Okefenokee Swamp, Georgia, 486f
Old Faithful Geyser, Yellowstone National Park, 336
Open Street Map project, 29
Orangutans, 518–519, 519f
Oregon, Wizard Island, 302
Organic sediment, 252f, 254–255
Organic soils, 478, 478f
Orogeny, 279–280
Orographic lifting, 116–117, 117f
Orographic precipitation, 116–117, 117f
Ostrich, 508f
Outlet glaciers, 428
Outwash plains, 438–439, 439f
Overgrazing, 527f
Overland flow
    discharge related to, 346, 347f
    surface water, 334, 342, 343f
Overthrust faults, 279, 283, 283f

Owens Valley, California, 315f
Ox-bow lakes, 379
Oxidation, rock weathering and, 315
Oxisols, 471f, 474, 474f
Oxygen gas, 48
Ozone, 49
Ozone hole, 57
Ozone layer, 49, 55, 57–59, 58f
Ozone monitoring sensors, 58f

P

Palos Verdes Hills, California, 325f
Pangea, 263, 263f
Parabolic dunes, 399f
Parallel faults, 284, 284f
Parallels, 12
Parasitism, 501
Parent material, soil, 456, 461
Pascal (Pa), unit of pressure, 130
Patterned ground, 443, 443f
Pawnee National Grassland, Colorado, 543f
Peat, 254
Peds, of soil, 464
Percolation, 335
Perennials, 520
Periglacial landforms, 440–444
    gelifluction and, 441–442, 441f
    ground ice and, 441–443
    human interactions with, 444
    ice wedges and, 442
    patterned ground and, 443, 443f
    permafrost and, 440–441, 440f
    pingo formation and, 442, 443f
    thermokarst and, 444, 444f
Perihelion, 42f
Permafrost, 440–441, 440f
    distribution of, 440
    processes and, 441
    tundra climate and, 226
Permanent gases, atmosphere, 48
Peru, Iquitos, climograph for, 207f
pH scale, soil fertility and, 467, 467f
Philippines, Mt. Pinatubo, 85
Photosynthesis, 488–489, 488f
Physical geography
    methods and tools and, 6–7, 6f
    science of geography, 4–5
    world of systems, 5, 5f
Physical weathering
    biological action and, 314
    exfoliation and, 314, 314f
    frost action and, 312, 312f
    salt-crystal growth and, 313, 313f
    thermal action and, 314
Picote Dam, Portugal, 352f
Piedmont glaciers, 427f
Pingo, formation of, 442, 443f
Pinnacles, 381
Pioneer stage, ecological succession and, 503
Plane map projections, 22f

Wind speed and direction *(continued)*
  geostrophic wind, 134, 134f
  high-pressure system and, 136
  isobars and, 132f, 133
  low-pressure system and, 136
  pressure gradients and, 132–133,
    132f, 133f
Wind vanes, 132, 132f
Wind waves, 401–403
Winkel Tripel projection, 24f, 25
Wisconsin
  Kettle Moraine State Park, 438f
  Madison, climograph for, 223f

Wisconsin Glaciation, 445
Wizard Island, Oregon, 302
Woodchucks, 535f
Woodlands, 521
World patterns, air temperature and,
    80–82
World Wind software system (NASA), 30
Wyoming, Grand Tetons National Park, 461

X

Xerophytic trees, 540

Y

Yakutsk, Siberia, 75f, 225f
Yardangs, 393, 393f
Yellowstone National Park, 336, 526
Yuma, Arizona, climograph for, 73f

Z

Zabriskie Point, Death Valley National
    Monument, California, 382f
Zone of ablation, glaciers, 430
Zone of accumulation, glaciers, 429–430